The Minerals, Metals & Materials Series

Shijie Wang · Michael L. Free
Shafiq Alam · Mingming Zhang
Patrick R. Taylor
Editors

Applications of Process Engineering Principles in Materials Processing, Energy and Environmental Technologies

An EPD Symposium in Honor of Professor Ramana G. Reddy

TMS

Springer

Editors
Shijie Wang
Rio Tinto Kennecott Utah Copper
Magna, UT
USA

Mingming Zhang
ArcelorMittal Global R&D
East Chicago, IN
USA

Michael L. Free
University of Utah
Salt Lake City, UT
USA

Patrick R. Taylor
Colorado School of Mines
Golden, CO
USA

Shafiq Alam
The University of Saskatchewan
Saskatoon, SK
Canada

ISSN 2367-1181 ISSN 2367-1696 (electronic)
The Minerals, Metals & Materials Series
ISBN 978-3-319-51090-3 ISBN 978-3-319-51091-0 (eBook)
DOI 10.1007/978-3-319-51091-0

Library of Congress Control Number: 2016961672

Printed on acid-free paper

This Springer imprint is published by Springer Nature
The registered company is Springer International Publishing AG
The registered company address is: Gewerbestrasse 11, 6330 Cham, Switzerland

Preface

The Professor Ramana Reddy Honorary Symposium on Applications of Process Engineering Principles in Materials Processing and Energy and Environmental Technologies has been proposed and designed as a tribute to Prof. Reddy and his distinguished career as professor, teacher, and researcher over a period of 35 years at the University of Nevada, Reno, and the University of Alabama, Tuscaloosa, Alabama. Professor Reddy has advised and worked with over 100 Ph.D. and M.S. graduate students, research scholars, and visiting scientists. He has published more than 378 research papers and 26 books (4 CDs) including one undergraduate student textbook in thermodynamics. He has also delivered more than 268 invited lectures and research presentations in 26 nations. As an ACIPCO Endowed Professor, Prof. Reddy holds many leadership positions and honors and awards as well. The extent of Prof. Reddy's contribution to energy, environmental, and materials science and process engineering is underlined in the biographical sketch provided in these proceedings.

The response to the symposium has been most gratifying. When a call for papers went out, 78 abstracts were submitted for inclusion in the conference program. The net result has been that the number of papers offered for presentation has exceeded the time allocation and a Poster Session has been included in the program and accommodated in the proceedings. The international nature of the symposium is attested to by the fact that the selected papers are drawn from 17 countries.

The symposium has been named "Applications of Process Engineering Principles in Materials Processing, Energy and Environmental Technologies: An EPD Symposium in Honor of Ramana Reddy." It is little bit longer but covers the most papers, which are presented at the symposium.

The symposium program encompasses 10 sessions with each session being preceded by a plenary lecture. The sessions cover a range of topics within materials processing, energy, and environmental:

(1) Plenary Session
(2) Electrometallurgy
(3) Hydrometallurgy

(4) Pyrometallurgy I
(5) Pyrometallurgy II
(6) Materials Processing and Plasma Processing
(7) Energy Storage and Engineering Issues
(8) Modeling and Simulation
(9) Thermodynamics and Kinetics
(10) Poster Session

Within these topic areas, the individual papers represent a diversity of subjects. What is evident is a thrust toward bridging the gap between theory and practice via contributions dealing with process modeling and simulation, process and equipment design, the development of novel processes, and the generation of new data for the better understanding of current process technology. The environmental challenges facing metallic and nonmetallic industries are also addressed.

The symposium should be a fitting salute to Prof. Reddy for his contributions to extractive metallurgy, materials science, and energy and environmental technologies over the breadth of his distinguished career. It is hoped that the symposium volume will provide not only a record of a meeting but also a basis for further excellent work in the applications of process engineering principles in materials processing and energy and environmental technologies.

This symposium will provide a forum where industrial, research institutes and university professionals can interact and exchange with other stakeholders to facilitate the advancement of materials processes and engineering.

Shijie Wang
Michael L. Free
Shafiq Alam
Mingming Zhang
Patrick R. Taylor

Contents

About the Editors

Shijie Wang received his B.Sc. in mineral processing from China and his masters and Ph.D. in metallurgical engineering from the University of Nevada, Reno. Dr. Wang has experience working at the General Research Institute for Non-Ferrous Metals (Beijing, China), American Smelting and Refining Co. (or ASARCO Inc.), and Phelps Dodge Corporation (now called Freeport-McMoRan Inc.) within many technical leading roles. He is currently the principal advisor at Rio Tinto Kennecott Utah Copper. Dr. Wang has been active in extractive metallurgy and has 30 years of expertise in hydro- and electrometallurgical process development, existing operation optimization, and troubleshooting. His work interests include metal recovery, operational efficiency, and profitability, as well as process safety management. Dr. Wang holds three US patents and two patent applications on copper anode slime hydrometallurgical processing and electrowinning/refining cell innovation. Dr. Wang has published more than 40 journal papers including nonferrous metals', precious metals', rare metals', and rare earth metals' resourcefulness and recoveries and has several technical papers to his credit. He is a copper PM/PGM/rare metals refining subject matter expert. Dr. Wang has been a TMS member since 1991 and is the former chair of the Hydrometallurgy and Electrometallurgy Committee of TMS from 2011 to 2013. He has been the lead and co-organizer of five symposia at the international conferences through TMS.

Michael L. Free is a professor of metallurgical engineering at the University of Utah in Salt Lake City, Utah. He has performed research and taught courses as a faculty member since 1996. His areas of expertise include hydrometallurgy, electrometallurgy, corrosion, and materials synthesis. He has been the principal investigator of 56 research projects funded by 16 companies, the Department of Defense, the Department of Energy, the Office of Naval Research, and the National Science Foundation. He has authored or co-authored more than 175 publications, including a hydrometallurgy textbook. He has consulted for 37 organizations. He received a B.S. degree in metallurgical engineering, an M.S. degree in chemical engineering, and a Ph.D. degree in metallurgical engineering from the University of Utah. He worked as a postdoctoral associate at the University of Florida in the Departments of Materials Science and Engineering and Chemical Engineering for two years before becoming a faculty member at the University of Utah.

Shafiq Alam is an associate professor at the University of Saskatchewan, Canada. In 1998, he received his Ph.D. degree in chemical engineering from Saga University, Japan. From 1999 to 2001, he was appointed as a postdoctoral research fellow at the University of British Columbia and the University of Toronto, Canada.

Dr. Alam has extensive experience in industrial operations, management, engineering, design, consulting, teaching, research, and professional services. Before joining the University of Saskatchewan in 2014, he was an assistant/associate professor at Memorial University of Newfoundland for approximately 7 years. Prior to starting his career in academia, he worked with many different companies, such as Shell, Process Research ORTECH Inc., Fluor Canada Ltd., and the National Institute of Advanced Industrial Science and Technology (AIST), Japan. Dr. Alam is highly experienced in the area of mineral processing and extractive metallurgy, and he possesses two patents and has more than 140 publications. He is the co-editor of four books and an associate editor of the

International Journal of Mining, Materials and Metallurgical Engineering (*IJMMME*). He is the winner of the 2014 TMS Extraction & Processing Division's Technology Award.

Dr. Alam is a registered professional engineer and has worked on projects with many different mining companies including Falconbridge, INCO (Vale), Barrick, Hatch, Phelps Dodge, Rambler, and Anaconda. He is an executive committee member of the Hydrometallurgy Section of the Canadian Institute of Mining, Metallurgy and Petroleum (CIM), and currently he holds the office of secretary (2013–2017). Dr. Alam is also the chair of the Hydrometallurgy and Electrometallurgy Committee of the Extraction and Processing Division (EPD) of The Minerals, Metals & Materials Society (TMS) for the period 2015–2017. He is a co-organizer of many symposia at the international conferences through CIM and TMS.

Mingming Zhang is currently a senior research engineer at ArcelorMittal Global R&D at East Chicago, Indiana. His main responsibility includes raw material characterization and process efficiency improvement in the mineral processing and iron-making areas. He also leads technical relationships and research consortia with university and independent laboratory members and manages the pilot pot-grate sintering test facility at ArcelorMittal Global R&D East Chicago. Dr. Zhang has over 15 years of research experience in the field of mineral processing and metallurgical and materials engineering. He obtained his Ph.D. degree in metallurgical engineering from the University of Alabama and his master degree in mineral processing from the General Research Institute for Non-ferrous Metals in Beijing, China. Prior to joining ArcelorMittal, he worked with Nucor Steel Tuscaloosa, Alabama, where he was a metallurgical engineer leading the development of models for simulating slab solidification and secondary cooling process. Dr. Zhang has conducted a number of research projects involving mineral beneficiation, thermodynamics and kinetics of metallurgical reactions, electrochemical processing of light metals, and energy-efficient and environmental cleaner technologies. He has published over 30 peer-reviewed research papers, and he is the recipient of several US patents. Dr. Zhang also serves as editor and reviewer for a number of prestigious journals including *Metallurgical Transactions A* and *B*, *JOM*, *Journal of Phase Equilibria and Diffusion*, and *Mineral Processing and Extractive Metallurgy Review*. Dr. Zhang has made more than 20 research presentations at national and international conferences including more than 10 keynote presentations. He is the recipient of 2015 TMS Young Leaders Professional Development Award. He has been invited by a number of international professional associations to serve as conference organizer and technical committee member. These associations include the Association for Iron & Steel Technology (AISTech) and The Minerals, Metals & Materials Society (TMS).

Patrick R. Taylor is a registered professional metallurgical engineer with over 39 years of experience in mineral processing and extractive metallurgy engineering, research, teaching, and consulting. He holds an endowed chair as the George S. Ansell Distinguished Professor of Chemical Metallurgy and is the director of the Kroll Institute for Extractive Metallurgy at the Colorado School of Mines. He has directed research for more than 100 graduate students and post-docs. He has taught extractive metallurgy and mineral processing university courses for the past 39 years. He has developed and taught 10 short courses to industry. He is active in many professional organizations including the Society for Mining, Metallurgy and Exploration (SME) (distinguished member), The Minerals, Metals & Materials Society (TMS), ASM International (fellow), and Mining and Metallurgical Society of America (MMSA). He has been very active in TMS since 1975 and has received the TMS Extraction & Processing Division (EPD) Distinguished Service Award and the TMS EPD Distinguished Lecturer Award. His areas of expertise include extractive and process metallurgy, mineral processing, recycling, waste treatment and minimization, and thermal plasma processing. Pat was educated at the Colorado School of Mines, where he received B.S. degrees in both mathematics and metallurgical engineering and a Ph.D. degree in metallurgical engineering.

Part I
Plenary Session

Applications of Process Engineering Principles in Materials Processing, Energy and Environmental Technologies—Contributions of Professor Ramana Reddy

Shijie Wang

Abstract Life time achievements of Professor Reddy's teaching, research and scholarship in the field of chemical and materials science and engineering are briefly described. He has devoted his entire career in the introduction and application of process engineering principles to the quantitative description of materials processing reactions, and industrial operations. He has formulated quantitative methodologies based on first principles of thermodynamics, phase equilibria and kinetics to: design of slags and fluxes for production and purification of metals and alloys; development of Reddy-Blander model for a priori prediction of impurities capacities of melts; development of novel ionic liquid electrolytes for materials processing; development of new theories for alloys and molten salts; design materials for fuel cells and capacitors; use of thermodynamic approaches to predict physical properties of materials; industrial energy waste separation and remediation; and thermal energy storage. He is a scholar and mentor, and has made numerous scientific and service contributions to the society and metallurgical and materials engineering education.

Keyword Professor Ramana Reddy

S. Wang (✉)
Rio Tinto Kennecott Utah Copper, 8362 West 10200 South, Bingham Canyon, UT 84006, USA
e-mail: shijie.wang@riotinto.com

S. Wang et al. (eds.), *Applications of Process Engineering Principles in Materials Processing, Energy and Environmental Technologies*, The Minerals, Metals & Materials Series, DOI 10.1007/978-3-319-51091-0_1

Biographical Overview

Dr. Reddy is American Cast Iron Pipe Company (ACIPCO) Endowed Professor and Professor of Metallurgical and Materials Engineering at The University of Alabama, Tuscaloosa, Alabama. He served as an Associate Director of Green Manufacturing Center, University of Alabama and visiting researcher at the University of California Lawrence Berkeley Laboratory, Berkeley; Indian Institute of Technology, Bombay; Argonne National Laboratory, Chicago and National Renewable Energy Laboratory, Golden, CO. He was appointed as the University of Alabama Coordinator for the National Space Science and Technology Center (NSSTC) and NASA and Council Member for the Alabama State Committee for Department of Defense-EPSCoR programs.

Professional level and status of people Professor Reddy mentored are outstanding. Professor Reddy has 35 years of teaching, research, service and academic administration experience. He has conducted projects involving thermodynamics and kinetics of metallurgical reactions, Phase equilibria and thermodynamic properties of metal systems, composite materials and nano-materials, radioactive and waste materials, plasma processing of materials, fuel cells and capacitors, energy efficient and environmental friendly processes and molten salts and fluxes, and energy storage materials. Dr. Reddy is also an excellent teacher. He has taught several metallurgical and materials engineering undergraduate and graduate courses. He understands the importance of excellence in teaching and earning the esteem of his students. He has advised and worked with over 100 Ph.D. and M.S. graduate students and visiting scientists. He is faculty advisor to several undergraduate students. He is an external Ph.D. dissertation examiner for the graduate students at the Universities in Australia, India and Sweden. Many students of his research group have received outstanding research awards. He has served on the advisory board of several scientific agencies and industries. In 2001, he delivered invited lectures at various society chapters through the SME Henry Krumb Lecturer award.

Dr. Reddy is among the world's top metallurgical and materials engineers. His seminal contributions in thermodynamic modeling and his creative application of the principles of thermodynamics, phase equilibria and kinetics to a number of contemporary materials problems of scientific and technological importance have led to his world-wide reputation as the leader in this field. He has succeeded in applying the principles of engineering to development of new materials and redefined the role of classical chemical metallurgy in materials engineering, leading to his theory (Reddy-Blander model) by which the impurities capacities of oxide melts in industrial systems can be predicted a priori with the fundamental structure and thermodynamic properties data of melts. The greatest impact of Dr. Reddy's work and achievements is in the introduction and application of process engineering principles to the quantitative description of materials processing reactions, and industrial operations. He has formulated quantitative methodologies based on the first principles of thermodynamics, phase equilibria and kinetics to design slags and fluxes for the production and purification of metals and alloys; and thermodynamic approaches to predict thermophysical properties of materials for industrial applications. He has written a definitive textbook on thermodynamics and taught and promoted engineering education. Professor Reddy has received numerous citations and awards from several professional societies.

Dr. Reddy is an Associate Editor of *Journal of Phase Equilibria and Diffusion*, ASM International, Materials Park, Ohio, USA; Associate Editor of *High Temperature Materials and Processes*, Walter De Gruyter, Germany; Board of Review Member, *Metallurgical and Materials Transactions*, TMS, USA; and Editor-in-Chief, *Journal for Manufacturing Science and Production*, Walter De Gruyter, Germany. Dr. Reddy has long been associated with the Minerals and Metallurgical industries and has made many significant contributions to the development of several industrial technologies. Dr. Reddy has received as PI and Co–PI for a total award for over $16.66 million in funding, of which his dollars share is over $10.64 million. He has received research grants from the National Science Foundation (five divisions, including DMR, DMII, ECS, EPS and INT), the Department of Energy (five national laboratories—ANL, ORNL, INEL, SRNL and NREL; ITP and CSP), the Department of Defense (ARO and DSWA), the Department of Transportation-UA (CAVT), and U.S. Bureau of Mines. Industrial sponsors included General Motors, BASF, BHP Minerals International, ADL, ALCOA, Honda R&D and ALSTOM Power.

Dr. Reddy's most unique contributions to professional development in his areas of expertise may lie in his work with industrial professionals and post-graduate professionals. In particular, Dr. Reddy has been closely involved with the development of industrial professional training through continuing education programs. He has taught short courses attended by technical professionals from industries all over the USA and abroad. He has presented several metallurgical courses at Union Miniere, Belgium (Extractive Metallurgy of Non-ferrous Metals and Phase Diagrams—Thermodynamics and Construction); IPMI, USA (Precious Metals Extraction and Refining from Secondary Sources), The Universidad de Concepción, Concepción, Chile (Advances en Metalurgia Extractiva), and TMS (Magnesium Metallurgy-Processing and Industrial Applications at the Annual Meeting in

San Diego). He has also organized 4 workshops including NASA/MSFC on Materials Sciences for Advanced Space Propulsion.

Dr. Reddy is an outstanding leader. The most important is his inherent ability to coalesce strong diverse opinions into an effective consensus. As the Head of the MTE Department, Dr. Reddy has implemented highly successful educational and research program. In 2008, under his leadership, the department of Metallurgical and Materials Engineering (MTE) successfully received the ABET accreditation. Also the MTE department received 2007 national ranking of #1 materials engineering in productivity. He has served as a Chair of the UA College of Engineering Dean Search Committee, Head of the Department of Metallurgical and Materials Engineering. He served on several TMS, SME, AIME, ASM International and University committees. Chair/member of more than 43 professional societies committees, 65 technical sessions, and 25 public and universities committees; consultant to more than 20 industries, federal institutions and universities. He has received the distinguished service award from TMS.

Dr. Reddy is an outstanding scientist/engineer in both research and education as evidenced by the several honors and awards bestowed upon him by professional societies and other organizations. These honors include: Who's Who in Technology Today, National Dean's list, and Phi Kappa Phi National Fraternity. Several of Dr. Reddy's students received outstanding student/research awards. Dr. Reddy has received the TMS 2002 Extractive and Processing Lecturer award, the 2005 Milton E. Wadsworth Extractive Metallurgy Award, 2006 ATA award for Excellence in Engineering, Chicago. He is the recipient of the 2009 EPD-TMS Science Award and also the 2012 TMS Distinguished Service Award. He has received distinguished professional achievement award from his *alma mater*. He is a follow of two societies, Distinguished Member (Fellow) of SME and Fellow of ASM International.

Significant Research Accomplishments

Dr. Reddy is a prolific writer and has published over 380 research papers and 26 books including one undergraduate student textbook in thermodynamics. He has also delivered more than 270 plenary and invited lectures at numerous national and international conferences, research centers and industries in 26 nations. Dr. Reddy's Google scholar citations for his scholar and scientific research are above 4200 and an H-index 31 and i10-index 83.

A brief summary of Dr. Reddy's research areas and achievements is as follows:

Research Area 1: Modeling of Impurity Reactions in Metal-Slag-Gas Systems
Although metals constitute about 5% of the construction and industrial minerals, they are important fuel for industrial growth and world economic development. The complexity in treating or eliminating many environmentally harmful impurities such as As, Sb, Bi from the base metal matte or metal in the smelting stage are

responsible for the high cost of the refining process. The impurity capacities (such as sulfur, arsenic, antimony and bismuth) of slags were calculated a priori using Reddy-Blander (RB) model. Dr. Reddy made the capacities predictions in a wide range of matte and slag compositions, and temperatures in copper, nickel and lead smelting conditions and sulfur in iron and steel making. The calculated impurities capacities results are in excellent agreement with the experimental and industrial slags data. The a priori knowledge of impurities is useful for reduction of energy consumption and enhanced environmental control in the current and future non-ferrous metal processes.

- Delivered the award lecture: Distinguished Lecturer Award, Extraction and Processing Distinguished Lecturer Award, TMS, 2002
- Received Henry Krumb Lecturer Award, SME. Through this award gave series of lectures at educational institutions and materials societies in USA and abroad.

Commercial Software Thermodynamic Simulator (TSIM) for thermodynamic calculations software was developed and distributed along with his text book—Thermodynamics by Gokcen and Reddy. Reddy-Blander Model was adopted and incorporated in the thermodynamic commercial software.

Accomplishments A novel thermodynamic model (Reddy-Blander model) a priori prediction of impurity capacities of oxide melts on the basis of the structure and thermodynamic properties of melt was developed.

Research Area 2: Phase Equilibria and Thermodynamic Properties of Metal Systems

Dr. Reddy's research in the field of phase equilibria and thermodynamic properties of Ti–Al based intermetallic materials relates to the study of high temperature stability of these materials. He has contributed to a major improvements towards developing new thermodynamic data base and resolving the problem of how to prevent high temperature oxidation of Ti–Al based intermetallic alloys by the addition of a ternary metallic element such as Nb, Ta or Mo by proposing a mechanism of multilayer oxidation in these alloys. His research article in this field received recognition and the best science paper award.

Metal alloy solutions Dr. Reddy is without a doubt an expert in the field of thermodynamics whose research has contributed greatly to the chemical processing of materials. He applies thermodynamics, phase equilibria, kinetics and transport phenomena to materials processing. His research has introduced the new concept of interaction parameters to predict the behavior and interaction among elements such as carbon, sulfur, oxygen and nitrogen on liquid iron and how the thermodynamic properties of liquid iron (which contains these elements) changes with their amounts and with temperature.

Dr. Reddy's major research work has been on the investigation of the effect of solute elements like carbon, sulfur, oxygen, on the thermodynamic properties of liquid iron during the iron making process. He developed theoretical derivation of the relevant equations that express the interaction of various solutes in liquid iron.

His studies on phase equilibria and thermodynamic modeling of metallic solutions using Margules equations are novel. In this, the modified Margules equations are expressed up to infinite order in the vicinity of solvent. Quaternary system of one solvent and three solute partials are derived. First order interaction coefficients of binary, ternary and cross-interaction parameters of quaternary system are used to evaluate the partial functions. In the original form of Margules equations, derived partial functions are convergent and same as Unified Interaction Parameter Formalism. But considering up to infinite order of the Margules coefficients, the functions are divergent, in order to overcome this restriction, he developed a new solution theory. Margules equations are modified to get consistent equations. The derived partial functions were thermodynamically consistent with Maxwell and Gibbs Duhem relations. The derived logarithmic activity coefficient of the solvent and solutes are consistent with the ternary systems. The calculated thermodynamic data, i.e., activity coefficient values and interaction coefficients are in excellent agreement with the experimental data for binary, ternary and quaternary Fe alloys systems. The use of interaction parameters in the bulk phase to express the thermodynamic properties such as activity coefficient, enthalpy, entropy, and Gibbs energy in various metallurgical systems is well established and known for several decades, but Dr. Reddy's the concept of interaction parameters for multicomponent solutions is new in the field of metallurgical thermodynamics.

Intermetallics Dr. Reddy's research has made significant contributions in several fields of material engineering, such as oxidation resistance of titanium aluminides, thermodynamics and phase equilibria studies in titanium and aluminum alloys. Further continuation of his research towards the determination of thermodynamic properties of Ti–Al is extremely important in the successful design of utility alloys for high temperature applications. The Ti–Al based intermetallic compound materials have great potentials as a new functional material due to their excellent properties such as lightweight, high strength and creep resistance. But despite their great potentials, these materials are not often applied in the industrial field since these compounds suffer from poor phase stability and oxidation resistance at high temperatures. Dr. Reddy's research in the field of phase equilibria and oxidation resistance of Ti–Al based intermetallic materials relates to the study of high temperature stability of these materials. Extensive research has already taken place to understand the oxidation behavior and find ways to improve the oxidation resistance. What makes Dr. Reddy's research so groundbreaking is that in this context, he has discovered the mechanism of oxidation resistance for these intermetallic materials (Ti_3Al, TiAl and $TiAl_3$) by the addition of elements as an alloying addition. He proposed a mechanism for the formation of multilayered oxide structure in Ti_3Al–Nb alloy which he observed during oxidation experiments in pure oxygen. The mechanism is based on the migration of oxygen ions which are generated by the placement of a higher valent Nb in a lower valent Ti site in the TiO_2 structure. Dr. Reddy contributions in this field are in three main areas: (1) Determination of thermodynamic properties of Ti–Al–X alloys. For this, he carried out solid state galvanic cell experiments to generate the activity data in various alloys and determined the Gibbs energy of mixing of different phases. (2) Determination of phases and phase equilibria in the Ti–Al–X systems. He carried out phase equilibria experiments to determine the stable phases. (3) Determination of high temperature phase stability of the Ti–Al–X alloys in gaseous environment. He studied phase equilibria and reaction kinetics of alloys in oxygen and sulfur by high temperature thermogravimetry technique and deduced the possible reaction mechanisms.

- Science Award: Received Extraction and Processing Division Science paper Award, TMS, 2009.

Research Area 3: Innovations in Processing of Metal Matrix Composites, Nano-powders and Materials

Dr. Reddy developed a new in situ processing technology for production of lightweight alloys matrix with ceramic particles reinforcements was developed. Successful in situ formation of Al alloy with ceramic particles (i.e. SiC, TiC, TiN) composites by bubbling reactive gas (i.e. methane, nitrogen, ammonia) into Al alloy and Mg alloy melts was discussed. Effect of processing parameters on the formation of AlN and SiC composites was investigated. Kinetic rate equations of in situ formation of composites were developed.

Thermal plasma processing of nanoscale metal matrix TiC and TIN composites was developed. The Fe–TiN composite was synthesized in a non-transferred arc D. C. plasma reactor from ilmenite ore concentrate using methane and nitrogen as the reactive gases. The standard Gibbs energy minimization method was used to calculate the equilibrium composition of reaction species. A mathematical model was developed to describe the plasma gas and particle dynamics and conversion yields. The model was used to study the thermal decomposition of ilmenite in the non-transferred arc plasma reactor. TiN an average dimension of about 300 nm was produced. The in situ production of composites by molten metal technology and production of composites powders by plasma technology are developed.

Accomplishments Research worked led to a novel in situ processing of metal matrix composites. Two types of technologies molten metal and thermal plasma were developed. Core-shell structures of nano-composite powders were developed. Research worked led to Pilot plant processing of nano-composites. *Patent*: 1 patent approved.

Research Area 4: Advancement of Radionuclides Materials Research Contributions

Based on Dr. Reddy's research works, a technologically potential flow sheet was proposed on hydrometallurgical and pyrometallurgical routes for the recovery, removal, immobilization of radionuclides and heavy metals from nuclear energy wastes. Studies were carried out for the solutional reactions over the concentrations found at the Department of Energy (DOE), Department of Defense (DOD) waste sites and weapons detonation areas. The experimental and theoretical work was conducted on treatment of energy materials containing radionuclides, and recovery of uranium. The studies involved in remediation, separation and storage of radionuclides and thermal energy storage materials are discussed below.

Remediation The bioremediation technology of nuclear waste containing the radionuclides (Rads) and uranium was evaluated. The microbial cultures and active strains of decontaminant bacteria were used to leach uranium from the waste. Thermodynamic model on bacterial-uranium oxide interfacial reaction was developed. From the combination of Young's equation and equation of state, the change in Gibbs energy of adhesion for bacterium-uranium oxide system was calculated. The influence of liquid and uranium contaminated waste surface tension on the change in Gibbs energy of adhesion was discussed. The results showed that the two strains of *T. ferrooxidans* adhesion to uranium oxide surface would be increased by decreasing the change in Gibbs energy of adhesion. Based on the thermodynamic model predictions, experiments were carried out on bioleaching of uranium waste using *Thiobacillus ferrooxidans.* Extraction of about 99% uranium from nuclear waste using sulfur-oxidizing bacteria, *Thiobacillus ferrooxidans,* was obtained. The results from thermodynamic model and experimental data showed that bioremediation was an effective technology to clean up the uranium waste.

Separation The resulting uranium in solution was desorbed by zeolites for eventual recovery or glassy vitrification processing. Modification of basic structure of zeolites specifically pore size, by changing the temperature, pH, concentrations were investigated. Uranium species were strongly adsorbed by chabazite. Maximum sorption, more than 98% uranium ions was reported for chabazite. The sorption also appeared to depend on Al_2O_3/SiO_2 ratio of the zeolite. Chabazite, a porous zeolite with low Al_2O_3/SiO_2 ratio demonstrated greater uranium adsorption than other zeolites. Adsorption of uranium by chabazite follows the Langmuir adsorption isotherm. Mechanism of uranium adsorption on chabazite was evaluated and activation energy 3.612 kJ/mol has been obtained for the adsorption process. The adsorption kinetics was discussed in terms of adsorption and intraparticle diffusion.

Storage The zeolites containing uranium were further processed to fix the uranium in the glassy matrix for safe disposal/storage. The immobilization of radionuclides and heavy metals from nuclear wastes fixing it in glassy matrix by vitrification technology was developed. The Zeolites, borosilicate glass, and Aluminates (spent potliners) and phosphates were tested for the immobilization of contaminants by vitrification process. Optimum amount of additions were evaluated for obtaining the lowest viscosity of the melt, lowest contaminant leachability and best glass formation characteristics for the most suitable media for the immobilization and also to meet the required properties for the contaminants containing nuclear wastes disposal and storage.

Thermal Energy Storage Molten Slats In addition to innovative thermodynamic modeling, Dr. Reddy has also made significant contribution to thermal energy storage research on molten salts phase diagram development, determination of multi-component systems melting point, Gibbs energy, enthalpy of fusion and entropy, and also development of heat treatment medium and thermal energy transfer fluids for solar thermal energy storage technology applications in industry. In 2012, Dr. Reddy developed a brand new binary $Mg(NO_3)_2$–KNO_3 system and calculated its thermodynamic properties and published. The phase diagram of Mg $(NO_3)_2$–KNO_3 binary system has two eutectics and one congruently melting solid phase, $2KNO_3 \cdot Mg(NO_3)_2$ (labeled as MgKN). The two eutectic points appear on either side of the congruently melting solid. The only available data for the Mg $(NO_3)_2$–KNO_3 system was its melting point which was calculated and published in 1942. No experimental data are available in the literature on the accuracy of this melting point. Thermodynamic properties such as heat capacity, enthalpy, and entropy and Gibbs energy were also not available in the literature. Dr. Reddy's research re-built the entire phase diagram for the MgKN system by developing the complete thermodynamics database. Dr. Reddy not only successfully measured the melting points for a wide range of different compositions of $Mg(NO_3)_2$–KNO_3 system but also found out the heat capacity values for this system from solid state to liquid state which is the most significant parameter for the thermal energy storage and heat transfer properties and also most critical data for deducing the thermodynamic parameter as function of temperature. Different from the data calculated more than 70 years ago, Dr. Reddy found the new MgKN system has approximately 30 K lower congruently melting temperature than what was believed before. This finding fill the 70-year gap in metallurgical thermodynamics. Given the significance of Dr. Reddy's research, his published findings are included in National Institute of Standards and Technology (NIST) database.

- Received the Energy Best Paper Award on thermal energy storage materials: Energy Committee, Light Metals Division Energy Paper Award, TMS, 2010.
- Energy Best Paper Award: Student award, Energy Committee, Light Metals Division, TMS, 2013.

Accomplishments Research worked led to a technologically potential flow sheet for radionuclides and heavy metals from energy waste materials. Dr. Reddy's research also led to pilot plant testing of aluminum spent pot liners.

Research Area 5: Fuel Cells and Capacitors for Portable Power Applications Successful application of fuel cell technology is key for the sustainable hydrogen economy. The widespread commercialization of the technology has still not been made possible due to high costs associated with the fuel cell components. One such component in the fuel cell stack is the bipolar/end plate. Dr. Reddy's research works concentrated on developments in the materials, design, and concepts for bipolar/end plates in the polymer electrolyte membrane fuel cell stack. Experimental results for use of Fe-based alloys for bipolar plate as an alternative to the expensive conventionally used graphite material are developed. The developments of the models for optimizing the design parameters in the gas flow-field of these plates are discussed. Based on these simulations results, some of the new concepts for these plates were urbanized. These include: use of metal foam in the gas flow-field and corrugated thin sheet bipolar/end plate. Experimental results with these new concepts are presented and compared with the model predicted results.

Applications of these new concepts in the development of commercial fuel cell stacks in the era of hydrogen economy are discussed.

Novel Capacitor electrode materials based on Mn-oxides, cobalt and vanadium oxide were developed. Experimental and electrochemical modeling were conducted to optimize the capacitor parameters. Coin Capacitors was designed, fabricated and tested for its long term performance. Dr. Reddy research works led to fabrication and testing of micro-fuel cells, portable fuel cells and large scale fuel cells. Novel low cost catalysts were developed.

- Delivered Defence Science Forum Lecture: 102-Invited lecture, DRDO, New Delhi, India, 2013.

Accomplishments Research work led to fabrication and testing of micro-fuel cells, portable fuel cells and large scale fuel cells. Novel low cost catalysts were developed.

Research Area 6: Addressing Energy and Environmental Concerns in Production of Light Metals Through Novel Ionic Liquid Electrolytes

Ionic liquids are organic salts which are liquid at room temperature and have wide operational temperature range. Ionic liquid has some special properties compared with aqueous solution and regular organic liquid, such as very low vapor pressure, wide temperature range for liquid phase (400 °C), broad electrochemical window ($\sim$4.5 V) and being environmentally benign. Many potential applications of ionic liquids in materials processing were assessed by Dr. Reddy. The example applications include reduction of metal oxide to metal at low temperature, extraction and refining of reactive metals, electrodeposition, recycling and recovery metals from composites, mixed scrap and waste materials, heat transfer and thermal energy storage fluids. Dr. Reddy's research works concentrates on application of engineering fundamentals and potential applications of ionic liquids as electrolytes in light metals production, refining and recycling with specific emphasis on energy and environmental concerns.

The ionic liquids were synthesized and characterized using NMR and TGA/DSC. Properties of these ILs such as thermal stabilities, heat capacities and thermodynamic properties were measured. The calculated storage densities for $[C_4min][Tf_2N]$ was more than 190 MJ/m^3. Feasibility of these ILs particularly $[C_4min][Tf_2N]$ as thermal storage and heat transfer fluids applications in intermediate temperature solar power generation was evaluated.

Dr. Reddy also developed a novel process for the high energy efficiency electrodeposition of metals from metal oxides and metal compounds by means of electrolysis in ionic liquids at low temperature. Electrochemical deposition processes of Zn from ZnO and Cu from Cu_2O, Pb from PbO using Urea ($(NH_2)_2CO$) and Choline chloride ($HOC_2H_4N(CH_3)_3^+Cl^-$) or (ChCl) were developed. Successful deposition of pure metal from metal oxide dissolved in Urea/ChCl (2:1) was accomplished with efficiency greater than 87%.

Dr. Reddy developed patented technology for an electro-winning and electrorefining of aluminum and titanium using ionic liquids at low temperatures.

The process is based on using ionic liquids of 1-butyl-3-methylimidazolium chloride with the addition of $AlCl_3$ and $TlCl_4$ in the temperature range of 50–140 °C, molar ratio ($AlCl_3$ to C_4mimCl) of 1.5–2.0 and voltage varied from 2 to 4 V. The cathode deposits obtained were dendrite free pure aluminium (>99.9% pure) and Al–Ti (up to 37 wt% Ti) alloy. Cathode current densities of up to 500 A/m^2 for electrowinning of aluminium were achieved. The surface modification of electrodes not only improves the cathode current density but also eliminates the dendritic deposition of metals. Use of glassy surface finish, aluminum cathode and aluminum alloy anode improved the current density of the process. The critical cathode overpotential and limiting current density at which aluminum dendritic growth begins was determined as −0.53 V and 518 A/m^2, respectively. Lower overpotentials (<0.42 V) prevent the dendrite formation in the aluminum deposits. It was concluded that the overpotential is the key in preventing dendrite formation.

Dr. Reddy developed a 3-D mathematical model (CFD) for the batch reactor electrodeposition of metals process. The optimum conditions for electrodeposition of metals and alloys were determined. The model results are in good agreement with the experimental data for the Al electrowinning process.

- Delivered the award lecture Milton E. Wadsworth Award for Extractive Metallurgy, Society for Mining, Metallurgical and Exploration Inc., SME-AIME, USA, 2005.

Accomplishments The advantages of the electrodeposition of metals using ionic liquids compared to industrial aqueous metals processes are low temperature, low energy consumption, and low pollutant emissions.

Patents 4 approved; 1 pending; The University of Alabama signed 1 patent license agreement with BASF.

PROFESSIONAL AWARDS AND HONORS RECEIVED

- **Fellow of ASM International**: American Society for Metals International, 1996.
- **Distinguished Member (Fellow)**: Society for Mining, Metallurgical and Exploration Inc., SME-AIME, 2008.
- **National Dean's List**: Named in the National Dean's List, The University of Utah, USA, 1979.
- **Who's Who in Technology Today**: Named in the Who's Who in Technology list, USA, 1986.
- **NATO Fellowship**: NATO Advanced Study Institute on Molten Salt Chemistry, held at Universita De Camerino, Italy, 1986.
- **ASM-IIM Fellowship Award**: ASM International, 1987.
- **Sterling Who's Who Member**: Sterling Executive Club, USA 1995.
- **Service Award**: Light Metals, TMS, 1996.

- **Service Award**: Light Metals, TMS, 1997.
- **Research Award**: J. Manufacturing Technology, 1999.
- **Research Award**: Light Metals Recycling Best Paper Award, TMS, 2001.
- **Eminent Engineer of the Society Honor**: Tau Beta Pi Eminent Engineer, The Tau Beta Pi engineering Honor Society, 2002.
- **Distinguished Lecturer Award**: Extraction and Processing Distinguished Lecturer Award (EPD), TMS, 2002.
- **Milton E. Wadsworth Award for Extractive Metallurgy**: Society for Mining, Metallurgical and Exploration Inc., SME-AIME, 2005.
- **Excellence in Engineering Award**: American Telugu association (ATA), USA 2006.
- **Henry Krumb Lecturer Award**: Society for Mining, Metallurgical and Exploration Inc., SME-AIME, 2006–2007.
- **Chief Guest**: Inauguration of Winter School, Advanced Materials: Concepts and Applications (AMCA), NIT, Warangal, India, 2008.
- **Excellence in Research and Education Award**: Osmania University Alumni Association, North America, USA, 2008.
- **Science Award**: Extraction and Processing Division Science Award, TMS, 2009.
- **Distinguished Service Award**: Extraction and Processing Division Distinguished Service Award, TMS, 2009.
- **Energy Best paper Award**: Energy committee, Light Metals Division Energy Paper Award, TMS, 2010.
- **TMS Distinguished Service Award**: Alexander Scott Distinguished Service Award, TMS 2012.
- **Energy Best paper Award**: Student award, Energy committee, Light Metals Division, TMS, 2013.
- **Distinguished Alumni Professional Achievement Award**: National Institute of Technology, India, 2016.

Status of the Development of Flash Ironmaking Technology

H.Y. Sohn, Y. Mohassab, M. Elzohiery, D.-Q. Fan and A. Abdelghany

Abstract The Flash Ironmaking Technology being developed at the University of Utah is aimed at producing iron directly from iron oxide concentrate. In this technology, the concentrate is reduced by H_2 and CO gas mixtures formed from the partial oxidation of natural gas in a flash reactor. Natural gas represents an economically and environmentally superior reductant/fuel for the flash ironmaking. The rate equations for the reduction kinetics by H_2, CO and H_2 + CO gas mixtures were determined in the temperature range 1150–1600 °C. These rate equations were applied to experimental results from a laboratory flash reactor using Computational Fluid Dynamics CFD simulation. A new mini-pilot reactor, which is capable of operating at 1150–1550 °C with a concentrate feeding rate of 2–5 kg/h, has been installed. Commissioning of the reactor with an emphasis on preheating of the reactor, production of reducing gas mixtures and the feeding and collection of iron ore concentrate and product particles has been completed.

Keywords Flash reaction · Magnetite concentrate · Reduction kinetics · Hydrogen reduction · Ironmaking · Natural gas

Introduction

Blast furnace (BF) produces over 90% of the world iron from iron ore. Ironmaking via BF is a multi-step process, involving sintering/pelletization, cokemaking, and decarburization, which consume energy and produce hazardous emissions and greenhouse gas CO_2. The Flash Ironmaking Technology being developed at the University of Utah [1–8] aims at producing iron directly from iron oxide concentrates by a gas-solid flash reaction utilizing natural gas as the reductant and fuel in the temperature range of 1200–1600 °C. The direct use of iron ore concentrate

H.Y. Sohn (✉) · Y. Mohassab · M. Elzohiery · D.-Q. Fan · A. Abdelghany
Department of Metallurgical Engineering, University of Utah, Salt Lake City, UT 84112, USA
e-mail: h.y.sohn@utah.edu

S. Wang et al. (eds.), *Applications of Process Engineering Principles in Materials Processing, Energy and Environmental Technologies*, The Minerals, Metals & Materials Series, DOI 10.1007/978-3-319-51091-0_2

(<100 μm) will bypass pelletizing and sintering steps, thus lowering energy consumption by 30% compared with the average blast furnace process [9].

The use of inexpensive and abundant natural gas will eliminate the cokemaking step, thus reducing the CO_2 gas emissions by ~40% compared to an average blast furnace [4]. As a part of developing this process, reduction kinetics of magnetite concentrates by hydrogen and carbon monoxide gas mixture in the temperature range of 1150–1600 °C was studied. In order to establish the reduction kinetics with the gas mixture, the reduction kinetics by the single gases were first determined. A laboratory-scale flash reactor was then built and used for producing iron directly from iron oxide concentrate using hydrogen or natural gas partially combusted with oxygen serving as a source of heating as well as reductant. Electric power was used to compensate for heat loss due to the high surface area to volume ratio in this small reactor. This was the first experimental realization of the flash ironmaking technology. CFD was used to apply the developed rate equation to the results of this lab scale flash reactor and good agreement was achieved. These promising results obtained by the lab scale reactor lead to designing a mini-pilot reactor that is heated completely by a natural gas flame. This reactor is able to operate in the temperature range of 1150–1550 °C with a concentrate feeding rate 2–5 kg/h.

Reduction Kinetics Determination

The reduction of iron ore concentrate by H_2 and CO was investigated previously by many researchers, but the investigations were mainly focused on large particles or pellets at temperatures lower than the temperature to be used in the flash ironmaking process (1250–1550 °C). Most of the kinetics studies were performed on hematite ore [10–13].

The magnetite concentrate to be used in the flash ironmaking process contains oxide gangues that were found experimentally to lower the melting point of the particles to 1350 °C. Therefore, the reduction mechanism in the solid state at temperatures below 1350 °C is different from that at higher temperatures. Thus, the reduction kinetics were determined below 1350 °C separately from those above 1350 °C.

Experiments were designed and rate equations were formulated for the reduction of magnetite by H_2, CO and $H_2 + CO$ mixtures in the temperature range 1150–1350 °C within which the particles remain solid and at 1350–1600 °C within which the particles experience fusion and melting. The samples produced in this work were analyzed by an ICP method, which was developed in this laboratory to determine the iron content in the samples. This method has high accuracy in determining iron content in the samples containing high amounts of oxide gangues as in the concentrate used in this work [14].

The formulated rate equations are important in developing the flash ironmaking process as they will be used for analyzing the data obtained from a laboratory scale flash reactor and also for designing an industrial flash reactor. In both reactors, natural gas is partially oxidized with industrial oxygen thereby forming a mixture of CO and H_2 gas, in which H_2 will be the main reducing agent. The reduction kinetics by each gas was determined before investigating the reduction rate with H_2 + CO mixtures.

Reduction Kinetics in Temperature Range 1130–1350 °C

For each experiment, the reduction degree was calculated based on the amount of removed oxygen from the sample before and after reduction. Magnetite reduction by hydrogen was determined to follow the nucleation and growth kinetics equation with the Avrami parameter $n = 1$. The dependence on the H_2 gas partial pressure was found to be of the first order with activation energy of 193 kJ/mol. The complete rate equation obtained for the reduction of magnetite with hydrogen in the temperature range 1150–1350 °C was

$$[-Ln(1-X)] = 8.65 \times 10^6 \times e^{\frac{-193{,}000}{RT}} \times \left[p_{H_2} - \frac{p_{H_2O}}{K_{H_2}}\right]_{lm} \times t \qquad (1)$$

where X is fractional removal of oxygen combined with iron, R is the gas constant 8.314 J/mol K, T is temperature in K, p is partial pressure in atm, K_{H_2} is equilibrium constant for the hydrogen reduction of wustite, t is time in s and lm stands for logarithmic mean.

The reduction rate by CO was much lower than by hydrogen. The CO reduction also followed the nucleation and growth model with the Avrami parameter $n = 0.4$. The Avrami parameter is typically greater than 1, but there are examples of reactions with n less than 1, in which the nuclei growth rate decreases with growth [15]. Wunderlich [16] has analyzed a large amount of data on the rates that follow the Avrami equation, and reported that the Avrami parameter can range from lower than 1 to greater than 6.

The dependence on CO partial pressure was found to be of first order and the activation energy was 479 kJ/mol. The complete rate equation obtained for the reduction of magnetite with carbon monoxide gas in the temperature range 1150–1350 °C was

$$[-Ln(1-X)]^{1/0.4} = 3.41 \times 10^{14} \cdot e^{\frac{-479{,}000}{RT}} \cdot \left[p_{CO} - \frac{p_{CO_2}}{K_{CO}}\right]_{lm} \cdot t \qquad (2)$$

in which all other symbols have been defined after Eq. (1) except that K_{CO} stands for the equilibrium constant for the reduction of wustite by CO.

In both cases, there was no effect of particle size on the reduction rate within the size range tested (20–53 μm). The most likely reason for the absence of size effect is that the particles experience thermal stress and develop cracks of similar

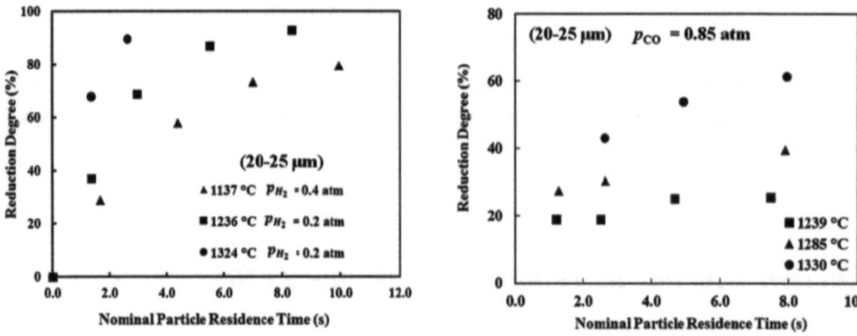

Fig. 1 Reduction degree of magnetite as a function of residence time by H_2 and CO singly under different conditions in the temperature range 1130–1350 °C

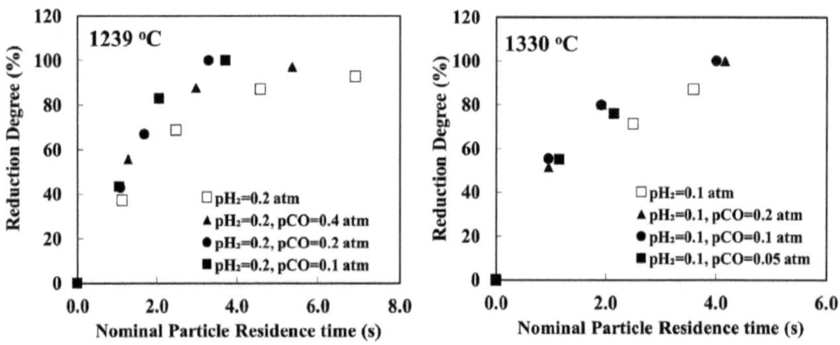

Fig. 2 Reduction degree of magnetite as a function of residence time by H_2 + CO mixtures under different conditions in the temperature range 1200–1350 °C

dimensions when rapidly heated as they are fed into the reactor at a high temperature. The reaction rate thus depends on the dimension of the solid between the cracks and no longer on the original size. Figure 1 shows some of the results obtained from the experiments.

These rate equations were used to develop a rate equation for the reduction kinetics by CO + H_2 gas mixtures. The presence of CO increased the reduction rate to be higher than the sum of the rates by individual gases. The results obtained are shown in Fig. 2.

Reduction Kinetics in Temperature Range 1350–1600 °C

The 1st order model in term of the fraction of solid remaining unreacted, equivalent to the nucleation and growth kinetics with $n = 1$, best describes the reduction kinetics of magnetite concentrate by H_2 in this temperature range. The reaction kinetics had a first-order dependence on the hydrogen partial pressure. Particle size had a significant effect on the reduction, as expected for the reaction of a molten/fused sphere. The activation energy was 177 kJ/mol. The complete rate equation obtained for the reduction of magnetite with hydrogen gas in the temperature range 1350–1600 °C is

$$[-Ln(1-X)] = 4.2 \times 10^7 \times e^{\frac{-177{,}000}{RT}} \times (d_p)^{-1} \left[p_{H_2} - \frac{p_{H_2O}}{K_{H_2}}\right]_{lm} \times t \quad (3)$$

where R is 8.314 J/mol K, T is in K, d_p is the particle size in μm, p is in atm, and t is in s.

In the case of reduction by CO, the nucleation and growth model also described the reduction kinetics. Similarly to the lower temperature range, an Avrami parameter $n = 0.5$ was found to best describe the results. The dependence on CO partial pressure was found to be of the first order and the activation energy was 49 kJ/mol. The results showed the same dependence on particle size as in the case of reduction by hydrogen, which is consistent with the fact that the size dependence originates from the geometrical factor. The complete rate equation obtained for the reduction of magnetite with carbon monoxide gas in the temperature range 1350–1600 °C is

$$[-Ln(1-X)]^{1/0.5} = 4.2 \times 10^2 \times e^{\frac{-49{,}000}{RT}} \times (d_p)^{-1} \left[p_{CO} - \frac{p_{CO_2}}{K_{CO}}\right]_{lm} \times t \quad (4)$$

where R is 8.314 J/mol K, T is in K, d_p is in μm, p is in atm, and t is in s.

Figure 3 shows some of the results obtained from the experiments. These rate equations were used to develop a rate equation for the reduction kinetics by $CO + H_2$ gas mixtures. The increase in reduction rate due to the presence of CO over the sum of the rates by individual gases was even greater in the higher temperature range than in the lower temperature range, as shown in Fig. 4.

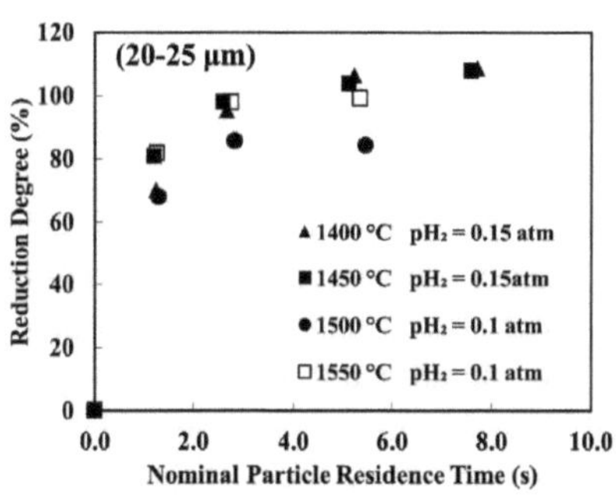

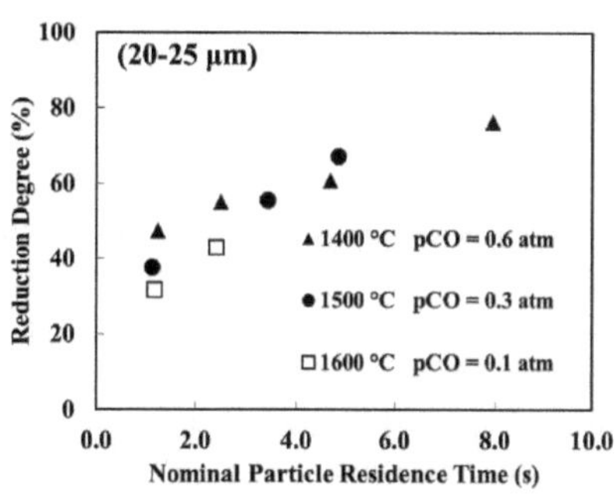

Fig. 3 Reduction degree of magnetite as a function of residence time by H_2 and CO singly under different conditions in the temperature range 1350–1600 °C

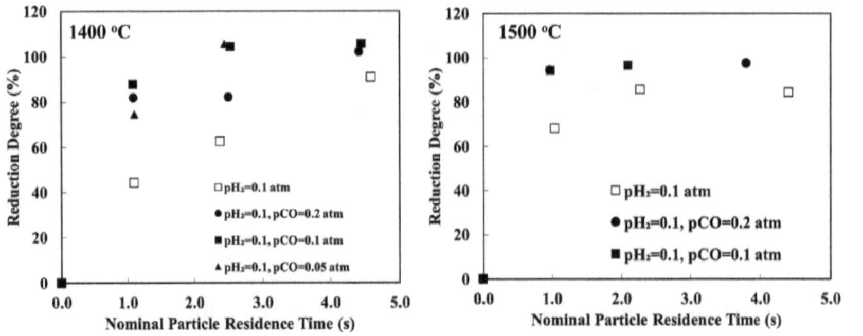

Fig. 4 Reduction degree of magnetite as a function of residence time by H_2 + CO mixtures under different conditions in the temperature range 1350–1600 °C

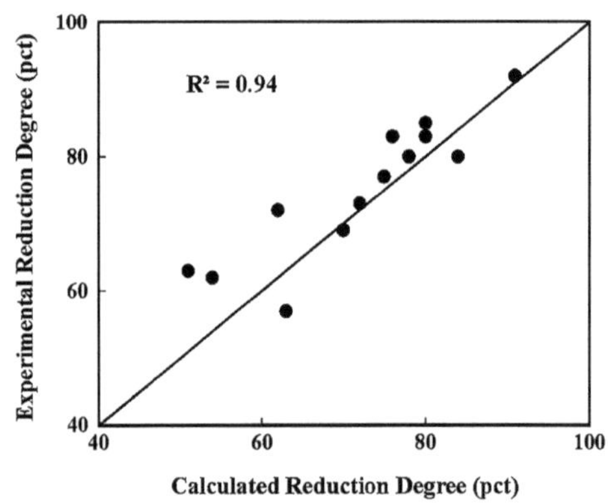

Fig. 5 Comparison between the calculated reduction degrees by CFD versus experimental results [17]

Laboratory Scale Flash Reactor

A laboratory scale flash reactor [17] was used to verify the rate expressions under an actual flash reaction condition. In this reactor, hydrogen or methane were used as the fuel/reductant by producing a non-premixed flame. Due to the high volume to surface ratio, electrical heating was used to compensate for the heat loss in the reactor. A three-dimensional computational fluid dynamics (CFD) model was developed to simulate the reduction of magnetite concentrate by hydrogen in which the fluid flow, heat transfer, and chemical reactions involved were incorporated. The governing equations for the gas phase were solved in the Eulerian frame of reference, while the particles were tracked in the Lagrangian framework. The partial combustion of H_2 injected through a non-premixed burner was simulated and the temperature profiles and reduction degrees obtained from the simulations had a satisfactory agreement with the experimental measurements, as shown in Fig. 5. The magnetite reduction by H_2 + CO mixtures produced from the partial oxidation of natural gas is now under investigation.

Mini-pilot Scale Flash Reactor

A mini-pilot flash reactor was built in the University of Utah for the flash reduction of magnetite concentrate. The new reactor has an inner diameter of 0.8 m and a reaction zone length of 2.3 m. This reactor has an 18 cm thick refractory layer to minimize heat loss. The reactor is preheated by full combustion of natural gas with oxygen through a preheat burner with a heating rate of 90–95 °C/h. The operating temperature range is 1200–1550 °C.

The new reactor is controlled completely with an automated system with a high built-in safety measures. The reactor vessel is designed to withstand a pressure up to 3.4 atm, and is equipped with an off-gas duct with a rupture desk that allows the gas to escape from the reactor if the pressure inside the reactor exceeds 0.7 atm. The concentrate feeding rate is 2–5 kg/h.

The reactor is equipped with a main burner by which natural gas is partially combusted with oxygen to produce a mixture of H_2 and CO reducing gases as well as H_2O and CO_2. This main burner produce a swirl in the produced gas for better distribution of the reducing gases and heat in the reactor as well as increasing the residence time of the particles. When the main burner is used, the off-gas is flowed through a flare stack where all the remaining H_2 and CO gases are completely burned before releasing to the atmosphere. An industrial gas analyzer is used to measure the contents of H_2, CO, CO_2, O_2 and CH_4 in the gas stream; to ensure the complete combustion during the preheat mode; and to check the H_2 and CO amounts produced during the experiment by the main burner.

The preheating and main burner components are water cooled to avoid any damage to them during the experiment. Alternative powder feeding ports are available on the roof of the reactor for different feeding modes.

Conclusions

The reduction kinetics of magnetite concentrate particles by H_2, CO and H_2 + CO gas mixtures was investigated in the temperature range of 1150–1600 °C. The rate equation of the magnetite reduction by each gas singly was obtained. The nucleation and growth model was used to describe the reduction kinetics by each gas in two temperature ranges of 1150–1350 °C and 1350–1600 °C. For magnetite reduction with CO gas, an Avrami parameter of less than 1 was determined, indicating a possible restriction on the nuclei growth by diffusional effects in them. These rate equations were used to formulate a rate equation for reduction by H_2 + CO mixtures. A laboratory scale flash reactor was used to verify the kinetics results under actual flash reaction conditions in which H_2 or natural gas is partially combusted with oxygen to produce the reductant and heat. CFD simulation was used for comparing the kinetics model to the experimental results and good agreement was achieved. A mini-pilot flash reactor that can operate in the temperature range 1200–1550 °C has been built and commissioned. The new reactor is preheated by complete combustion of natural gas with a heating rate of 90–95 °C/h. The magnetite concentrate feeding rate ranges from 2 to 5 kg/h. Experimental tests using this mini-pilot reactor is scheduled to begin in the near future.

Acknowledgements The authors acknowledge the financial support from the U.S. Department of Energy under Award Number DEEE0005751 with cost share by the American Iron and Steel Institute (AISI) and the University of Utah.

Disclaimer This report was prepared as an account of work sponsored by an agency of the United States Government. Neither the United States Government nor any agency thereof, nor any of their employees, makes any warranty, express or implied, or assumes any legal liability or responsibility for the accuracy, completeness, or usefulness of any information, apparatus, product, or process disclosed, or represents that its use would not infringe privately owned rights. Reference herein to any specific commercial product, process, or service by trade name, trademark, manufacturer, or otherwise does not necessarily constitute or imply its endorsement, recommendation, or favoring by the United States Government or any agency thereof. The views and opinions of authors expressed herein do not necessarily state or reflect those of the United States Government or any agency thereof.

References

1. H.Y. Sohn, Suspension ironmaking technology with greatly reduced energy requirement and CO_2 emissions. Steel Times Int. **31**, 68–72 (2007)
2. H.Y. Sohn et al., Suspension reduction technology for ironmaking with low CO_2 emission and energy requirement. Iron Steel Technol. (AIST Trans.) **6**, 158–165 (2009)
3. M.E. Choi, H.Y. Sohn, Development of green suspension ironmaking technology based on hydrogen reduction of iron oxide concentrate: rate measurements. Ironmak. Steelmak. **37**, 81–88 (2010)
4. H.K. Pinegar, M.S. Moats, H.Y. Sohn, Flowsheet development, process simulation and economic feasibility analysis for novel suspension ironmaking technology based on natural gas: Part 1—Flowsheet and simulation for ironmaking with reformerless natural gas. Ironmak. Steelmak. **39**, 398–408 (2012)
5. M.Y. Mohassab Ahmed, Phase equilibria between Iron and slag in $CO/CO_2/H_2/H_2O$ atmospheres relevant to a novel flash ironmaking technology. Ph.D. dissertation, The University of Utah, 2013
6. H.K. Pinegar, M.S. Moats, H.Y. Sohn, Flowsheet development, process simulation and economic feasibility analysis for novel suspension ironmaking technology based on natural gas: Part 3—Economic feasibility analysis. Ironmak. Steelmak. **40**, 44–49 (2013)
7. Y. Mohassab, H.Y. Sohn, Effect of water vapor on sulfur distribution between liquid Fe and MgO-saturated slag relevant to a flash ironmaking technology. Steel Res. Int. **86**, 753–759 (2014)

8. F. Chen et al., Hydrogen reduction kinetics of hematite concentrate particles relevant to a novel flash ironmaking process. Metall. Mater. Trans. B **46**, 1133–1145 (2015)
9. H.Y. Sohn, M. Olivas-Martinez, Methods for calculating energy requirements for processes in which a reactant is also a fuel: need for standardization. JOM **66**, 1557–1564 (2014)
10. E.G. Davis et al., Flash reduction of iron ore. Report of investigations. United States Department of the Interior, Bureau of Mines; 7627 (U.S. Department of Interior, Bureau of Mines, Washington, 1972), 10 p
11. N.J. Themelis, W.H. Gauvin, Reduction of iron oxide in gas-conveyed systems. AIChE **8**, 437–444 (1962)
12. W.M. McKewan, Kinetics of iron oxide reduction. Trans. Am. Inst. Min. Metall. Pet. Eng. **218**, 2–6 (1960)
13. R.J. Fruehan, Y. Li, L. Brabie, E.J. Kim, Final stage of reduction of iron ores by hydrogen. Scand. J. Metall. **34**, 205–212 (2005)
14. Y. Mohassab et al., Determination of total iron content in iron ore and DRI: titrimetric method versus ICP-OES analysis. *In 2016 EPD Congress* (Wiley, 2016), pp. 125–133
15. P.G. Hedmark et al., Cold-crystallization in liquid crystalline poly(p-hydroxybenzoic acid-co-ethylene terephthalate). Polymer **30**, 2068–2073 (1989)
16. B. Wunderlich, *Macromolecular Physics* (Academic Press, New York, 1976), pp. 115–347
17. D.-Q. Fan et al., Computational fluid dynamics simulation of the hydrogen reduction of magnetite concentrate in a laboratory flash reactor. Metall. Mater. Trans. B **47**(6), 1–12 (2016)

Innovations and Insights in Fluid Flow and Slime Adhesion for Improved Copper Electrorefining

Weizhi Zeng, Michael L. Free and Shijie Wang

Abstract Copper electrorefining is an old technique that is generally performed in electrolytic cells with a bottom inlet and top outlet that are convenient for electrolyte flow control. Nevertheless, this configuration cannot effectively direct electrolyte flow to the regions between electrodes, which results in weak convection and inadequate delivery of chemical species and additives. In this article, we discuss the innovations in electrorefining cell design to improve electrolyte flow for better cathode quality and simulation of fluid flow field and particle movements based on finite element modeling method. The resulting fluid flows in cells can transport suspended slime particles to the cathode, where they can be incorporated as impurities. The concentration of suspended slime particles is inversely related to slime adhesion, which is strongly influenced by slime particle sintering, and sintering is dependent on temperature and slime particle composition. Thus, the combination of innovative fluid flow and slime particle sintering can play important roles in improving cathode quality.

Keywords Fluid flow · Slime particle movement · Innovative cell · Slime adhesion · Sintering and coalescence

Introduction

Electrorefining and electrowinning is widely used to produce Grade A copper (99.99% purity). They are performed in large electrolytic cells in industry. Most refineries are still using traditional electrolytic cells that have a bottom inlet and a top outlet on sidewalls. This configuration has been employed since the first copper

W. Zeng (✉) · M.L. Free
Department of Metallurgical Engineering, University of Utah,
Salt Lake City, UT 84112, USA
e-mail: weizhi.zeng@utah.edu

S. Wang
Rio Tinto Kennecott Utah Copper, Magna, UT 84044, USA

S. Wang et al. (eds.), *Applications of Process Engineering Principles in Materials Processing, Energy and Environmental Technologies*,
The Minerals, Metals & Materials Series, DOI 10.1007/978-3-319-51091-0_3

electrolysis operation in Europe in 1865 [1], and is convenient for electrolyte level control. Nevertheless, it has many drawbacks, which significantly affect impurity control in both electrorefining and electrowinning.

Under such configuration, electrolyte inflow enters the cell from the bottom inlet, with most of its kinetic energy dissipated on the way to the inter-electrode domains, where most electrochemical reactions, species transfer, and slime particle transport in copper electrolysis occur. As a result, two significant problems can be encountered here: electrolyte additives and chemical species can not be delivered to the regions near electrodes efficiently; convection between the electrodes is poor, as electrolyte flows here are almost only driven by density gradients [2, 3]. Consequently, cathode surface quality is affected; larger electrolyte copper concentration is required for electrowinning; operating current density is limited.

For copper electrorefining, this configuration raise further problems. According to previous studies [4], most cathode contaminations in copper electrorefining originates from suspended slimes, rather than settled slimes or impurity ions. The configuration featuring a bottom inlet and a top outlet makes upward electrolyte flows dominant in electrolytic cells, which is unfavorable for slime particle settling [2, 3, 5]. Therefore, the concentration of suspended slime particles can not be reduced, and cathode deposit is usually under high risk of impurity contamination in traditional electrolytic cells.

Consequently, innovations in copper electrolytic cell designs would be of great help in improving productivity, efficiency, and purity in copper electrorefining. Some refineries have been doing research on new cell designs and have even installed their innovative electrolytic cells [6, 7]. Nevertheless, most copper refineries are still using traditional electrolytic cells, and literatures concerning the design, mechanism and effects of innovative electrolytic cells can be barely found. Thus, this paper will give a description of one of our innovative electrorefining cells [8], along with the simulation results of fluid flow field and slime particle movements.

While we can use innovative electrolytic cells to help improve cathode purity by generating dominant downward flows and settling most released slime particles, it can be even better if we can improve the adhesion of slime particles and increase the sizes of released particles. In our recent studies, we found that slime particle sintering/coalescence have significant effects on anode slime adhesion and the sizes of slime particles [4, 9]. Furthermore, we can control the extent of slime particle sintering/coalescence through cell temperature and anode composition [4]. Some results and suggestions will be presented in this paper.

Therefore, this paper shows two approaches to improved copper electrorefining. One is the design and application of innovative electrolytic cells and the other one is the control of the adhesion and sizes of slime particles.

Design and Simulation Results of an Innovative Electrolytic Cell

The designed innovative electrolytic cell has a configuration of top outlet, bottom inlet, and shortened distance from the inlet to the inter-electrode gap, which are considerably different from traditional electrolytic cells. The effects and advantages of such configurations can be evaluated by modeling and simulation of the innovative cell. The geometry of the cell is shown in a 3D Cartesian coordinate system in Fig. 1.

As shown in Fig. 1, the cell has dimensions of 0.1 m × 0.075 m × 0.12 m, with one pair of electrodes sit in the middle, one inlet at the top, and one outlet at the bottom. The heights of the anode and the cathode are 0.09 and 0.095 m, respectively. The widths of the electrodes are 0.08 m. Note that only the front sides of the electrodes have electrochemical reactions. The distance between the electrodes is 2.5 cm, and the distance between the inlet and the anode is 2.4 cm.

The domain of the innovative cell was discretized into tetrahedral elements and the geometric faces were discretized into triangular elements. By applying governing equations for current distribution, fluid flow, and particle motion [8], the electrolyte flow field and the slime particle movements can be simulated under specific boundary conditions, which are 11 ml/min flow rate, 50 °C cell temperature, and 300 A/m^2 current density in this study. The resulting electrolyte flow field is shown in Fig. 2a, b.

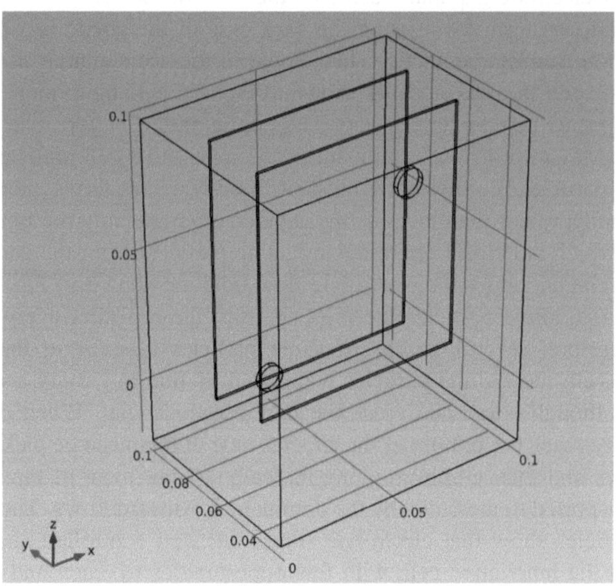

Fig. 1 The geometry of the innovative electrolytic cell (units in meters) [8]

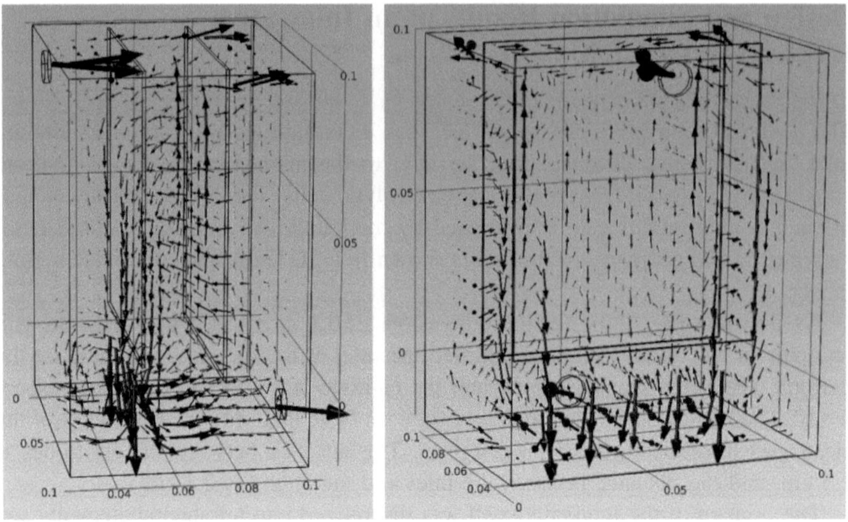

(a) (b)

Fig. 2 The electrolyte flow field in the innovative cell from different angles [8]

As shown in Fig. 2a, looping fluid flows (downward flows along the anode and upward flows along the cathode) are generated in the inter-electrode gap, due to electrolyte density gradients caused by electrochemical reactions [2, 3]. In addition, electrolyte inflows enter the cell through the top inlet. Most of the inflows cross over the inter-electrode gap and reach the other side of the cell behind the cathode. Part of the inflows turn down along the back side of the anode and create strong circulating flows under the anode. Then, some of these circulating flows enter the inter-electrode gap from its sides, as shown in Fig. 2b, leading to more convections and better species transfer in the gap. The configuration of top inlet and bottom outlet makes downward flows dominant in the innovative cell and can help settle down slime particles. Furthermore, most of the downward flows along the anode turn to the outlet rather than towards the cathode after reaching the bottom domain of the cell, which is helpful for removing slime particles from the cell. The slime particle movements under the electrolyte flow field are shown in Fig. 3a–d.

In Fig. 3a–d, slime particles are injected from the top inlet every 500 s from t = 0 s. Therefore, several groups of slime particles co-exist in the cell. After entering the cell, most slime particles begin to drop due to gravity and the downward flows, though some can cross the inter-electrode gap. When the dropping slime particles reach the bottom of the anode, some of them can be picked up by the looping flows and then climb up along the cathode, but most of them are settled down or transported to the outlet by the dominant downward flows. The movements of slime particles show that electrolyte slimes are well controlled under the fluid flow field in the innovative cell, with few opportunities to reach and contaminate the cathode.

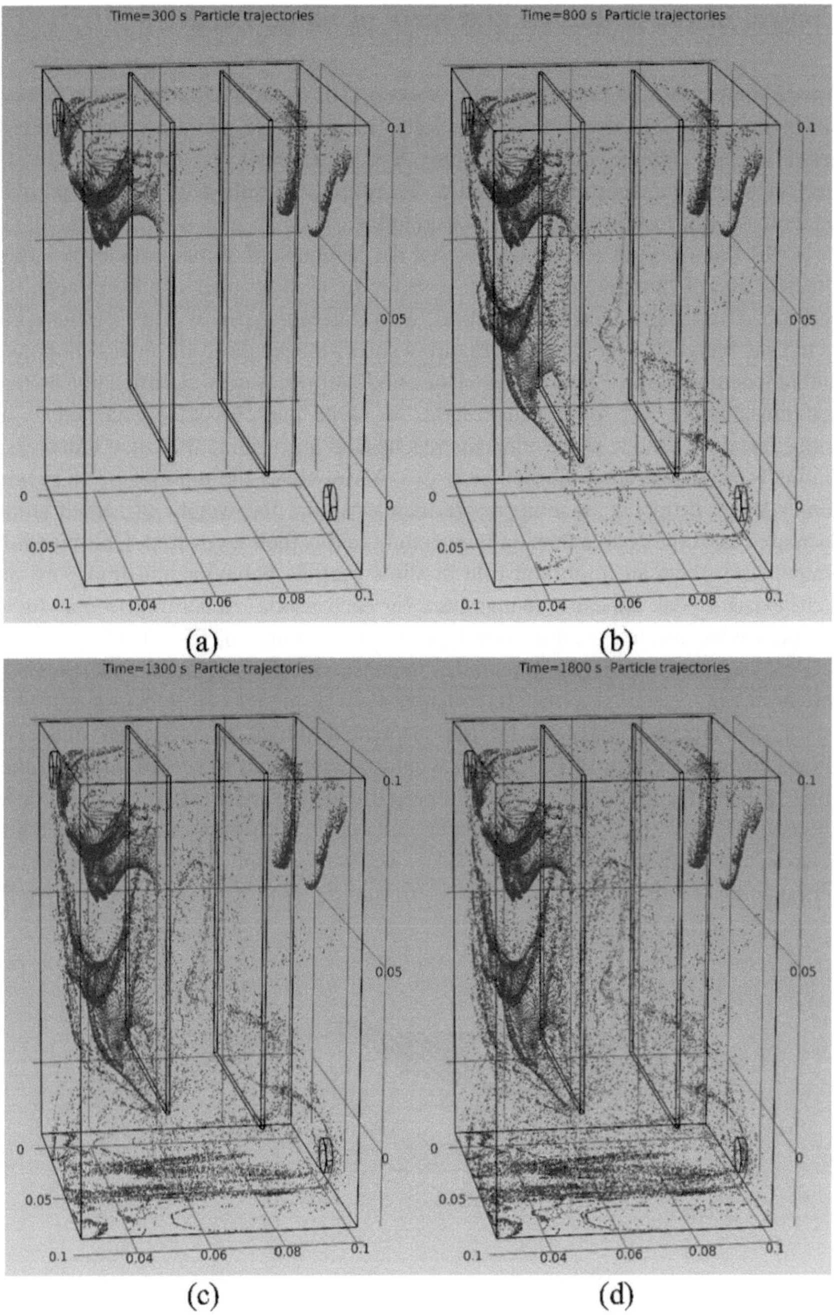

Fig. 3 Slime particle movements in the innovative cell at different time points after released from the inlet [8]

Control of the Adhesion and Sizes of Slime Particles

Anode slime adhesion has significant effects on slime particle behavior and cathode purity in copper electrorefining. Generally, larger anode slime adhesion leads to better copper cathode quality, because less slime particles are released to the electrolyte and transported to cathodes. Therefore, the adhesion of slime particles needs to be improved in copper electrorefining.

It was found in our recent studies that the behavior of slime particles is largely affected by cell temperature [4]. In a series of tests, copper electrorefining was performed at different cell temperatures, with different types of slime particles collected and weighed after each test. Figure 4 shows the resulting slime distributions at different temperatures. The adhesion of anode slimes varies significantly as temperature changes. At room temperature, no slime particles were attached to the anode; at 65 °C, anode slime adhesion reaches the maximum; after that, it decreases gradually as temperature further increases. Meanwhile, the amount of suspended slimes keeps dropping as temperature increases, and the weight of settled slimes decreases until the peak adhesion temperature and is then recovered. Consequently, temperature plays an important role in slime particle behavior and transport, and there exists a peak adhesion temperature for each anode. According to our studies [4], peak adhesion temperatures are typically in the range of 55–70 °C.

Besides, the sizes of slime particles increase as cell temperature increases. As shown in Fig. 5, settled slimes have larger sizes at higher cell temperatures. Only minor size increase is shown for the suspended slimes, because that suspended slimes are smaller than 9–10 μm and would be converted to settled slimes if their sizes become too large [4]. Larger slime particles are more likely to settle down without affecting the purity of copper cathodes. With further studies in slime behavior at high temperatures [4], it was found that slime particle sintering/coalescence occurs when the surrounding temperature is raised above the

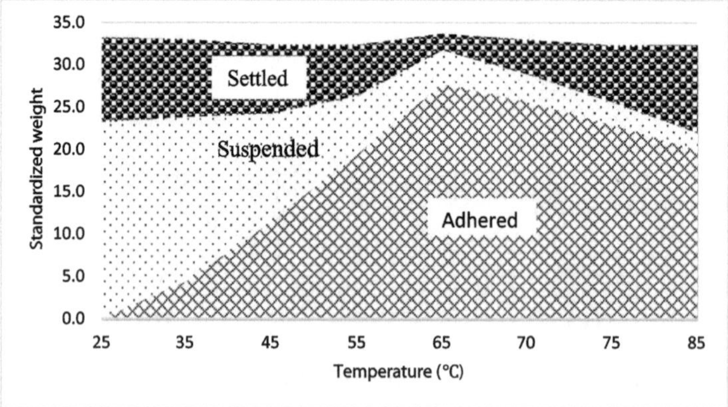

Fig. 4 Slime distributions in copper electrorefining at different temperatures

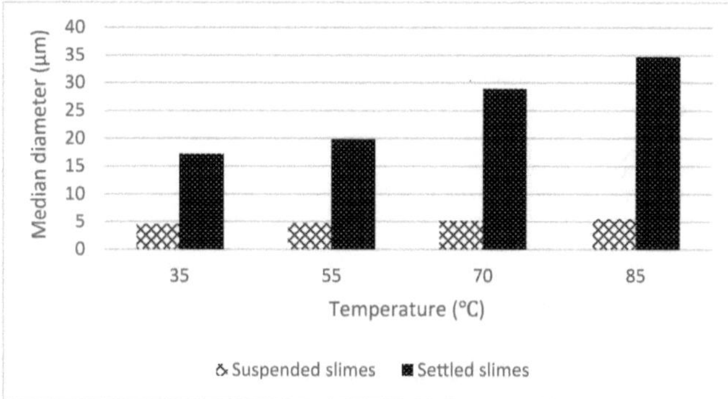

Fig. 5 Relationship between the slime particle median diameter D_{50} and the cell temperature [4]

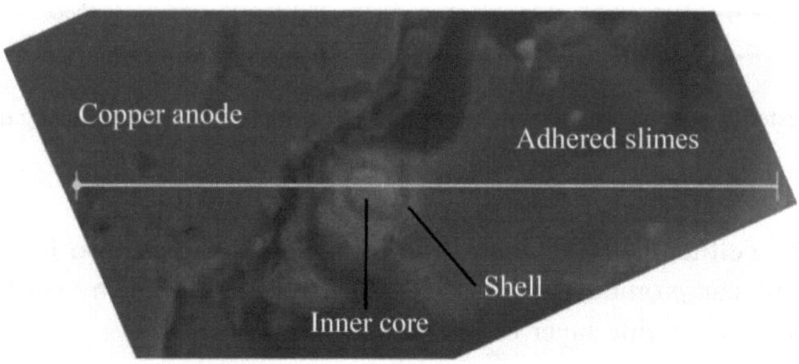

Fig. 6 SEM image of the anode slimes layer structure in front of the anode surface of a harvested anode [4]

sintering temperature, which is affected by the composition of slime particles/anodes. Generally, slime particles from high impurity anodes have lower sintering temperatures [4]. As a result, higher cell temperatures can induce more slime particle sintering/coalescence, which leads to improved anode slime adhesion and larger sizes of released slime particles. Figure 6 shows the adherence of slime particles on the anode by sintering/coalescence.

Improved slime particle adhesion and larger released particle sizes can effectively reduce the concentration of suspended slimes. As a result, impurity concentrations in copper cathodes can be lowered. Figure 7 presents the bismuth levels in the harvested copper cathodes at different cell temperatures, which were determined by DC Arc. Bismuth levels are above 0.5 ppm when cell temperatures are below 45 °C. At higher temperatures, the resulting copper cathodes have much lower bismuth levels. It is clear that cell temperature has larger effects on the high impurity anodes, which have slime particles with lower sintering temperatures and are more sensitive to temperature increase.

According to the results, anode slime adhesion and the sizes of released slime particles are largely affected by slime particle sintering/coalescence, which can be

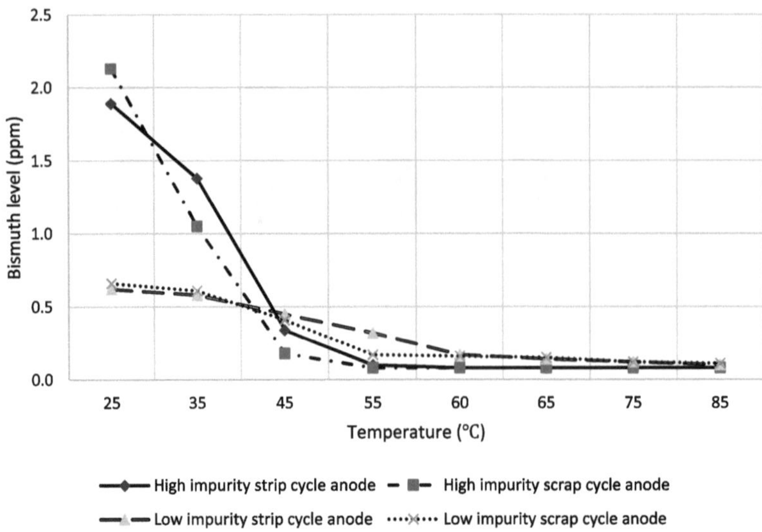

Fig. 7 Harvested copper cathode bismuth levels at different cell temperatures using different types of anodes [4]

controlled by cell temperature and anode composition. It was also found that high current density can promote slime particle sintering/coalescence by generating more heat in front of the oxide layer on the anode [9].

Conclusions

Based on the results, the innovative cell can provide strong convection in the inter-electrode gap that can facilitate species transfer, as well as dominant downward flows that can help settle down slime particles. Also, the adhesion and sizes of slime particles can be controlled by temperature, current density, and anode composition. As a result, we can reduce the number of released slimes by improving slime adhesion, increase the settling velocities of slime particles by increasing their sizes, and facilitate slime particle settlement by improving electrolyte flow field.

References

1. S. Wang, Novel electrowinning technologies: the treatment and recovery of metals from liquid effluents. JOM. **60**, 41–45 (2008)
2. W. Zeng, M.L. Free, J. Werner, S. Wang, Simulation and validation studies of impurity particle behavior in copper electrorefining. J. Electrochem. Soc. **162**, E338–E352 (2015)

3. W. Zeng, S. Wang, M.L. Free, Experimental and simulation studies of electrolyte flow and slime particle transport in a pilot scale copper electrorefining cell. J. Electrochem. Soc. **163**, E111–E122 (2016)
4. W. Zeng, M.L. Free, S. Wang, Studies of anode slime sintering/coalescence and its effects on anode slime adhesion and cathode purity in copper electrorefining. J. Electrochem. Soc. **163**, E14–E31 (2016)
5. W. Zeng, J. Werner, M.L. Free, Experimental studies on impurity particle behavior in electrolyte and the associated distribution on the cathode in the process of copper electrorefining. Hydrometallurgy **156**, 232–238 (2015)
6. A. Filzwieser, I. Filzwieser, S. Konetschnik, New technology for electrorefining of copper. J. Miner. Met. Mater. Soc. **64**, 1290–1295 (2012)
7. C. Wenzl, A. Filzwieser, S. Konetschnik, METTOP-BRX-technology—industrial application. in *T.T. Chen Honorary Symposium on Hydrometallurgy, Electrometallurgy and Materials Characterization*, Wiley, Hoboken, 2012, pp. 63–76
8. W. Zeng, S. Wang, M.L. Free, A comparative study of electrolyte flow and slime particle transport in a newly designed copper electrolytic cell and a laboratory-scale conventional electrolytic cell. J. Miner. Met. Mater. Soc. (2016). doi:10.1007/s11837-016-2076-x
9. W. Zeng, M.L. Free, S. Wang, Simulation studies of electrolyte flow and slime particle transport in a newly designed copper electrorefining cell. in *229th ECS Meeting*, San Diego, California, 1 June 2016
10. W. Zeng, S. Wang, M.L. Free, Experimental studies of the effects of anode composition and process parameters on anode slime adhesion and cathode copper purity by performing copper electrorefining in a pilot-scale cell. Metall. Mater. Trans. B **47**, 3178–3191 (2016)

Molten Flux Design for Solid Oxide Membrane-Based Electrolysis of Aluminum from Alumina

Uday Pal, Shizhao Su and Thomas Villalon

Abstract Compared with the current Hall–Héroult process for smelting aluminum, the Solid Oxide Membrane (SOM)-Based electrolysis process brings various advantages such as simplified design, lower cost, lower energy use, and many other environmental benefits. However, the state-of-the-art SOM electrolysis process is limited by the solid oxide membrane's stability and its compatibility with the molten salt and the electrode materials. This work will identify cost-effective molten salt compositions and membrane materials/architecture along with electrodes and current collectors that are all chemically compatible and provide optimum electrochemical properties for efficient aluminum oxide electrolysis.

Keywords Solid oxide membrane · Yttria stabilized zirconia · Aluminum production · Flux

Introduction

The Solid Oxide Membrane-based Electrolysis process is a cost-effective and energy efficient clean metals production technology that utilizes an oxygen-ion-conducting solid oxide membrane (SOM) to directly electrolyze the desired metal oxide dissolved in a non-consumable molten salt. During the SOM electrolysis process, the desired ions of the metal such as magnesium, aluminum, silicon, or a rare earth are reduced at the cathode while oxygen ions migrate through the membrane and are oxidized at the anode (Fig. 1). It provides a general route for directly reducing various metal oxides to their respective metals, alloys, or inter-

U. Pal (✉) · S. Su · T. Villalon
Division of Materials Science and Engineering,
Boston University, Brookline, MA 02446, USA
e-mail: upal@bu.edu

U. Pal
Department of Mechanical Engineering, Boston University,
Brookline, MA 02215, USA

S. Wang et al. (eds.), *Applications of Process Engineering Principles in Materials Processing, Energy and Environmental Technologies*,
The Minerals, Metals & Materials Series, DOI 10.1007/978-3-319-51091-0_4

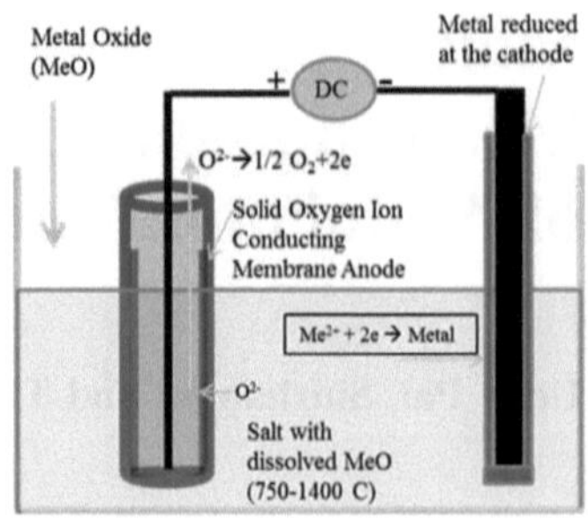

Fig. 1 Schematic of the generic SOM electrolysis process

metallics. Compared with current metal production processes, such as the chloride-based electrolysis process for magnesium production and the Hall–Héroult process for smelting aluminum, the SOM process brings various advantages such as simplified design, lower cost, lower energy use, and many other environmental benefits. However, the state-of-the-art SOM process is limited by solid oxide membrane's stability and its compatibility with the molten salt and the electrode materials. This work will attempt at identifying cost-effective molten salt compositions and membrane materials/architecture along with electrodes and current collectors that are all chemically compatible and provide optimum electrochemical properties for efficient aluminum oxide electrolysis.

Background

Today's metals industry is a major consumer of carbon-based energy and generates significant amount of pollutants including greenhouse gases (GHG). The societal impetus to lower energy consumption and GHG emissions provides motivation for the development of clean technologies that could reshape the metals industry and enable environmental sustainability. To meet this challenge, (SOM)-based electrolysis process for metals production is being developed. It is a cost-effective, energy-efficient, and potentially zero direct emission process. The SOM-based electrolysis technology produces metals (Me) from their respective oxides (MeO_x) by the direct electrolysis reaction: $MeO_x \rightarrow Me$ (cathode) + $x/2\ O_2$ (anode). Specifically, in the metal product value chain from mined ores to finished products, the most energy-intensive step is usually the oxide to metal conversion process. For example, in primary aluminum production via the Hall–Héroult process, Al_2O_3 is directly electrolyzed, the electricity use is more than double the reaction enthalpy due to heat losses through frozen cryolite sidewalls. In addition, the hot CO_2 gas emitted from the graphite anodes contains not only HF, which renders heat recovery from exiting gas impractical, but also perfluorocarbons, powerful GHG with global warming potential that is 5000–10,000 times that of CO_2. Compared with the traditional methods of extractive metallurgy, the SOM process brings various advantages such as simplified design, lower cost, lower energy use, and possibly zero direct emissions. The state-of-the-art SOM process features the utilization of an oxygen-ion-conducting membrane, such as the yttria-stabilized zirconia (YSZ), for directly electrolyzing metal oxides. Since the late 1960s, researchers have recognized and investigated the use of zirconia solid electrolyte for metals production and refining; however, limited progress was achieved due to the poor stability of the zirconia membrane, low current density, and lack of an appropriate current collector assembly. Pioneering work in our laboratory identified new molten salt compositions with increased solubility of many oxides, decreased zirconia membrane corrosion rate, and helped in developing an inert-anode current collector. It has generated new interest in the solid oxide membrane-based electrolysis process for metals production. The technology has been applied on a laboratory scale to reduce various metal oxides or oxide compounds to the respective metals or alloys, including Mg, Al, Ti, Ca, Fe, Cu, Ta, Cr, Nb, Yb, Nd, Pr, Dy, Si, $CeNi_5$, and $La_xCe_{1-x}Ni_5$, Ti–Fe alloy, and Ti–Si intermetallics [1, 2]. However, the major impediment towards scale-up design and commercialization of this process for primary production of metals is identification of compatible cell components (molten salt composition, solid oxide membrane, cathode, and anode) that have the required properties under the chosen operating conditions (temperature and environment) for energy efficient and cost-effective operation. This paper targets aluminum production and provides a strategy to determine possible compatible membrane materials, salt compositions, and electrodes having the required properties. The major risk is materials instability under the operating conditions, however, several alternate materials can be considered for risk mitigating strategy. If successful, the resulting process will reduce aluminum oxide at lower cost, use at least 30% less electricity compared to the Hall–Héroult process, and not directly generate greenhouse gas (GHG) emissions.

New Research Strategy

New research must focus on the selection of membrane materials/architecture, salt systems, and electrodes for cost-effective and energy efficient aluminum production via the solid oxide membrane based electrolysis process. Solid Oxide Membrane stability during oxide electrolysis in molten salts is affected by both salt composition and the applied electric potential [3]. Based on our experience in modeling such systems, the design criteria and the property requirements for the membrane material, salt systems, and electrodes are outlined below.

- Selected salt must have high ionic and negligible electronic conductivity. A possible candidate flux system for 1000–1100 °C operation is MeF_2–CaO–Al_2O_3 system where ‘Me’ is one or more of Ca, Mg or other alkaline earth metals. A possible candidate system for 700–900 °C operation is MF–CaO–Al_2O_3 system where ‘M’ is one or more of Na, Li, or other Group I elements. The objective here is to design the conducting flux to have the required oxide solubility, low volatility, and also match its basicity with the membrane for stability.

- Selected membrane can have a one-end-closed tubular structure (Fig. 1). It can either be a stand-alone structure or coated over a porous support tube of the same material. It must have high oxygen ion conductivity (see table below) and preferably some (5–25%) electronic conductivity for low anodic charge transfer resistance and enable application of larger potentials (up to 5 V) for increased metal production rate without membrane dissociation.
- Selected membrane must be compatible with the salt (must have similar oxide basicity), reducing environment (stable under cathodic conditions) and the oxidizing gas environment on the anodic side (oxygen).
- Membrane Material and Design Option 1: Bi-layer membrane such that one layer is compatible with the cathodic (salt) side and the other layer is compatible with the anodic side (e.g. acceptor doped ceria and zirconia films on the cathodic and anodic sides, respectively).
- Membrane Material and Design Option 2: One-layer of composite membrane (e.g. acceptor doped ceria or zirconia with an electronic or mixed conducting perovskite such as $LaCrO_3$ or $La_{1-x}Sr_xCo_{1-y}Fe_yO_{3-x-y}$). These membranes will be stable and have the required oxygen ion and electronic conductivities under both cathodic (salt) and anodic (oxygen) environments. This will allow application of higher potentials without membrane dissociation.
- Membrane Material and Design Option 3: A single-phase mixed oxygen ion-electron conducting perovskite membrane layer that is stable under both cathodic (salt) and anodic (oxygen) environments (e.g. $La_{1-x}Sr_xCo_{1-y}Fe_yO_{3-x-y}$). Single phase brownmillerite and pyrochlore phases that are similarly mixed conducting may also be considered.
- Anode: Porous composite of acceptor doped ceria or zirconia with a p-type electronic or mixed conducting perovskite such as $La_{1-x}Sr_xMnO_3$ or $La_{1-x}Sr_xCo_{1-y}Fe_yO_{3-x-y}$, and silver. Previously developed inert anode current collector [4] can also be used with such anodes.
- Cathode: Liquid Al as cathode with C or TiB_2 serving as current collector (prior work) [4].

The above mentioned materials system must be screened to meet the following property requirements:

SOM cell component	Conductivity (S/cm)	Ionic transference number	Electronic transference number	Volatility ($\mu g/cm^2$-s)	Stability	Viscosity (Pa-s)	Oxide content (w%)
Molten salt (supporting electrolyte)	>2	~ 1	<0.1	<1	Match basicity with membrane	<0.1 Improve ion diffusivity	>5

SOM cell component	Oxygen-ion conductivity (S/cm)	Other conducting characteristics	Stiffness GPa Strength Gpa	Corrosion rate in molten salt (μm/hour)	Stability (no composition or phase change)
Oxide membrane	>0.01	Mixed ionic-electronic conductivity for stability and charge transfer	>100 >0.1	<0.2	Under both anodic and cathodic conditions

SOM cell component	Electronic conductivity (S/cm)	Other characteristics	Volatility ($\mu g/cm^2$-s)	Stability	Sustained current density at 3–5 V
Cathode	>100	Low charge transfer resistance	<0.01	Compatible with molten salt, current collector, and metal (aluminum) produced	>1 amp/cm^2 (Comparable or exceed Hall–Héroult process)
Anode	>100	Low charge transfer resistance; for solids, match expansion with membrane (within 12%)	<0.01	Compatible with membrane and current collector under anodic conditions (oxygen atmosphere)	>1 amp/cm^2 (Comparable or exceed Hall–Héroult process)

Component materials having the required properties need to be screened and run in trials of SOM electrolysis for Al production. The results of the electrolysis process can be analyzed using equivalent circuit modeling (Fig. 2) to quantify all the efficiency (polarization) losses [5]. The results will enable selection of the most cost-effective and optimum combination of salt chemistry, membrane material, electrodes and current collectors for the scale-up trial. Successful outcome of the work will make SOM electrolysis emerge as a promising process for Al production and also demonstrate it as an attractive and viable technology that can replace other state-of-the-art metals production processes.

Results

Aluminum oxide (alumina) is dissolved in a molten salt flux and electrolyzed employing an oxygen-ion-conducting yttria-stabilized zirconia. The above strategy was employed to select and design the molten salt flux. The base flux chosen was a eutectic composition of CaF_2–MgF_2. The more acidic the fluoride (e.g. MgF_2 in CaF_2–MgF_2 system) the more it favors forming complex ions; acidity is inverse of

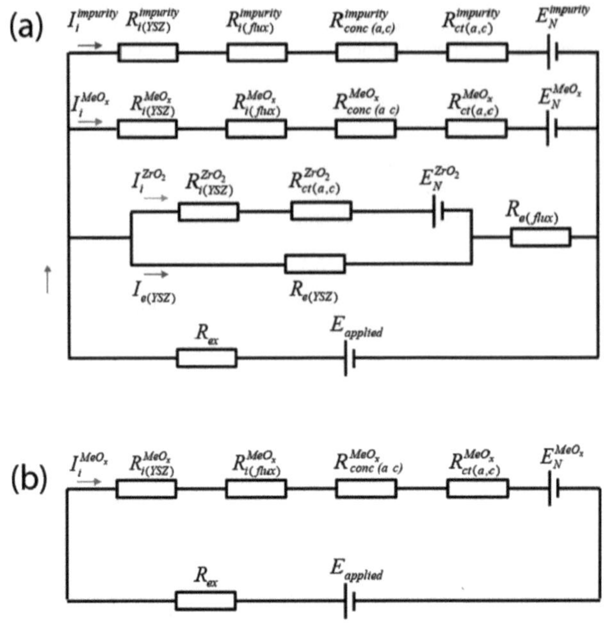

Fig. 2 Equivalent circuit of the SOM electrolysis cell with a traditional zirconia membrane: **a** General case and **b** Ideal case

basicity. When an oxide with a higher melting point, having no common specie and dissimilar size and structure is added to the fluoride flux (e.g. when Al_2O_3 is added to CaF_2–MgF_2) it is generally expected to have a very limited increase in the entropy of mixing and any limited solubility is expected to be due to complex aluminum oxyfluoride ion formation resulting in a decrease in the enthalpy of mixing. The acidic oxide (Al_2O_3) added will have a greater tendency to react with a more basic oxide, such as the ZrO_2 membrane, and thus result in the membrane attack. This phenomena has been experimentally observed in our laboratory (see Fig. 3).

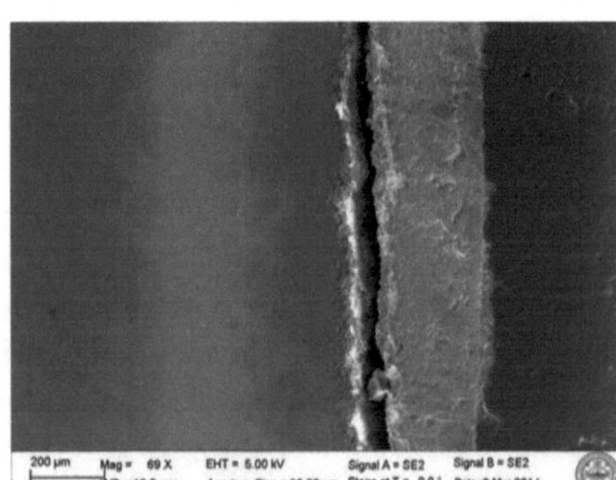

Fig. 3 YSZ membrane attack due to Al_2O_3 dissolved in MgF_2–CaF_2 flux (SEM image, YSZ exposed to molten flux for 8 h at 1175 °C)

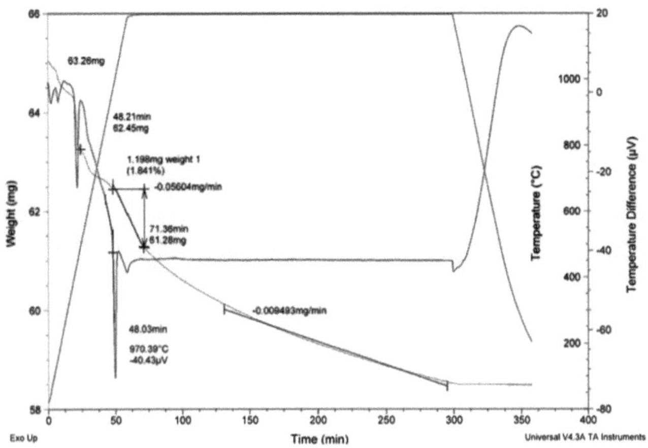

Fig. 4 Thermogravimetric analysis of 45 wt% MgF_2–55 wt% CaF_2 with 6 wt% CaO, 4 wt% YF_3 and 2 wt% AlF_3. Aluminum oxyfluoride is highly volatile

On the other hand if AlF_3 is added to the CaF_2–MgF_2 flux, more alumina (Al_2O_3) may dissolve in the flux resulting in a higher concentrations of the aluminum oxyfluoride complex ion. However, the formation of oxyfluoride species would likely compete with the reaction between the alumina and the zirconia membrane (membrane attack). The oxyfluoride species in general also have a higher vapor pressure and so even if it is the stable specie it may be lost if the temperature of the system is high. Laboratory experiments have indeed shown that aluminum oxyfluoride species volatilize at temperatures above 1000 °C (see Fig. 4).

When the AlF_3 is added to the flux, it volatilizes as aluminum oxyfluoride taking the oxygen from the basic CaO; the resulting flux had lower aluminum, oxygen and fluorine contents. To overcome these difficulties a more basic oxide than zirconia was added to the CaF_2–MgF_2 flux. The basic oxide is selected considering the mixing and bonding behaviors between the alumina and the basic oxide. For instance if CaO is selected for addition with alumina it is known that the CaO–Al_2O_3 system has a lower melting eutectic following similar consideration between the entropy and enthalpy of mixing; the complex alumina ions that form get depolymerized by the oxygen ions from the basic oxide. Therefore, when basic oxide CaO is added to CaF_2–MgF_2 flux, it dissolves due to increase in the entropy of mixing and the resultant flux is capable of dissolving the alumina (Al_2O_3) forming complex ions without forming the volatile oxyfluoride species (see Fig. 5).

Thus when CaO is added to the eutectic CaF_2–MgF_2 system, it dissolves the alumina likely not as an oxyfluoride but as a complex alumina ion that is associated with calcia (CaO); extent of polymerization is determined by the oxygen ions in calcia. Thus the more basic calcia interacts with the alumina and prevents it from attacking the zirconia membrane. For chemical inertness with the membrane, it was found necessary to tune the flux composition of the CaF_2–MgF_2–CaO–Al_2O_3

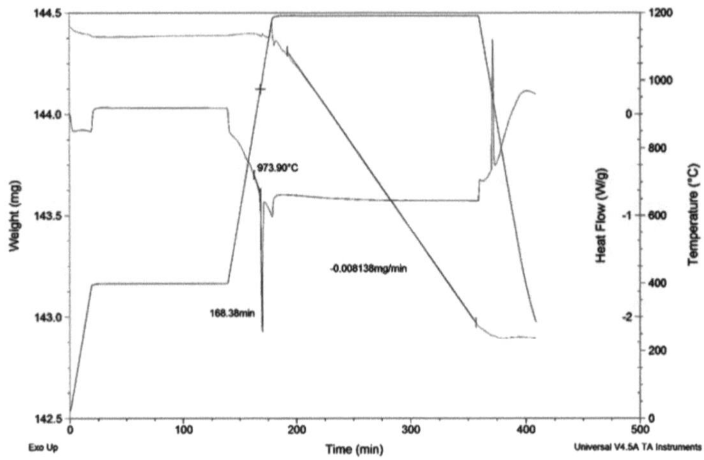

Fig. 5 Thermogravimetric analysis of 45 wt% MgF_2–55 wt% CaF_2 with 12 wt% CaO, 9 wt% Al_2O_3 and 4 wt% YF_3

system such that it matched the optical basicity of the flux composition represented by the eutectic CaF_2–MgF_2 system containing 10 wt% MgO (optical basicity 0.7–0.8). The latter composition was the flux used for magnesium oxide electrolysis using the yttria-stabilized zirconia membrane [6]. Also, if the zirconia membrane is stabilized with an oxide such as Yttria (Y_2O_3), then appropriate additions of yttrium fluoride and/or yttria must be added to the flux to eliminate any chemical potential gradient of yttria between the flux and the zirconia membrane (see Fig. 6).

The viscosity can be adjusted by tailoring the amounts of the eutectic fluoride system, the basic oxide, and the acidic oxide. Increasing the amount of the fluoride system and the basic oxide relative to the acidic oxide will decrease the viscosity (alumina is a network former and the basic oxide is a network breaker). The electrical conductivity can also be similarly adjusted. Increasing the amount of the fluoride system relative to the oxides increase the electrical conductivity of the system. The volatility can be tailored through selection of the fluoride system. Group II fluorides in general have lower volatility than Group I fluorides.

Based on the above-mentioned strategy we engineered a flux system for alumina electrolysis and successful aluminum production employing an oxygen-ion conducting yttria-stabilized zirconia membrane [7]. The flux composition that was successfully engineered was a eutectic MgF_2–CaF_2 (45 wt% MgF_2–55 wt% CaF_2) system having 3–9 wt% Al_2O_3, 6–12 wt%CaO, and 4 wt%YF_3. The flux optimization process is shown in a flow chart (Fig. 7). The physico-chemical properties of the flux system were measured and it met all the property requirements (specified above), such as electrical conductivity, transference numbers, melting point, volatility, membrane stability, viscosity, and alumina solubility [7].

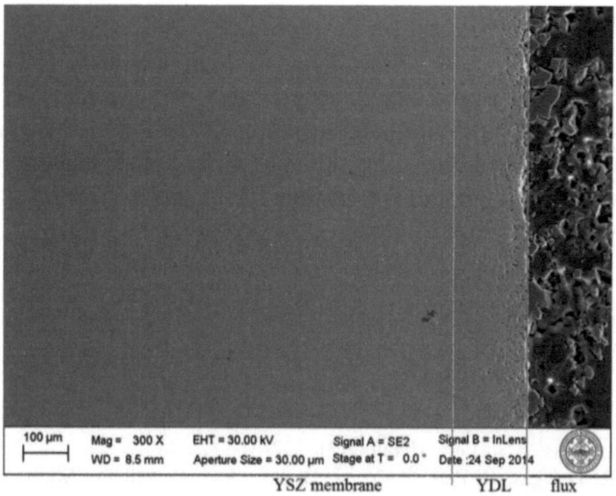

Fig. 6 The YSZ membrane after exposed to the flux (45 wt% MgF_2–55 wt% CaF_2 with 3 wt% Al_2O_3, 6 wt% CaO and 4 wt% YF_3) at 1200 °C for 500 h, showing negligible corrosion and limited yttria depletion layer (YDL) growth

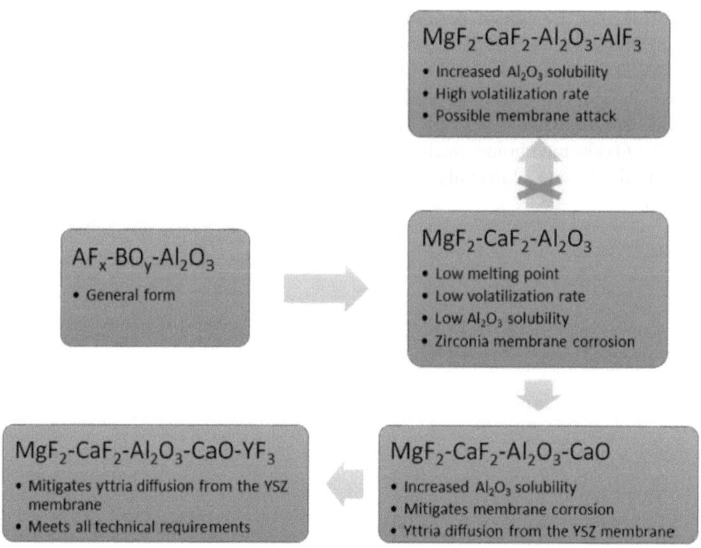

Fig. 7 Flux optimization flow chart

Conclusion

A design strategy and some experimental results are presented to identify and evaluate molten salt compositions and membrane materials/architecture along with electrodes and current collectors that are chemically compatible and provide optimum electrochemical properties for efficient aluminum oxide electrolysis employing the solid-oxide-membrane based electrolysis process.

References

1. X. Guan, U.B. Pal, Y. Jiang, S. Shizhao, Clean metals production by solid oxide membrane electrolysis process. J. Sustain. Metall. **2**(2), 152–166 (2016)
2. M. Suput, R. Delucas, S. Pati, G. Ye, U.B. Pal, A.C. Powell IV, Solid oxide membrane technology for environmentally sound production of titanium. Min. Process. Extractive Metall. **2**(2), 152–166 (2008)
3. X. Guan, U.B. Pal, A.C. Powell, Environmentally friendly solid oxide membrane electrolysis process for magnesium oxide reduction: experiment and modeling. Metall. Mater. Trans. E **1**(12), 132–144 (2014)
4. X. Guan, U. Pal, S. Gopalan, A. Powell, LSM ($La_{0.8}Sr_{0.2}MnO_{3-\delta}$)–inconel inert anode current collector for Solid oxide membrane electrolysis. J. Electrochem. Soc. **160**(11), F.1179 (2013)
5. X. Guan, U.B. Pal, Design of optimum solid oxide membrane electrolysis cells for metals production. Prog. Nat. Sci. **25**(6), 591–596 (2015)
6. E.S. Gratz, J.D. Milshtein, U.B. Pal, Determining yttria-stabilized zirconia (YSZ) stability in molten oxy-fluoride Flux for the production of magnesium with the SOM process. J. Am. Ceram. Soc. **96**(10), 3279–3285 (2013)
7. S. Su, Solid Oxide membrane electrolysis process for clean efficient aluminum production. Doctoral Thesis, Boston University, April 2016

Effect of Slag Phase on Mixing and Mass Transfer in a Model Creusot Loire Uddeholm (CLU) Converter

A. Chaendera and R.H. Eric

Abstract Gas jet was blown through bottom nozzles of a 1/5th scale CLU model in the presence and absence of slag. Kerosene and water were used to simulate the slag and metal phases respectively. Mixing time increased with increasing bath height and with a decrease in the gas flow rate. It was related to bath weight, bath height and gas flow rate. Slag increased mixing time by about 16% and the mass transfer values by about 32%. Mass transfer rates were higher near the bath surface and in the gas-liquid plume region. They increased with increasing gas flow rate and with a decrease in bath height. A relationship showing the dependence of derived mass transfer coefficient on the gas injection rate was established. Turbulence characteristics inside the bath liquid were established to vary with location, vessel geometry and gas injection rate.

Keywords CLU converter · Mixing time · Mass transfer coefficient

Introduction

Gas stirring plays very important role in steelmaking processes. The gases purged into the converter bring about thorough mixing of the bath thus accelerating the physical and chemical reactions taking place in the vessel. Knowledge of the slag impact on mixing and mass transfer rates in the converter operation helps to accurately determine the necessary gas blowing rate for efficient stirring of the bath. Therefore, a thorough understanding of the slag phase influence on mass transport in the bath is vital for efficient operation of the CLU-converter.

A. Chaendera · R.H. Eric (✉)
School of Chemical and Metallurgical Engineering,
University of the Witwatersrand, Johannesburg, South Africa
e-mail: rauf.eric@wits.ac.za

R.H. Eric
Department of Materials Science and Engineering, Aalto University, Espoo, Finland

S. Wang et al. (eds.), *Applications of Process Engineering Principles in Materials Processing, Energy and Environmental Technologies*,
The Minerals, Metals & Materials Series, DOI 10.1007/978-3-319-51091-0_5

Nyoka et al. [1, 2] studied the process of mass transport in a CLU converter using one-fifth scale water model experiments. Their investigations were conducted in the absence of a simulated slag phase. The bath was stirred by air injection through five bottom-placed nozzles. The mixing conditions in the bath were evaluated in terms of the gas flow rate (Q) and bath weight (W) through Eq. (1).

$$T_{mix} = 1.08Q^{-1.05}W^{0.35} \quad (1)$$

The bath surface and the gas-liquid plume regions in the purged bath were identified as the zones exhibiting the highest mass transfer rates and turbulence effects. Lehner et al. [3] in their investigations immersed graphite rods into argon stirred melts of known composition for specified times. The dissolution pattern was found to be non-uniform, with the maximum dissolution rates occurring in the vicinity of the melt surface. The mass transport coefficients were largely independent of the bath depth away from the bath surface but showed a small increase towards the bath bottom. The results of several studies [1, 6, 7] showed that the rate of mixing in metal processing operations depends on the rate of energy input or energy dissipation. Investigations by Wright [4], showed that the dissolution rates of the rods outside the plume region were lower than the rates in the gas-liquid plume region at the same gas flow rate. The mass transfer rates were found to increase with gas injection rates.

Akdogan and Eric [5] established that the presence of a simulated slag layer caused an increase in the mixing time. That was explained in terms of the simulated slag layer tending to dissipate some of the input energy. In gas-stirred metallurgical vessels the turbulence caused by the gas-liquid plume rising through the bath facilitates the increase in mixing. Several researchers [6, 7] have come up with different mathematical relationships to explain the mass transport phenomenon in the reactors. Asai et al. [6], presented the results of earlier studies and their own cold model work in terms of the values of exponent n, by the relationship in Eq. (2) where K is mass transfer coefficient and n is the stirring factor of the gas flow rate (Q).

$$K \propto Q^{n} \quad (2)$$

Numerous values of n have been reported in various plant and pilot plant studies on the desulphurisation of steel as well as in many water model experiments.

Szekely et al. [8] investigated mixing, flow phenomena and mass transfer in argon stirred ladles. The mass transfer measurements obtained for the dissolution of carbon rods in argon stirred melts provided semi-direct confirmation of predictions made regarding the turbulence patterns within the system. The mass transfer rate was observed to increase with an increase in the stirring gas flow rate. They also found that the local turbulence exhibited appreciable spatial variations. The vicinity of the plume recorded high levels of turbulence especially in the upper regions.

Experimental Measurements

The experimental set-up consisted of a clear cylindrical PVC tank that was one-fifth the size of the CLU-converter that was being simulated. Water and kerosene (10% by volume) were adopted to simulate the metal and the slag phase respectively. The top oxygen lance was not simulated in the investigation. The omission of the top lance was considered acceptable, since it is not used continuously and for a long period of time during the actual blowing process in the commercial CLU-converter. Total bath heights (including the slag layer) of 0.56, 0.61, 0.67, 0.72 and 0.78 m were investigated. The water bath heights before the addition of kerosene were 0.50, 0.55, 0.60, 0.65 and 0.70 m. Gas flow rates in the range between 0.010 and 0.023 m^3/s were adopted for the investigations. The gas injection rates between the model and the real CLU-converter were related through the modified Froude number, the value of which was kept 242 that was calculated for the industrial vessel.

Bath Mixing Intensity Measurements

The location for the pH-temperature probe in the bath and the point for injecting the tracer solution were fixed. The experimental conditions that were adopted by Nyoka et al. [1] were also used for these investigations. This was done to allow for a clear comparison of results obtained in the absence and in the presence of the simulated slag phase. The experimental runs were 20 min long. The average of at least eight values obtained from different experimental runs was adopted as the mixing time under the investigated conditions. The mixing time was calculated at 99.7% bath homogeneity. The effects of bath height, gas flow rate and the presence of slag phase on the mixing behaviour of the bath were measured.

Mass Transfer Experiments

Solid-liquid mass transfer measurements were performed by the mass loss technique. The investigation involved measuring the mass losses of the solid benzoic acid specimens suspended at various radial and vertical locations in the gas purged bath for 15 min. Instantaneous mass loss measurements were reduced to equivalent mass transfer coefficients according to Eq. (3) in which $m = \rho\pi R^2 L$, $A = 2\pi RL$, and ΔC is the change in concentration. The term k is the mass transfer coefficient, R is the radius, L is the length, and ρ is the density of the specimen (Fig. 1).

$$(dm/dt) = kA\Delta C \tag{3}$$

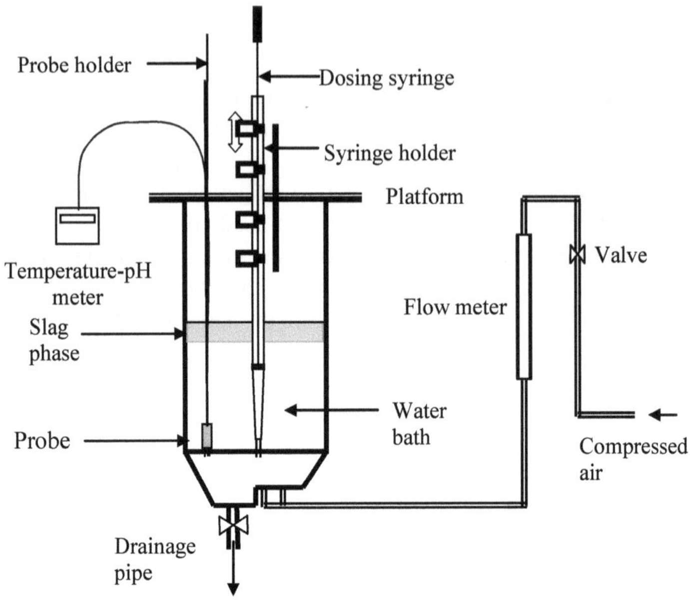

Fig. 1 The general set-up of the apparatus used for mixing time measurements [9]

Taking the benzoic acid concentration in the bulk liquid to be zero, the mass transfer coefficient is estimated from Eq. (4) where C_s is the saturation concentration of benzoic acid in water.

$$k = (dR/dt)\rho/C_s \tag{4}$$

Measuring the radii of the specimens that had undergone uneven dissolution during bath purging posed a challenge. Consequently, average radii were calculated when necessary from the weight losses recorded on the specimens. Experimental results were used to explain the effect of bath height and gas flow rate on the mass transfer coefficients in the bath, which were calculated from the obtained data. The experiments were carried out at gas flow rates of 0.010, 0.015 and 0.023 m^3/s and water bath heights of 0.4, 0.5 and 0.65 m. The spatial positioning of the specimens in the experiments has been presented elsewhere [2].

Results and Discussion

The results obtained were discussed in terms of the bath height and gas flow rate. The effect of these operating variables on mixing intensity and solid-liquid mass transfer was analyzed.

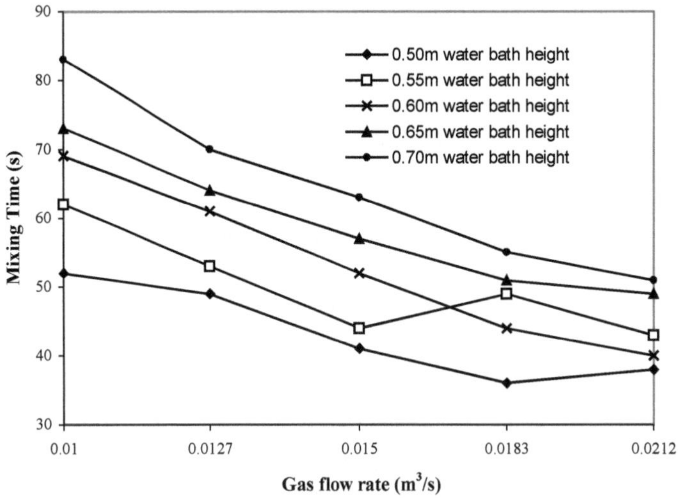

Fig. 2 Effect of gas flow rate on the mixing time at different bath heights

Mixing Rates

Generally, the mixing time was observed to decrease with increasing gas flow rate. Figure 2 presents the effect of gas flow rate on the mixing time. This trend was explained in terms of the gas-liquid plume velocity changes that resulted from increasing the gas flow rate. At a constant bath height, as the gas flow rate increased the centre line plume velocity increased resulting in rapid recirculation of the bath that enhanced mixing. In earlier investigations [5, 10], it was noted that as the gas flow rate increased the plume cone angle increased thus increasing the plume radius at the surface and effectively increasing the plume volume in the bath. The enlarged high turbulence region enhanced the bath liquid circulation and mixing in the bath. The largest decrease in mixing time was observed at the highest bath height investigated, 0.78 m. The mixing time decreased by about 40% at this bath height but only by 31% at 0.56 m bath height. The observed changes in mixing time showed the possible depreciating influence of gas flow rate at higher levels of specific energy input.

Many researchers [1, 4, 5, 11] have also observed the inconsistent behaviour of the mixing time shown in Fig. 2 at 0.50 and 0.55 m water bath heights. The seemingly consistent occurrence has been attributed to various reasons. The most convincing explanation was given by Mazumdar et al. [11]. They attributed the anomalies to zone shifting in the bath. In the current study, the probe was positioned in a generally 'slow moving zone' that was determined under specific conditions adopted in the preliminary experiments. Mixing rates in the various regions of the bath vary according to bath liquid activity or turbulence. Due to probable zone shifting, the probe was situated in a less active zone when the mixing time increased with an increase in gas flow rate. After this anomaly, it was observed that the mixing time at the 0.55 m water bath height resumed the normal trend when it decreased after the gas flow rate was increased to 0.0212 m^3/s. It was therefore concluded that the new zone of activity was now determining the time taken for bath homogeneity. The existence of minimum mixing time values could have resulted from a probable swirling of the bath. Observations by Krishna et al. [10], of the two-phase plume and flow conditions in the bath revealed that the plume swirled above a certain gas flow rate resulting in enhanced mixing rates in the bath. The minimum mixing times recorded in this study were therefore attributed to probable swirling of the plume that increased the rates of mixing in the bath.

Above a certain value of specific energy input, the level of tracer dispersion considered for the mixing time value to be determined could have been altered. Under conditions of very high stirring employed it was possible that, the partially affected zones in the bath were reached by the mixing power of the purging gases. The result was an increase in the volume of the bath liquid in the bulk flow regime and a consequent shifting of the equilibrium concentrations of acid in the two flow regimes. Consequently, the mixing time increased against an increase in gas flow rate and as a result the mixing time values recorded tended to be longer than was expected. It was therefore deduced that under these conditions the transfer of acid in the radial direction was lowered as the gas-liquid plume velocity in the axial direction increased.

Using the experimental results obtained, a correlation that could be used to relate the mixing time (T) to gas flow rate (Q) and bath weight (W) under the investigated conditions was established as,

$$T_{mix} = 2.87Q^{-0.73}W^{0.24} \tag{5}$$

The established correlation shows a greater influence of the gas flow rate on the mixing time than the weight of the load in the model CLU-converter. To get a clearer picture the effect of the bath height on mixing time was also investigated.

For the gas flow rates investigated, the mixing time increased with an increase in the bath height as presented in Fig. 3. The decrease in specific stirring energy input resulted in increased dead volumes at the bottom of the vessel and decreased bubble formation as the bath height increased thus, the reduction in mixing intensity observed. The bath height effect on the mixing time was greatest at the lowest gas flow rate as significant specific energy input changes were recorded here. An inconsistent variation of the mixing time and bath height was also observed.

The occurrence of peaks in Fig. 3 at gas flow rates 0.0183 and 0.0212 m^3/s was attributed to zone shifting in the bath as explained above. A peak was observed when a quieter zone occupied the point where the probe was located. Assuming this was the case then, the mixing times recorded were bound to be the opposite of what was expected since mixing rates in the various regions of the bath vary according to activity or turbulence.

The mixing pattern at total bath heights above 0.61 m could be explained in terms of the columnar effect becoming pronounced with an increase in bath height.

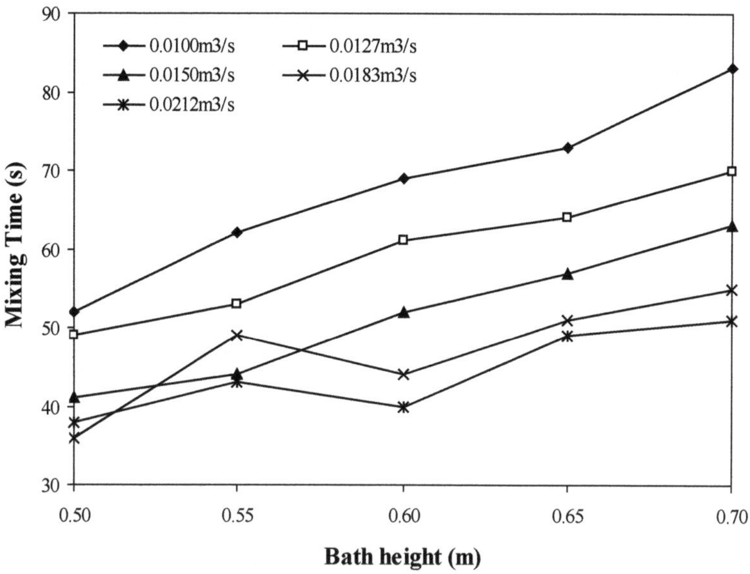

Fig. 3 Effect of bath height on the mixing time

Castillejos and Brimacombe [12] and Asai et al. [6] discussed that at high ratios of bath height to diameter of the vessel, dead volumes that formed near the bottom of the vessel had the effect of increasing mixing time. Krishna et al. [10] explained this phenomenon by means of a multiple circulation model, in which the bath consists of several circulation cells in the axial direction that make mixing by bulk flow ineffective. The formation of the circulation cells in the bath inhibits the thorough mixing of the bath solution. Thus, longer periods of mixing are observed under conditions of increased bath load as more circulation cells are formed. When the formation of circulation cells in the bath occurs at lower bath heights, the mixing time that is recorded tends to be longer than the expected value. The multiple circulation cells model could also be used to explain the large mixing time values recorded at the 0.61 m bath height and gas flow rates of 0.0183 and 0.0212 m^3/s. The deviation from the 'normal' depends on the extent of circulation cells formation. Using this model, the formation of circulation cells in the bath was therefore more pronounced at the gas flow rate of 0.0183 m^3/s than at 0.0212 m^3/s.

The bath height in the vessel that had a diameter of 0.5 m was observed to have the largest effect on the mixing time. Using the experimental results obtained, a correlation that could be used to relate the mixing time (T) to gas flow rate (Q), bath weight (W) and bath height (H) was established as,

$$T_{mix} = 4.39Q^{-0.73}W^{0.24}H^{1.12} \quad (6)$$

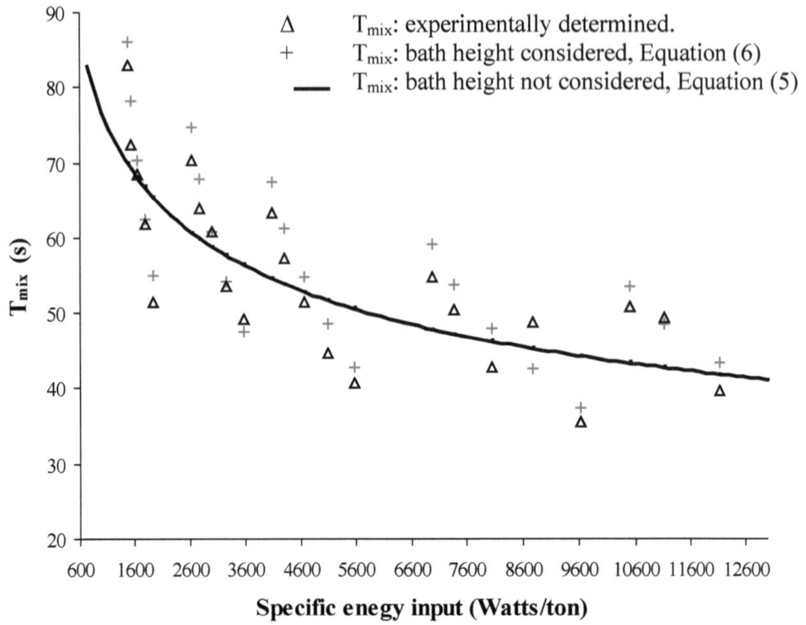

Fig. 4 A comparison of the deviations of the estimates from the experimentally determined mixing time values

The mixing time values estimated from this correlation had an average deviation of ±3% from the experimentally determined values. Figure 4 presents a comparison of the mixing time estimates obtained using the two correlations.

The crosses in Fig. 4 above represent the sample points estimated by Eq. (6) and the solid line by Eq. (5). The sample points that were determined taking the bath height into consideration, i.e. +, are closer to the experimentally determined values represented by the triangles, Δ. These two corresponding values can be identified in Fig. 4 by checking points falling on the same specific energy input line. The solid line in Fig. 4 was plotted from the points calculated from the relationship that did not take the contribution of the bath height into account.

The effect of slag was established by comparing the results of the current study to those obtained by Nyoka et al. [1]. Longer mixing time results were recorded because of the increased load coming from the incorporated slag phase. The influence of tracer partitioning between the two liquids constituting the bath also prolonged the mixing time further. Since the incorporated slag phase constituted only 10% of the bath, it was argued that the reduced mass transfer rates of the tracer in the bath liquid had a huge influence.

Investigations by Akdogan and Eric [5] centred on measuring the rate of transfer of a tracer from the water phase to the kerosene phase, i.e.; liquid-liquid mixing and mass transfer. They reported that the time taken for the tracer to distribute into equilibrium partitioning ratios was a variable of gas flow rate, bath height and

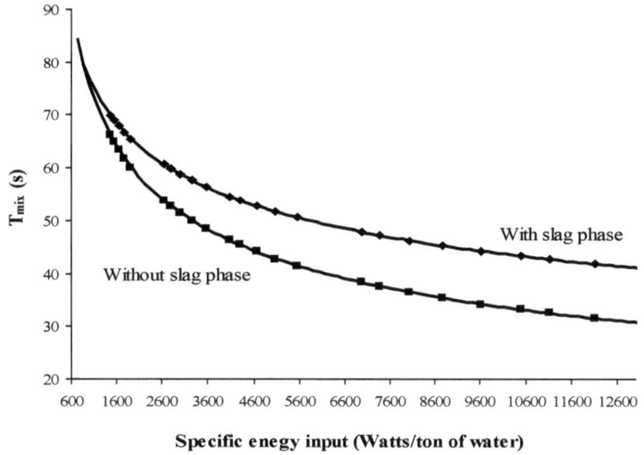

Fig. 5 Effect of the slag phase on the mixing time values predicted for model converter operation

nozzle configuration. As a result of the density differences of the kerosene and water, the purged liquid at the bottom of the bath where dosing was effected had a greater proportion of water than kerosene. The consequence of that was the prolonging of the mixing time as an equilibrium portion of tracer had to be transferred to the larger part of the slag phase in the dynamic fluid system.

The mixing time values predicted using the established correlations with and without the slag phase were also compared on the basis of their variations with specific energy input to the water bath. Figure 5 presents a comparison of predicted mixing time values when the baths were at the same water bath height but had different weights because of the slag included in the other case. It is apparent from the graphs in Fig. 5 that the mixing time increased with the inclusion of the slag phase. It took a longer period for the same energy input to cause the same level of mixing on an increased bath volume. The slag phase dissipated some of the input energy to the bath. Thus, higher mixing time values were recorded. The effect of the slag phase increased with the specific stirring energy density. Presence of slag reduced the mean gas velocity in the bath. Thus, an increased change in mixing time value was observed when the specific energy input increased. These observations indicated the significance of kinetic energy at high specific stirring energy density. Under conditions of high bath weight, the kinetic energy factor of input energy has relatively less effect on mass transfer rates that are mainly controlled by buoyancy energy. Other factors therefore become more significant in influencing the mixing behavior of the bath.

At lower stirring rates, the bath took long periods to homogenize. The extended period observed during bath homogenization allowed the tracer enough time to partition itself, in equilibrium amounts, between the two phases making up the bath liquid. This occurred in almost the same time as it would have in a single phase. This was the reason why the mixing time change with the presence of the slag phase

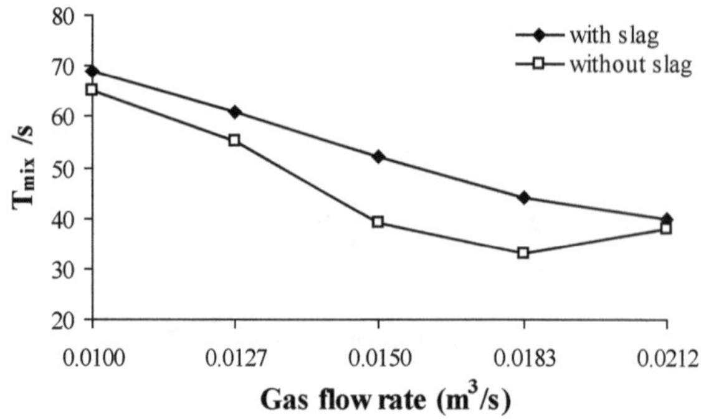

Fig. 6 Mixing time variation with gas flow rate at a water bath height of 0.60 m

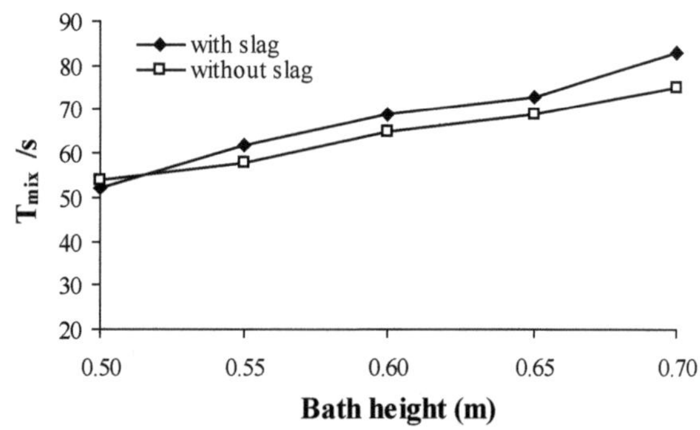

Fig. 7 Mixing time variation with bath height at a gas flow rate of 0.010 m^3/s

was small at lower specific energy input than at high specific energy input. Under the latter conditions, bath homogeneity was quickly achieved in a single phase system. Thus, the presence of a second phase tended to prolong the mixing time as the tracer distributed itself in a bath that was not homogenous. The different kerosene-water proportions at the bottom and surface was responsible for the inhomogeneity of the bath liquid. In the absence of stirring or at very low stirring rates, mixing of the tracer is mainly controlled by convective forces that are less affected by a second phase. Therefore, the mixing time values approached each other as the specific energy input was lowered. An average increase in mixing time of 16.3% was calculated from the correlations used for the conditions investigated in the presence of slag.

The effect of the presence of slag was clearly shown by graphical comparisons of the results in Fig. 5. The graphs in Figs. 6 and 7 illustrate further the effect of the simulated slag phase under the conditions investigated.

It is apparent that the inclusion of a second phase had the effect of increasing the mixing time. The difference in mixing time values was lesser at lower bath heights where the specific input rates were lower and vice versa. An average value of the percentage change in mixing time was calculated from these results to be 17%. The value compared quite well with the 16.3% value obtained from Fig. 5 presentation. The close similarity of these percentage changes in T_{mix} values is important. The similarity exhibited supported the soundness of the developed mathematical relationships. It was therefore concluded that the analysis presented provided a true picture of the influence of the slag phase on mixing in the model bath. The comparison of the two investigations was therefore considered successful.

Mass Transfer Rates

The solid-liquid mass transfer rates in different regions of the bath were represented by calculated mass transfer coefficients. Figure 8 presents an example of the effect of increasing gas flow rate on mass transfer coefficients in the bath. S2, S3, S4 and S7 are individual solid benzoic acid samples suspended at different locations.

The increased stirring power imparted by the purging gases was responsible for increasing the energy needed for the dissolution of the benzoic acid specimens. At a constant bath height, as the gas flow rate increased, the center line plume velocity increased resulting in rapid recirculation of the bath. The plume cone angle was observed to increase with gas flow rate thus effectively increasing the plume volume in the bath. The effect of the enlarged high turbulence region was the enhancement of circulation, mixing and solid-liquid mass transfer. Significant mass transfer rates were observed near the bath surface. The weight of the much reduced overlying liquid near the bath surface promoted the multiplication of the rising air bubbles in the buoyancy energy dominated region. Under these conditions more stable gas bubbles were formed, thereby promoting turbulence and mass transfer. The wave action, splashing and eddies caused by the gases leaving the bath at high velocity markedly enhanced the dissolution rates of benzoic acid specimens in the region.

Mass transfer coefficients were almost constant inside the bath but increased slightly at the bottom. Samples S2 and S7 were located inside the bath but outside the highly turbulent gas-liquid plume. Sample S3 was in the eye, transition region between the turbulent gas plume region that was originating from the nozzles and the recirculation liquid from the bath surface. The kinetic energy of the bath liquid and bubble formation were lower in these zones. The presentation in Fig. 8 shows that there was almost uniform change in the mean mass transfer rates as the gas flow rate increased. Comparable gradients of the plots indicated an almost constant effect of increased stirring power on the mass transfer coefficients inside the bath. The gradient of the graph observed for sample S4 is slightly steeper indicating the influence of more turbulence that existed in the region close to the bath surface as the gas flow rate was increased. The results obtained were consistent with the observations made in earlier investigations. The mass transfer coefficient was highest close to the bath surface, lower and almost uniform inside the bath before increasing slightly at the bottom of the vessel. The spatial positioning of the benzoic acid specimens in the three regions influenced the observed mass transfer characteristics. The geometry of the bath also modified the mass transfer characteristics of the bath in certain locations, especially close to the step at bottom.

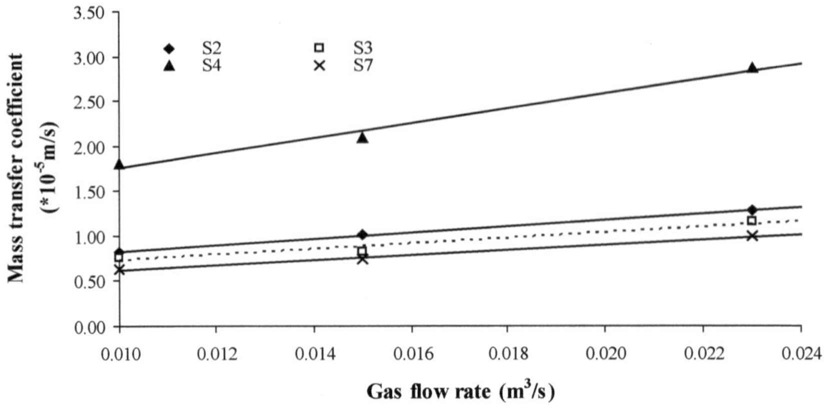

Fig. 8 Effect of gas flow rate on mass transfer coefficient at 0.44 m bath height

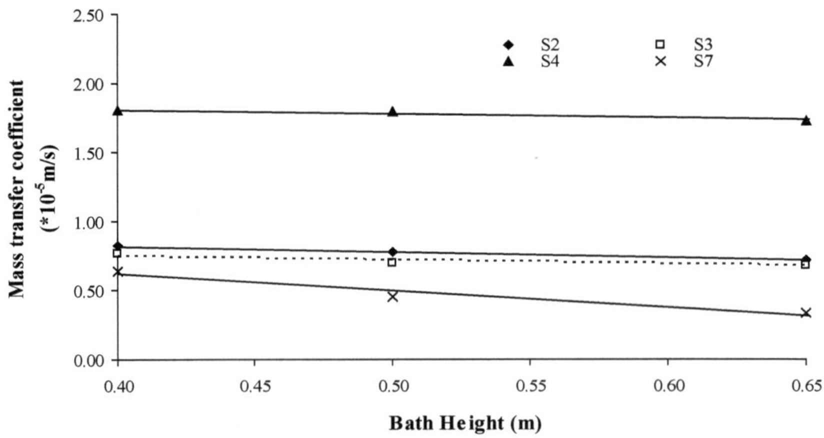

Fig. 9 Effect of bath height on mass transfer coefficient at 0.01 m^3/s gas flow rate

Increasing bath height reduced the mass transfer properties in the model converter. Figure 9 presents the effect of bath height on the mass transfer coefficient. The negative gradients of the graphs resulted from reductions in specific bath energy as the bath load was increased at a constant gas flow rate. The change in mass transfer coefficients was gradual for most locations at 0.010 m^3/s gas flow rate. However, the gradients of the graphs became steeper as the gas flow rate increased. The sensitivity of mass transfer coefficients at higher gas flow rates was attributed to a much larger specific energy input drop as the bath height was increased. As the load increased, input energy dissipation also increased hence the reduction in the specific input energy to the bath. The exponential relationship that exists between mass transfer and the gas injection rate was also responsible for greater sensitivity of the mass transfer coefficients at higher gas flow rates.

Turbulence in gas-stirred metallurgical vessels was established to be a result of the gas-liquid plume rising through the bath by Akdogan and Eric [5]. An increase in mass transfer rates was attributed to increased turbulence in the bath. At very high gas blowing conditions used in the CLU-converter, the effect of gas flow rate on mass transport becomes less pronounced. The observations made in this study is supported by the work done by Wright [4] and Chou et al. [13]. In these studies of mass transport, it was established that the mass transfer rate increased with gas flow rate. The mass transfer coefficients initially increased strongly with the gas flow rate, but this effect diminished with further increases and lower stirring factors were observed at higher gas flow rates. The dissolution rates were observed to be significantly higher in the plume region than in the other parts of the bath. Chou et al. [13] also attributed this to better mixing of oil and water in the cold model. As the gas flow rate approached 0.00067 m^3/s the improvement in the mass transfer rate with gas flow rate became less obvious.

A comparison of mapping results obtained in the absence and presence of the simulated slag phase was done to establish the effect of the simulated slag phase. Comparisons of the respective mass transfer coefficients obtained in the model experiments are presented in bar charts. The results of the experiments showed that the mass transfer values were lowered by the presence of the slag phase. This was attributed to an increased bath load reducing the specific stirring energy under the given conditions. Although a lesser factor, due to increased viscosity of the bath when slag was present, the dissolving capacity of the bath was reduced. The decrease in the mass transfer coefficients in the presence of slag were not constant but averaged 19.3% for all the gas flow rates investigated. The decreases were about 26, 14 and 18% for the gas flow rates 0.010, 0.015 and 0.023 m^3/s respectively. The changes in mass transfer coefficients were highest at low gas flow rates. This was explained by the fact that, when stirring energy is low any opposing energies or factors result in a large change in mass transfer coefficients. The effect was minimised as the gas flow rate increased, resulting in increased stirring energies, whereas the retarding factors remained almost unchanged.

Figure 10 presents the mass transfer coefficient in the bottom region of the converter in the presence and absence of the slag phase. Mass transfer coefficients were generally higher in the regions close to the nozzles where samples S3, S4 and S5 were located. The step at the bottom of the model converter was responsible for increased turbulence in its vicinity thus, the higher mass transfer rates observed on sample S3. Mass transfer rates in the bottom region increased with gas flow rate.

However, the mass transfer rates in this region were generally low because the samples were placed at the end of the bath liquid re-circulating loop. The liquid in this region had lost much of its kinetic energy. Turbulence was therefore limited as gas bubbles were only confined to regions in the vicinity of the nozzles.

The presentation in Fig. 11 is an example of comparisons of the mass transfer coefficients near the cone region. The samples were located above the nozzles, where the cylindrical section of the vessel started. A greater influence of the gas-liquid plume in this region became apparent. As a result of the close vicinity of

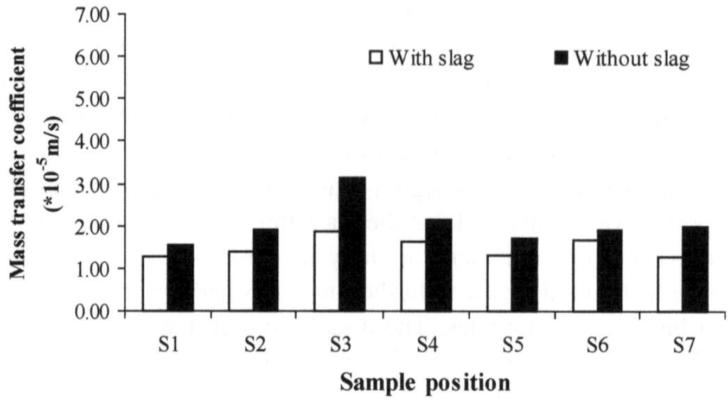

Fig. 10 Comparison of mass transfer coefficient values at 0.010 m^3/s gas flow rate in the bottom region. S1 to S7 are individual solid benzoic acid sample positions

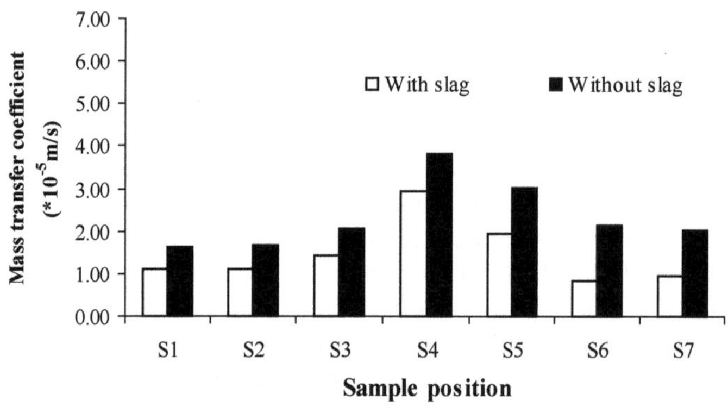

Fig. 11 Comparison of mass transfer coefficient values at 0.010 m^3/s gas flow rate near the cone region

the nozzles to the region, samples S4 and S5 that were located inside the high turbulent gas-liquid plume experienced much increased mass transfer rates.

Generally, the mass transfer rates were higher in the bottom region than near the cone region for positions that did not fall inside the plume. The mass transfer rates followed the turbulence patterns. Two distinct regions were observed in the near cone region. There was a region that fell inside the plume and another one that was extending from the vessel walls to the interface between the re-circulating liquid and the plume region. Existence of dead volumes in the latter region was quite probable.

When the gas flow rate was low slow moving slag (kerosene) droplets were observed in this region for quite long periods of time before they broke up into minute particles and became indistinguishable which was an indication of the existence of dead volumes resulting in low turbulence levels and low mass transfer coefficients.

The average change in mass transfer coefficients was 37.7% in the cone region. The reductions in mass transfer coefficients in the presence of the slag phase were about 38, 39 and 36% for the gas flow rates 0.010, 0.015 and 0.023 m^3/s respectively. A greater reduction in the mass transfer coefficient values near the cone region than in the bottom region was observed when the slag was present. This probably resulted from a combination of two factors. Firstly, the less exposure of the samples in the bottom region to the aggressive gas-liquid plume turbulence meant that the samples experienced marginal changes originating from a reduced bath stirring power that resulted from the presence of slag. The major driving force for dissolution in the bottom region was eddies produced by the bath liquid bouncing off the bottom of the vessel. These eddies were most likely less affected by the reduction in specific stirring energy of the bath. Secondly, in the near cone region, mass transfer processes were mainly driven by gas stirring power related to turbulence. The reduced influence of the bottom step in this region led to reduced gas stirring power. Thus, the reduced specific energy input resulted in less effective mass transfer in the presence of slag.

Samples near the bath surface region recorded the highest mass transfer rates. As explained earlier, the increased turbulence resulting from a much wider plume region plus the splashing action of the liquid was responsible for the high mass transfer rates. The sudden pressure drop at the bath surface also added to the turbulence in this region. Eddies produced in the liquid by the gas leaving the bath at high velocity were responsible for the increase in the dissolving power of the liquid in the vicinity of the bath surface.

Figure 12 shows the effect of slag on mass transfer coefficients near the bath surface. Samples at location S4 recorded maximum mass transfer coefficients. The plume effect at this location was marginal only in the bottom region where the sample was not positioned inside the plume volume. The average change in mass transfer coefficients was 40% near the bath surface region. The reductions in the mass transfer coefficients with slag inclusion were about 33, 39 and 48% for the gas flow rates 0.010, 0.015 and 0.023 m^3/s respectively. Comparable average changes

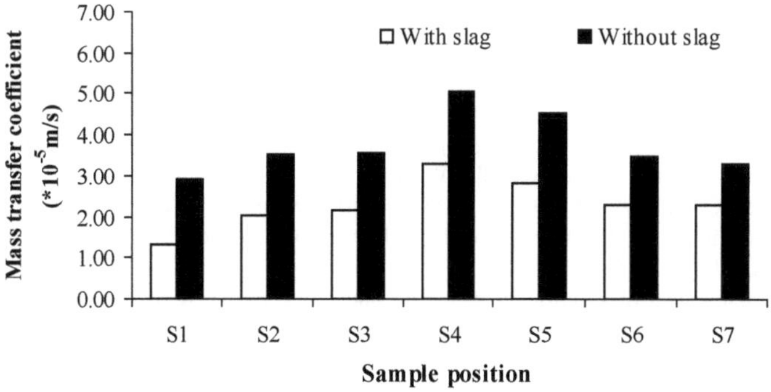

Fig. 12 Comparison of mass transfer coefficient values at 0.015 m^3/s gas flow rate near the water bath surface region

in mass transfer coefficient values near the cone region and water bath surface regions show that the presence of the slag had similar effect in the two regions. This was to be expected as mass transfer processes in both regions were mostly influenced by turbulence in the bath liquid resulting from bath purging.

Conclusions

Under conditions of a constant slag proportion in the bath, the mixing time was found to increase with bath height and decrease with gas flow rate. The mixing time data obtained was related to the specific energy dissipation rate. The correlation, $T_{mix} = 4.39Q^{-0.73}W^{0.24}H^{1.12}$, was established to estimate the mixing time in the model vessel to an acceptable degree of accuracy. Presence of the simulated slag phase resulted in the mixing time increasing by an average of about 16%. The mixing time was significantly prolonged at high levels of specific energy input to the bath and only marginally at very low specific energy levels.

Mass transfer coefficients were experimentally determined at varying gas flow rates and bath heights using dissolution of cylindrical solid benzoic acid samples suspended at various positions in the model reactor in the presence of slag. It was established that the bath liquid is in continuous circulation in an anti-clockwise motion with the leading edge above the gas purging nozzles. This motion provided for bulk mixing and mass transfer of the bath liquid. Turbulence characteristics inside the bath liquid varied with location and gas injection rate. The bath surface region, the gas-liquid plume region and the bath bottom region displayed higher mass transfer characteristics. The step at the bottom of the vessel increased turbulence by impeding the smooth bulk liquid flow. As a result, solid-liquid mass transfer characteristics of the bath increased in its locality.

References

1. M. Nyoka, G. Akdogan, R.H. Eric, N. Sutcliffe, Mixing and solid-liquid mass-transfer rates in a Creusot-Loire-Uddeholm vessel: a water model case study. Metall. Trans. **34B**, 833–842 (2003)
2. M. Nyoka, *Mixing and Mass Transfer in a Creusot Loire Uddeholm Converter*. M.Sc. (Eng) Dissertation, University of the Witwatersrand, Johannesburg, 2001
3. T. Lehner, G. Carlsson, T. Hsiao, in *On fluid flow and metallurgical reactions in gas stirred melts*. Scaninject II, 2nd International Conference on Injection Metallurgy, Lulea, Sweden, 1980, vol. 1, pp. 22.1–22.34
4. J.K. Wright, Steel dissolution in quiescent and gas stirred Fe/C melts. Metall. Trans. **20B**, 363–374 (1989)
5. G. Akdogan, R.H. Eric, Model study on mixing and mass transfer in ferroalloy refining processes. Metall. Trans. **30B**, 231–239 (1999)

6. S. Asai, M. Kawachi, I. Muchi, in *Mass transfer rate in ladle refining processes*. Scaninject III, 3rd International Conference on Refining of Iron and Steel by Powder Injection, Lulea, Sweden, 1983, no. 12, 12.1–12.29
7. R.J. Fruehan, L.J. Martonik, The rate of absorption of nitrogen into liquid iron containing oxygen and sulphur. Metall. Trans. **11B**, 615–621 (1980)
8. J. Szekely, T. Lehner, C.W. Chang, Ironmaking Steelmaking **6**, 285–293 (1979)
9. A. Chaendera, R.H. Eric, in *Effects of a simulated slag phase on mixing and mass transfer rates in a 0.2-scale Creusot-Loire Uddeholm converter model*, ed. by F. Kongoli, R.G. Reddy. Sohn International Symposium on Advanced Processing of Metals and Materials; Thermodynamics and Physicochemical Principles: Iron and Steel Making, TMS, Warrendale, USA, 2006, vol. 2, 263–270
10. G.G. Krishna Murthy, A. Ghosh, S.P. Mehrotra, Characterization of two-phase axisymmetric plume in gas stirred liquid bath-a water model study. Metall. Trans. **19B**, 839–850 (1988)
11. D. Mazumdar, S.K. Kajani, A. Ghosh, Mass transfer between solid and liquid in vessels agitated by bubble plume. Steel Res. **61**(8), 339–346 (1991)
12. A.H. Castillejos, J.K. Brimacombe, Physical characteristics of gas jets injected vertically upward into liquid metal. Metall. Trans. **20B**, 595–601 (1989)
13. J.M. Chou, M.C. Chuang, M.H. Yeh, W.S. Hwang, S.H. Liu, S.T. Tsai, H.S. Wang, Effects of process conditions on mixing between molten iron and slag in smelting reduction vessel via water model study. Iron Steelmaking **30**(3), 195–202 (2003)

Part II
Electrometallurgy

Modeling of Aluminum Electrowinning in Ionic Liquid Electrolytes

Mingming Zhang and Ramana G. Reddy

Abstract A 3-D mathematical model was developed for the batch reactor of low temperature aluminum electrowinning using ionic liquid electrolytes. This model describes the deposition process by incorporating the mass transport of participating ionic species, homogeneous chemical reactions within the diffusion layer, and the associated electrochemical kinetics. Processing parameters, current and potential distribution, species concentration, fluid flow distribution, and electrode spacing were evaluated for the optimal reactor performance. The results indicated that the electrode spacing significantly affects the electrolyte fluid flow and current density distribution. The parallel electrode configuration (in line with electrolyte inlet) improved the convection and resulted in uniform current density distribution and electrolyte fluid flow. However, for this electrode configuration, electroactive species distribution was most favorable between the electrodes. Perpendicular configuration of electrodes resulted in a more non-uniform fluid flow within electrolyte domain, and it has a potential to cause non-uniform deposits. Aluminum electrowinning experiments were conducted using batch reactor at 80 °C, electrolyte flow rate of 5–20 ml/min, and applied cell voltage of 3–3.5 V. Good agreement was obtained between the model and the batch aluminum electrowinning experimental results.

Keywords Aluminium · Electrowinning · Computer fluid dynamics (CFD) · Ionic liquid

M. Zhang (✉)
ArcelorMittal Global R&D, East Chicago, USA
e-mail: Mingming.Zhang@ArcelorMittal.com

R.G. Reddy
Department of Metallurgical and Materials Engineering, The University of Alabama, Tuscaloosa, AL 35487-0202, USA

S. Wang et al. (eds.), *Applications of Process Engineering Principles in Materials Processing, Energy and Environmental Technologies*,
The Minerals, Metals & Materials Series, DOI 10.1007/978-3-319-51091-0_6

Introduction

The low temperature aluminum electrowinning process in ionic liquids has been shown to possess a number of advantages over traditional organic solvents and molten salts. It is considered to be a low energy consumption process, since the operating temperatures are low (<100 °C), and the emission of pollutants is negligible due to low vapor pressure [1]. Particularly chloroaluminate ionic liquids made by mixing imidazolium chloride and $AlCl_3$ have been investigated by many groups as possible electrolyte for aluminum production [1–3]. The $C_n$mimCl-$AlCl_3$ melts showed high conductivity, produced high purity, and thick adherent deposits [3]. By varying the molar ratio of the organic salt to $AlCl_3$, basic, neutral and acidic melts can be formed. System with a molar ratio of $AlCl_3$:RCl greater than 1 is acidic, which contains R^+, $Al_2Cl_7^-$ and $AlCl_4^-$ ions. Aluminum electrodeposition is feasible only in acidic melts. The cathode reaction can be expressed as:

$$4Al_2Cl_7^- + 3e^- = Al + 7AlCl_4^- \quad (1)$$

The reaction taking place at the anode is:

$$4AlCl_4^- = 2Al_2Cl_7^- + Cl_2 + 2e^- \quad (2)$$

The overall electrolysis reaction can be obtained by combining Eqs. (1) and (2):

$$2Al_2Cl_7^- = 2Al + 2AlCl_4^- + 3Cl_2 \quad (3)$$

A parallel plate electrochemical reactor is the most convenient cell design for an electrowinning process. Such design helps to obtain the current density, energy consumption and efficiency of aluminum deposition process data. These process parameters depend on the input of the cell potential, electrolyte composition and flow rate. Pickett [4] has given cell design methods for parallel plate cells. But they were based on the assumption that the mass transfer controlled electrode reaction is dominant. The models developed by Parrish and Newman [5] and Caban and Chapman [6] approaches are referred to as thin diffusion layer models. They assumed that the reactant concentration is constant in the bulk solution and differs only within the thin diffusion layer of the electrodes. White et al. [7, 8] presented a mathematical model for a parallel plate electrochemical reactor and associated recirculation system. They assumed that the distance between the electrodes is much smaller than the dimensions of the electrodes and the axial diffusion and migration were ignored. Georgiadou [9] further refined these models considering exponential kinetics of heterogeneous electrode reactions under steady state laminar flow. All these models mentioned above were studied for aqueous electrolytes and their PDE equations were solved using either finite difference (FD) or finite volume (FV) methods in two dimensions (2-D).

Wahnsiedler [10] developed the isothermal electrochemical model for industrial Hall cells. The authors have developed a general three-dimensional (3-D) model mainly applicable to Hall industrial aluminum electrowinning cell. Their model uses fluid dynamics codes based on concentration, heat and momentum conservation. A macro-homogeneous approach has been developed by Newman [11], to describe the potential variations in the electrolytes. The material balance in the electrolyte was described using the concentrated solution theory and in the solid phase using a Fickian diffusion equation in Cartesian coordinates. Later, Pals and Newman [12], Song and Evans [13] and Wang et al. [14] extended this model by including the energy balance terms in order to predict cell temperature. This electrochemical-thermal model using local heat generation terms has shown to be more accurate and includes all the features of the isothermal model developed by Wahnsiedler [10]. The electrochemical-thermal model in its entirety or in partial was used by various researchers for modeling single electrodes and or full electrochemical cells to simulate various operating cell conditions at high temperatures. The problems encountered with aluminum electrowinning systems such as gas evolving, thermal runaway, and occurrence of undesired side reactions were addressed using this model.

The objective of this paper is to provide design information for the laboratory/batch scale electrowinning process using ionic liquid electrolytes at near room temperatures. In conjunction with various scale experimental efforts, a 3-D numerical modeling was developed for understanding and controlling the fluid dynamics and electrochemical reactions at the electrode-electrolyte interface during the low temperature aluminum electrowinning in ionic liquids. The approaches developed in this paper can be extended for the large scale production of aluminum using ionic liquids.

Experimental Methods

The ionic liquid electrolytes were prepared according to the procedures reported earlier [2]. The ionic liquid (C_6mimCl) was mixed with anhydrous $AlCl_3$ in the electrolyte reservoir with a constant stirring in glove box (deaerated with argon). After the temperature of the electrolytic mixture was stabilized, the reservoir was removed from glove box to ventilated cabinet and connected to the electrolytic cell. Electrolysis begins when the flow rate and temperature of electrolyte reaches pre-determined values. The ceramic piston pump and hot plate were used to control the flow rate and temperature of the electrolyte.

The electrolysis experiments were conducted using an electrolytic cell (500 ml cylindrical beaker) as shown in Fig. 1. Cathode was copper sheet (5.5 $\times$ 3.0 $\times$ 0.1 cm) and anode was graphite plate (5.5 $\times$ 3.0 $\times$ 0.4 cm). A Kepco ABC programmable power supply (Flushing, N.Y.) was used as AC power supply and a Keithley 2000 Multimeter was used to measure the cathode potential. A Corning stirrer/hotplate Model PC-320 (Acton, MA) was used to maintain the temperature of electrolyte and the flow rate of electrolyte was controlled by a Cole Palmer Series Q ceramic valveless piston pump (Vernon Hills, IL).

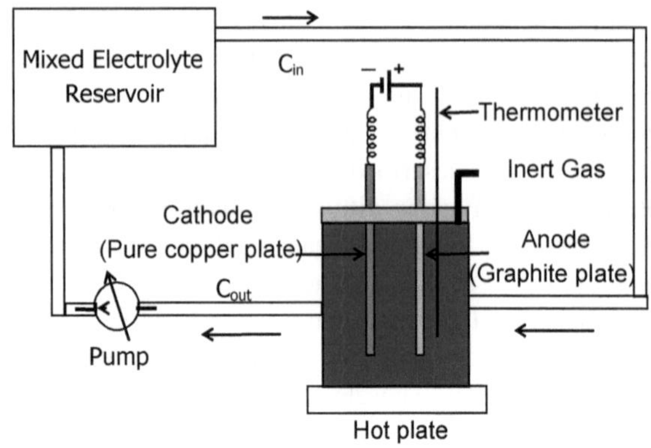

Fig. 1 Schematic for the batch recirculation electrowinning

Table 1 Electrowinning parameters

Electrolyte composition	$C_6mimCl + AlCl_3$ (Al concentration: 5 mol/L)
Temperature	80 °C
Cathode	Copper
Anode	YBD extrude carbon
Electrolyte flow rate	5–20 ml/min
Avg. current density	150–250 A/m^2
Applied cell Volt.	3.0 V versus Al(III)
Electrode area	5.5 $\times$ 3.0 cm^2

Table 1 shows the conditions employed in the electrowinning experiments. All experiments were conducted under a constant applied cell voltage condition. Inert gas (Argon) was purged through electrolysis cell during all experiments. After the experiment, the pump was stopped and electrodes were removed from the cell. The deposits were cleaned and weighted. Further analysis was performed to determine current efficiency and microstructures of the deposits.

Mathematical Formulation

Electrical Potential Modeling

The electrical potential distribution in 2-D electrowinning domain results was presented in our earlier paper [15]. In the present paper, the same methodology was extended to 3-D domain by solving the Poisson equation which describes the potential distribution within the bulk electrolyte. The Poisson equation relates the charge density to the Laplacian of the electric potential, as shown in Eq. (4).

However, the use of both Poisson's equation and electroneutrality could render inconsistence especially when it needs to determine the behavior of each species in the electrolyte near the electrodes. This is because the charge density in the electric double layer near the electrodes cannot be neglected. This double layer region has very large electric field and thus the electroneutrality may not be valid. The detailed equations describing this region are addressed in section Electrode-Electrolyte Interface.

The Poisson equation for Φ is solved with constant values of Φ at the electrodes, and zero flux at the other boundaries.

$$\nabla \cdot (\nabla \phi) = \nabla^2 \phi = -\frac{\rho'}{\varepsilon} \tag{4}$$

With the help of the Ohm's and Coulomb's Laws, it is possible to numerically solve the Poisson equation in 3-D domain;

$$i = -\kappa \nabla \Phi \tag{5}$$

$$\nabla \cdot \Phi = 0 \tag{6}$$

The partial current densities can be evaluated once the conductivity and potentials are known. The conductivity of the electrolyte can be calculated by the following equation [16].

$$\kappa = F^2 \sum_{i=1}^{2} m_i C_{i,ref} \tag{7}$$

Fluid Flow Modeling

The fundamental physical laws applicable to the electrolyte fluid flow system are the Navier-Stokes equations. The expression for the electrolyte fluid motion is given by:

$$\rho \frac{\partial u}{\partial t} + \rho (u \cdot \nabla) u = -\nabla P + \mu \nabla^2 u + \rho g \tag{8}$$

The continuity equation which expresses conservation of the mass is given by:

$$\nabla \cdot (\rho u) = 0 \tag{9}$$

For the electro-hydrodynamic flow which is the case in electrowinning process, it is necessary to simultaneously solve for the transport of space charge and the electric field, in addition to the momentum equations and the mass conservation.

Mass Transport in Bulk Electrolyte

The governing equations for species involve a mass transport for species *i* and the electroneutrality condition. The governing equations within the electrolyte domain include differential material balances for all species considered in the model. The conservation of species expression is given by:

$$\frac{\partial C_i}{\partial t} = D_i \nabla^2 C_i + \frac{z_i}{|z_i|} m_i \nabla \cdot (C_i \nabla \phi) \tag{10}$$

where, t denotes time, D_i is the diffusion coefficient, z_i is the electrical charge number and m_i is the ionic mobility of species i.

The electroneutrality condition in the electrolyte domain is given by:

$$\sum_i z_i C_i = 0 \tag{11}$$

The current density is determined by combining the Eqs. (10) and (11). The equation is given by:

$$i = -F \sum_i z_i D_i \nabla C_i - F \sum_i z_i^2 D_i \nabla \Phi \tag{12}$$

Electrode-Electrolyte Interface

Because electrode reactions which occur at the electrode-electrolyte interface play a major role in determining the electrochemical kinetics, knowledge of these reactions is critical in evaluating the process parameters such as current efficiency and current density. Species charges and solvent dipoles are arranged at the interface between electrode and electrolyte solution. The arrangement in the inter-phase region at the boundary of an electrode-electrolyte is termed as "electric double layer". Several models were proposed to describe electric double layer. These are: (a) Helmholtz compact layer model, (b) Gouy-Chapman diffuse layer model, and (c) Stern model. A more comprehensive model addressing the inter-phase phenomena in three-dimension is the Bockris-Devanathan-Muller (BDM) model. This triple layer model incorporates all the above three models. Because of the non-linear variation in the charge density within the electric double layer, the potential distribution is no longer governed by Laplace's equation. The charged species and potential distribution equations are coupled together. They have to be solved simultaneously in order to obtain the current density and potential distribution within the interface.

$$\rho' = \sum q_i c_j = \sum z_j e c_j \tag{13}$$

$$c_j = c_j^{\infty} \exp\left[-\frac{z_j e \phi_0}{k_B T}\right] \tag{14}$$

$$\nabla^2 \phi = -\frac{\rho'}{\varepsilon} = -\frac{1}{\varepsilon} \sum z_j e c_j^{\infty} \exp\left[-\frac{z_j e \phi_0}{k_B T}\right] \tag{15}$$

Due to the complexity involved in solving the above Eqs. (13)–(15), it was assumed that $i = i_{0k}$. The conductivity k was calculated using Eq. (7). The solution potential ϕ_o (at the cathode surface ϕ_{oc} and anode surface ϕ_{oa}) was calculated by substituting i and k into the Eq. (5). The equilibrium open-circuit potential E_k is given by Eq. (16). Then the reaction current density was calculated by Butler-Volmer Eq. (17).

$$E_k = E_k^0 - \frac{RT}{n_k F} \ln \prod_i \left(\frac{C_{i\infty}}{\rho}\right)^{s_{ik}} \tag{16}$$

$$i_k = i_{0k}^{ref} \prod_i \left(\frac{C_{i0}}{C_{iref}}\right)^{\gamma_{ik}} \times \left(\exp\left(\frac{\alpha_{ak} n_k F}{RT}(V_M - \Phi_{0a} - E_k)\right) - \exp\left(-\frac{\alpha_{ck} n_k F}{RT}(V_M - \Phi_{0c} - E_k)\right)\right) \tag{17}$$

Boundary Conditions

The boundaries of the studied electrowinning cell include inlet, outlet, electrode surfaces and inner cell walls. The implementations of these boundary conditions are shown in Fig. 2. The boundary conditions were evaluated assuming that the fluid flow is incompressible laminar flow (Re < 1000). The grid was extended to store the values of the physical properties at the boundaries. The boundary conditions applied in the discretized equation via source term are described in the following sections.

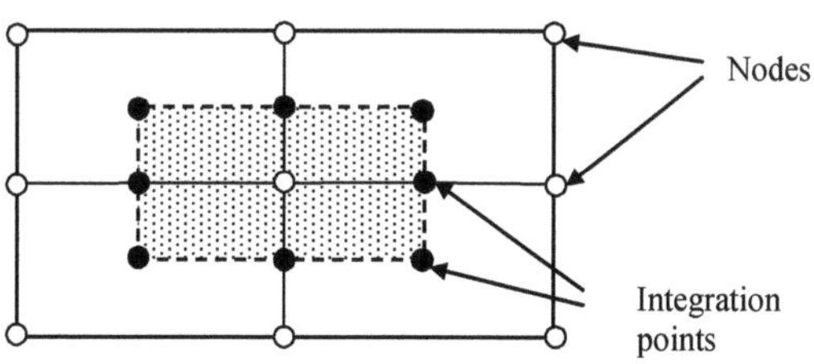

Fig. 2 Finite volume based discretization

Inlet Boundary Conditions

The distributions of all flow variables, except pressure, were specified at the inlet to the electrowinning cell. The flow direction was top nozzle to the bottom nozzle on the opposite side of cell. Uniform initial conditions were assumed for C_i and T,

$$C_i = C_\infty \tag{18}$$

$$T = T_0 \tag{19}$$

On the fluid inlet boundary,

$$u_i = u_0 \tag{20}$$

$$C_i = C_{i\infty} \tag{21}$$

$$\frac{\partial \Phi}{\partial n} = 0 \tag{22}$$

Outlet Boundary Conditions

At the outlet boundary, the flow is directed out of the domain. The hydrodynamic boundary condition (i.e., mass and momentum) involves constraints on the boundary static pressure, velocity or mass flow. For all other transport equations, the outlet value of the variable is part of the solution.

In our model, zero gradients for all variables, except pressure, in the normal direction were assumed.

On the fluid outlet boundary,

$$\frac{\partial C_i}{\partial n} = 0 \tag{23}$$

$$\frac{\partial \Phi}{\partial n} = 0 \tag{24}$$

Wall Boundary Conditions

In this modeling work, non-slip and non-penetrating wall conditions were assumed. The electrolyte velocity in the additional grid line was set to be zero (the grid line used for implementation wall boundary conditions). Since there is no mass transfer of species through the cell wall, the gradient of the species concentration was set to zero, ($\partial C_w/\partial n = 0$). This was implemented by setting the values of the species concentration at the wall as the nearby cells.

On insulating wall boundaries:

$$u_i = 0 \tag{25}$$

$$\frac{\partial C_i}{\partial n} = 0 \tag{26}$$

$$\frac{\partial \Phi}{\partial n} = 0 \tag{27}$$

Electrode Surface Boundary Conditions

On a cathode (+) or anode (−) surfaces, the interfacial flux of certain species should be equal to the generation or consumption of the species by the associated electrochemical reactions, i.e. the normal flux of reacting species is a sum of reaction currents (i_K), whereas for non-reacting species is zero. The conditions are given by:

Reacting species:

$$-D_i \frac{\partial C_i}{\partial n} - z_i D_i C_i \frac{\partial \Phi}{\partial n} = (\pm 1) \sum_k \frac{s_{ik} i_k}{n_k} \tag{28}$$

Non-reacting species:

$$-D_i \frac{\partial C_i}{\partial n} - z_i D_i C_i \frac{\partial \Phi}{\partial n} = 0 \tag{29}$$

The kinetic and thermodynamic parameters, species diffusion coefficients, and reference species concentrations used to solve this model are listed in Table 2 and Table 3. The governing equations and associated boundary conditions were numerically solved using finite volume method which is discussed in the following section.

Table 2 Standard potentials for reactions at cathode and anode [1, 2, 16]

Reactions	n_k	γ_{ik}	α_{ak}	α_{ck}	i_{0k}^{ref} (A/m^2)	E_k^0 (Volts)
(1)	3	0.7	0.5	0.5	10^{-5}	−0.5
(2)	2	0.7	0.5	0.5	10^{-5}	2.50

Table 3 Transport properties of aluminum ions in C_6mimCl-$AlCl_3$ mixture (at 25 °C) [3, 15]

Species/Anions	C_6mim^+	$AlCl_4^-$	$Al_2Cl_7^-$	Cl^-
Reference concentration, $C_{i,\,ref} \times 10^3$(mol/ml)	1	0.5	0.5	0.1
Diffusion coeff., $D_i \times 10^5(cm^2s^{-1})$	0.144	1.065	0.61	9.321
Mobility ($m_i = D_i/RT$) $\times 10^9$, $cm^2\ mol\ J^{-1}s^{-1}$	0.581	4.30	2.46	37.6
Transport number, $t^{+/-}$	0.71	0.23	0.06	–
Density, g/m^3	–	1.294	1.389	1.234
Molecular weight, g/mol	–	279.96	413.22	–
Molar volume, V_{mol}, cm^3/mol	–	216.4	297.54	–
Conductivity, mS/cm	–	22.6	14.5	–

Results and Discussion

Electrical Field Distribution

The results of the computation of the potential distribution and the electrical field is shown in Fig. 3 for V_{app} = 3 V versus Al (III) in the xy-plane at z = 0.5, 1.0, 3.0 and 9.0 cm. As seen from these figures, the electrical potential is distributed intensively between two separated electrodes and the main gradient exists around the edges of electrodes and induces electrolyte to flow from anode to cathode. Therefore, electric field intensity is high between the electrodes and almost zero in the cross region between electrodes and cell wall.

Concentration Profile

During the aluminum electrowinning in chloroaluminate ionic liquid electrolyte, the current is transported through the bath almost exclusively by the $Al_2Cl_7^-$ and $AlCl_4^-$ ions, whereas the cation C_nmim^+ acted as supporting electrolyte. Figure 4 shows 3-D examples of concentration profiles of $Al_2Cl_7^-$ within the electrowinning cell. Concentration gradients were found to decrease from the upper part of cell to the inactive bottom part. The iso-concentration contour becomes steeper near the center of the electrodes. The higher concentration gradients on the outside of electrode surface are likely due to the presence of large inactive surface. This surface allows partial replenishment of the boundary layer by the diffusion of the electroactive species. It is also observed that the contours are smooth and even at the low Reynolds number. In addition, the development of the concentration profile with increasing Reynolds number was as expected. At the low Reynold number, the concentration changes in the electrodes realm were mainly due to diffusion and migration. At greater Reynolds number, the convection effects dominated and

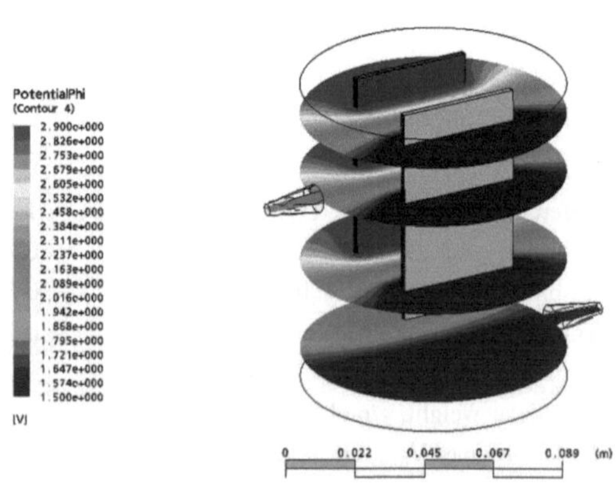

Fig. 3 Iso-potential contour with aspect ratio 1:1 at applied cell potential 3.0 V versus Al(III)

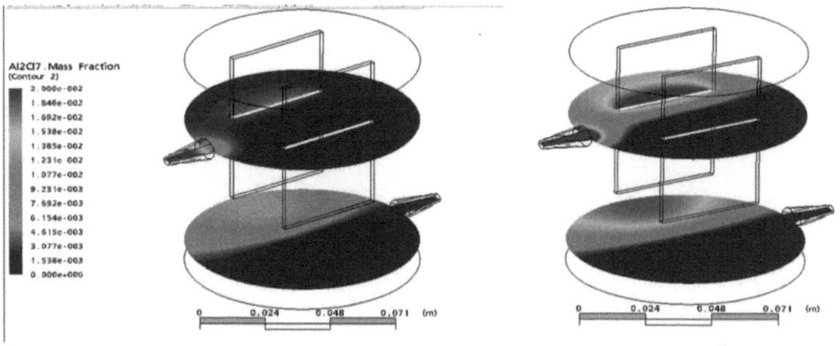

(a) (b)

Fig. 4 Iso-concentration contours of $Al_2Cl_7^-$ with aspect ratio 1:1 under laminar flow for Re = 7.31 (**a**) and for Re = 731 (**b**)

concentration changes were confined into a smaller distance from the electrode surfaces, since the thickness of the mass transfer boundary layer was progressively reduced.

Current Density Distribution

The current density modeling are based on the electrical field and current density distribution calculation. Figure 5 shows the computed current density vector plots. As seen in the current density vector plot, current flow from the anode through the electrolyte to the cathode, while it is nearly zero everywhere else. The streamtraces which are actually the paths of ions moving in the current field within the electrolyte correspond to the current line. The current density near the edges of

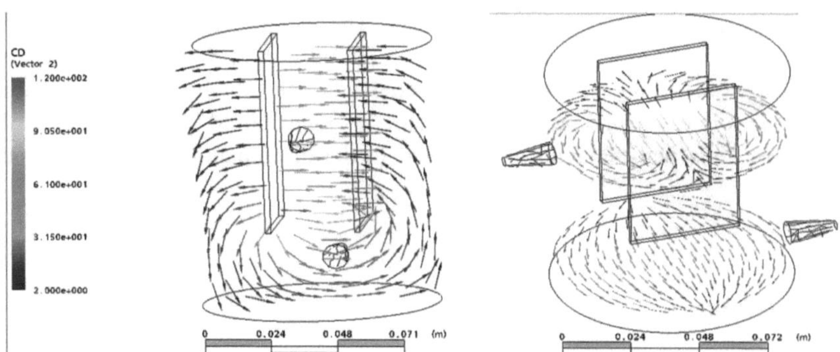

(a) (b)

Fig. 5 Distribution of dimensionless primary current density with aspect ratio 1:1 at applied potential 3.0 V versus Al(III)

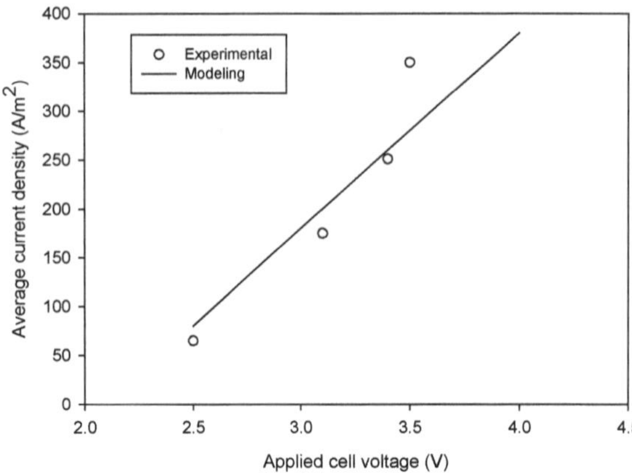

Fig. 6 Voltage versus current curves, experimental and modeling results

electrode is generally more intensive than elsewhere within the cross section. With varying electrode distance, the mean current density values Ψ within the cross section electrolyte domain changes. Under optimum electrode distance the mean current density (dimensionless) becomes maximum corresponding to the applied minimum cell voltage, where the electrolyte concentration, temperature and geometry of cell are set to constant. Our early modeling results on cylindrical electrowinning cell showed the relation between applied cell voltage and electrode distance with mean current density as a parameter when electrolyte temperature was 90 °C and concentration was 50 mol% of $AlCl_3$. The optimum electrode distance was determined to be ~ 1.5 cm when maximum mean current density was achieved with minimum applied cell voltage [15].

Based on the results of 3-D electrical field and current density modeling on electrowinning cell, the experimental data were compared with modeling results as shown in Fig. 6. The modeling results were in good agreement with experimental data below 3.5 V, but there was significant difference with applied cell voltages above 3.5 V. This may be due to the cathodic side reactions are significant when cell voltage increases to higher than 3.5 V. The electrochemical window of the studied ionic liquid electrolyte is between 2.2 and 4.0 V. When the applied cell voltage approaches the maximum electrochemical cell voltage (window) of ionic liquid electrolyte, electrochemical properties of ionic liquid electrolyte may also experience abnormal behavior as observed in the discrepancy between our modeling and experimental results.

Conclusions

A 3-D mathematical model for the low temperature aluminum electrowinning in ionic liquid electrolytes is developed. A number of ionic liquid electrowinning process parameters were evaluated for optimal reactor performance, including current and potential distributions, species concentration profiles, fluid flow distribution, and electrode spacing. The results indicated that the electrode configuration significantly affects the electrolyte fluid flow and current density distribution. The parallel electrode configuration (in line with electrolyte inlet) improved the convection and resulted in homogenous current density distribution and electrolyte fluid flow. The Electroactive species distribution was most favorable between the electrodes in the case of parallel electrode configuration. Perpendicularly configured electrodes resulted in a more complex fluid flow within electrolyte domain.

The specific conclusions derived from present study are as follows:

(1) The electric field intensity is high between the electrodes and almost zero outside cross section of the electrodes.
(2) The current density and fluid flow for two different electrode configurations were evaluated. Higher current densities are obtained with the parallel configuration.
(3) Aluminum electrowinning experiments were conducted using batch reactor. The current density modeling results are in good agreement with experimental data below applied cell voltage of 3.5 V.

Acknowledgements The authors would like to thank the financial support from the U.S. Department of Energy, The University of Alabama and the computer software assistance from Alabama Supercomputer Center (ASC).

Appendix

List of Symbols

C_i	Concentration of specie i, mol/m^3, $C^*_i = C_i/C$
C_i^0	Surface concentration of specie i, mol/m^3
C_i^{ref}	Reference concentration of specie i, mol/m^3
C_i^{∞}	Bulk concentration of specie i, mol/m^3, $C^*_{i\infty} = C_{i\infty}/C$
D_i	Diffusion coefficient of specie i, m^2/s
E_k	Equilibrium open-circuit potential of reaction k, Volts
E_k^0	Standard potential of reaction k, Volts
F	Faraday's constant, 96,487 C/mol
g	The acceleration of gravity, 9.8 m/s
i	Primary current density, Am^{-2}
i_{0k}	Reference current density of reaction k, Am^{-2}
i_{0k}^{ref}	Exchange current density of reaction k at an interface, A/m^2
i_k	Current density of reaction k, A/m^2
k_B	Boltzmann constant, 1.380×10^{-23} J/k; 8.617×10^{-5} electron-volt/k
m_i	Mobility of specie i, cm^{2}mol J^{-1}s^{-1}, $m_i = D_i/RT$
n	Outward normal direction to a boundary surface

$n_{i,j}$ Components of the outward normal surface vector
n_k Number of electrons transferred in reaction k
P Pressure, Pascal
q_i Electron charge of specie i, 1.602×10^{-19} C per electron
R Universal gas constant, 8.314 J/mol K
S^{ik} Stoichiometric coefficient of specie i in reaction k
S_θ Source term in scalar equation
T Temperature, K
t Time, second
u Velocity of the electrolyte, m/s
u_i Velocity of the specie i, m/s
$[u_i]$ i Component mean velocity, m/s
V_M Electrode potential, Volts
V_{app} Applied cell potential, Volts
$x_{i,j}$ Components of the Cartesian co-ordinate
z_i Charge number of specie i
α_{ak} Anodic transfer coefficient of reaction k
α_{ck} Cathodic transfer coefficient of reaction k
ε Permittivity of electrolyte, Faradays/meter
ρ' Charge density, ev/m^3
γ_{ik} Constant order of reaction k
Γ_θ Scalar diffusion coefficient term
Φ Electric potential, Volts
Φ_0 Solution potential in electrolyte adjacent to electrode surface, Volts
κ Conductivity of the electrolyte, S/m or $\Omega^{-1}m^{-1}$
ρ Density, kg/m^3
θ Scalar variable
μ Kinetics viscosity of the electrolyte, cm^2/sec

References

1. M. Zhang, V. Kamavaram, R.G. Reddy, New electrolytes for aluminum production: ionic liquids. JOM **11**, 54–57 (2003)
2. M. Zhang, V. Kamavaram, and R.G. Reddy, Application of fluorinated ionic liquids in the extraction of aluminum, in *Light Metals 2004* (2004), pp. 315–319
3. M. Zhang, V. Kamavaram and R.G. Reddy, Aluminum electrowinning in ionic liquids at low temperature, in *Light Metals 2005* (2005), pp. 583–588
4. D.J. Pickett, *Electrochemical Reactor Design*, 2nd edn. (Elsevier Scientific Publishing Co., New York, 1979)
5. W.R. Parrish, J.S. Newman, Current distributions on plane, parallel electrodes in channel flow. J. Electrochem. Soc. **117**(1), 43–48 (1970)
6. R. Caban, T.W. Chapman, Rapid computation of current density distribution by orthogonal collecation. J. Electrochem. Soc. **123**, 1976 (1036)

7. T.V. Nguyen, C.W. Walton, R.E. White, A mathematical model for a parallel plate electrochemical reactor, CSTR, and associated recirculation system. J. Electrochem. Soc. **133** (6), 1130–1138 (1986)
8. R.E. White, M. Bain, M. Raible, Parallel plate electrochemical reactor model. J. Electrochem. Soc. **130**(5), 1037–1042 (1983)
9. M. Georgiadou, Modeling current density distribution in electrochemical systems. Electrochim. Acta **48**, 4089–4095 (2003)
10. W.E. Wahnsiedler, *Mathematical Modeling of Materials Processing Operations*. (Metallurgical Society, Inc., 1987)
11. J.S. Newman, *Electrochemical Systems* (Prentice Hall, Englewood Cliffs, NJ, 1991)
12. C.R. Pals, J. Newman, Thermal modeling of the lithium/polymer battery, I. Discharge behavior of a single cell. J. Electrochem. Soc. **142**(10), 3274–3281 (1995)
13. L. Song, J.W. Evans, Electrochemical-thermal model of lithium polymer batteries. J. Electrochem. Soc. **147**(6), 2086–2095 (2000)
14. Z.H. Wang, W.B. Gu, C.Y. Wang, Extended abstracts. in: *Proceedings of the 196th ECS Meeting*, Honolulu, Hawaii. (The Electrochemical Society Inc., Pennington, NJ, 1999) vol. 96, pp. 17–22
15. M. Zhang, R.G. Reddy, Electrical field and current density distribution modeling of aluminum electrodeposition in ionic liquid electrolytes. ECS Trans. **1**(16), 47 (2006)
16. J.F. Yan, P.S. Fedkiw, C.G. Law, A model for the Kolbe reaction of acetate in a parallel-plate reactor. J. Appl. Electrochem. **26**(2), 175–185 (1996)

Electrochemical Processing of Rare Earth Alloys

Karen Sende Osen, Ana Maria Martinez, Henrik Gudbrandsen, Anne Støre and Ole Kjos

Abstract The light rare earth metals Nd, Pr, La, Ce as well as some alloys with Fe, are today produced in China by electrolysis in molten fluorides using oxide raw materials. A major challenge is to obtain a good cell operation without de-composing the electrolyte leading to emissions of perfluorinated carbon (PFC) green-house gases to the atmosphere. This work is focused on understanding the fundamental requirements to run the electrolysis cells for DyFe alloy production in an efficient and environmental friendly way. Electrolysis experiments was carried out in DyF_3-LiF melts at 1050 °C. A Fe rod was used as consumable cathode and the (consumable) anode was made of graphite. To establish at which anode potential PFC occurred and thus enabling optimisation of the oxide batch feed rate, analysis of the anode gases was performed with Fourier Transform Infrared Spectrometer (FTIR). The produced DyFe alloy was characterised by X-ray Diffraction (XRD), Scanning Electron Microscopy (SEM) and Energy Dispersive X-ray Spectroscopy (EDS).

Keywords Dysprosium electrolysis · Gas analysis · Perfluorinated carbon · Fluoride melt

Introduction

The light rare earth metals Nd, Pr, La, Ce as well as some alloys with Fe, are today produced in China by electrolysis in molten fluorides using oxide raw materials. A major challenge is to obtain a good cell operation without de-composing the electrolyte leading to emissions of perfluorinated carbon (PFC) green-house gases to the atmosphere.

The "Chinese technology" relies on an electrolytic process using a vertical electrode set-up cell, with graphite anodes and molybdenum or tungsten as inert

K.S. Osen (✉) · A.M. Martinez · H. Gudbrandsen · A. Støre · O. Kjos
SINTEF Materials and Chemistry, Richard Birkelands vei 2b, 7465 Trondheim, Norway
e-mail: Karen.S.Osen@sintef.no

S. Wang et al. (eds.), *Applications of Process Engineering Principles in Materials Processing, Energy and Environmental Technologies*,
The Minerals, Metals & Materials Series, DOI 10.1007/978-3-319-51091-0_7

cathode materials, where the Rare Earth element (RE) is deposited in a liquid form [1]. Consumable cathodes of transition elements are also used to obtain the corresponding RE-based alloy. The most common case is the use of Fe as cathode material, enabling use of the alloy product as a master alloy for permanent alloy manufacturing. The electrolyte typically consists of an equimolar REF_3-LiF mixture and RE_2O_3 is used as raw material at a temperature of operation of ca. 1050 °C [2]. Despite the fact that there is limited reliable information on the actual electrolysis technology used in China, it seems that the typical operational anodic current densities are around 1 A/cm^2 with CO reported as the main anode off-gas product, and that the process releases large amounts of PFC gases (mainly CF_4 and C_2F_6) almost continuously [3]. It is believed that conventional electrolytic production of Nd contaminates the atmosphere with PFC as much as the entire aluminium industry if the off-gases were not treated. PFCs are qualified as one of the gases with highest global warming potentials (GWP) for ozone-depleting substances. The tabulated data for 100-year GWP relative to CO_2 are 6500 and 9200 for CF_4 and C_2F_6, respectively (values referred to the GWP of CO_2). The life time is established to be 10,000 and 50,000 years for CF_4 and C_2F_6, respectively [4].

Reports on Dy–Fe electrolysis are scarce, but the off gas composition during electrodeposition of Nd have been studied by several. Keller [3] studied the off gas during Nd electrolysis in NdF_3–CaF_2–LiF (60-20-20 wt%) with 1 wt% Nd_2O_3. At 915 °C, below a cell voltage of 3.5 V, only CO/CO_2 was evolved. The content of both CF_4 and C_2F_6 became significant at 4.5 V. CF_4 levelled off at higher voltages (5–7 V), but the C_2F_6 continued to rise and became the predominant PFC component. Guihua et al. [5] studied the off gas products in NdF_3–LiF with only residual oxide content during polarisation at 1273 and 1173 K, and found that at the lower temperature C_2F_6 and CF_4 were produced under stable operating conditions without leading to anode effect (massive increase in cell potential and current drop due to blockage of active anode sites by the formed PFC).

A main challenge is to run a process with no PFC gas evolution and at the same time maintain a good cell operation in terms of avoiding sludge formation due to over feeding of oxide. To avoid PFC evolution, a sufficient supply of oxide is continuously needed to match the operational amperage. The relatively low oxide solubility and slow dissolution rate gives a narrow operational window in this respect. The objective of this work is to understand some fundamental requirements on how to carry out DyFe alloy production without PFC emission. This was done by studying the gas composition by on line gas analysis during electrolysis experiments in equimolar DyF_3–LiF melts at approximately 1050 °C.

Experimental

A vertical electrode set up was used for the electrolysis cell. This included graphite anodes surrounding the iron cathode rod. Such a set up would ensure an even current distribution. The molten DyF_3–LiF (50:50 molar basis) electrolyte was contained in a glassy carbon crucible (Mersen Nordic AB) and the experiment was run in a gas tight tubular furnace at 1050 ± 5 °C under Ar 5.0 atmosphere. Sintered alumina tubes were used both for feeding the Dy_2O_3 material and gas sampling. The Dy_2O_3 raw material was stored in a desiccator.

Electrolysis experiments were performed by applying constant current using an HP 6031A System Power Supply. The cell voltages as well as the temperature variations were logged during the course of the experiment by means of a multi-channel Keithley 2000 Multimeter. The temperature was monitored continuously using a thermocouple type S (Pt–Pt 10% Rh). A Fourier Transform Infrared Spectrometer (FTIR) apparatus (Protea ProtIR 204m) was used for online analysis of the off-gases (CO, CO_2, C_2F_6 and CF_4) of the cell during the electrolysis. Existing models were used in order to transform the FTIR readings to actual gas concentrations, upon calibration of the apparatus.

Approximately 580 g of equimolar DyF_3–LiF had been mixed, pre-melted and cooled in a graphite crucible. The obtained chunk of salt was then placed in the glassy carbon crucible, and the furnace was sealed off with the electrode set up in place and heated to ca. 1060 °C. After melting, around 2 wt% Dy_2O_3 (anhydrous 99.9% (REO), Strem Chemicals Inc.) was added and left to dissolve until the morning after. Then a melt sample was taken and immediately analyzed with respect to oxide content by carbothermal method using LECO TC-436 DR (Leco Corp., USA) to verify that everything had dissolved. When this was confirmed, the electrolysis experiment was run galvanostatically at anodic current density of around 0.15 A/cm^2, and initial cathodic current density of around 1.5 A/cm^2. The argon flow through the furnace was kept constant during the whole experiment to 500 mL/min, and the out gas was led through the FTIR via the alumina out gas tube situated above the melt "inside" the anode. Subsequent amounts of oxide were fed batch wise during the course of the electrolysis. The oxide additions could be carried out without opening the cell to the atmosphere, so the gas analysis could take place during the whole experiment without disturbance. The cell voltage, current and temperature were recorded, and in addition, the glassy carbon crucible was used as a "quasi reference electrode" so that the potential between the anode and the crucible and cathode and crucible could be monitored.

After typically 5–6 h of electrolysis, the furnace was slowly cooled under argon to room temperature. The next day the cell was opened, salt crushed, and any metal produced recovered. The composition of the Dy–Fe alloy obtained at the cathode is expected to have a dysprosium content of ca. 80 wt%, according to the reported phase diagram [6].

Results and Discussion

Figure 1 shows electrolysis data, and Figs. 2 and 3 the results from the gas analysis plotted together with the anode voltage for one of the experiments.

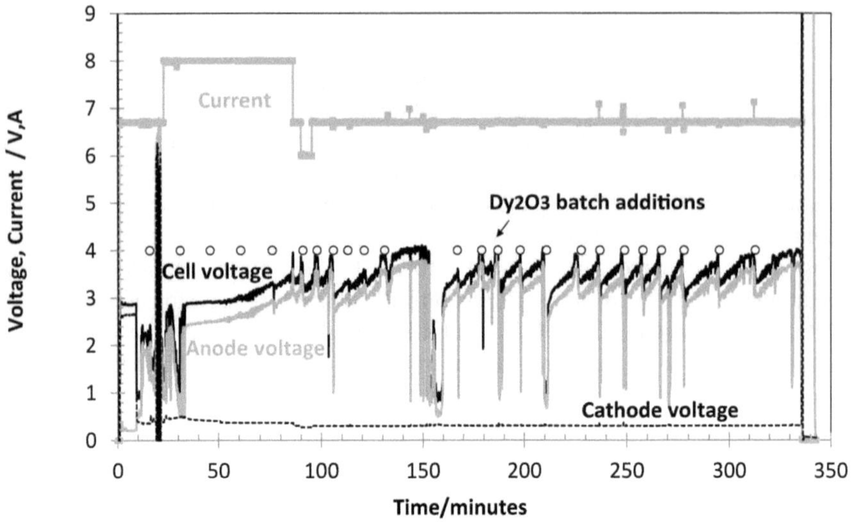

Fig. 1 Example of cell voltage, anode and cathode voltages, current and temperature recorded during the course of one of the experiments. The *red dots* indicate the point in time for the oxide feedings

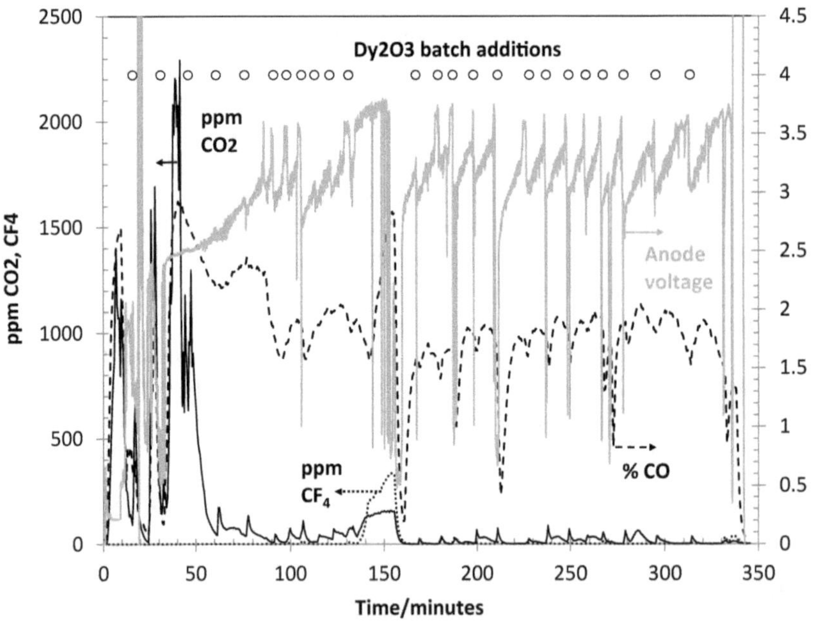

Fig. 2 Anode voltage and concentrations of CO (g), CO_2 (g) and CF_4 (g) in the off gas recorded during the electrolysis experiment shown in Fig. 1

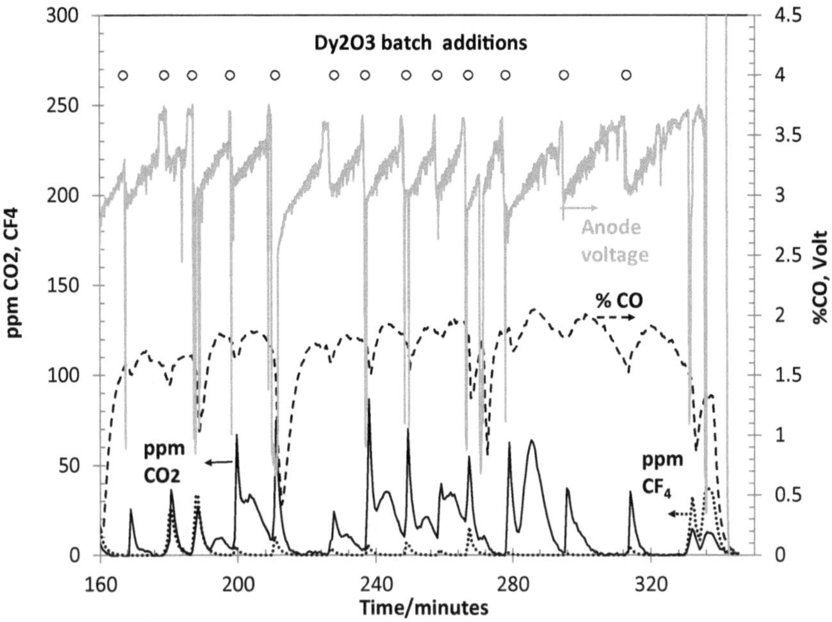

Fig. 3 Same as data shown in Fig. 2 for the last three hours of the experiment: anode voltage and concentrations of CO (g), CO_2 (g) and CF_4 (g) in the off gas

The characteristic periodic sudden increases in the anode voltage (spikes) from around 3.4 to 3.6–3.7 V that can be seen in the figures, correlates with the onset of CF_4 gas (shown most evidently in Fig. 3). Upon oxide feeding, the anode potential dropped to around 3 V, and started to climb slowly again until the sudden increase to 3.6–3.7 V, accompanied by CF_4 evolution. As the experiment went along, this feature was deliberately used to govern the oxide addition rate. An optimal feed rate for Dy_2O_3 in terms of batch addition frequency that would prevent CF_4 onset could be established by feeding just before the abrupt anode potential rise occurred.

This behaviour of the anodic voltage is due to the depletion of dissolved oxides at the anode interface and subsequent discharge of fluoride ions from the electrolyte. The reoccurring potential jumps accompanied by CF_4 formation probably represents an incipient anode effect that would eventually end in a massive rise in potential and drop of current if no action had been taken (oxide addition). However, it seems possible to sustain a stable operation at higher anode potentials and lower oxide contents, with co-evolution of CF_4. Due to a temporarily blocking of the oxide feeder tube, no oxide was fed to the bath for about 30 min (time period from ~125 to 160 min), and from Fig. 1 it can be seen that after jumping up to 3.6–3.7 V, the potential remained there and no current block occurred. Similar PFC released previous to a full anode effect has also been observed during aluminium electrolysis, either as so called "Low voltage" anode effects, or as continuous back ground emissions [7].

On average, more than 99% of the off gas was CO, but in the first period of the electrolysis, also significant amounts of CO_2 was observed. However, the CO_2 concentration dropped, but smaller amounts seemed to evolve simultaneously as CF_4.

The reason for the mentioned decline in the CO_2 content in the beginning might be that the primary anode product when oxide is discharged in fact is CO_2 as in aluminium electrolysis, but as more dysprosium iron alloy is formed as the electrolysis proceeds, some of the dysprosium dissolves into the melt and reacts back with CO_2 to form Dy_2O_3 and CO. Some of the CO_2 formed might also react with the carbon in the graphite anode and be converted to CO according to the Boudouard reaction. However, it cannot be ruled out that CO could be the result of a primary reaction.

That the evolution of CO_2 also to coincides with CF_4 might be due to formation of the former as some of the CF_4 reacts with dissolved Dy_2O_3 in the melt.

No C_2F_6 gas was detected, however, the cell potential never exceeded 4 V.

Upon opening the cell after cooling, metal lumps consisting of a Dy–Fe alloy could be recovered from the bottom of the crucible, shown in Fig. 4a.

The iron cathode, shown in Fig. 4b, had evidently been consumed. The deposited material around the Fe cathode situated just above the electrolyte surface was analysed by X-ray Diffraction (XRD). The analysis of the deposit showed mainly Dy fluoride salt ($LiDyF_4$), and some amounts of dysprosium oxyfluoride, LiF and Dy metal.

Analysis of the cathode product by XRD, demonstrated the presence of $DyFe_2$ and Dy, and this is expected after slow cooling of the cathode product, according to the phase diagram.

Figure 5 shows back scattered Scanning Electron Microscopy (SEM) micrograph analysis and Energy Dispersive X-ray Spectroscopy (EDS) analysis of the Dy–Fe alloy obtained. The hexagonal phase is rich in Fe, the continuous phase rich in Dy. The relatively high carbon content of 2 wt% might be due the reaction of the liquid alloy and the bottom of the cell reactor (glassy carbon crucible).

By weighing the recovered metal and assuming a Dy content of 80 wt%, and knowing the total amount of Dy_2O_3 added as well as the oxide concentrations in the melt before and after electrolysis, an oxygen balance could be carried out. The

a

b

Fig. 4 **a** Dysprosium-iron lump recovered after the experiment, **b** iron cathode consumed

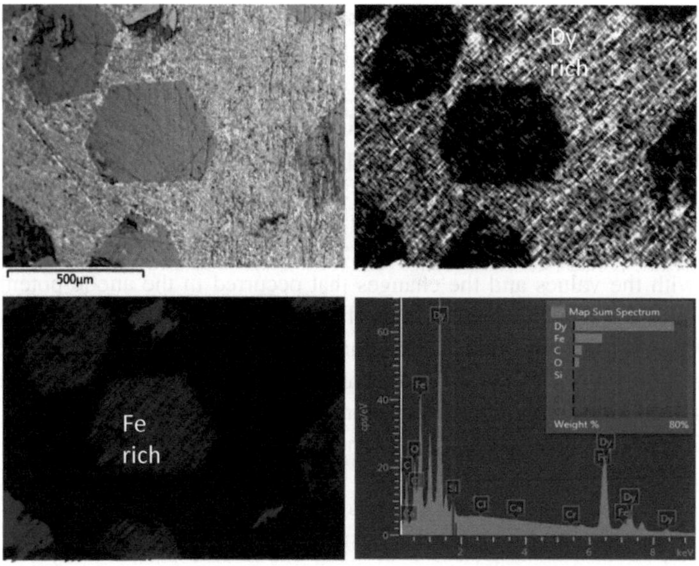

Fig. 5 Back-scattered SEM micrograph analysis of the Dy–Fe alloy product and EDS elemental analysis

oxide yield was found to be ~32%, meaning that around 68% is "not accounted for". This has probably ended up as sludge at the bottom of the cell. XRD of the three layers of salt that can be observed in Fig. 4a indicated a higher content of dysprosium oxy fluoride in the lowest layer. All though the average Dy_2O_3 addition rate was kept at around 60% of the total charge imposed, the single batch sizes seemed to have been too large for all of it to have time to dissolve before it settled in the cell bottom. Dysinger and Murphy reported on similar challenges in Nd electrolysis, and commented that once settled in the cell bottom the oxide will not dissolve in an appreciable rate [8]. Continuous feeding would probably help to utilize more of the oxide added.

The current efficiency of the process was determined to be approximately 19% by weighing the cathode product.

Oxide yield and current efficiency calculated by using the gas analysis to estimate the total amount of oxygen that exit the cell yielded numbers in the same range as the weighing method.

A significant part of the loss in current is most likely due to back reactions between the anode and cathode products in the small laboratory cell. Other current losses might be due to Dy metal dissolution and co deposition of Li, perhaps giving rise to electronic conduction. The back reaction may be diminished by increasing the anode cathode distance. It is in general also more probable to obtain higher current efficiencies in larger scale electrolysis cell.

Summary and Concluding Remarks

The objective of this work was to understand some fundamental requirements on how to avoid PFC emission during DyFe alloy electrolysis. Electrolysis experiments in equimolar DyF_3–LiF melts at approximately 1050 °C and simultaneous on line gas analysis was carried out.

The optimal Dy_2O_3 batch feed rate for avoiding PFC formation could be established by following the cell voltage and gas analysis and correlating the onset of PFC with the values and the changes that occurred in the anode potential.

On-line analysis of the cell off-gasses by FTIR showed that the main component in the off gas is CO, but significant amounts of CO_2 were observed in the beginning of the electrolysis. If CO_2 is the primary anode product, back reaction between CO_2 and dissolved Dy yields CO as a reaction product.

Dy–Fe alloy was produced, and the phases $DyFe_2$ and Dy of the solidified lump was verified by XRD.

The oxide yield and current efficiency determined by weighing the cathode product was approximately 32 and 19% respectively.

Acknowledgements This work has received funding from the European Union's 7th Research and Innovation Program under Grant Agreement No. 603564.

Zhaohui Wang at SINTEF is gratefully acknowledged for having performed the SEM and XRD analysis.

References

1. J. Lucas, P. Lucas, T. Le Mercier, A. Rollat, W. Davenport, *Rare Earths. Science, Technology, Production and Use* (Elsevier, Netherlands, 2015)
2. S. Pang, S. Yan, Z. Li, D. Chen, L. Xu, B. Zhao, Development on molten salt electrolytic methods and technology for preparing rare earth metals and alloys in China. Chin. J. Rare Met. **3**(35), 440–450 (2011). (in Chinese language)
3. R. Keller, Electrolytic production of neodymium, with and without emissions of greenhouse gases. in *Proceedings of the Volume Energy and Electrochemical Processing for a Cleaner Environment*. Joint ECS/ISE Meeting Paris, France, 1997
4. http://www.climatechangeconnection.org/
5. W. Guihua, W. Xiangcheng, Z. Hongmin, Electroanalytical study of electrode processes on carbon anode in lithium fluoride and neodymium fluoride melt. J. Rare Earths **25**, 533 (Spec Issue Dec. 2007)
6. T.B. Massalski et al. (eds.), *Binary Alloy Phase Diagrams*, vol. 2 (ASM International, Ohio, 1990)
7. D.S. Wong, P. Fraser, P. Lavoie, J. Kim, PFC emissions from detected versus nondetected anode effects in the aluminum industry. JOM **2**(67), 342–353 (2015)
8. D.K. Dysinger, J.E. Murphy, Electrowinning of Neodymiuh, from a molten oxide-fluoride electrolyte. Report Bureau of Mines 1994, ISSN 1066-5552

Effect of Cobalt and Iron Concentration on the Potential for Oxygen Evolution from Pb–Ca–Sn Anodes in Synthetic Copper Electrowinning Electrolytes

C.E. Abbey and M.S. Moats

Abstract It is well known that the addition of cobalt to copper sulfate—sulfuric acid electrolytes decreases the overpotential for oxygen evolution and decreases the rate of corrosion of Pb–Ca–Sn anodes. This effect, however, has not been adequately quantified in the presence of iron and manganese in this type of electrolyte. This work provides quantifiable data on the effect of cobalt concentration in the range of 0–0.6 g/L in synthetic electrowinning electrolytes with and without the presence of iron. The effect of cobalt on anode potential was determined using 2 and 24 h chronopotentiometry experiments. As expected, Pb–Ca–Sn potentials increased with decreasing cobalt concentration over the range of 0–0.6 g/L Co with and without the presence of Fe. Two regression models were developed to allow plant operations the ability to predict anode potentials as a function of cobalt concentration with or without iron in the electrolyte.

Keywords Anode potential · Copper electrowinning · Cobalt · Iron · Manganese

Introduction

Electrowinning is an imperative determinant of the quality of copper produced [1]. Approximately 30% of the energy consumed during electrowinning is used for anodic reactions, especially oxygen gas evolution [1]. A decrease in the oxygen overpotential and anode oxidation would improve the profitability of copper electrowinning operations. In line with efforts at increasing profitability, cell house design modifications [2] to include more automation, lead alloy variations [3] and cobalt addition [4] have been the foci of electrowinning optimisation research in recent years. These options have been proposed to improve the future of copper

C.E. Abbey (✉) · M.S. Moats
Materials Research Center, Missouri University of Science and Technology, Rolla, MO, USA
e-mail: ceag28@mst.edu

S. Wang et al. (eds.), *Applications of Process Engineering Principles in Materials Processing, Energy and Environmental Technologies*,
The Minerals, Metals & Materials Series, DOI 10.1007/978-3-319-51091-0_8

electrowinning [5]. Reducing anode potential, energy consumption and anode corrosion is vital to profitability.

Several reviews have compared different lead alloy [3], coated titanium alloy, and mesh on lead [6–9] anodes. Other non-lead alloys containing titanium have also been compared with lead alloy anodes [10]. Non-lead anodes provide reduced anode potentials but are expensive; hence multi-directionally rolled Pb–Ca–Sn anodes [11] remain the most widely used electrode for many industrial operations.

According to the review by Clancy et al. [4], the Pb in the anode reacts with sulfates in the electrolyte to form $PbSO_4$, which has low conductivity, leading to increasing potential under constant current operation. Oxygen evolution occurs after the $PbSO_4$ reacts with H_2O to form either α-PbO_2 [12] or β-PbO_2 [13] depending on the conditions. The overpotential needed to drive the oxygen evolution reaction is high on lead anodes partly due to the potential needed to form and sustain PbO_2.

Cobalt is known to reduce oxygen overpotential and power consumption, promote higher quality copper and protect lead anodes resulting in longer lifespans. This is true whether it appears in the electrolyte [4, 14–16] or as an alloy in the anode [3]. X-ray Diffraction (XRD) and X-ray Photoelectron Spectroscopy (XPS) characterization of anode surfaces show no Co compounds and indicate that the presence of Co ions do not form a Co film that protects the anode surface from oxidization [14]. This suggests that Co plays a catalytic role in reducing the porosity of the PbO_2 (whether rhombic or tetragonal) layer. This less porous layer is responsible for hindering the oxidation of $PbSO_4$ to PbO_2 [14].

The effect of the Co in the presence of Fe or Mn in the electrolyte has not been sufficiently quantified. The advantages of having about 0.5 g/L Co in the electrolyte have been reported [15] and some have even attempted to explain the mechanism of anode corrosion reduction [15, 17]. Central African ores can produce electrolytes with high [Co] and significant concentrations of manganese and iron. This paper reports quantifiable data on the effects of 0.1 g/L increases in [Co] from 0–0.6 g/L Co on the anode potentials in the presence or absence of Fe in the electrolyte. These conditions are similar to an industrial operation in central Africa.

Experimental

All experiments were conducted in an electrolyte (250 mL) with a starting concentration of 170 g/L H_2SO_4, 0.6 g/L Mn, 20 mg/L Cl^- at a temperature of 50 °C ± 1 °C. A second batch of tests was performed with the addition of 0.6 g/L Fe (III). In both conditions, cobalt concentrations examined were 0.0, 0.1, 0.2, 0.3, 0.4, 0.5 and 0.6 g/L. Mn and Cl concentrations were not controlled during the experiment.

A three electrode cell was used with a Pb–Ca–Sn working electrode with a surface area of ~1 cm^2 placed 3.8 cm from a platinum mesh counter electrode. A mercury-mercury sulfate reference electrode (0.64 V vs. standard hydrogen electrode (SHE), at room temperature, 0.63 V vs. SHE at 50 °C) was inserted directly into the cell equidistant from the working and counter electrodes. Electrode potentials are reported versus SHE.

Chronopotentiometry was performed for 2 or 24 h to measure the anode potential of the Pb–Ca–Sn anode as a function of cobalt concentration and to grow anode scale (presumably PbO_2 and MnO_2). A 300 A/m^2 current density was applied based on the measured surface area of each working electrode. Anode potential data was collected every 3 s. All experiments were repeated and all measurements presented are averages of at least two runs. There was less than 1% variation between replicates.

Results and Discussion

Pb–Ca–Sn electrodes were operated at 300 A/m^2 in 170 g/L H_2SO_4, 0.6 g/L Mn, and 20 mg/L Cl^- at 50 °C ± 1 °C. Figure 1 shows a plot of the 24 h chronopotentiograms with 0.6 g/L Fe. The 2 and 24 h tests and with (MnFeCo) and without Fe (MnCo) reveal that anode potential increases with decreasing cobalt concentration as observed in [15] for different [Co]. The anode potential data produced by Tjandrawan [1] at 0.15 g/L Mn and 0.15 g/L Co are within the variation of the data at 0.1 and 0.2 g/L [Co] from this study. Generally, the presence of Co and Fe reduced the anode potentials as shown in Figs. 2 and 3. From the data with and without Fe, equations of the form shown in Eq. 1 were used to produce a reasonable empirical model [18].

$$y = a + b[Co]^n \quad (1)$$

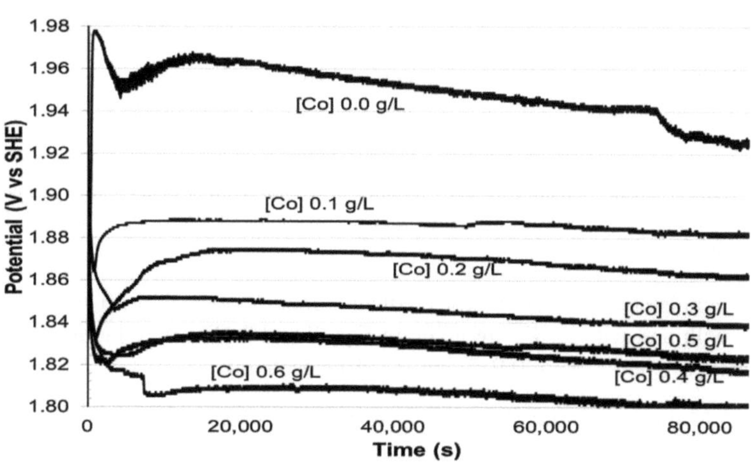

Fig. 1 Chronopotentiometry data for Pb–Ca–Sn electrode operated at 300 A/m^2 in 170 g/L H_2SO_4, 0.6 g/L Mn, 0.6 g/L Fe, 20 mg/L Cl^- at 50 °C at various cobalt concentrations for 24 h

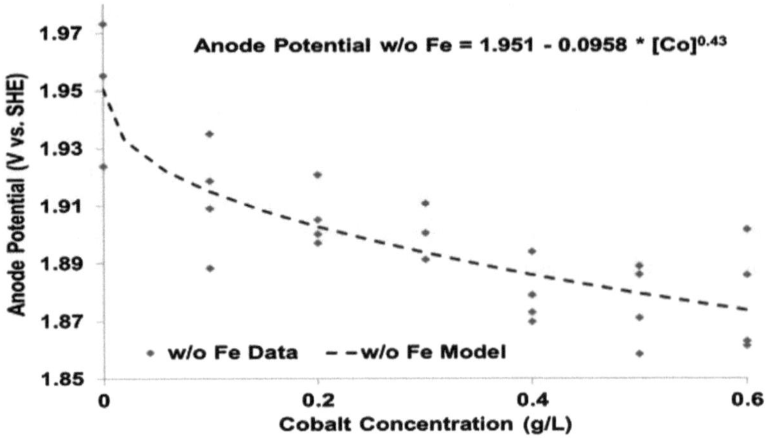

Fig. 2 Average anode potential for Pb–Ca–Sn over the last 10 min of a 24 h experiment and mathematical model for electrolyte versus cobalt concentration at 300 A/m^2 in 170 g/L H_2SO_4, 0.6 g/L Mn, 20 mg/L Cl^- at 50 °C without Fe

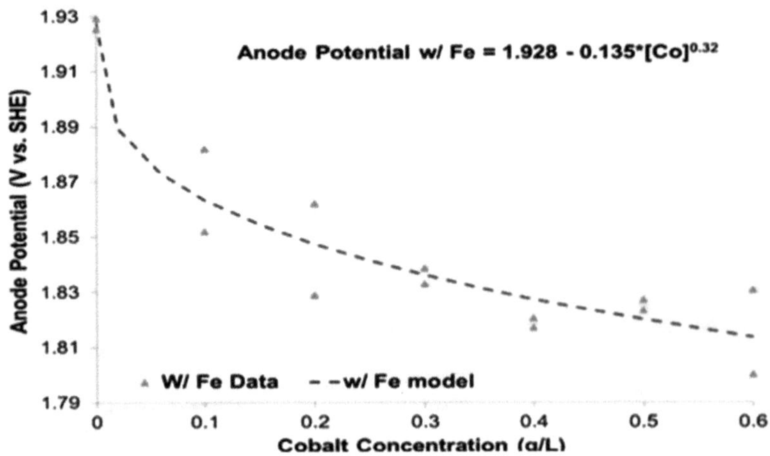

Fig. 3 Average anode potential for Pb–Ca–Sn over the last 10 min of a 24 h experiment and mathematical model for electrolyte versus cobalt concentration at 300 A/m^2 in 170 g/L H_2SO_4, 0.6 g/L Mn, 20 mg/L Cl^- at 50 °C with 0.6 g/L Fe

The mathematical equations for predicting anode potential at 50 °C in 170 g/L H_2SO_4 with and without Fe are given in Eqs. 2 and 3. The correlation coefficient, R^2, for Eqs. 1 and 2 were 0.915 and 0.715 respectively.

$$\text{MnFeCo} \quad \textit{Anode Potential}\,(V\,vs.\,SHE) = 1.928 - 0.135[\text{Co}]^{0.32} \tag{2}$$

$$\text{MnCo} \quad \textit{Anode Potential}\,(V\,vs.\,SHE) = 1.951 - 0.096[\text{Co}]^{0.42} \tag{3}$$

The models can be used to predict anode potential for [Co] over the ranges of 0–0.6 g/L Co with or without 0.6 g/L Fe. The data also indicates that Co and Fe individually and jointly reduce the anode potential with the combination giving the highest anode potential decrease. This indicates the possibility of a synergistic interaction between iron and cobalt. While iron appears to reduce the anode potential in these tests, increasing iron concentration also reduces cathodic current efficiency [19]. Even so, copper electrowinning electrolytes often are maintained with a 8:1 to 10:1 Fe to Mn ratio to avoid the production of high oxidation state manganese species [20]. Reducing the [Co] in the electrolyte increases the anode potential and increases the Mn oxidation rate (cyclic voltammetry data not shown here due to space constraint).

The empirical formulae are valid over the range of 0.0–0.6 g/L cobalt concentrations. Hence, the anode potential can be estimated to evaluate electrical costs as a function of cobalt concentration to make informed decisions about operating a tankhouse. Tables 1 (with Fe) and 2 (without Fe) illustrate the effect of reducing [Co] from a baseline concentration (600 mg/L) on power requirements and costs based on the assumptions of a tankhouse with 44,000 A, 156 cells, $0.15/kWh, and a 95% online factor. These values are vital for an operation considering lowering

Table 1 Added electrical power requirement and power costs projected to be caused by lowering the cobalt concentration in electrolyte from 600 mg/L with 0.6 g/L Fe

Cobalt Concentration (mg/L)	Added power requirement (kW)	Added power cost ($/year)
600	0 (base case)	0 (base case)
500	44	56,000
400	96	119,000
300	156	195,000
200	233	290,000
150	281	351,000
100	343	428,000

Table 2 Added electrical power requirement and power costs projected to be caused by lowering the cobalt concentration in electrolyte from 600 mg/L without 0.6 g/L Fe

Cobalt Concentration (mg/L)	Added power requirement (kW)	Added power cost ($/year)
600	0 (base case)	0 (base case)
500	40	50,000
400	85	105,000
300	136	170,000
200	198	248,000
150	237	296,000
100	283	354,000

their cobalt concentration. The cost involved can be compared with profit from extra cobalt recovered sales to make a decision.

Conclusions

Results of experiments examining the effect of cobalt concentration on Pb–Ca–Sn anode potential under typical copper electrowinning conditions have been provided. The results show that the anode potential increases with decreasing cobalt concentration with or without the presence of 0.6 g/L Fe for the [Co] tested. Empirical equations were developed to allow the prediction of anode potentials as a function of [Co] between 0–0.6 g/L when there is Fe in the electrolyte or not. The results indicate that Co, Fe and Mn are likely interacting at or near the anode surface. This interaction should be studied further.

Acknowledgements The financial support and permission to publish these results given by the AMIRA Project P705C is highly appreciated. We also immensely appreciate the help of Margaret Scott and Kevin Foster for their experimental assistance.

References

1. V. Tjandrawan, The role of manganese in the electrowinning of copper and zinc, in *School of Chemical and Mathematical Science*, Murdoch University (2010), p. 153
2. T. Robinson, K.C. Sole, M.S. Moats, F.K. Crundwell, M. Moritmitsu, L. Palmu, in *Developments in Base Metal Electrowinning Cellhouse Design*. TMS Annual Meeting (2012)
3. A. Hrussanova, L. Mirkova, T. Dobrev, Electrochemical properties of Pb–Sb, Pb–Ca–Sn and Pb–Co_3O_4 anodes in copper electrowinning. J. Appl. Electrochem. **32**(5), 505–512 (2002)
4. M. Clancy, C.J. Bettles, A. Stuart, N. Birbilis, The influence of alloying elements on the electrochemistry of lead anodes for electrowinning of metals: a review. Hydrometallurgy **131**, 144–157 (2013)
5. M. Moats, M. Free, A bright future for copper electrowinning. JOM **59**(10), 34–36 (2007)
6. K. Kawaguchi, G.M. Haarberg, M. Morimitsu, Nano-architecture on the mud-cracked surface of IrO_2-Ta_2O_5 binary system. ECS Trans. **25**(33), 67–73 (2010)
7. K. Kawaguchi, M. Morimitsu, Effects of oxide composition on structure, surface morphology, and oxygen evolution behaviors of IrO_2-Ta_2O_5/Ti anodes prepared at a high temperature. Electrochemistry **83**(4), 256–261 (2015)
8. M. Morimitsu, Performance and commercialization of the smart anode, MSA™, for environmentally friendly electrometallurgical process. Electrometallurgy **2012**, 49–54 (2012)
9. T. Zhang, M. Morimitsu, A novel oxygen evolution anode for electrowinning of non-ferrous metals. Electrometallurgy **2012**, 29–34 (2012)
10. L. Cifuentes, A. Montes, G. Crisóstomo, Corrosion behaviour and catalytic effectiveness of Pb–Ca–Sn, RuO_2–IrO_2/Ti and IrO_2–Ta_2O_5/Ti anodes for copper electrowinning. Corros. Eng. Sci. Technol. **46**(6), 737–744 (2011)
11. J. Yang, B. Chen, H. Hang, Z. Guo, S. Wang, Effect of rolling technologies on the properties of Pb-0.06 wt% Ca-1.2 wt% Sn alloy anodes during copper electrowinning. Int. J. Miner. Metall. Mater. **22**(11), 1205–1211 (2015)

12. I. Ivanov, Y. Stefanov, Z. Noncheva, M. Petrova, T. Dobrev, L. Mirkova, R. Vermeersch, J.P. Demaerel, Insoluble anodes used in hydrometallurgy: Part I. Corrosion resistance of lead and lead alloy anodes. Hydrometallurgy **57**(2), 109–124 (2000)
13. W. Zhang, G. Houlachi, Electrochemical studies of the performance of different Pb–Ag anodes during and after zinc electrowinning. Hydrometallurgy **104**(2), 129–135 (2010)
14. T. Nguyen, A. Atrens, Influence of lead dioxide surface films on anodic oxidation of a lead alloy under conditions typical of copper electrowinning. J. Appl. Electrochem. **38**(4), 569–577 (2008)
15. T. Nguyen, N. Guresin, M. Nicol, A. Atrens, Influence of cobalt ions on the anodic oxidation of a lead alloy under conditions typical of copper electrowinning. J. Appl. Electrochem. **38**(2), 215–224 (2008)
16. P. Yu, T.J. O'Keefe, Evaluation of lead anode reactions in acid sulfate electrolytes II. Manganese reactions. J. Electrochem. Soc. **149**(5), A558–A569 (2002)
17. A. Gendron, V. Enel, S., Abe, Effect of cobalt added to electrolyte on corrosion rate of Pb-Sb anodes in copper electrowinning. Can. Metall. Q. (2013)
18. G. Miller, Methods of managing manganese effects on copper solvent extraction plant operations. Solv. Extr. Ion Exch. **29**(5–6), 837–853 (2011)
19. Y. Khouraibchia, M.S. Moats, Evaluation of the effect of copper electrowinning parameters on current efficiency and energy consumption using surface response methodology. ECS Trans. **28**(6), 295–306 (2010)
20. G. Miller, The problem of manganese and its effects on copper SX-EW operations. Paper from COPPER 95, vol. III, Electrorefining and hydrometallurgy of copper, international conference held in Santiago, Chile, 26–29 Nov, Papers, 1995, pp. 649–663

Cobalt Electrodeposition from Cobalt Chloride Using Urea and Choline Chloride Ionic Liquid: Effect of Temperature, Applied Voltage, and Cobalt Chloride Concentration on Current Efficiency and Energy Consumption

Andrea R. Kim and Ramana G. Reddy

Abstract Electrodeposition of cobalt from cobalt chloride using the urea and the choline chloride ionic liquid (2:1 molar ratio) was studied to search the optimized parameters for the higher current efficiency. The cyclic voltammetry was tested with 0.2 M $CoCl_2$ using various scan rates to determine the reduction potential in Urea/ChCl at 323 K. Based on the data from cyclic voltammetry test, the transfer coefficient and the diffusion coefficient were calculated as 0.22 and 3.38×10^{-6} cm^2/s, respectively. The parameters for electrodeposition of cobalt were various temperatures (323–383 K), applied potentials (2.4–3.3 V), and concentration of $CoCl_2$ (0.2–0.5 mol/L). Current efficiency and energy consumption were calculated to investigate the optimal condition for the electrodeposition of cobalt. The highest current efficiency (95%) was obtained under the temperature 323 K, the applied voltage 2.7 V, and concentration 0.5 mol/L of $CoCl_2$. Scanning electron microscope (SEM) and X-ray diffraction (XRD) were used for the characterization of cobalt deposits.

Keywords Cobalt · Electrodeposition · Ionic liquid · Current efficiency · Energy consumption · Morphology

Introduction

Cobalt is the 33rd most abundant element in the earth's crust. The cobalt is formed with other metal ores such as arsenides, sulfides, and oxide. Cobalt can be used in various applications with producing cathodes in lithium ion batteries, super alloys,

A.R. Kim (✉) · R.G. Reddy
Department of Metallurgical and Materials Engineering, The University of Alabama, Tuscaloosa, AL 35487-0202, USA
e-mail: arkim@crimson.ua.edu

S. Wang et al. (eds.), *Applications of Process Engineering Principles in Materials Processing, Energy and Environmental Technologies*,
The Minerals, Metals & Materials Series, DOI 10.1007/978-3-319-51091-0_9

magnets, inks, catalysts for chemical industry [1–6]. The electrodeposition of cobalt alloy from aqueous solution is not an easy process due to the narrow electrochemical potential window of water [7]. The ionic liquids are good solutions for electrodeposition because of large electrochemical windows comparing to non-ionic liquid solutions. The ionic liquid has the several advantages as an electrolyte: wide potential windows, high solubility of metal salts, high conductivity, thermal stability, non-flammability, and low vapor pressure [8–11]. The Deep Eutectic Solvent (DES) is a new type of solvent and urea and choline chloride (ChCl) mixture (2:1 molar ratio) is one of significant DES [12]. Comparing to the common ionic liquid, the mixture of choline chloride and urea has several advantages: low price, easy storage, easy preparation, and non-toxic. In addition, its physic-chemical properties are really similar to the typical ionic liquids [13, 14].

Thus, the reline (commercial name of mixture of Urea/ Choline chloride) was used for the extraction and the electrodeposition of metal [15]. Yang and Reddy [16, 17] investigated the solubility of ZnO in the eutectic mixture of Urea/ChCl with the FTIR spectroscopy, the cycle voltammetry, the calculations of the current efficiency and the energy consumption in various temperatures, applied voltages, and concentrations of $[BMIM]HSO_4$ addition for the electrodeposition of Zn with ZnO. They [18] also studied the solubility of PbO in reline with the same method (FTIR spectroscopy) and successfully synthesized the pure lead on the copper substrate. Cojocaru et al. [19] studied the electrochemical impedance spectroscopy of selenium behavior to present the cathodic process from the electrodeposition of Se films in reline. They reported that electrochemical impedance measurements showed that the cathodic process of selenium ions in ChCl-urea and ChCl-EG ionic liquids had almost similar characteristics. Ana-Maria et al. [20, 21] discussed the electrochemical reaction in the electrodeposition of copper using reline and CuCl. Their group calculated the diffusion coefficient in CuCl and reline system. They also studied the electrochemical behavior of $CuCl_2$ in reline using the cyclic voltammetry and the electrochemical impedance spectroscopy. Ali et al. [22] published the electrodeposition of copper from reline with $CuCl_2$. They successfully synthesized and investigated the dense, smooth, and continuous copper deposition using scanning electron microscopy and X-ray diffraction. The current efficiency of their deposition of pure copper is about 97%. Chu et al. [23] proved to form zinc-cobalt alloys from choline chloride and urea solvent mixture. They investigated that the size of grain cluster of Zn–Co deposition increases as the negative value of potential increases. Abbott group [24] deposited zinc-tin alloy in reline on the platinum electrode. Their group showed that zinc and tin could be deposited as individual metals and as an alloy. Gómez et al. [25] were successful in electrodepositing platinum on the vitreous carbon electrode. They discussed that the deposition condition depended on the nature of solvent and the deposited species because of the change of electrochemical window which is the factor to improve the current efficiency and minimize the damage of the coating. Costovici et al. [26] electrochemically synthesized the ZnO nano powder using the platinum cathode and zinc anode with at least 85% efficiency at 20–30 °C. Their cyclic voltammogram showed the continuous increasing current caused by the dissolution of Zn and the oxidation of peroxide ion in the anodic scan. At the same time, the precipitated ZnO was synthesized. Anicai et al. [27] studied the electrodeposition of Ni on Cu substrate with some cracks due to the corrosion behavior on it. They also performed the cyclic voltammetry to investigate that the limit cathodic peaks for Ni (II) reduction at 30 and 80 °C were located at −1.05 V and −0.75 to −0.88 V, respectively. Huynh et al. [28] used choline chloride/ urea (1:1 molar ratio) to synthesize Al from $AlCl_3$ on glassy carbon, iron, and copper cathodes, whose current efficiencies were 30.8, 64.45, and 29%, respectively. They also analyzed the purities of Al depositions on each cathode, which were 92.42% for glass carbon, 70.33% for iron, and 82.63% for copper, respectively. Golgovici et al. [29] performed the electrodeposition of Cd and CdTe on the platinum cathode from TeO_2 and $CdCl_2$. They observed only one couple of reduction/oxidation peaks from Cd^{2+} ion and the complex shape due to Cd^{2+}/Te^{4+} ions using the cyclic voltammetry. Li et al. [7] reported the electrodeposition of Co from $CoCl_2$ on copper cathode. They used the cyclic voltammetry to study that two irreversible reactions occurred from Co^{2+} to Co with one-step two electron process. They also calculated the diffusion coefficient at 373 K, the current efficiency at various temperatures, and the reduction potentials.

This work is focused on investigating the experimental parameters which lead the optimized consequences with high current efficiency and low energy consumption in Urea /ChCl–$CoCl_2$ mixture. The transport coefficient and the diffusion coefficient were calculated at low temperature as well using cyclic voltammetry.

Experimental Procedure

Preparation of Urea and Choline Chloride Ionic Liquid

Choline chloride ($HOC_2H_4N(CH_3)_3Cl$) (98%) and Urea ($Co(NH_2)_2$) (99.3%) from Alfa Aesar were dried for more than 2 h to remove residual moisture in the salts. These two chemicals were mixed in the molar ratio 1:2 (46.15% urea and 53.85% choline chloride) to synthesize the right amount of ionic liquid. The mixture of urea and choline chloride was heated at 363 K to obtain a homogeneous clear solution with constant stirring. Cobalt chloride in the right amount for each study (Alfa Aesar, 99.7%) was dissolved in the ionic liquid mixture for more than 12 h.

Cyclic Votammetry

Tungsten wire, silver wire, and platinum wire were used as a working electrode, a reference electrode, and a counter electrode, respectively. All electrodes were cleaned by SiC paper until it is shiny to obtain better results by eliminating oxide on the surface and washed with deionized water. The electrodes were fixed in a rubber cap with a thermometer and were covered on a beaker filled in ionic liquid. The CV test was performed using pure reline and reline with 0.2 mol/L $CoCl_2$ at 323 K. The scan rate range used was 20–80 mV/s to determine reduction potential of reline with 0.2 M $CoCl_2$ at 323 K.

Electrodeposition of Cobalt

The copper sheet was the working electrode and the graphite sheet was the counter electrode. Both working and counter electrodes were cut to appropriate size (2.5 cm × 5 cm) to fit into the beaker. The copper sheet was polished by SiC paper for the same reason in the CV test. The electrodes were fixed in Teflon plate using bolts and nuts. The insulator is shaped to fit into the beaker. The experiment of electrodeposition of cobalt from cobalt using urea and choline chloride had been performed for 2 h with variation of temperatures, cell voltages, and concentrations of $CoCl_2$. The temperature variation rage was 323–383 K (50–110 K), the applied voltage range was 2.4–3.3 V, and the concentration of $CoCl_2$ range was 0.2–0.5 mol/L.

EG&G PARC model 273A Potentiostat/Galvanostat was used for cyclic voltammetry and electrodeposition experiment using chronoamperometry. The deposited cobalt on cathode was rinsed with deionized water and ethanol, then dried to analyze using SEM and XRD.

Characterization of the Deposited Cobalt

The characterization of deposited cobalt on cathode was analyzed using Scanning Electron Microscopy (SEM, JEOL 7000, Japan) and X-ray diffraction pattern (Bruker D8 XRD) with Co kα radiation. SEM was used to analyze the morphology of deposited cathode surface and XRD was used to examine the phase and structure of the deposit.

Result and Discussion

Cyclic Voltammetry

To study the electrode reaction such as the electron transportation in the ionic liquid, the cyclic voltammetry test was performed at 323 K. As shown in Fig. 1, the cathodic limit which generates the hydrogen cathodic gas [30] is about −1.2 V at

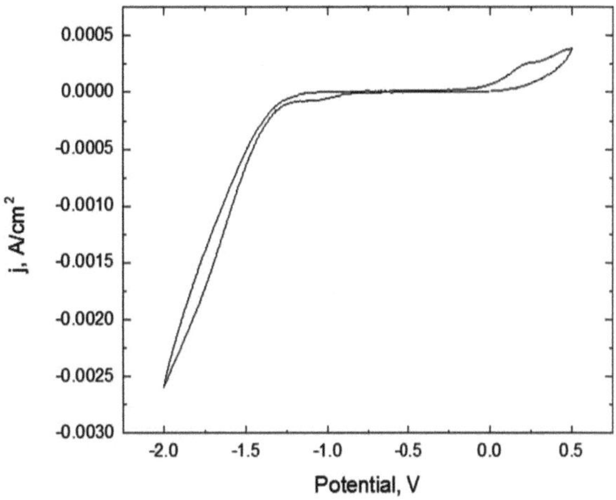

Fig. 1 Cyclic voltammogram of Urea/ChCl (2:1 ratio) with 50 mV/s scan rate at 323 K

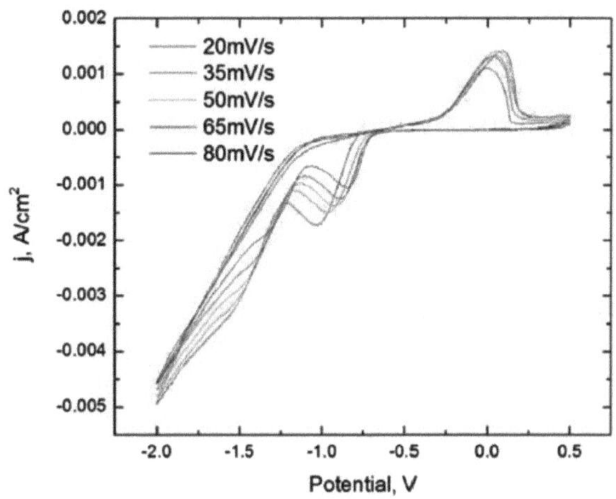

Fig. 2 Cyclic voltammogram of 0.2 mol/L $CoCl_2$-Urea/ChCl as function of various scan rates at 323 K

323 K. Figure 2 shows the CV curve of 0.2 mol/L $CoCl_2$ Urea/ChCl mixture according to the different scan rates. As the scan rate increases, the cathodic reduction potential becomes more negative, which indicates the irreversible process takes place. For the irreversible process, the transfer coefficient of $Co^{2+} + 2e^{-} \rightarrow Co$ can be calculated by the following equation [31]:

$$\left|E_p - E_{p/2}\right| = 1.857\,RT/\alpha nF \tag{1}$$

where E_p is the cathodic potential in V, $E_{p/2}$ is the half of cathodic potential in V, R is the gas constant which is 8.314 J/K mol, T is the temperature in K, α is the transfer coefficient, n is the number of exchanging electrons, and F is the Faraday constant, 96,500 C/mol. Using the equation above and the data from Fig. 2, the average transfer coefficient is 0.22.

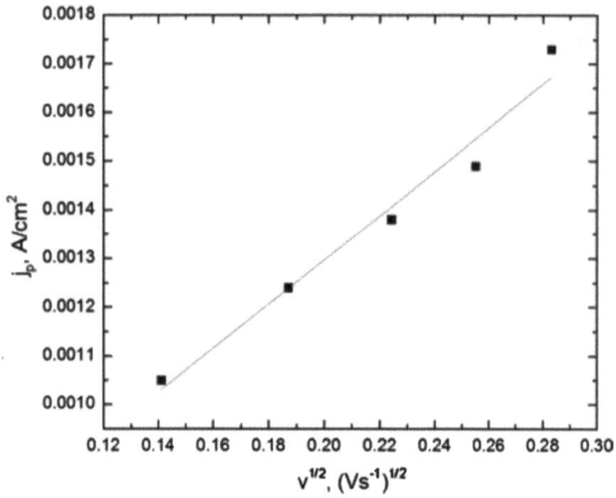

Fig. 3 Cathodic current density peaks from cyclic voltammogram of 0.2 mol/L $CoCl_2$-Urea/ChCl as a function of various scan rates at 323 K

Table 1 The transfer coefficient and the diffusion coefficient in various ionic liquid systems

System	T (K)	α	D (cm^2/s)	Reference
$CoCl_2$ in Urea/ChCl	323	0.22	3.38×10^{-6}	This work
$CoCl_2$ in Urea/ChCl	373	0.36	1.7×10^{-6}	[7]
$CoCl_2$ in urea-NaBr-KBr	373	0.45	2.5×10^{-6}	[33]
$CoCl_2$ in urea-NaBr-acetamide	353	0.26	2.83×10^{-7}	[34]
$CoCl_2$ in urea-NaBr-acetamide	353	0.23	2.24×10^{-7}	[35]
$Co(BF_4)_2$ in BMIMBF4	353	0.3	7.6×10^{-8}	[36]
$Co(TFSA)_2$ in BMPTFSA	300	–	7.2×10^{-8}	[37]
$CoCl_2$ in $AlCl_3$-EMIC	295	–	4.40×10^{-7}	[38]
ZnO in Urea/ChCl	373	0.2	1.89×10^{-8}	[17]
PbO in Urea/ChCl	363	0.52	1.42×10^{-7}	[18]
$AlCl_3$ in BMIC	363	–	2.2×10^{-7}	[39]
$AlCl_3$ in EMIC	363	–	9.1×10^{-7}	[40]

T temperature, *α* transfer coefficient, *D* diffusion coefficient

To calculate the diffusion coefficient of cobalt ion in the electrolyte at 323 K, the following equation and the slope from Fig. 3 [32] can be used:

$$I_p = 0.4958nAc^i\left(\frac{\alpha nF}{RT}\right)^{1/2}D^{1/2}v^{1/2} \tag{2}$$

where I_p is the cathodic peak current in A, A is the electrode area in cm^2, c^i is the initial concentration in mol/cm^3, α is the transfer coefficient, D is the diffusion coefficient in cm^2/s and v is the scan rate in V/s. The calculated diffusion coefficient is 3.38×10^{-6} cm^2/s using Fig. 3. Table 1 lists the transfer coefficient and the diffusion coefficient in various ionic liquid systems. According to Table 1, the calculated values are reasonable as compared with the literature data in the same ionic liquid system with different temperatures. The data in Table 1 shows that the transfer coefficient is generally greater at high temperature in the system. This may be due to the faster ion transfer rate at high temperature. The diffusion coefficient depends on the concentration of electrolyte according to Eq. (2). Thus, higher concentration results in the higher diffusion coefficient.

Electrodeposition of Cobalt

To determine the effect of temperature on the electrodeposition of cobalt on the copper cathode, four experiments were performed under the fixed cell voltage, concentration of $CoCl_2$, and time. Based on the data from the experiments, the current density plot, the current efficiency, and the energy consumption were calculated.

The current efficiency (η) of electrodeposition can be calculated by the following equation:

$$\eta = \frac{m_1}{m_2} \times 100\% \tag{3}$$

where m_1 is the weight of deposited cobalt on cathode in gram, its value can be calculated from the weight difference of the cathode before and after the electrodeposition. m_2 is the theoretical weight of cobalt in grams which can be calculated using Faraday's law [26]:

$$m_2 = \frac{Ita}{nF} \tag{4}$$

where I is the applied current in A, t is the deposition time in s, a is the molecular weight of cobalt in grams, and F is Faraday constant.

The energy consumption (E) is the measure of consumption of power and can be calculated by:

$$E = V\frac{Q}{\eta} \tag{5}$$

where V is cell voltage used in this experiment, Q is total charge required for a certain amount deposit of cobalt on cathode, and η is the current efficiency on the cathode.

The Effect of Temperature on the Electrodeposition of Co

Figure 4 shows that the current density decreases rapidly at the initial 40 s and begins to be stable for the rest of the time. The current density drop is due to the

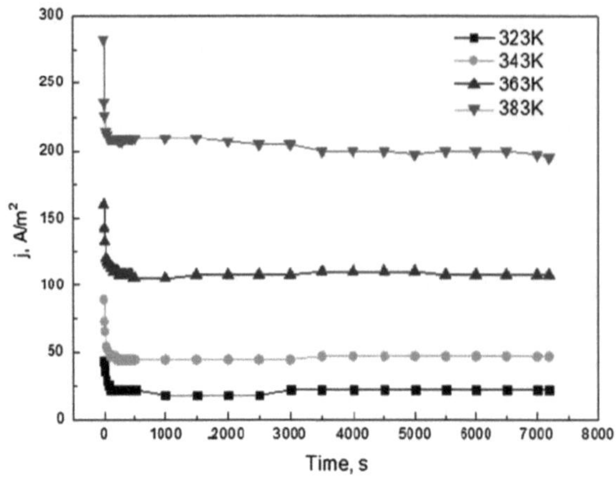

Fig. 4 Current density as a function of time at various temperatures under the applied voltage 3.0 V and concentration 0.2 mol/L of $CoCl_2$

concentration polarization on the cathode. As the heating temperature increases, the current density increases because the diffusion rate increases with temperature, which increases the ion transportation near the copper substrate. The lower viscosity of the electrolyte with increasing temperature would also cause higher current density at higher temperature.

The current efficiency and the energy consumption at different temperatures are summarized in Table 2. The current efficiency as a function of the temperature is ranged from 76.78 to 78.55%. Thus, there exists no significant difference in the current efficiency with temperature. The more amount of the cobalt deposited was obtained at higher temperature because of increase in diffusion rate. However, higher current density due to higher temperature also causes a result in the more theoretical weight from the calculation using Eq. (4). Thus, the current efficiency which is the ratio of actual and theoretical weight of the deposit may not be greater at high temperature. The current efficiency at 323 K is greater than other values and thus the energy consumption at 323 K is the lowest.

The surface morphology of the cobalt deposited on the copper cathode at different temperatures was studied using SEM. In Fig. 5, all micrographs show the clustered particles that are uniformly layered on copper substrates. The size of cobalt particles at 323 K is relatively small compared with that of the other depositions at higher temperatures. As the temperature increases, cobalt particles become larger but remain spherical. Figure 6 shows the cross-sectional Co deposits at 323 and 383 K under the concentration of 0.2 mol/L, 3.0 V and 2 h. It can be seen that the cobalt deposition is uniformly distributed on the copper cathodic electrode. Also, the dependence of the particle size in the cross-sectional area on the temperature (Fig. 6) is similar to that in the surface morphology of the cobalt deposition (Fig. 5).

Table 2 Summary of the current efficiency and the energy consumption as function of temperature (concentration of $CoCl_2$ = 0.2 mol/L, Cell voltage = 3.0 V)

Temperature (K)	η (%)	E (kWh/kg)
323	78.58	3.47
343	76.78	3.55
363	78.22	3.49
383	77.62	3.52

η current efficiency, *E* energy consumption

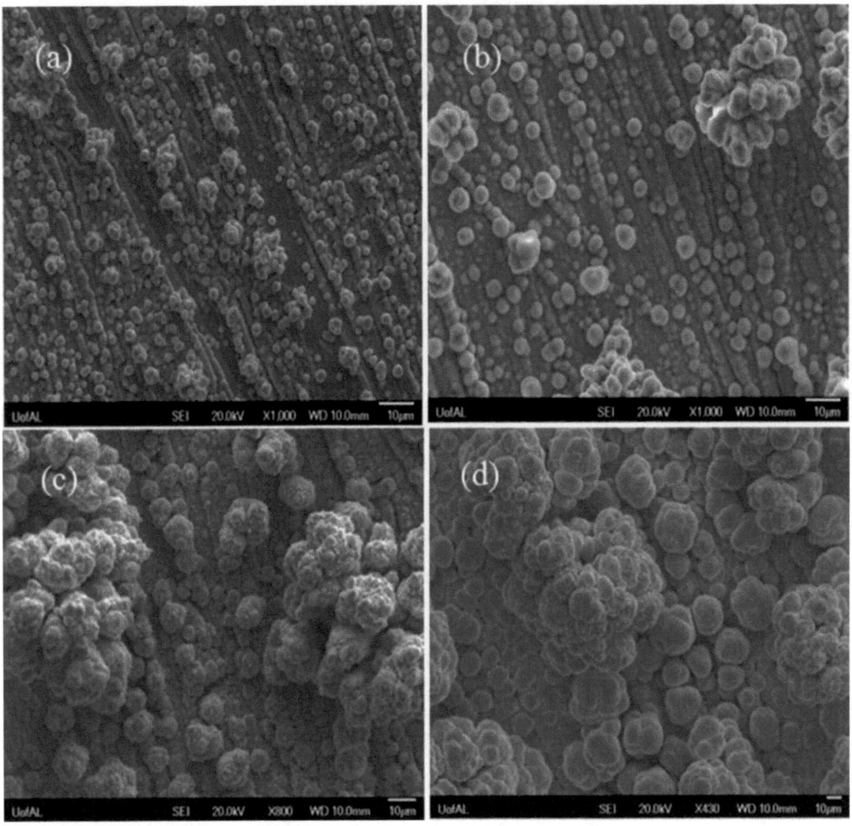

Fig. 5 SEM micrographs of Co electrodeposits at **a** 323 K, **b** 353 K, **c** 363 K, **d** 383 K under the applied voltage 3.0 V, concentration 0.2 mol/L of $CoCl_2$ and 2 h

Effect of the Applied Voltage on the Electrodeposition of Co

The effect of the cell potential on the cathode for the electrodeposition of cobalt was studied. The current density, calculating the current efficiency and the energy consumption, and the surface morphology using SEM and XRD patterns were determined. The current density as a function of time at 353 K, 0.2 mol/L of $CoCl_2$ for 2 h with different applied voltage was plotted as shown in Fig. 7. The highest

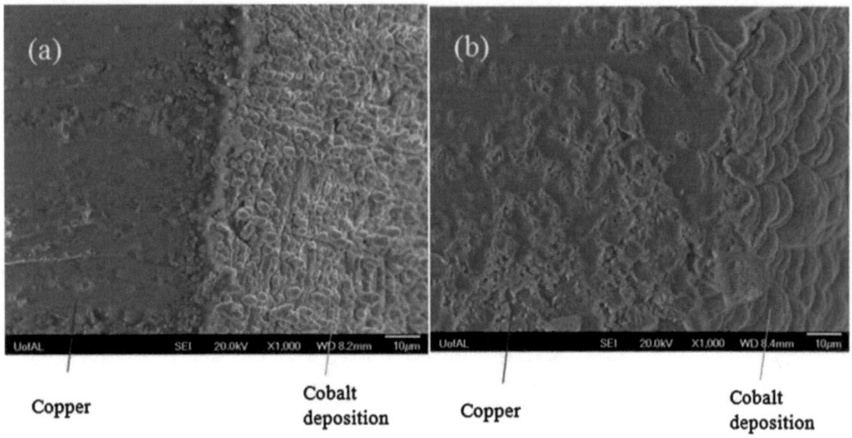

Fig. 6 SEM micrographs of cross-sectional Co electrodeposits at **a** 323 K and **b** 383 K under the applied voltage 3.0 V, concentration 0.2 mol/L of $CoCl_2$, and 2 h

Fig. 7 Current density as a function of time at the different applied voltage under the temperature 323 K and concentration 0.2 mol/L of $CoCl_2$

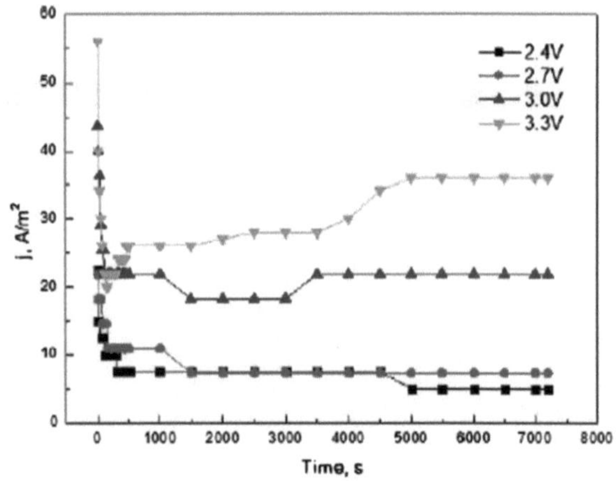

current density is observed at 3.3 V while the current density at 2.4 V is the lowest. As the larger cell potential is applied, the higher current density is obtained. This might be due to the larger reduction potential on the cathode surface as the larger cell voltage is applied.

The current efficiency and the energy consumption according to the cell voltage are summarized in Table 3. The current efficiency increases as the cell voltage increases from 2.4 to 2.7 V. However, there is a huge current efficiency drop observed at 3.0 V because the ionic liquid was decomposed during the experiment. Also, the dendrite formation would be another reason for decreasing current efficiency at the higher cell voltage. The energy consumption at 2.4 V is only

Table 3 Summary of the current efficiency and the energy consumption as function of reduction potential. (concentration of $CoCl_2$ = 0.2 mol/L, Temperature = 323 K)

Cell voltage (V)	η (%)	E (kWh/kg)
2.4	83.64	2.61
2.7	87.92	2.79
3.0	78.58	3.47
3.3	73.54	4.08

η current efficiency, E energy consumption

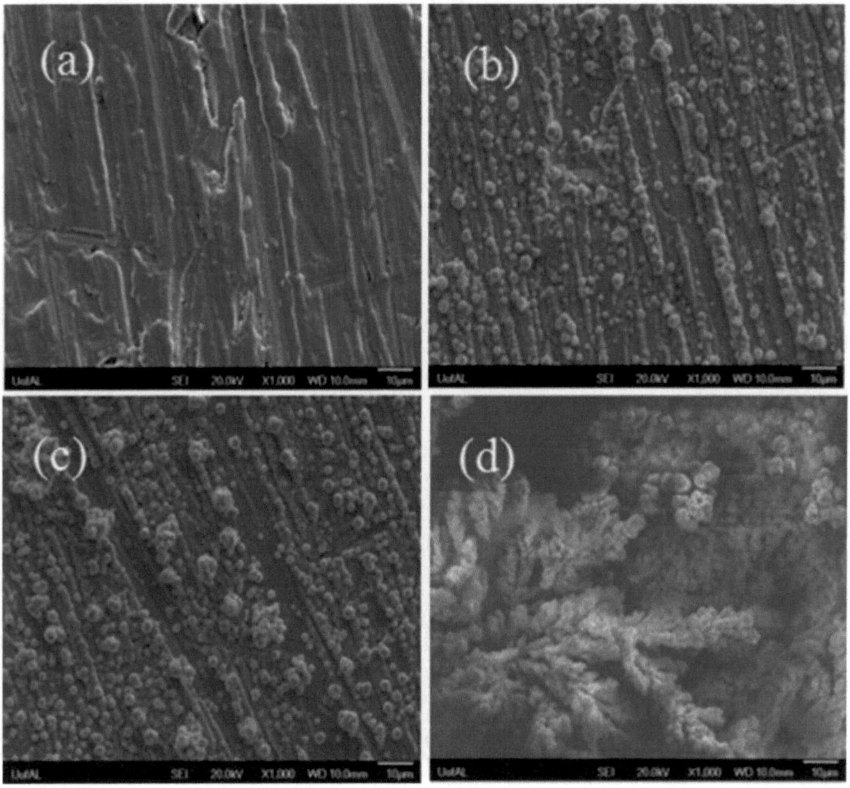

Fig. 8 SEM micrographs of Co electrodeposits at the applied voltages **a** 2.4 V, **b** 2.7 V, **c** 3.0 V, **d** 3.3 V under the temperature 323 K, concentration 0.2 mol/L of $CoCl_2$ and 2 h

2.61 kWh/kg, which is the lowest value compared with others. The energy consumption increases significantly at 3.0 V.

The surface morphology on the copper cathode with the applied voltage was examined using SEM in Fig. 8. The deposition at the applied voltage 2.4 V is compact and uniformly layered without particles and clusters. At 2.7–3.0 V, the micrographs show the nodular shape of particles and clusters and of the similar particle size. However, the cluster size at 3.0 V is larger than that at 2.7 V. At

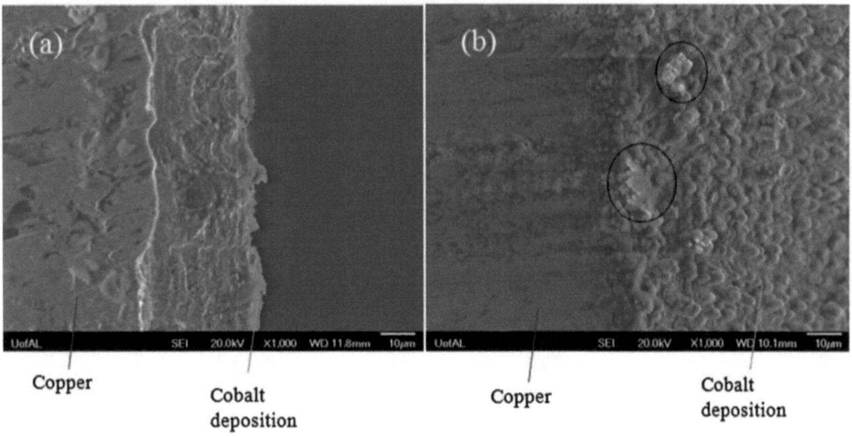

Fig. 9 SEM micrographs of cross-sectional Co electrodeposits at the applied voltages **a** 2.4 V and **b** 3.3 V under the temperature 323 K, concentration 0.2 mol/L of $CoCl_2$ and 2 h

Fig. 10 SEM micrographs of the single dendrite of Co electrodeposits under the temperature 323 K, concentration 0.2 mol/L of $CoCl_2$, the applied voltage 3.3 V, and 2 h

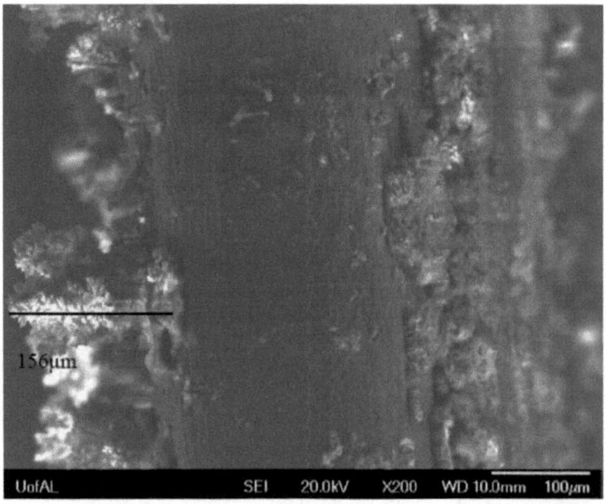

3.3 V, the dendrite forms on the uniform and dense cobalt deposition as a background. The cross-sectional morphologies of Co deposition at 2.4 and 3.3 V are shown in Fig. 9. No particle is shown on the copper substrate. However, the boundary between the copper and the cobalt deposits shows that the cobalt was layered uniformly on the copper electrode at 2.4 V. At 3.3 V, a few dendrites in the circle in Fig. 8b are observed but not uniformly distributed. This resulted from the loss of the dendrites when the deposited sample was cut to prepare the specimen for the SEM. However, the dendrites were observed in other spots far from the cut area, as shown in Fig. 10, where the average size of the cobalt dendrites was measured as 156 μm.

Effect of the Concentration of $CoCl_2$ on the Electrodeposition of Co

The effect of the concentration on the electrodeposition of cobalt with different concentration of $CoCl_2$ was investigated. As shown in Fig. 11, there is not much difference between results from 0.2 to 0.5 M of $CoCl_2$. However, the current density at 0.5 mol/L of $CoCl_2$ is still higher than lower concentration values. The trend in this current density plot shows that the current density increases at higher concentration. This might be due to more reducible ions in the higher concentration electrolytes.

The calculated current efficiency and energy consumption are listed in Table 4. As the concentration of $CoCl_2$ increases from 0.2 mol/L to 0.3 mol/L, a small increase in the current efficiency was observed. The current efficiency increases significantly at 0.5 mol/L $CoCl_2$. The saturation of $CoCl_2$ was observed when 0.6 mol/L $CoCl_2$ was added to Urea/ChCl. This indicates maximum solubility of $CoCl_2$ is about 0.5 mol/L $CoCl_2$. That might cause the sudden increase of current efficiency at 0.5 mol/L which resulted in lower energy consumption which is relatively lower than other data. The low temperature and the low applied voltage lead

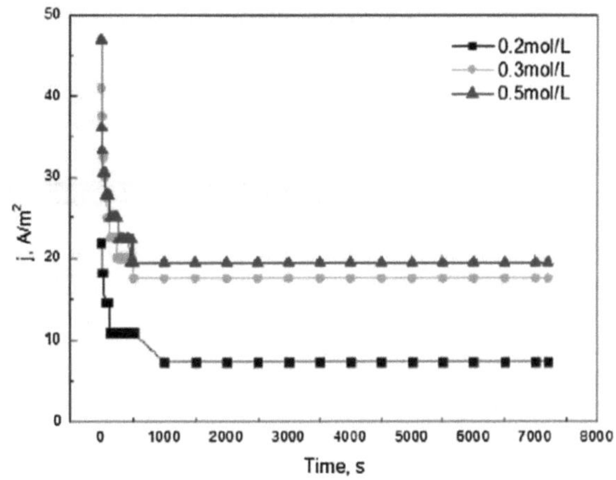

Fig. 11 Current density as a function of time at the different concentrations of $CoCl_2$ under the applied voltage 2.7 V and temperature 323 K

Table 4 Summary of the current efficiency and the energy consumption as function of concentration of $CoCl_2$ (Temperature = 323 K, Cell voltage = 2.7 V)

Concentration of $CoCl_2$ (mol/L)	η (%)	E (kWh/kg)
0.2	87.92	2.79
0.3	85.06	2.89
0.5	95.00	2.59

η current efficiency, *E* energy consumption

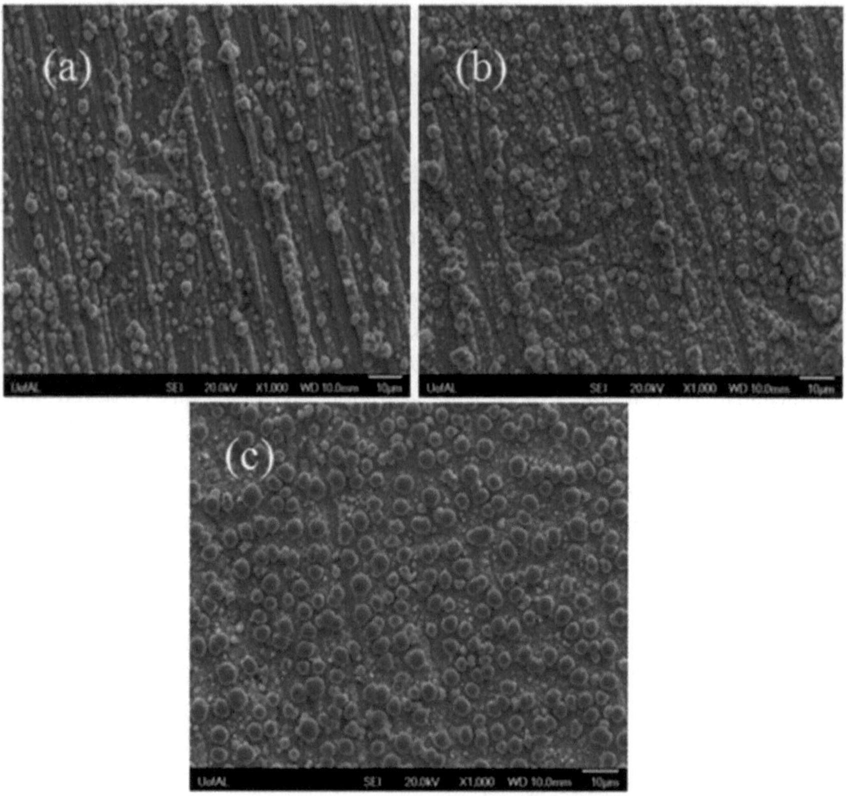

Fig. 12 SEM micrographs of Co electrodeposits at concentration of CoCl2 **a** 0.2 mol/L, **b** 0.3 mol/L, **c** 0.5 mol/L under the applied voltage 2.7 V, the temperature 323 K and 2 h

to the smaller total charge for the lower energy consumption. The optimized energy consumption in this work is 2.59kWh/kg at 0.5 mol/L.

The microstructures of the cobalt deposition on the copper substrate with different $CoCl_2$ concentrations were analyzed using SEM, as shown in Fig. 12. The shapes of the depositions at the $CoCl_2$ concentrations 0.2 and 0.3 mol/L are really similar to each other although more particles and clusters are appeared under the 0.3 mol/L concentration. In Fig. 13, XRD peaks from the cobalt deposit at 0.2 mol/L $CoCl_2$ is shown from the copper cathode to test the purity of Co deposition. All peaks match with the pure cobalt metal (ICDD File #04-001-3273). This proves that highly pure Co deposits are obtained without any other elements through the experiments. In the surface morphology at 0.5 mol/L, particles are rounded with slightly larger size. The cross-sectional morphology is determined, as shown in Fig. 14, for the concentration of electrolytes under 2.7 V, 323 K, and 2 h. In Fig. 14a of 0.3 mol/L $CoCl_2$, the spherical particles are observed on the copper electrode. In Fig. 14b of 0.5 mol/L $CoCl_2$, the shape of particles in the formation of

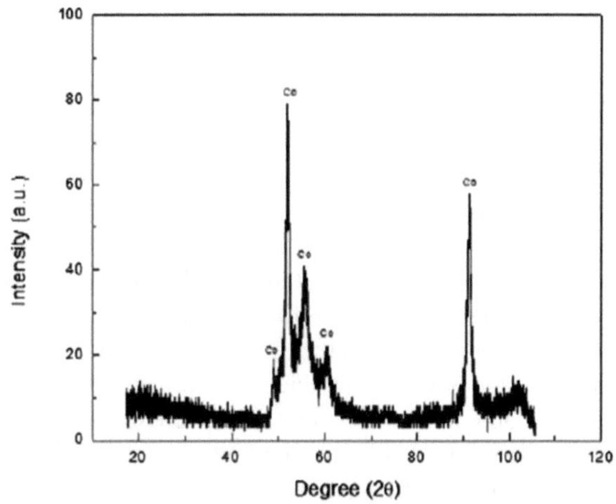

Fig. 13 XRD pattern of the Co electrodeposit on copper cathode

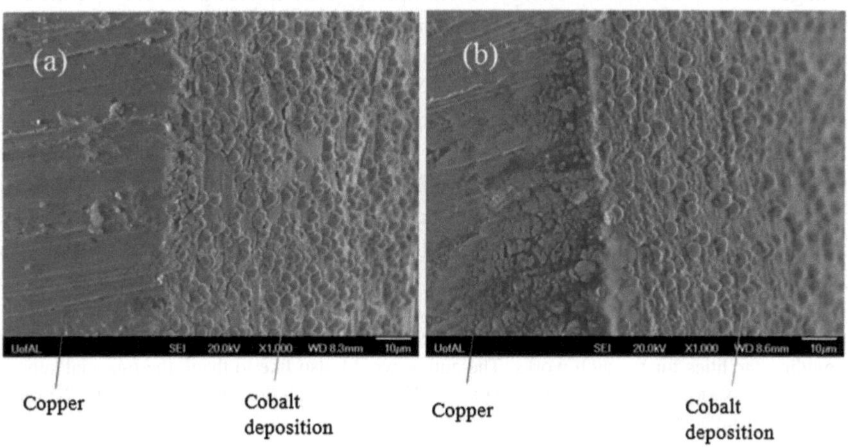

Fig. 14 SEM micrographs of the cross-sectional Co electrodeposits at the concentration of $CoCl_2$ **a** 0.3 mol/L and **b** 0.5 mol/L under the applied voltage 2.7 V, temperature 323 K and 2 h

cobalt deposition is spherical which are similar to the particles in Fig. 14a. However, the Co deposition layer without the particles at 0.5 mol/L (Fig. 14b) might be thicker than that at 0.3 mol/L (Fig. 14a). As the concentration of the ions increases at the cathode, the growth of the cobalt deposited layer forms the large grains by diffusion.

Conclusion

To obtain the high purity cobalt deposit with high efficiency, the effect of temperature, cell voltage, and $CoCl_2$ concentration on the electrodeposition of cobalt from $CoCl_2$ in Urea/ChCl (2:1 ratio) was investigated using current density, current efficiency, and energy consumption. The cyclic voltammetry was conducted with 0.2 mol/L $CoCl_2$ using various scan rates to determine the reduction potential in Urea/ChCl at 323 K, which was the temperature of the lowest energy consumption. The curve of cyclic voltammetry shows that the reduction of Co^{2+} to Co is an irreversible process with one step, 2 electrons transport. Based on the data from cyclic voltammetry test, the transfer coefficient of Co^{2+} and the diffusion coefficient were calculated as 0.22 and 3.38×10^{-6} cm^2/s respectively, which were reasonable as compared with the literature. Under the applied voltage 3.0 V and concentration 0.2 mol/L of $CoCl_2$, the current efficiency with the temperature range of 323–383 K didn't change significantly but still the highest current efficiency (78.58%) was observed at 323 K with the lowest energy consumption (3.47 kWh/kg). Under the temperature 353 K and concentration 0.2 mol/L of $CoCl_2$, the change of the applied voltage showed the lowest energy consumption (2.61 kWh/kg) at 2.4 V with current efficiency 83.64%. The higher applied voltage caused the lower current efficiency due to the formation of dendrites during the cobalt deposition. Under the temperature 323 K and cell voltage 2.7 V, increasing the concentration of $CoCl_2$ showed the highest current efficiency (95%) with the lowest energy consumption (2.59 kWh/kg) at the concentration of 0.5 mol/L $CoCl_2$, which was about saturation concentration in the present system. SEM images of the surface morphology of cobalt deposits on the copper cathodes according to the different parameters were obtained. The XRD analysis of the cobalt deposit confirmed the deposition of pure cobalt.

Acknowledgements The author would like to acknowledge the University of Alabama for providing facilities for research works. The author would also like to thank the financial support from the National Science Foundation (NSF) under the grant number NSF-No. 1310072 throughout this research work.

References

1. M. Hawkins, Why we need cobalt. Appl. Earth. Sci.: Trans. Inst. Min. Metall., Sect. B **110**, 66–71 (2001)
2. M.R. Ali, A. Nishikata, T. Tsuru, Electrodeposition of Co-Al alloys of different composition from the $AlCl_3$-BPC-$CoCl_2$ room temperature molten salt. Electrochim. Acta **42**, 1819–1828 (1997)
3. M.J. Donachie, S.J. *Donachie, Superalloys: A Technical Guide* (ASM International, 2002)
4. R. Michel, M. Nolte, M. Reich, F. Löer, Systemic effects of implanted prostheses made of cobalt-chromium alloys. Arch. Orthop. Trauma Surg. **110**, 61–74 (1991)

5. R. Michel, M. Nolte, M. Reich, F. Löer, Systemic effects of implanted prostheses made of cobalt-chromium alloys. Arch. Orthop. Trauma Surg. **110**, 61–74 (1991)
6. J.A. Disegi, R.L. Kennedy, R. Pilliar, *Cobalt-base Alloys for Biomedical Applications* (ASTM International, 1999), pp. 62–70
7. Min Li, Zhaowen Wang, Ramana G. Reddy, Cobalt electrodeposition using urea and choline chloride. Electrochim. Acta **123**, 325–331 (2014)
8. A.P. Abbott, K.S. Ryder, U. Konig, Electrofinishing of metals using eutectic based ionic liquids. Trans. Inst. Met. Finish. **86**, 196–204 (2008)
9. C.P. Fredlake, J.M. Crosthwaite, D.G. Hert, S.N.V.K. Aki, J.F. Brennecke, Thermophysical properties of imidazolium-based ionic liquids. J. Chem. Eng. Data **49**, 954–964 (2004)
10. S. Forsyth, J. Pringle, D. Macfarlane, Ionic liquids—an overview. Aust. J. Chem. **57**, 113–119 (2004)
11. C. Chiappe, D. Pieraccini, Ionic liquids: solvent properties and organic reactivity. J. Phys. Org. Chem. **18**, 275–297 (2005)
12. A.P. Abbott, G. Capper, D.L. Davies, R.K. Rasheed, V. Tambyrajah, Novel solvent properties of choline chloride/urea mixtures. Chem. Commun., 70–71 (2003)
13. Q. Zhang, K.D.O. Vigier, S. Royer, F. Jérôme, Deep eutectic solvents: syntheses, properties and applications. Chem. Soc. Rev. **41**, 7108–7146 (2012)
14. M. Pal, R. Rai, A. Yadav, R. Khanna, G.A. Baker, S. Pandey, Self-aggregation of sodium dodecyl sulfate within (Choline Chloride + Urea) deep eutectic solvent. Langmuir **30**, 13191–13198 (2014)
15. R.G. Reddy, Ionic liquids: how well do we know them? J. Phase Equilib. Diff. **27**, 210–211 (2006)
16. H. Yang, R.G. Reddy, Electrochemical deposition of zinc from zinc oxide in 2:1 urea/choline chloride ionic liquid. Electrochim. Acta **147**, 513–519 (2014)
17. H. Yang, R.G. Reddy, *Electrodeposition of zinc from zinc oxide using urea and choline chloride mixture: effect of [BMIM]HSO4, temperature, voltage on current efficiency, energy consumption, and surface morphology* (TMS, Warrendale, PA, USA, 2014), pp. 21–26
18. H. Yang, R.G. Reddy, Fundamental studies on electrochemical deposition of lead from lead oxide in 2:1 urea/choline chloride ionic liquids. J. Electrochem. Soc. **161**, 586–592 (2014)
19. A. Cojocaru, M. Sima, Electrochemical investigation of the deposition/dissolution of selenium in choline chloride with urea or ethylene glycol ionic liquids. Rev. Chim. (Bucharest) **63**, 217–223 (2012)
20. A.M. Popescu, A. Cojocaru, C. Donath, V. Constantin, Electrochemical study and electrodeposition of Copper(I) in ionic liquid-reline. Chem. Res. Chin. Univ. **29**, 991–997 (2013)
21. A.M. Popescu, V. Constantin, A. Cojocaru, M. Olteanu, Electrochemical behaviour of Copper (II) chloride in choline chloride-urea deep eutectic solvent. Rev. Chim. (Bucharest) **62**, 206–213 (2011)
22. M. Rostom Ali, Md Ziaur Rahman, S. SankarSaha, Electrodeposition of copper from a choline chloride based ionic liquid. J. Electrochem. **20**, 139–145 (2014)
23. Q. Chu, J. Liang, J. Hao, Electrodeposition of zinc-cobalt alloys from choline chloride–urea ionic liquid. Electrochim. Acta **115**, 499–503 (2014)
24. A.P. Abbott, G. Capper, K.J. McKenzie, K.S. Ryder, Electrodeposition of zinc–tin alloys from deep eutectic solvents based on choline chloride. J. Electroanal. Chem. **599**, 288–294 (2007)
25. E. Gómez, E. Vallés, Platinum electrodeposition in an ionic liquid analogue. Solvent stability monitoring. Int. J. Electrochem. Sci. **8**, 1443–1458 (2013)
26. S. Costovici, A. Petica, C.-S. Dumitru, A. Cojocaru, L. Anicaia, Electrochemical synthesis of ZnO nanopowder involving choline chloride based ionic liquids. Chem. Eng. Trans. **41**, 343–348 (2014)
27. L. Anicai, A. Florea, T. Visan, *Studies regarding the nickel electrodeposition from choline chloride based ionic liquids* (Applications of Ionic Liquids in Science and Technology, Middle Tennessee State University, USA, 2011), pp. 261–286

28. T.-C. Huynh, Q.P.D. Dao, T.-N. Truong, N.-G. Doan, S.-L. Ho, Electrodeposition of aluminum on cathodes in ionic liquid based choline chloride/urea/AlCl3. Environ Pollut **3** (2014), 59–69
29. F. Golgovici, T. Visan, Electrodeposition behavior of cadmium telluride from chlorine chloride-urea ionic liquids. Chalcogenide Lett **9**, 165–174 (2012)
30. D. Yue, Y. Jia, Y. Yao, Y. Sun, Y. Jing, Structure and electrochemical behavior of ionic liquid analogue based on choline chloride and urea. Electrochim. Acta **65**, 30–36 (2012)
31. A.J. Bard, L.R. Faulkner, *Electrochemical Methods: Fundamental and Applications* (Wiley, New York, 1980)
32. F. Golgovici, T. Visan, Electrodeposition behavior of cadmium telluride from chlorine chloride-urea ionic liquids. Chalcogenide Lett **9**, 165–174 (2012)
33. Q.Q. Yang, K.R. Qiu, D.R. Zhu, L.C. Sa, Electrodeposition of cobalt and rare earth-cobalt in urea-NaBr-KBr melt. Electrochemistry **1**, 274–277 (1995)
34. C.W. Xu, W.J. Pan, D.S. Yuan, Y.X. Tong, G.K. Liu, Preparation of Gd-Co alloy film in acetamide-urea-NaBr melt. T. Nonferr. Metal. Soc. **12**, 1007–1010 (2002)
35. J.C. Wang, C.W. Xu, P. Liu, Y.X. Tong, G.K. Liu, Preparation of Y-Co alloy film in acetamide-urea-NaBr melt. T. Nonferr. Metal. Soc. **12**, 1191–1194 (2002)
36. M. Li, Z. W. Wang, R. G. Reddy, "Electrodeposition of cobalt from air and water-stable ionic liquid 1-butyl-3- methylimidazolium tetrafluoroborate," EPD Congress, San Diego, (2014), pp. 241–250
37. R. Fukui, Y. Katayama, T. Miura, The effect of organic additives in electrodeposition of Co from an amide-type ionic liquid. Electrochim. Acta **56**, 1190–1196 (2011)
38. R.T. Carlin, H.C. De Long, J. Fuller, P.C. Trulove, Microelectrode evaluation of transition metal-aluminum alloy electrodepositions in chloroaluminate ionic liquids. J. Electrochem. Soc. **145**, 1598–1607 (1998)
39. D. Pradhan, R.G. Reddy, Mechanistic study of Al electrodeposition from EMIC-AlCl3 and BMIC-AlCl3 electrolytes at low temperature. Mater. Chem. Phys. **143**, 564–569 (2013)
40. Q. Liao, W.R. Pitner, G. Stewart, C.L. Hussey, G.R. Stafford, Electrodeposition of aluminum from the aluminum chloride-1-methyl-3-ethylimidazolium chloride room temperature molten salt + benzene. J. Electrochem. Soc. **144**, 936–943 (1997)

METTOP-BRX Technology—Eliminating Concerns and Highlight Potentials of the Concept of Tankhouse Optimization

A. Filzwieser, I. Filzwieser, S. Wallner and M.B. Hanel

Abstract Whenever thinking about increasing tankhouse performance two major aspects has to be considered, the technological feasibility as well as the economical impact. For meeting the requirement of an increased current density leading to an increased productivity, Mettop has developed the METTOP-BRX Technology, which allows the introduction of fresh electrolyte between each pair of anode and cathode. As an example of a greenfield approach the installation at Xiangguang Copper Co., Ltd will be described. Since the start up in 2011, Xiangguang Copper Co., Ltd operates one of their two copper refining tankhouse with a current density of 420 A/m^2 and a current efficiency above 98% in average. At Montanwerke Brixlegg Ag, as an example of a brownfield installation, the annual production could be increased by 50%. Although for many years a total of two industrial applications are in operation, the successful implementation of this technology that result in significant operational and economic advantages has not been fully accepted within the industry due to misunderstandings, incorrect interpretations and lack of awareness. It is intended with this paper to create a better understanding about the parallel flow technology and help to eliminate worries and concerns.

Keywords Tankhouse optimization · Copper hydrometallurgy · Parallel flow device

Introduction

Whenever thinking about increasing tankhouse performance two major aspects have to be considered, the technological feasibility as well as the economic impact. For an optimized tankhouse performance, not only in copper industry but also for zinc electrolysis, the following points are of interest:

A. Filzwieser (✉) · I. Filzwieser · S. Wallner · M.B. Hanel
Mettop GmbH, Peter Tunner Str. 4, 8700 Leoben, Austria
e-mail: andreas.filzwieser@mettop.com

S. Wang et al. (eds.), *Applications of Process Engineering Principles in Materials Processing, Energy and Environmental Technologies*,
The Minerals, Metals & Materials Series, DOI 10.1007/978-3-319-51091-0_10

- High tankhouse capacity
- High current efficiency
- High chemical cathode quality
- Low operating cost
- Lower footprint
- Low bound capital.

Increasing tankhouse capacity can generally be done in two ways: Either by increasing the cathodic deposition area (i.e., adding more cells or increasing the number of electrodes in the cell) or by increasing the current density. The former possibility is relatively cost-intensive, as new buildings and infrastructure are required, and sometimes it is not possible due to limited space and/or other restrictions. Increasing the current density is generally a matter of the actual electrical system but it is also limited as current efficiency and cathode quality tend to decrease with increasing current density, due to problems with both cathodic deposition and anodic dissolution. Consequently, the current densities that are used in electrorefining are far below the theoretical limiting current density.

According to physical and chemical laws, the only way to increase the current density in a way that ensures that the electrodeposition is still pure and that the metal sheets are smooth and of a high quality, is to minimize the diffusion boundary layer and/or to influence the electro crystallization by an optimized inhibitor distribution. This can be achieved by increasing the flow directly at the cathode, for example by using a manifold inlet to achieve a parallel electrolyte flow and introduce the fresh electrolyte with the active inhibitors directly at the cathodes.

For meeting all those requirements, Mettop has developed together with Montanwerke Brixlegg AG, Austria a new type of manifold inlet: the METTOP-BRX Technology [1–3]. Described simply, the METTOP-BRX Technology increases the productivity of a conventional electrorefining cell by allowing the operator to increase the current density up to 400–420 A/m^2. This is accomplished by increasing the electrolyte flow rate and introducing it into the cell parallel to the electrode surfaces. Introduction of the electrolyte in this way, in contrast to introducing it at the end wall of the cell, aids the transfer of Cu^{2+} from anode to cathode by enhancing the natural convection of electrolyte between the two opposing electrodes, minimizing the diffusion boundary layer at the cathode surface, and improving the homogeneity of the electrolyte temperature and chemistry within the cell.

The parallel electrolyte flow can be realized by either using a Parallel Flow Device (PFD), namely a separate device that is installed in the cell, or a Parallel Flow Plate (PFP), namely an integrated system in the polymer concrete cell.

Theoretical Background and Functional Principle

To increase current density in a copper refining electrolysis the limiting current density has to be increased. The only physical quantity which can be influenced in a wider range is the diffusion boundary layer δN as shown in Eq. 1. The concentration in the bulk electrolyte is limited to 45–50 g/l because above 50 g/l the movement of the Cu ions will be negatively affected. The diffusion coefficient could be influenced via the temperature but the temperature is limited to 65 °C.

$$i_{lim} = n \cdot F \cdot D \cdot \frac{c^0}{\delta_N} \tag{1}$$

i_{lim} limiting current density [A/m^2]
n valency of ions [–]
F Faraday constant (F = 96,485 [As/mol])
D diffusion coefficient [m^2/s]
c^0 concentration in bulk electrolyte [mol/m^3]
δ_N thickness of diffusion layer [m].

Decreasing the diffusion boundary layer means increasing the relative movement of electrolyte in front of the cathode surface. The limiting current density is theoretically the highest current density which can be used in a given system. At the limiting current density the copper ions concentration in front of the cathodes is zero.

However, the current density used in industrial practice amounts only 30–35% of the theoretical limiting current density. That ratio can be increased by an excellent operation practise. A perfect electrode lining, homogeneous copper concentration and electrolyte temperature distribution as well as a perfect inhibitor distribution (directly in front of the cathode!) are of importance.

To increase the limiting current density and the industrial used current density the following has to be done:

Decreasing the diffusion boundary layer by increasing the relative electrolyte velocity between electrolyte and cathode surface.

Distribution of all inhibitors directly in front of the cathodes

The electrolyte flow within an electrolysis cell can be divided in a forced convection and a natural convection. The forced electrolyte flow in a conventional electrorefining cell is shown on the left side in Fig. 1. The electrolyte enters the cell at the bottom of the front and leaves it at the upper opposite end. One major disadvantage of this conventional arrangement of inlet and outlet is that the fresh electrolyte, and hence also the additives like inhibitors, is not inserted directly in front of the active cathode surface, where it is needed, and a part of it also leaves the cell by bypassing the electrodes on the sides. By using a conventional forced convection the relative movement between electrolyte and cathode is hardly influenced by the forced convection. The diffusion boundary layer is mainly influenced by the natural convection.

Figure 1 (right side) shows the natural convection in the cell, which occurs during electrorefining due to differences in electrolyte density near the electrodes. Close to the anode, the density of the electrolyte increases due to anodic dissolution and hence higher concentration of metal ions, whereas near the cathode a local decrease of density occurs due to electrodeposition. This natural convection was proven and determined quantitatively by measurements of fluid flow velocity between anode and cathode, as shown in Fig. 2.

Again, to increase the limiting current density the diffusion boundary layer has to be decreased and therefore the relative movement (velocity) between electrolyte and cathodes surface has to be increased. This means higher electrolyte velocities enable higher limiting current densities in electrorefining. These higher velocities can be realized by implementing a parallel electrolyte flow between the electrodes. The development of the Parallel Flow Device (PFD) enables this conditions. However, the position of the electrolyte inlet has to be chosen accurately, as a high electrolyte velocity at the anode can also cause problems with the anode slime,

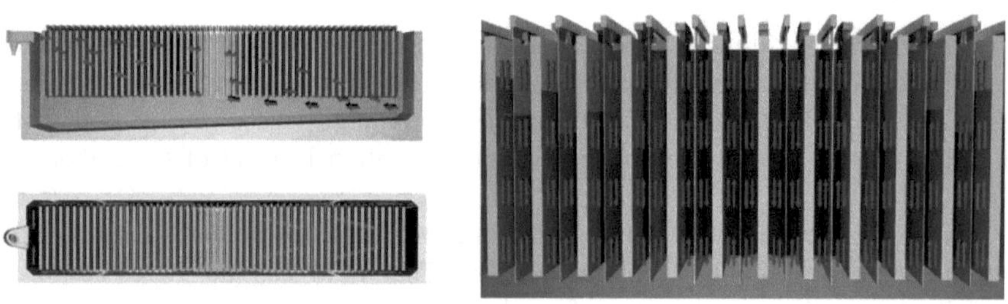

Fig. 1 Forced electrolyte flow in a conventional electrorefining cell (*left*) and natural convection during electrorefining (*right*)

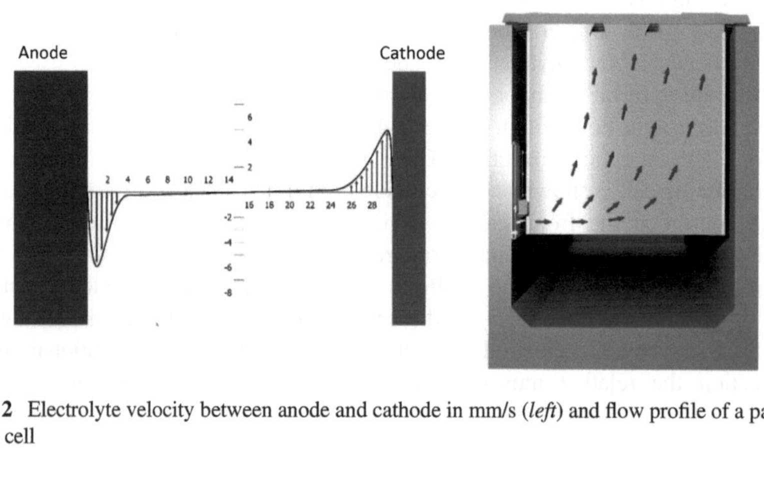

Fig. 2 Electrolyte velocity between anode and cathode in mm/s (*left*) and flow profile of a parallel flow cell

which may be detached from the anode surface, transferred to the cathode, and cause dendrites and buds there.

The flow profile of a parallel flow cell is shown on the right side of Fig. 2.

Due to a higher electrolyte flow (up to three times higher than the standard flow) the velocity in front of the cathode is much higher and therefore the diffusion boundary layer is decreased. Which means a higher limiting current density and therefore a higher industrial used current density is achieved using the parallel flow concept.

Inhibitor Distribution

To produce a very smooth copper cathode and to operate the tankhouse on a high current efficiency following inhibitors have to be added:

Glue: 50–100 g/t copper
Thiourea: 50–100 g/t copper
HCl: 30–70 mg/l.

The inhibitors are added to the electrolyte before the fresh electrolyte is entering the cell. In a conventional cell the inhibitors are not in front of the cathode (where they have to be; see Fig. 1).

Using a parallel flow concept the inhibitors are introduced directly in front of the cathodes and therefore smooth copper cathodes are produced even at very high current densities (up to 450 A/m^2). The inlet of the fresh electrolyte has to be close to the bottom of the cell (see Fig. 3 left). By doing so, the upwards flow of the natural convection can be used and the fresh electrolyte (and all inhibitors) are directly in front of the cathode surface. The density of the fresh electrolyte is lower as the used electrolyte and this leads also to an upwardly flow.

Using the METTOP-BRX Technology a current density of up to 450 A/m^2 at a current efficiency of above 98% can be achieved.

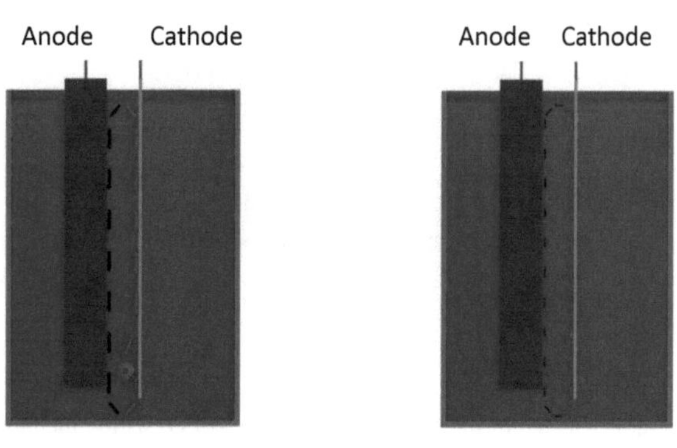

Fig. 3 Comparison of electrolyte flow profile using the METTOP-BRX Technology (*left*) with the natural convection at a conventional system (*right*)

Overall, the impact of introducing a parallel flow can be summarized [4–6]:

- Due to a higher electrolyte flow rate (up to three times higher than a standard flow rate) and a flow directly in front of the cathode leads to a decreased hydrodynamic and diffusion boundary layer. A direct introduction of the inhibitors near the active cathode surface lead to a more homogeneous distribution of inhibitors on the cathode surface and furthermore increase the product quality in terms of surface quality
- A simultaneous introduction of the inhibitor glue guarantees the same glue activity all over the cell lengths
- Since the distance between the anode and cathode is precisely fixed, the current density distributions is more homogenous
- The direct electrolyte introduction leads furthermore to a more homogenous electrolyte temperature distribution.

METTOP-BRX Technology—Hardware for Different Kinds of Application

There are two different approaches for optimization of the tankhouse performance, either when installing a perfectly new tankhouse (greenfield approach) or an upgrading of an existing tankhouse (brownfield approach).

Brownfield Approach—Parallel Flow Device (PFD) for Upgrading Existing Tankhouses

The core of the METTOP-BRX Technology is a Parallel Flow Device (PFD; Fig. 4 left side), which introduces the fresh electrolyte between each pair of electrodes

Fig. 4 Sketch of the parallel flow device (PFD) within the tankhouse (*left*) and magnified view of the nozzles and cathode positioning system (*plastic rectangles*, *right side*)

within the tankhouse. Special positioning devices—so-called cathode spacers as shown in Fig. 4—provide an accurate electrode positioning, as well as a defined relative position of cathode and nozzles. The nozzles, which are designed for each individual tankhouse, direct the fresh electrolyte in an upwards flow in front of the cathode surface, which only enhances the already existing flow resulting from natural convection. The PFDs are customer-tailored to guarantee optimum results for each specific tankhouse.

Greenfield Approach—Parallel Flow Plate for Optimized Tankhouse Performance

The Parallel Flow Plate (PFP) is a combination of PFD and the cell itself: In new tankhouses—or when replacing cells in an existing tankhouse—the METTOP-BRX Technology can be integrated in the polymer concrete cells as shown in Fig. 5. The main idea is to cast the cell in a way that the volume for the electrolyte distribution system is already included. Therefore, only a stainless steel front plate with nozzles and cathode spacers—the Parallel Flow Plate—is attached to the cell. Implementing PFPs minimizes the installation effort, as only the front plate has to be mounted. Furthermore, these systems can be preassembled at the cell supplier and the customer can install the complete cells just like normal cells.

In both application cases the, the basic equipment can be summarized:

- A welded stainless steel construction (PFD) or a combination of the polymer concrete cell and a stainless steel plate (PFP) for distributing the fresh electrolyte evenly along the entire cell

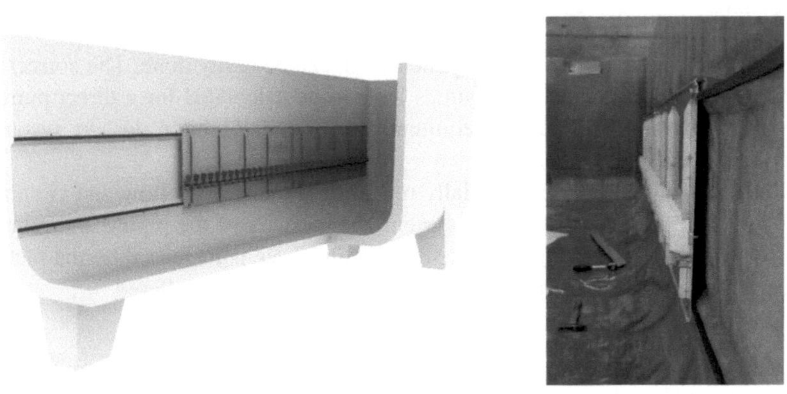

Fig. 5 Sketch of the parallel flow plate (PFP, *right*) and during installation at Montanwerke Brixlegg AG, Austria

- Stainless steel nozzles for introduction the fresh electrolyte precisely in between the anodes and cathodes
- Cathode positioning systems for ensuring the accurate spacing within the entire cell.

Due to the fact that a higher electrolyte flow rate is required (the outlet velocity of the electrolyte per nozzle has to be calculated individually for each tankhouse) the overall tankhouse design has to be adapted and again can be divided between greenfield and brownfield projects.

Considerations for Installation of the METTOP-BRX Technology

The use of high current density and the METTOP-BRX Technology requires an adequate tankhouse design. Optimum performance can only be achieved with a comprehensive design of the overall process and complete tankhouse equipment, comprising electrical and electrolyte system, as well as the mechanical components.

For installation of the METTOP-BRX Technology the following requirements have to be taken into account:

- Electrolyte system for high flow: Adequate piping diameters, as well as direct pumping instead of overhead tanks
- Electrical system for high current density: Suitable rectifier, transformer, and bus bars
- A good electrode quality/geometry is essential for exact positioning
- Polymer concrete cells—for smooth walls (PFD installation) and system integration (PFP)
- Crane positioning system to guarantee exact electrode positioning
- Stainless cathodes or smooth copper mother sheets.

For a greenfield project the adequate design can be easily done. The correct pipe dimensions as well as the right pumps have to be calculated for a direct pumping system. There are no specific requirements for the inhibitor dosage system or filtering system.

For a brownfield project especially regarding the increased flow and circulation rate compared to the conventional system, the capacity of the piping and pumping systems will be the limiting factors, since the flow rate is increased by the factor of three. To overcome these problems individual solutions are offered.

Solutions like:

- Creating internal loops for two third of the electrolyte flow and using the existing piping for one third

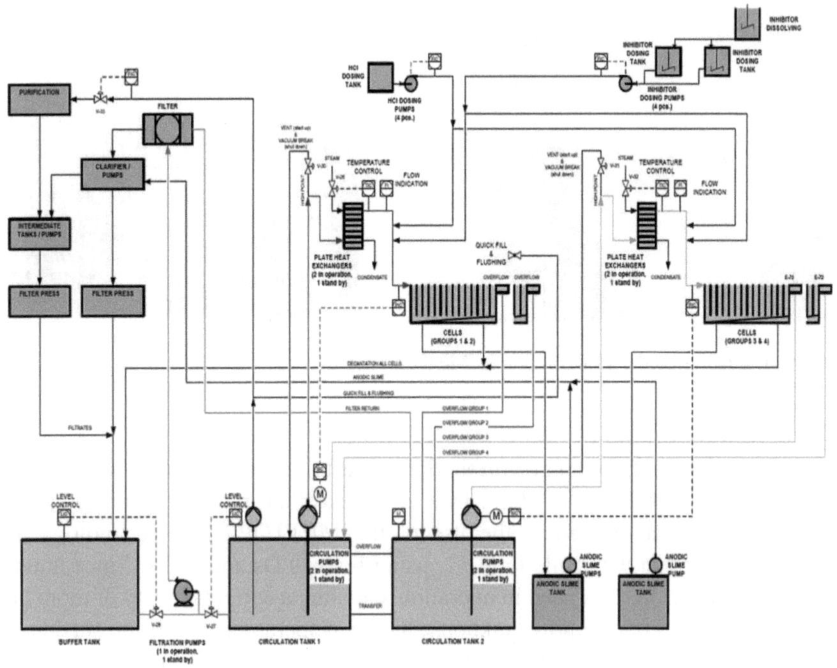

Fig. 6 Process flow sheet of a greenfield approach as realized in Xiangguang Copper Tankhouse No. 2

- Increasing the current density by taking out anodes and cathodes and creating the higher flow using pumps installed into the calls
- Replacing partly the existing pipes and change to a direct pumping system (Fig. 6).

References of the Technology in Industrial Scale Use

In both cases, brownfield and greenfield application, Mettop has industrial scale references for proving the applicability of the METTOP-BRX Technology. In the following sections a basic description of the installed system as well as the major benefits will be given.

Montanwerke Brixlegg AG, Austria—Upscaling an Existing Tankhouse [3, 7, 8]

The longest-standing METTOP-BRX Technology application is found at Montanwerke Brixlegg AG, a secondary copper producer in Austria, where—after

Fig. 7 Installation of the PFD on site at Montanwerke Brixlegg AG and electrolyte flow after installation

years of developments and high current density tests—52 cells of the new tankhouse were equipped with PFDs in 2007. Due to the excellent results regarding cathode quality and current efficiency, the use of the METTOP-BRX Technology was extended by another 104 cells in 2011, using both PFDs and PFPs. Since autumn 2011, the new rectifier has been in operation, enabling a current density of more than 420 A/m^2, and half of the entire tankhouse has been operating at high current density.

Montanwerke Brixlegg AG has made the following observations and conclusions since the introduction of the METTOP-BRX Technology:

- The technology allows using higher current density and therefore increasing production.
- With additive adjustments the current efficiency can be increased by two to three percent.
- In case of bad anode quality, current efficiency can be kept the same as when using good anode quality—or can even be higher by using the METTOP-BRX Technology.
- A better cathode quality due to more homogeneous additive distribution can be achieved.
- Using the technology results in shorter down times when filling the groups due to higher flow and consequently faster filling. Furthermore, the higher flow enables a faster heating-up of the anodes (for new anodes) in the cold season.
- Fewer shorts occur due to accurate cathode positioning/guiding system (Fig. 7).

Xiangguang Copper, China—Installation of a New Tankhouse

The first application of the METTOP-BRX Technology in an entire new tankhouse (720 cells) is Xiangguang Copper, a primary copper producer in China. The tankhouse engineering for their second tankhouse, which had been started up in May 2011, was done by Mettop. This tankhouse was constructed in the same way as the existing tankhouse, but now 300,000 t/a cathode copper, compared to the existing tankhouse with 200,000 t/a production capacity, can be produced by using the METTOP-BRX Technology. This increase in production rate of 50% is caused by the much higher current density, namely an increase from 280 A/m^2 in the old tankhouse up to 420 A/m^2.

As Xiangguang Copper decided from the very start that the new tankhouse will be equipped with the METTOP-BRX Technology and the high current density technology, Mettop designed the whole tankhouse accordingly. The tankhouse design included:

- Latest generation of PFDs (pyramid cathode spacers, hanging system, cleaning openings, handling hooks)
- Pumps for direct pumping (i.e., no overhead tanks) for approximately 100 l/min per cell
- Design of electrical system for 420 A/m^2 (2 electric circuits)
- Design of heat exchangers
- Good coordination of electrical and electrolyte circuits under consideration of potential differences
- Continuous addition of inhibitors.

This first greenfield METTOP-BRX tankhouse started up in summer 2011. In December 2011, the average current density of the new tankhouse was higher than 98% (98.51%) at current density of 410 A/m^2, and the quality of the produced cathodes was excellent as shown in Fig. 8. In 2012 Xiangguang Copper used 420 A/m^2 and achieved an average current efficiency of 98.3%. Today Xiangguang Copper reports current efficiency of >99% at 420 A/m^2 [9].

Fig. 8 Tankhouse with METTOP-BRX Technology at Xiangguang Copper (*left*) and excellent surface quality of the cathodes

Economical Benefits of the METTOP-BRX Technology

For the two industrial scale applications, either brownfield application as well as greenfield application, detailed economical calculations regarding operational expenditures, capital expenditures, savings and revenues have been conducted. Due to different power costs as well as different interest rate levels an economical calculation or evaluation has to be done for each customer individually. Just for an overview only a few points which are significant are given for a production of 400,000 t copper cathodes per year.

For a new tankhouse following CAPEX decrease can be achieve using 420 A/m^2 (by using the METTOP-BRX Technology) compared to a standard technology (310 A/m^2).

- Lower overall tankhouse footprint ($\sim$20%)
- Less cells (26%)
- Less cathode plates (26%)
- Less electrolyte and edge strips
- Less anodes in the tankhouse (25–30 Mio US$ bound capital reduction)!

The costs for machines like anode preparation machine, stripping machine, etc. and for the electrical installation in terms of generator, rectifier, bus bar, etc. will stay the same assumed the same production for both systems.

For the OPEX calculation the two following parameters are of importance:

- Interest savings due to a much lower CAPEX costs
- Increase in power consumption due to a higher current density.

Both industrial sites, where the METTOP-BRX Technology is used, report that the overall OPEX are decreased [9–11] and all economic calculations are supporting these statements even in countries with high power costs.

Summary and Conclusion

Today's most advanced tankhouse technology—the METTOP-BRX Technology—allows increasing the current density and productivity, respectively, by up to 50% compared to conventional copper electrorefining tankhouses. This technology can either be installed in existing facilities—in order to increase production—or considered in new plants—in order to reduce the footprint of the entire tankhouse.

The present paper illustrates the potential economic benefit of equipping copper electrorefining facilities with the METTOP-BRX Technology. Both for a greenfield project as well as for a brownfield upgrade, the economic value is shown by industrial inspired examples.

Right now the benchmark for a copper refinery is Xiangguang Copper, China, running their tankhouse no. 2 at 420 A/m^2 at a current efficiency >99% using the METTOP-BRX Technology.

References

1. Europäische Patentschrift EP 2 183 409 B1: Method for operating copper electrolysis cells
2. A. Filzwieser, I. Filzwieser, S. Konetschnik, New technology for electrorefining of copper. JOM **64**(11), 1290–1295 (2012)
3. C. Wenzl et al., High Current Density Tankhouse and METTOP-BRX Technology at Montanwerke Brixlegg AG. Erzmetall **64**(6), 5–1 (2011)
4. A. Filzwieser, K. Hein, G. Mori, Current density limitation and diffusion boundary layer calculation using computational fluid dynamics. JOM **54**(4), 28–31 (2002)
5. A. Filzwieser, Modelling of the processes close to cathodes in copper electrolysis (Modellierung der kathodennahen Vorgänge in der Kupferelektrolyse), Ph.D. thesis, Montanuniversität Leoben, Chair of Non-Ferrous Metallurgy, Leoben, Austria, 2000
6. I. Filzwieser, The analysis and mathematical modelling of the parameters influencing cathodic deposits in copper refining electrolysis, Ph.D. thesis, Montanuniversität Leoben, Chair of Non-Ferrous Metallurgy, Leoben, Austria, 2005
7. A. Filzwieser, J. Pesl, A. Anzinger, Ch. Kapfinger, Optimisation of tankhouse operation at Montanwerke Brixlegg, in *Proceedings EMC*, vol. 2, 28th June–1st July Innsbruck, Austria, 2009, pp. 715–726
8. A. Filzwieser, I. Filzwieser, S. Konetschnik, A. Anzinger, Economic benefits of operating a copper electro-refining tankhouse at high current density using the Mettop-BRX Technology, in *Proceedings of the Conference of Metallurgists COM*, 28th Sep–1st Oct, Vancouver, Canada, 2014
9. S. Zhou, The jet of parallel—a new technology of copper electrolysis, in *Proceedings of Copper 2016* (in print)
10. A. Filzwieser, I. Filzwieser, S. Konetschnik, A. Anzinger, Cost comparison between a conventional ER tankhouse and a high current density ER tankhouse using the METTOP-BRX Technology, in *Proceedings of the Copper 2013*, 1–4 Dec, Santiago, Chile, 2013
11. S. Zhou, Copper electrolytic refining technology operating at high current density, in *Proceedings of Copper 2013*, vol. V, pp. 333–344, 2013

Mathematical Modeling of Molten Salt Electrolytic Cells for Sodium and Lithium Production

Donghui Li, Kinnor Chattopadhyay, Lei Gao, Boyd Davis, Rüdiger Schwarze, Amjad Asad and Christoph Kratzsch

Abstract Sodium (Na) and Lithium (Li) are produced using molten salt electrolysis. The electrochemistry of the electrolyte is well-researched; however, there are benefits to understanding the melt flow and implications on it for cell design modifications. The basic configuration of alkali metal cells is the Downs cell. This consists of a central anode surrounded by a cathode, and this geometry was the basis for this mathematical modeling study. The behavior of gas bubbles in molten electrolyte was studied in both Na and Li cells through the use of computational fluid dynamics (CFD) techniques. The distance between the anode and the cathode was varied in the CFD model to ascertain whether strong circulatory flows would change significantly in the cell. The standard *k*-*ε* turbulence model was used to mimic turbulent flow, and a two-way coupled Discrete Phase Model (DPM) was adopted to simulate flotation behavior of chlorine bubbles and liquid metal droplets. The liquid metal distribution on the free surface was predicted using the Volume of Fluid (VOF) multi-phase model.

Keywords Downs cell · Sodium · Lithium · Electrolysis · CFD modeling

---

D. Li · K. Chattopadhyay (✉)
Process Metallurgy and Modelling Group, University of Toronto,
184 College Street, Suite 140, Toronto, ON M5S3E4, Canada
e-mail: kinnor.chattopadhyay@utoronto.ca

D. Li · K. Chattopadhyay
Innovative Extractive Technologies Ltd., 145 St. George Street,
Suite 406, Toronto, ON M5R2M1, Canada

L. Gao
State Key Laboratory of Complex Non-Ferrous Metal Resources Clean Utilization,
Kunming University of Science and Technology, Kunming 650093, China

B. Davis
Kingston Process Metallurgy Inc., 759 Progress Avenue Kingston,
Kingston, ON K7M 6N6, Canada

R. Schwarze · A. Asad · C. Kratzsch
TU Bergakademie Freiberg, Lampadiusstrasse 4, Freiberg 09599, Germany

S. Wang et al. (eds.), *Applications of Process Engineering Principles in Materials Processing, Energy and Environmental Technologies*,
The Minerals, Metals & Materials Series, DOI 10.1007/978-3-319-51091-0_11

Introduction

The use of molten salt electrolysis for light metal production is the norm for aluminum, sodium, and lithium extraction and is used outside of China for magnesium production. In order to produce sodium metal, a mixture of salts containing NaCl are used in the Downs cell [1]. In the molten salt, NaCl is present as Na^+ and Cl^- ions. When an electric current is passed through the molten salt, Na^+ ions are reduced at the cathode to form sodium metal. Cl^- ions are oxidized to make chlorine bubbles on the anode surface. For sodium production, a steel mesh membrane is used as a partition between the anode and cathode in order to keep the sodium away from chlorine, and prevent any chemical reaction. Because of the difference in density between the lighter metal and its salt bath, the molten sodium droplets formed in the cathode compartment rise to the top and coalesce to form a liquid metal layer. The chlorine bubbles produced on the anode surface float up out of the molten salt bath, and are subsequently collected by an overhead hood. This is the same for lithium, except that membrane is not always used since lithium partially wets the cathode.

In this paper, a central anode surrounded by a cathode was modeled using Computational Fluid Dynamics (CFD) technology. The nature of the molten salt flow and mass transport in the Downs cell was investigated as to see if the formation of bubbles and metal along the electrodes could be modeled to give electrolyte flows that are estimated in these types of cells.

Dimensions of the Anode/Cathode Pair

Figure 1 displays the schematic view of the modeled electrode pairs. The solid graphite anode was located at the centre of the Downs cell, surrounded with a cylindrical steel cathode. The gap between the anode and the cathode was set as 30 mm, to form a channel for the flotation of both chlorine bubbles and metal droplets. The anode and the cathode were submerged, and suspended in the molten

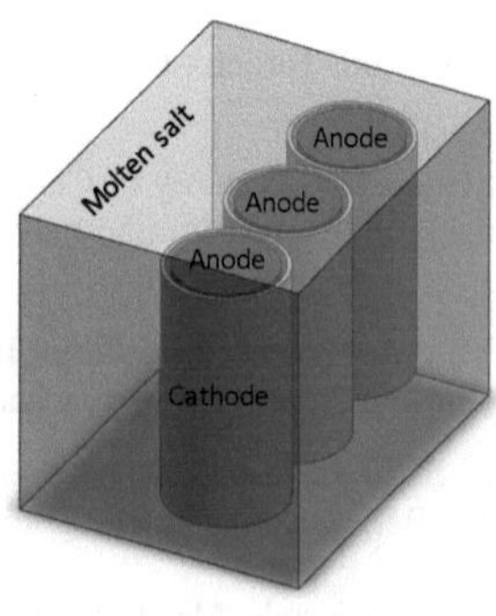

Fig. 1 Sketch of the anode/cathode pairs

Table 1 Dimensions of the Downs cell used in the CFD modeling

Item (m)	Value
Diameter of anode	0.45
Inner dia. of the cathode	0.51
Thickness of the circular steel cathode	0.02
Height of the electrodes	2.13
Height of the Downs cell	2.80

salt. The produced chlorine gas escaped from the top surface of the molten salt, and was collected by overhead hood. The dimensions of the anode/cathode pair in the CFD modeling are given in Table 1, and are not meant to be based on any commercial design.

Mathematical Model

The molten salt flow in the Downs cell is driven by the drag force as a result of the flotation of both chlorine bubbles and liquid metal droplets. The behavior of the chlorine bubbles and metal droplets were predicted by using the Lagrangian discrete phase method coupled with a two equation turbulence model.

Standard k-ε Turbulence Model

The standard *k-ε* model by Launder and Spalding was adopted to model the turbulent flow inside the cell [2]. Here, ε is the rate of turbulence energy dissipation, while *k* stands for the kinetic energy of turbulence per unit mass.

$$\frac{Dk}{Dt} = \frac{\nu_t}{\sigma_k}\nabla^2 k + G_k - \varepsilon \tag{1}$$

$$\frac{D\varepsilon}{Dt} = \frac{\nu_t}{\sigma_\varepsilon}\nabla^2 \varepsilon + \frac{\varepsilon}{k}(C_1 G_k - C_2\varepsilon) \tag{2}$$

The parameter G_k is the rate of production of k and is calculated by the equation of $G_k = \nu_t\left[\frac{\partial u_i}{\partial x_j} + \frac{\partial u_j}{\partial x_i}\right]\frac{\partial u_i}{\partial x_j}$. Finally, the turbulent and the effective viscosities are calculated by the equations $\mu_t = \frac{C_\mu \rho k^2}{\varepsilon}$ and $\mu_{eff} = \mu + \mu_t$.

The values of the constants in the standard *k-ε* model were recommended by Launder and Spalding: C_1 = 1.44, C_2 = 1.92, C_μ = 0.09, σ_k = 1 and σ_ε = 1.3, were used in the present work.

Discrete Phase Model (DPM)

In the discrete phase modeling procedure, the fluid phase is treated as a continuum by solving the Navier-Stokes equations, while the dispersed phase is solved by tracking a large number of bubbles, or droplets, through the calculated flow field in a Lagrangian frame of reference. The drag force, gravity, buoyancy force and virtual mass force were considered in the DPM model. There can be momentum exchange of the dispersed phase with the fluid phase, by two-way coupled calculating procedure. The basic equations involved in discrete phase modeling are as follows:

$$\frac{d\overrightarrow{u_p}}{dt} = \frac{18\mu C_D \mathrm{Re}_P}{24\rho_p d_p^2}\overrightarrow{u_{rel}} + \frac{g(\rho_p - \rho)}{\rho_p} + \frac{1}{2}\frac{\rho}{\rho_p}\frac{d}{dt}\overrightarrow{u_{rel}} \tag{3}$$

here, $\mathrm{Re}_P = \frac{\rho d_p \left|\overrightarrow{u_{rel}}\right|}{\mu}$, $\overrightarrow{u_{rel}} = \overrightarrow{u} - \overrightarrow{u_p}$, and $C_D = a_1 + \frac{a_2}{\mathrm{Re}} + \frac{a_3}{\mathrm{Re}^2}$, a_1, a_2 and a_3 are constants that apply over several ranges of particle Reynold number, given by Morsi and Alexander [3].

It is also important to consider the dispersion of bubbles and droplets due to presence of a highly turbulent flow in the continuous phase. A stochastic method known as 'Discrete Random Walk' (DRW) model is used to determine the instantaneous fluid velocity [4, 5], depending on the calculated turbulent kinetic energy.

Electrolytic Reaction Model

The following electrolytic reactions occur on the anode and cathode surfaces:

$$\text{On the cathode surface:}\, M^+ + e^- \rightarrow M\,(\text{metal droplets}) \tag{4}$$

$$\text{On the anode surface:}\, Cl^- \rightarrow \frac{1}{2}Cl_2 + e^-\,(\text{gas bubbles}) \tag{5}$$

here M is either Na or Li. According to the current density and the *Avogadro constant* (6.02×10^{23}), the chlorine gas and metal generation rate can be deduced. For both the lithium and sodium models, no membrane was used—CFD modeling was used to predict the mixing process of the bubbles and droplets.

Table 2 Physical properties used in the CFD modeling

Item	Value
Operating temperature of Li production (°C)	450
Operating temperature of Na production (°C)	500
Density of Cl_2 (kg/m^{3}at 450 °C)	1.25
Density of Cl_2 (kg/m^{3}at 500 °C)	1.13
Density of Na droplets (kg/m^3)	920
Density of Li droplets (kg/m^3)	490
Density of NaCl–$CaCl_2$ salt (kg/m^3)	1600
Density of LiCl–KCl salt (kg/m^3)	1646
Viscosity of NaCl–$CaCl_2$ salt (Pa S)	0.002
Viscosity of LiCl–KCl salt (Pa S)	0.00146
Current density (A/m^3)	10,000

Parameters and Computing Method

Table 2 gives the physical properties adopted in the CFD model. Half of a single anode/cathode pair was considered to exploit the symmetrical geometry. The central plane and the vertical plane were considered as a symmetrical boundary condition; the top surface was set as free surface with zero shear stress while all other boundary surfaces were set as wall surfaces. When the discrete bubbles or droplets reached the top surface of the cell in the DPM model, they were considered as escaped or trapped, and thus removed out of the subsequent calculation.

Very fine structured mesh was used in the channel region between the anode and the cathode, and the top region where the liquid metal/molten salt interface existed. Hexahedral control volumes were used. The other regions of the cell were not important for the present study consequently divided into unstructured computational domains. The computational domain was meshed into 1.6 million control volumes.

The SIMPLE algorithm [6, 7] for pressure-velocity coupling and second order upwind scheme was adopted for solving the momentum, k and ε equations. The body force weighted discretization scheme was used for the pressure equation. When liquid metal distribution on the top surface was calculated, the VOF method [8] was adopted to simulate the liquid metal/electrolyte interface. ANSYS Fluent 16.2 software package was used to solve the mathematical modeling [9].

Results and Discussion

Bubble Behavior and Flow Characteristics in LiCl Cell

The typical bubble distribution in the LiCl electrolytic cell at different time instants is displayed in Fig. 2. The color map for the bubbles was based on the bubbles'

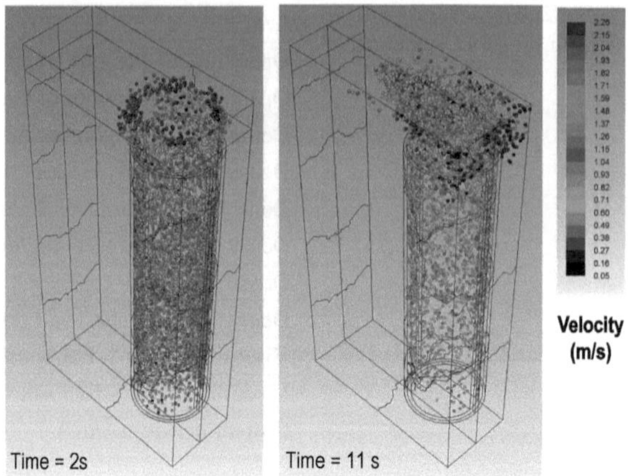

Fig. 2 Bubble distribution in the lithium electrolytic cell

local velocity magnitude. At the start of the electrolysis process, the chlorine bubbles are generated on the surface of the graphite anode. Because of the buoyancy force, the bubbles begin to rise. This flotation causes an upswing (gas lift) in the molten salt resulting in a circulatory flow in the bath. Steady state was reached in about 10 s. It is also important to note that more than one million bubbles were tracked in the molten salt bath using this CFD model; however, for visualization purposes, only 5% of the total bubbles were displayed in Fig. 2.

The predicted velocity field of the molten salt on the top free surface and the middle symmetrical plane in a single cell is shown in Fig. 3. In this case, the bubble diameter was set as 5 mm in the model. The average velocity magnitude between the anode and cathode was around 1.07 m/s. Since the diameter of the chlorine bubbles in the electrolytic cell is unknown, different bubble sizes, such as 4 and

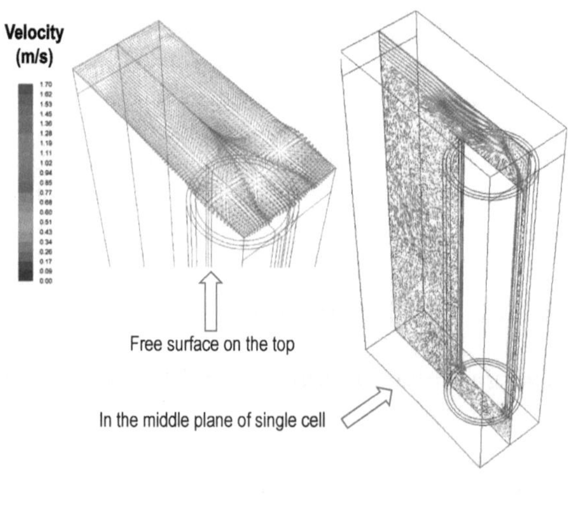

Fig. 3 Velocity field of the molten salt on the *top* free surface and *middle* symmetrical plane in lithium electrolytic cell

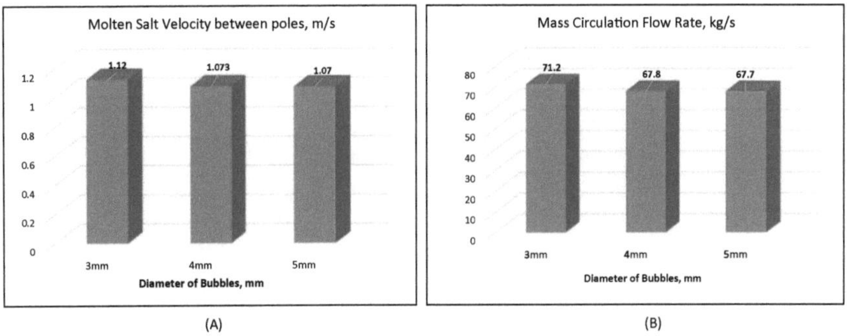

Fig. 4 Average velocity and mass flow rates flowing through the gap of one electrode pair using different chlorine bubble sizes in the CFD modeling

3 mm, were also tested. The consequent velocity fields were analyzed and compared. The average velocity magnitude of the molten salt in the gap between the electrodes for different bubble diameters is depicted in Fig. 4a. The mass flow rate through these electrode pairs are shown in Fig. 4b. Although the smaller bubbles had less buoyancy, the total number of small bubbles was much more than that of the large bubble. This was expected as the total volume production rate of chlorine gas remained unchanged for the same current density. For example, the total number of bubbles generated on the anode surface was 1.38×10^5/s for a bubble diameter of 5 mm. This number would increase to 6.40×10^5/s if the bubble diameter was reduced to 3 mm. As a result, the average molten salt velocity magnitude flowing through the gap between the electrode pairs became similar, in the range of 1.07–1.12 m/s. Therefore, there is no appreciable impact on the circulating velocity for a bubble size ranging from 3 to 5 mm. Accordingly, the mass flow rates through the gap of the electrode pair were in the range of 67–71 kg/s. This flow rate would be the amount of recycle electrolyte provided for the electrolytic process. This was much higher than the electrolytic reaction rate (given that the metal chlorides (Na or Li) form a significant fraction of the overall electrolyte) and therefore would help maintain a near constant electrolytic concentration between the electrodes.

Lithium Droplet Trajectories

In the LiCl electrolytic process, liquid lithium is produced as fine droplets on the inner cathode surface. Because the density of lithium is lighter than the electrolyte, the droplets will float up with the molten salt flow, and eventually reaching the top surface. At the top surface, the lithium droplets are trapped due to the surface tension. A liquid lithium layer was ultimately formed (assuming the lithium was not trapped in a bell) as more and more lithium accumulated on the surface. In the CFD

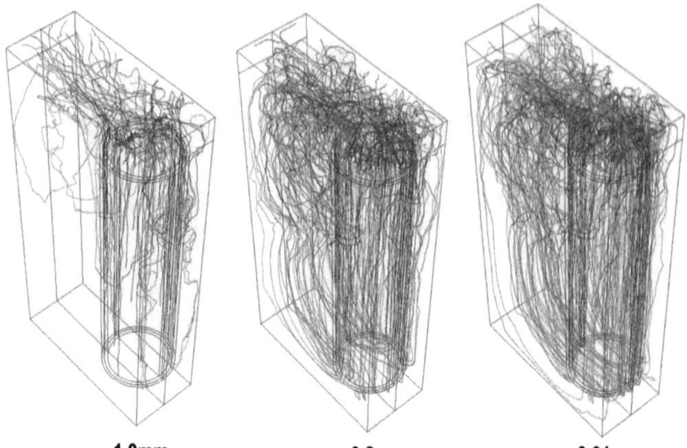

Fig. 5 Trajectories of lithium droplets in the molten salt bath (different droplets' trajectories were identified by *colors*)

modeling approach, the trajectories of the lithium droplets in the bath were predicted using the Lagrangian Discrete Phase Model. The trajectories of different sized droplets (Diameter = 1 mm, 0.2 mm and 0.01 mm respectively) were predicted by the model. At time t = 0 s, 12,800 droplets were released on the cathode surface. These droplets were tracked along with the molten salt flow. The droplet trajectories in the molten salt bath are shown in Fig. 5. In order to display the droplet trajectory clearly, only 5% of the total bubble population are given in Fig. 5. Different colors were used to distinguish the droplet ID number. It could be seen that the larger bubbles would quickly float up to the top free surface, and migrate downwards in quick succession. For a smaller size (Dia. = 0.2 mm), the buoyancy force was much smaller, resulting in more droplets entrapped towards the bottom of the bath by the circulating melt flow, and thus the residence time of small droplets staying in the molten salt bath increased significantly.

Aspects of Lithium Distribution on the Top Surface

To demonstrate the extension of the model after lithium reaches the surface, the distribution of liquid lithium on the free surface could be predicted using a VOF model coupled with DPM and turbulent flow. The results obtained from the lithium-electrolyte distribution in the electrolytic cell are depicted in Fig. 6. For this purpose, three single cells (in the same configuration as Fig. 1) are aligned together into the molten salt bath. The red region represents the electrolyte and the blue area represents the liquid lithium. Because of the turbulent circulatory flow caused by the chlorine bubbles, the accumulated liquid lithium on the top surface was pushed away from the center of the electrodes. Most of the liquid lithium accumulated in

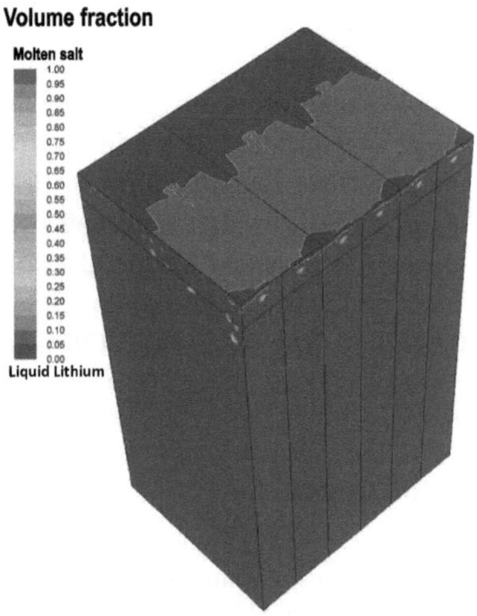

Fig. 6 Lithium distribution on the top surface in the electrolytic cells at time = 800 s

the rear of the bath resulting in a layer measuring 10 mm at the moment of time t = 800 s. A small portion of the lithium metal was trapped in the front of the bath. The circulatory flow also entrapped some of the lithium in the molten salt bath. When the time increased to t = 1600 s, the thickness of lithium layer at the rear surface would increase to around 20 mm, while the size of the exposed electrolyte eyes decreased.

Electrolytic Process in NaCl Cell

Similar modeling procedures were followed for the electrolytic process in the sodium bath, to simulate the sodium production using the same Downs cell type configuration. Because the gap between the anode and the cathode is only 30 mm, as the chlorine bubbles floated up, the chlorine bubbles become homogenized along with the newly formed sodium droplets in the electrode gap (without any porous membrane). However, this would lead to rapid chemical reaction between sodium droplets and chlorine bubbles forming NaCl again, because the sodium, once formed, enters into the electrolyte as a fog. Therefore, using a porous membrane to separate the chlorine bubbles from mixing with sodium droplets is an absolute necessity.

Conclusion

In the present study, it was found that the molten salt flow driven by flotation of bubbles could be predicted by the CFD model using the Lagrangian Discrete Phase Method coupled with two equation turbulence models. It is necessary to use a porous membrane for separating chlorine bubbles from sodium droplets in a sodium electrolytic cell, otherwise the chlorine bubbles will mix and react with sodium droplets and form NaCl again. In the lithium cell, the size of chlorine bubbles in the range of 3–5 mm has a minor effect on the velocity magnitude of the molten salt; the mass flow rate driven by the dragging of chlorine bubbles in the lithium cell was about 70 kg/s for the conditions used in this paper. This was the amount of recycled electrolyte provided for the electrolysis process by the circulating molten salt flow. The distribution of lithium metal layer on the top free surface could be predicted by CFD modeling using the VOF technique.

References

1. J.C. Downs, Electrolytic process and cell, US Patent, No. 001501756, 1924
2. B.E. Launder, D.B. Spalding, The numerical computation of turbulent flows. Comput. Methods Appl. Mech. Eng. **3**(2), 269–289 (1974)
3. S.A. Morsi, A.J. Alexander, An investigation of particle trajectories in two-phase flow systems. J. Fluid Mech. **55**(2), 193–208 (1972)
4. R. Schwarze, F. Obermeier, D. Janke, Numerical simulation of fluid flow and disperse phase behaviour in continuous casting tundishes. Model. Simul. Mater. Sci. Eng. **9**, 279–287 (2001)
5. A.D. Gosman, E. Ioannides, Aspects of computer simulation of liquid-fueled combustors. J. Energy **7**(6), 482–490 (1983)
6. S.V. Patankar, D.B. Spalding, A calculation procedure for heat, mass and momentum transfer in three-dimensional parabolic flows. Int. J. Heat Mass Transf. **15**(10), 1787–1806 (1972)
7. H.K. Versteeg, W. Malalasekera, *An introduction to computational fluid dynamics: the finite volume method* (McGraw-Hill, Loughborough, 1995)
8. C.W. Hirt, B.D. Nichols, Volume of fluid (VOF) method for the dynamics of free boundaries. J. Comput. Phys. **39**(1), 210–225 (1981)
9. ANSYS-fluent 15.0 theory guide (Southpointe, 2013)

Part III
Hydrometallurgy

P-CAC, A Unique Separation Technology for PGM Recovery

Shijie Wang and Tracy Morris

Abstract Preparative Continuous Annular Chromatography, or P-CAC, is an innovative separation technology, a unique approach to high-performance chromatography for the separation of PGM component mixtures from solutions. P-CAC consists of inlet/distribution head and outer/inner cylinders, and bottom plate/fraction-collection as well. P-CAC has been applied in biotechnological and pharmaceutical downstream processing. This was the first time to explore an efficient PGM separation and recovery at the Copper Refinery. Laboratory trials with using a chloride solution for raffination for PGM are described and the tests results are presented in this paper.

Keywords PGM recovery · P-CAC · Platinum · Palladium

Introduction

The Spanish named platinum *platina*, which means little silver. They regarded it as an unwanted impurity in the silver collected from the new world. The platinum group metals (PGM) are a family of six members consisting of ruthenium, rhodium, palladium, osmium, iridium, and platinum, which can be gained by examining the position they occupy in the periodic table, as shown in Fig. 1. Due to their high resistance to oxidation and corrosion, the PGM, like gold and silver, are classified as noble metals, and widely used in chemical laboratories, i.e., crucibles or combustion boats, thermocouples, electrodes, instrument components. For many years in automotive industry, platinum and palladium have been installed as oxidation

S. Wang (✉)
Rio Tinto Kennecott Utah Copper, 11500 West 2100 South, Magna, UT 84044, USA
e-mail: shijie.wang@riotinto.com

T. Morris
Groupo Mexico Asarco LLC, 7001 State Highway, Amarillo, TX 79106, USA

S. Wang et al. (eds.), *Applications of Process Engineering Principles in Materials Processing, Energy and Environmental Technologies*,
The Minerals, Metals & Materials Series, DOI 10.1007/978-3-319-51091-0_12

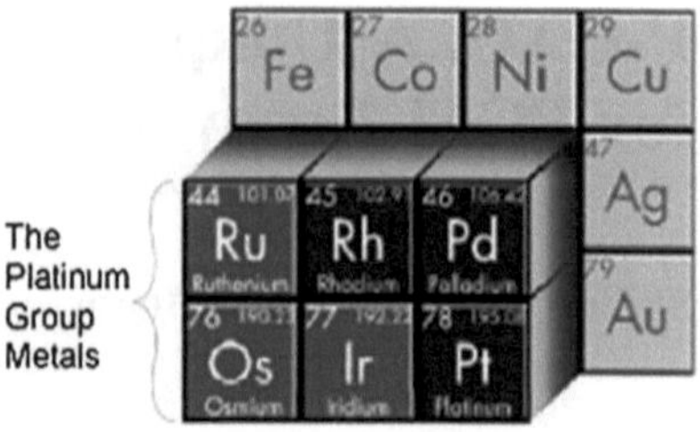

Fig. 1 Periodic table: the platinum group metals [1]

catalysts in the catalytic converters to treat automobile exhaust emissions. A widely range of PGM alloy compositions is also used in low voltage, low energy, and other contracts; thin- and thick-film circuits; in jewelry and decorations; and for dental and medical uses as well.

Sources of the platinum-group metals can be classified in two folds; the world newly mined, or primary PGM, as shown in Fig. 1, flow production chart at Rustenberg Platinum Mines for treatment of ore up to the concentrate stage [2]; and secondary PGM in which are recovered by refining precious metals slimes, used equipment, and spent catalysts. Most refining procedures take advantage of the ready solubility of gold, platinum, and palladium in aqua regia and the ease with which gold can be reduced to the metallic form from the chloride solution by solvent extraction technology [3]. Figure 2 schematically illustrates the aqua regia refining process [4], and Fig. 3 provides a detailed block flow chart of typical PGM recovery at a copper refinery.

Platinum is first removed by precipitation with ammonium chloride. Gold is then reduced with iron(II) sulfate to its metallic state. Finally, Palladium is oxidized at 80 °C with nitric acid to the +4 oxidation state and precipitated with ammonium chloride. Since it takes place in the presence of base metals, the entire process is lengthy, costly, and complicated, especially, an intermediate product, or so called "yellow cake", between the platinum and palladium precipitations, is unstable that needs to be prevented, a new separation technology—P-CAC, therefore, is selected and trialed for the replacement of the traditional process.

P-CAC Technology

Preparative Continuous Annular Chromatography, or P-CAC, is featured by an annulus with synchronically rotating inner and outer cylinders, packed with stationary phase. It consists of the following main parts:

1. Inlet and Distribution Head—It includes seven inlet ports to pump in the feed and eluent solutions. One inlet port is reserved for the ventilation of the column and the pressure indicator and pressure relief valve. One inlet port is reserved for the top eluent which floods the entire annulus.

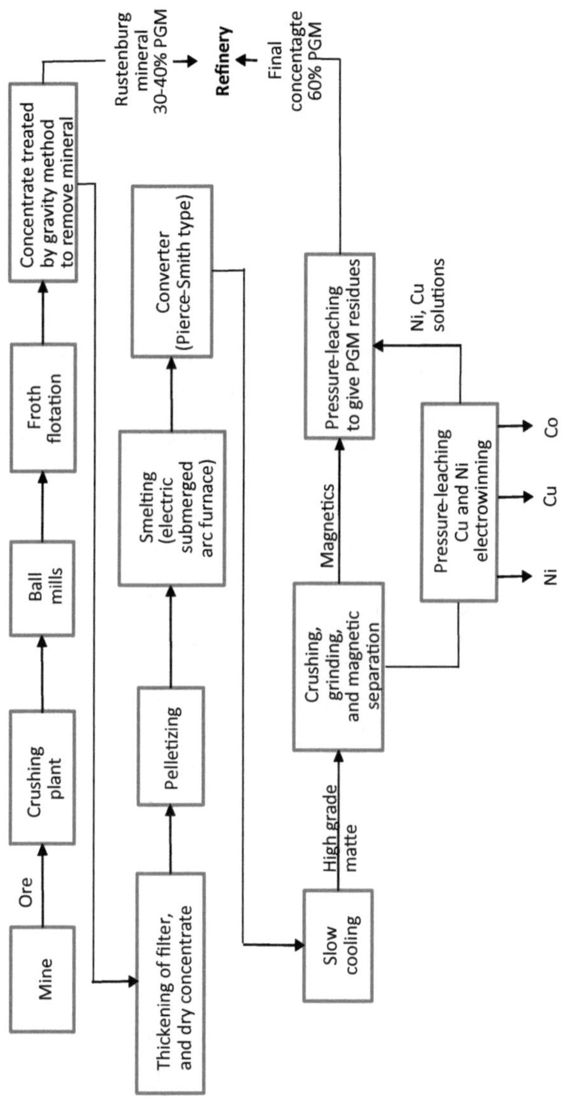

Fig. 2 Flow production chart at Rustenberg platinum mines for treatment of ore up to the concentrate stage

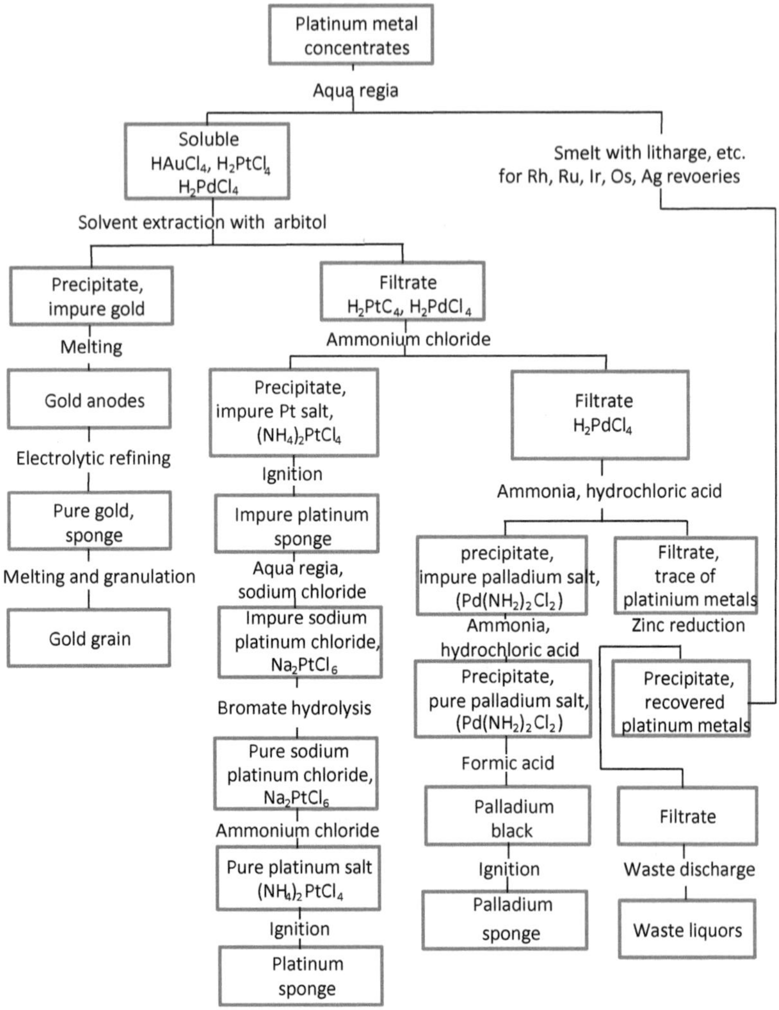

Fig. 3 Process route at international nickel acton refinery

The remaining five inlets allow adoption and optimization for processes such as:

- Multiple feed-inlet ports.
- Techniques such as step-elution, gradient-elution, displacement-eluent, as well as washing and recycling steps such as those used in ion-exchange chromatography.

2. Top-plate—This includes a rotating device and a bearing to keep the column aligned during rotation. The top-plate connects the distribution head and the outler cylinder of the annular column. The top-plate is made of polypropylene (PEEK optional) and can withstand pressures in the column of up to 10 bar.

3. Outer Cylinder—This outer cylinder is made of high precision glass, which can be used up to 3 bar operating pressure. A pressure relive valve installed in the distribution head, prevents the column from failing due to over pressure. The glass cylinder, in its basic design, allows the column to be packed to a total bed height of between 380 and 420 mm. Other outer cylinders to give different bed heights are available.
4. Inner Cylinder—The inner cylinder is made of titanium and is designed to withstand 10 bar pressure. The basic design allows the column to be packed to a total bed height 420 mm. Inner cylinders to give different bed heights are available.
5. Thermojacket—The inner cylinder serves at the same time as a thermojacket to cool or to heat the annular column. The design of the column allows to vary the temperature between 0 and 95 °C.
6. Bottom-plate—The bottom-plate contains 180 exit holes in the annulus and is made of PEEK and is designed to withstand pressure up to 10 bar. The holes are covered by a filterplate to prevent the stationary phase from entering the holes. The bottom-plate is connected to the drive through a driving-pin at the axis. The axis is made of titanium.
7. Fraction–collection—This is regulated by a stationary glide-ring system. The glides-ring system consists of 180 chambers each associated with one of the 180 exit holes in the bottom-plate. This ensures an easy and precise way to collect fractions by just connecting the required exit ports of the glide-ring with fluid supporting modules, which collect the fluid from the chambers. Thus a constant product quality with no cross contamination is ensured. The whole fraction-collection represents a closed system. Unused eluent can be reused. The glide-ring system is made of PTFE.
8. Drive—High precision step motor including control panel and software. The software allows three different operation modes—production, compression, and filling modes. Housing—The housing is made of coated steel. The housing protects the drive as well as the electronic parts from the surrounding environment.

Experimental Tests

For the bench experimental tests, a laboratory size P-CAC device with control system was set up as shown in Fig. 4. The feed solution was collected from the operation circuit at the Refinery and directly pumped on to the stationary phase using a fixed, internal nozzle. The main eluent is evenly distributed throughout the entire annulus, see Fig. 5. At the bottom of the column the separated components were simultaneously collected, as shown in Fig. 6, each at a given rotation angle.

Fig. 4 Laboratory size P-CAC set up

Fig. 5 The main eluent is evenly distributed throughout the entire annulus

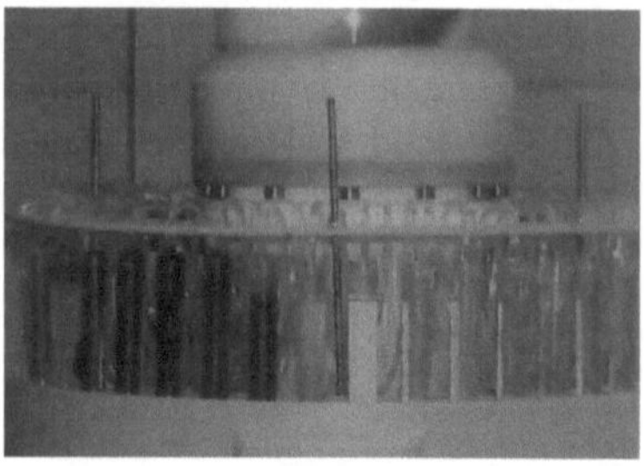

Fig. 6 The separated components were simultaneously collected at the bottom of the column

The localized separation permits continuous operation and simultaneous removal of all components. All collected samples were rearranged (Fig. 7a) by different chambers from the exit holes and the fraction-collection samples, which represent closed systems with different color (Fig. 7b), were then submitted for chemical analyses in the Central lab.

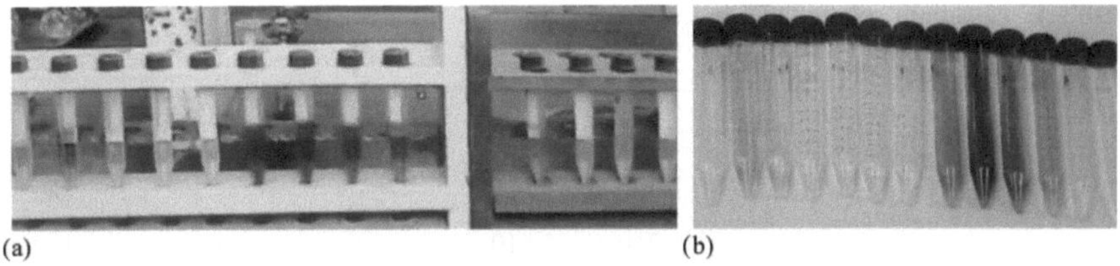

Fig. 7 **a** All collected samples were rearranged; **b** the fraction-collection samples represent each closed system with *different color*

Results and Discussion

Separation resolution depends on the affinity of the components to the stationary phase, as well as the throughput of the feed and eluents, and the speed of rotation. As a result, a colorful separating flow pattern on the outer glass cylinder is shown in Fig. 8; where the yellow (1st line), brown (2nd line, wider), and green (3rd line) represent platinum, palladium, and copper/silver, respectively.

All test results are tabulated in Tables 1 and 2. Test 1 was run with the Palladium solution after the gold is removed. Clean fractions of Pd and Pt are produced from the batch column. As shown in Table 1, the Pd/Pt ratios in feed and product solutions are 37 and 7040, respectively. After P-CAC separation, as shown in Table 2, a significant separation of Pd and Pt is demonstrated again in where the metals come out. Further work needs to be done to fine tune the process, i.e., Pd and Pt separation between AR-A7 and AR-A9.

From the plant practice, a 99.95% Palladium metal will be obtained when the Pd: Pt ratio equals or greater than 2000:1; the Pd:Pt ratio of 7040:1 in the palladium fraction, obtained from the experimental test, guarantees a very pure Palladium metal can be produced by employee the P-CAC Technology.

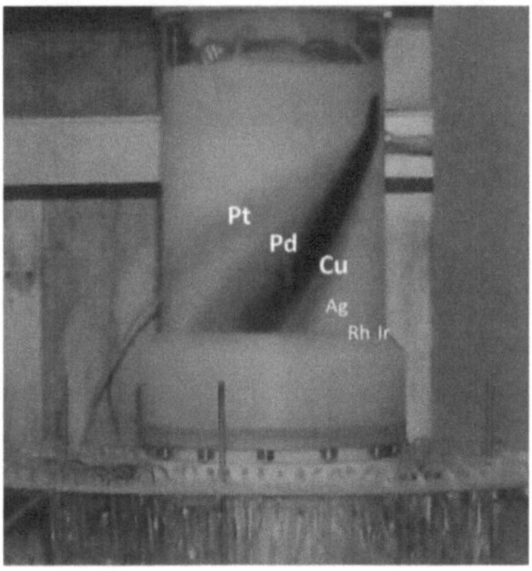

Fig. 8 A colorful separating flow pattern on the outer glass cylinder is shown

Table 1 Palladium solution and results for P-CAC tests 1 and 2

Sample ID	Cu	Ag	Pd	Pt	Au
Initial concentration (g/L)	61.6	0.40	125.6	4.40	0.10
Test 1 results (mg/L)					
1–7 A	7080	<10	<10	<5	<10
8–9 A	760	10	30	<5	<10
10 A	94	22	1080	<5	<10
11 A	24	36	7360	<5	<10
12–22 A	<10	18	8760	<5	<10
23–27 A	<10	<10	1360	<5	<10
28–36 A	<10	<10	<10	462	<10
Test 2 results (mg/L)					
1–7 B	7720	<10	<10	<5	<10
8–9 B	2160	12	1800	<5	<10
10 B	6760	50	11,800	<5	<10
11 B	5520	60	11,920	<5	<10
12–22 B	1320	42	12,320	<5	<10
23–27 B	16	<10	3880	58	<10
28–36 B	12	<10	<10	514	<10
Combined final solution (mg/L)	44	<5	35.2[a]	44	

[a]In g/L

Table 2 Palladium solution and results for P-CAC tests 3 and 4

Sample ID	Cu	Ag	Pd	Pt
Initial concentration (g/L)	112.8	0.84	134.4	3.20
Test 3 results (mg/L)				
AR-A1	<1	<1	<1	<1
AR-A2	7150	<1	5	<1
AR-A3	3450	<1	<1	<1
AR-A4	4	<1	<1	<1
AR-A5	4	<1	<1	<1
AR-A6	12	<1	<1	<1
AR-A7	15	<1	1	<1
AR-A8	8	<1	1650	60
AR-A9	<1	<1	7600	170
AR-A10	<1	<1	3600	7
AR-A11	<1	<1	750	1
AR-A12	<1	<1	16	<1
AR-A13	<1	1	7	<1
Test 4 results (mg/L)				
AR-B1	1400	<1	<1	<1

(continued)

Table 2 (continued)

Sample ID	Cu	Ag	Pd	Pt
AR-B2	8650	20	<1	<1
AR-B3	2	8	<1	<1
AR-B4	<1	<1	<1	<1
AR-B5 to B7	<1	<1	3	<1
AR-B8	<1	<1	4350	<1
AR-B9	<1	<1	4200	<1
AR-B10	<1	<1	2350	<1
AR-B11	<1	<1	1100	<1
AR-B12	<1	<1	222	<1
AR-B13	<1	<1	1	<1
AR-B14 to B16	<1	<1	1	<1
AR-B17 to B23	<1	<1	<1	<1

Conclusions

1. All test results have demonstrated that clean fractions of Pd and Pt are produced from the batch column. Using the direct collected solution feed and dilute HCl as eluent, the Pd fraction is of high purity. The Pd:Pt ratio in palladium fraction is 7080.
2. The first industrial trial is very successful; batch column tests have approved that P-CAC technology can be applied in the PGMs separation and production for process safety, efficiency, and profitability. Using data obtained from the batch column experiments for scale up the P-CAC to production size would be the next focus in path forward.

Acknowledgements Sixty-seven years has passed since the first concept of a rotating annular chromatograph for industrial applications was suggested by Nobel Prize winner A.V.P. Martin in 1949. In his paper [5], Martin wrote "Chromatography can also be a continuous process, provided that the developing solvent returns the adsorbent to its initial state within a reasonable period. Imaging the chromatogram to be packed within a narrow space between two concentric cylinders. The upper surface of the chromatogram is flooded with solvent and at one point the solution to be separated is fed on slowly. The annular chromatogram is slowly and uniformly rotated with the result that different zones will form helices of characteristic angle which can be collected at various fixed points around the bottom of the chromatogram".

It will be amazing to list all scientists and engineers and their efforts, of cause including all failures, to develop the P-CAC to date. The authors are very grateful to Drs. A. Prior and Y. Shang of PRIOR TECHNOLOGY GMBH for their technical support, which helped completing the first P-CAC industrial trial on PGM separation and made Dr. Martin and the other pioneers' dream come true.

References

1. Chem230 Wiki, *Platinum group metals* (2016). Retrieved from http://chem230.wikia.com/wiki/Platinum_Group_Metals
2. U.S. Bureau of Mines, Minerals yearbook, 1978–79, vol. 1 (U.S. Bureau of Mines, Washington, DC, 1979)
3. R.T. Jones, JOM world nonferrous smelter survey, part II: platinum group metals. JOM **56**, 59–63 (2004)
4. R.J. Dowsing, Met. Mater. **32** (1980)
5. A. Uretschager, A. Jungbauer, Preparative continuous annular chromatography (P-CAC), a review. Bioprocess Biosyst. Eng. **25**, 129–140 (2002)

The Physical Characteristics of Electrorefined Copper Starter Sheet Material

Daniel Majuste, Paul Laforest and Michael Moats

Abstract The interaction of anode and electrolyte chemistry in copper electrorefining is complex. This is especially true with regard to the behavior of Group 15 elements (As, Sb, and Bi). To better understand this system, laboratory electrorefining experiments were conducted using commercial anodes with As/(Sb + Bi) molar ratios of 0.54 and 3.8 and an electrolyte collected from a commercial refinery. The effect of adding thiourea during plating was also examined. Twenty-one hour copper deposits were produced in the laboratory to simulate starter sheet production. The mechanical properties of the electrodeposited copper as measured by an industrially relevant empirical bend test and a one-point bend test developed by UFMG are reported. The crystal structures of the samples were also examined.

Keywords Copper · Electrorefining · Starter sheet · Ductility · Bending

Introduction

Approximately 25% of the world's copper electrorefineries produce and use starter sheets for their operations [1]. While starter sheet technology has been employed for many decades, situations arise where their production can be problematic [2]. Specifically, the ductility of the starter sheet material can become compromised which leads to loop breakage.

The ductility of starter sheet material is measured in a plant setting using a simple bend test [2]. In a bend test, material is placed in a vice and bent, evenly, at 90° in each direction. Each motion is counted as one bend, and the total number of

D. Majuste (✉)
Department of Metallurgical and Materials Engineering, Universidade Federal de Minas Gerais, Belo Horizonte, MG 31270-901, Brazil
e-mail: daniel.majuste@demet.ufmg.br

P. Laforest · M. Moats
Materials Research Center, Missouri University of Science and Technology, Rolla, MO 65409, USA

S. Wang et al. (eds.), *Applications of Process Engineering Principles in Materials Processing, Energy and Environmental Technologies*,
The Minerals, Metals & Materials Series, DOI 10.1007/978-3-319-51091-0_13

bends to failure is recorded. A higher number of bends indicates a sample with more ductility. This method was used by O'Keefe and Hurst [3] when they showed that antimony in electrolyte produced brittle copper electrodeposits but chloride and glue additions counteracted this effect. Laforest and Moats [4, 5] also employed a bend test while examining the effects of anode chemistry and thiourea on deposit ductility and electrolyte chemistry.

While ductility of copper starter sheet material is important, the reporting of physical characterization using quantified and reproducible methods was not found in the open literature. Ductility of electrodeposited copper for other applications has been measured, but the copper deposits are thinner than those produced in this study [6–9]. Therefore, the ductility of copper starter sheet materials was measured using a one-point bending test. Comparisons of the one-point bending test results to crystal structure and bend test values were made.

Experimental

Starter Sheet Materials

Two sets of starter sheet materials were evaluated by bend testing and the UFMG one-point bending test. The first set was commercial starter sheet samples. The second set was laboratory starter sheet samples. The commercial starter sheet samples were produced by two facilities and selected to provide a range of bend test values. The laboratory starter sheet samples were produced during an earlier investigation with details reported elsewhere [4, 5, 9].

One-Point Bending Test

A one-point bending test was developed by researchers at Universidade Federal de Minas Gerais to simulate the stress that would be imposed upon copper and zinc electrodeposits produced by electrowinning during removal from permanent cathodes. The test was developed as part of the AMIRA P705 series of projects.

Cu samples (25 mm width × 55 mm length) were subjected to a one-point bending test in a customized device, which reproduces the bending of the metal sheet during the stripping stage. The device was placed on a universal testing machine (INSTRON 5582) using a 100 N load cell, where a controlled vertical force is applied to the punch of the device. One end of the sample is fixed by a lock device that applies a compression force, while the punch moves vertically at a constant speed of 0.2 mm s^{-1}, bending the sample. The horizontal distance between the punch and the support was kept constant during the tests at 10 mm. The punch's weight was considered in the calculations. This approach takes into account all geometric factors related to the experimental set-up. The variables measured here were the vertical displacement of the punch and the vertical force. In order to calculate the angle θ between the bent sample and its original position, the geometry of the test system was simplified: the sample was assumed to be straight.

X-ray Diffraction

The crystalline structure of the Cu deposits was analyzed by X-ray diffraction (XRD), using a PANalytical (Empyrean) X-ray diffractometer, with Cu Kα1 (λ = 1.5406 Å) radiation. The XRD patterns were measured in 2θ range from 30 to 100° using a step size of 0.02°. The patterns were identified using an ICDD (International Centre for Diffraction Data) file as reference (04-0836). The texture of the electrodeposited metal was discussed semi-quantitatively on the basis of the relative peak intensities.

Results and Discussion

Commercial Starter Sheets

Four commercial starter sheet samples were provided by an industrial partner. Selected impurity concentrations for the samples are provided in Table 1 along with the average number of bends to failure. The data indicate no correlation between these impurities and the bends to failure.

Specimens of the commercial samples were measured using the UFMG one-point loading test. Force versus deformation curves for the four commercial samples are shown in Fig. 1. No correlation was observed. The commercial samples were of differing thickness as indicated in Table 1. Once the force was normalized to account for the deposit thickness, a correlation between the average number of bends to failure and the maximum normalized deformation force was observed as shown in Fig. 1. It is not totally unexpected that samples which were stronger were also less ductile.

Laboratory Starter Sheets

To understand the effect of anode composition and thiourea on starter sheet ductility, laboratory electrorefining experiments were conducted. Sections of commercial anodes and cathodes were employed along with commercial electrolyte. Details of the electrorefining experiments were previously reported [4, 5, 9].

Table 1 Selected concentrations and average bends to failure for commercial starter sheet samples

Sample	As (ppm)	Bi (ppm)	Sb (ppm)	Pb (ppm)	Ag (ppm)	S (ppm)	Ave bends to failure	Thickness (mm)
Comm 1	3.0	0.1	0.7	0.2	1.7	23.0	2.6	0.60
Comm 2	6.7	0.2	1.2	0.6	3.3	15.3	4.0	0.39
Comm 3	1.1	<0.1	0.3	0.1	6.6	3.1	5.6	0.57
Comm 4	0.8	<0.1	0.1	0.3	2.4	18.4	11.5	0.22

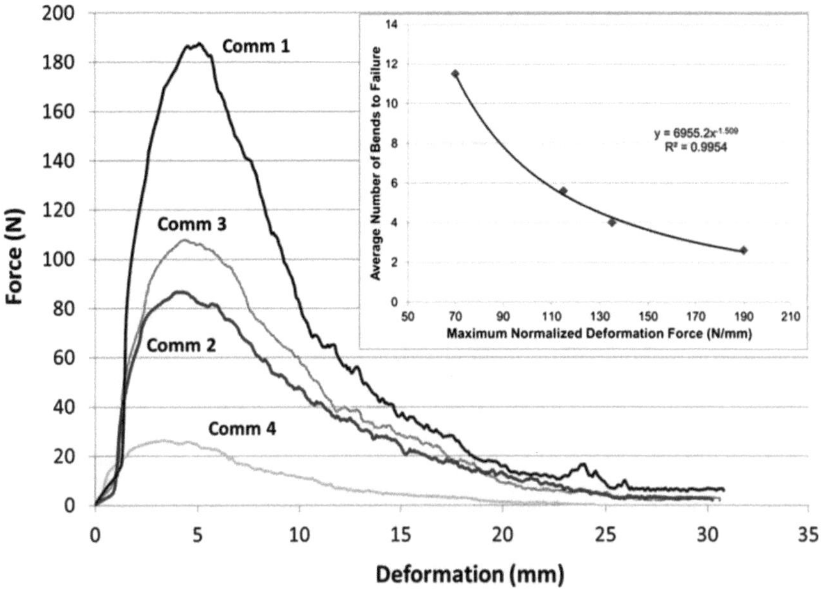

Fig. 1 Force versus deformation curves from one-point loading test for commercial samples. *Insert* Average number of bends to failure versus maximum normalized deformation force

Three experimental series were conducted. The conditions for each series and the average bends to failure are provided in Table 2. With 95% confidence, Series 2 (high molar ratio anodes) produced more ductile starter sheets than Series 3 (low molar ratio anodes with thiourea), which in turn was more ductile than Series 1 (low molar ration anodes).

Fourteen or fifteen electrorefining tests were conducted in each series. Four electrodeposits were produced in each test. Two deposits were subjected to the bend to failure test. Two were available for other characterization. From these extra samples, three specimens were selected from each series and sent to UFMG one-point bending measurement. The samples are identified as Lab with the series number and an arbitrary letter (A through C).

Figure 2 shows normalized load (N) versus deformation (mm) curves obtained for each lab produced Cu sample. Load was normalized based on differences in sample thickness. The range of thickness was 1.627–1.95 mm. An initial fracture (not complete breakage) was observed for samples Lab 1A, Lab 1B, Lab 3A, and Lab 3B during the one point bending test, thereby indicating that such samples are less ductile. This correlates well with the bend test values where Series 1 and 3 were less ductile than Series 2.

Moment (M) versus bending angle curves were plotted up to a bending angle of 60° (Fig. 3). The maximum angle of the test is likely greater than the typical wedge angle during stripping. It can be seen that the curves exhibit similar behavior: the bending moment increases parabolically with angle, and then displays increasing

Table 2 Summary of each series experimental conditions and average bend test results

Lab series	Anode As/(Sb + Bi) molar ratio	Daily glue (g/tonne)[a]	Avitone (g/tonne)	Thiourea (g/tonne)	Average bends per mm thickness	Std dev. of bends/mm
1	0.54	240	27	0	4	2
2	3.82	240	27	0	8	4
3	0.54	240	27	240	6	2

[a]Prior to each plating cycle 1800 g/tonne of top glue was added to the electrolyte

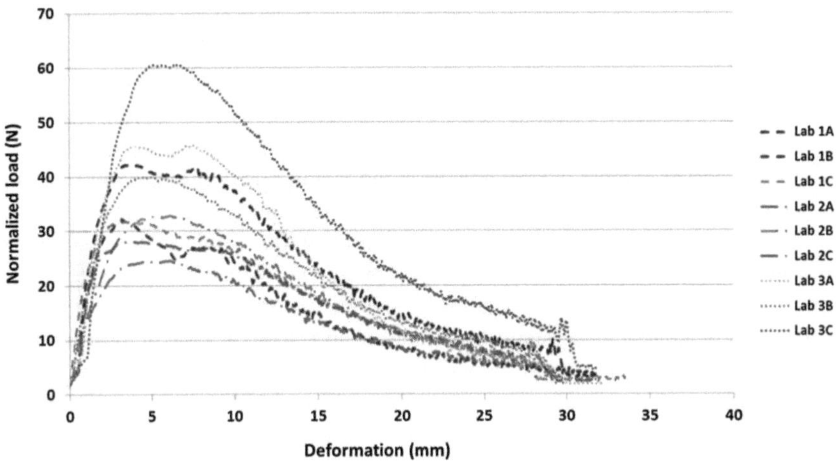

Fig. 2 Normalized load versus deformation curves obtained for the lab Cu samples

oscillations as the testing angle increases. These oscillations, which were observed for angles above about 40°, are probably associated with the friction between the punch and the sample, as the horizontal component of the bending force increases with the bending angle. The surface roughness has a direct influence on the friction between the punch and the specimen, making it difficult to assign a specific meaning to these oscillations. Figure 3 also reveals that the curve obtained for samples Lab 1A, Lab 1B, and Lab 3B reached a peak at about 48°, 51° and 59°, respectively, and then decreased, which was caused by the fracture of the corresponding sample. Again, this indicates that these Cu deposits present a relatively lower ductility and, therefore may present problems during loop making.

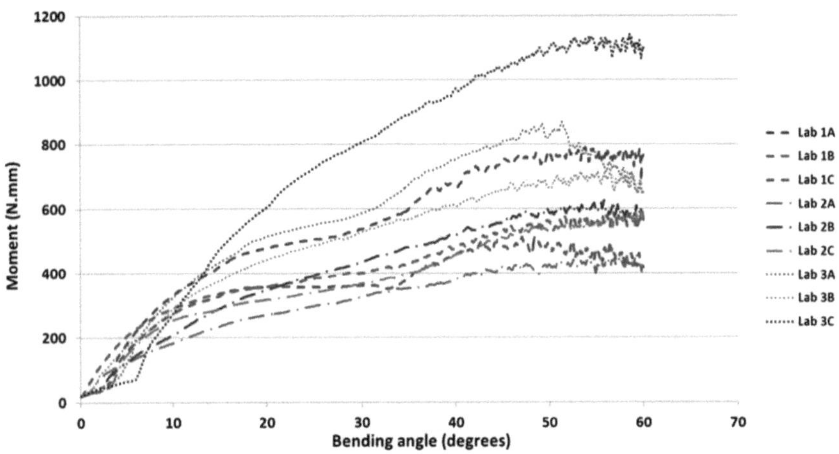

Fig. 3 Moment versus bending angles curves obtained for the Cu samples plotted with normalized data

All bent samples were submitted manually to further bending (about 180°)—one cycle, in an attempt to simulate the preparation of loops. The samples Lab 1B, Lab 3A, and Lab 3B broke, while a severe fracture was observed for samples Lab 1A and Lab 1C. This again correlates well with the bend test values where Series 1 and 3 were less ductile than Series 2.

It is difficult to explain the differences in the magnitude of the moment measured for the various samples (or net force F that causes bending), since it depends closely on the material strength that is affected by the crystallographic texture, grain size, thickness and stress concentrators in the specimens. Considering the testing of non-porous Cu samples (as visualized), the behavior observed here would be probably related to the crystallite orientation and size. A careful polishing and metallographic analysis should be performed to generate additional information.

Comparison of the results from the industrial bend test and one-point bending test did not produce a strong correlation as seen with the commercial samples. On average, the Lab 2 samples which had the highest number of bends to failure needed the least amount of force to deflect the samples. The best commercial

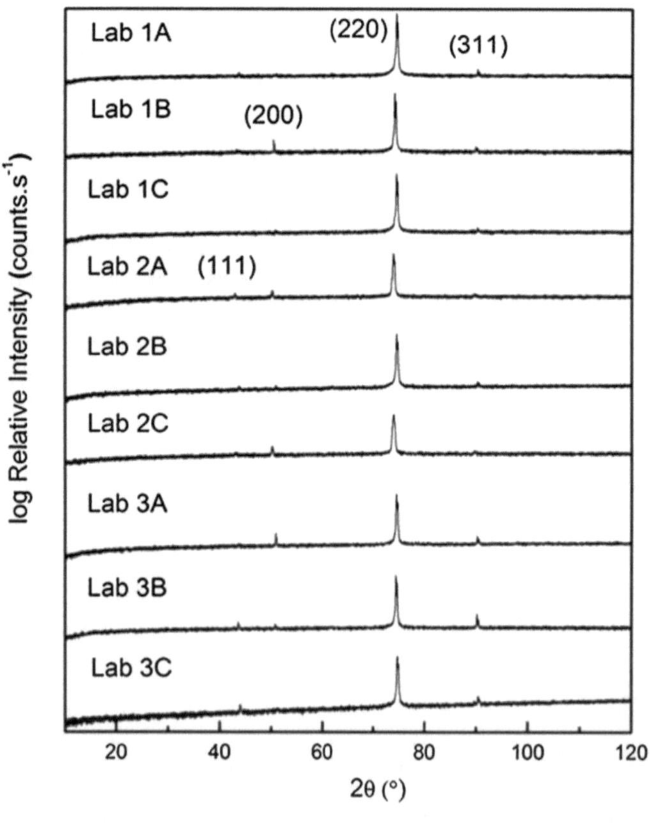

Fig. 4 XRD patterns for laboratory starter sheet deposits

Table 3 Relative intensity of the XRD peaks detected for the lab samples

Sample	Relative intensity			
	(111)	(200)	(220)	(311)
Lab 1A	–	–	1	0.008
Lab 1B	–	0.017	1	0.010
Lab 1C	–	–	1	–
Lab 2A	0.034	0.043	1	–
Lab 2B	–	0.011	1	0.016
Lab 2C	–	0.061	1	–
Lab 3A	–	0.035	1	0.028
Lab 3B	0.016	0.014	1	0.038
Lab 3C	0.014	–	1	0.031

sample also needed the least amount of force. The one-point bending test seems to indicate that starter sheets need to be weaker and more ductile to perform better in the industrial bend test.

In an attempt to understand the root causes of the bend and one-point bending measurements, XRD was performed on the laboratory Cu deposits. Figure 4 shows the XRD patterns. The relative intensity of the peaks is given in Table 3. Only peaks ascribed to Cu crystallites were detected. As can be seen, the preferred orientation detected in all samples was the (220). It can also be observed that the relative intensity of orientation (200) was magnified for samples Lab 1B, Lab 2A, Lab 2B, Lab 2C, Lab 3A and Lab 3B. The increased presence of the (200) plane was also detected during our previous investigation [4, 5, 9] and correlated to improved bend test results. These XRD data appear to confirm this correlation.

It therefore appears that Sb(V) in the electrolyte promotes the orientation of the deposit in favor of the (220) plane parallel to the surface. Change in anode chemistry decreases Sb(V) and promotes the growth of grains not completely orientated with the (220) plane. Thiourea at the dosage used also promotes this trend, but not as effectively as the change in anode chemistry. As stated previously, a detailed investigation on the crystallite size and type of growth (via cross-section analysis) may help us to further explain the observed trends.

Conclusions

Physical characterization of commercial and laboratory produced copper starter sheet materials was conducted. One point bending tests provided complimentary data about the strength and ductility of the electrodeposits. The one point bending test data revealed that, in general, samples that perform well in an industrial bend test require less force for deformation.

XRD analysis confirmed previous observations that samples produced with a higher molar ratio anode or in the presence of thiourea exhibits grains with more (200) orientation. The number of bends appears to be a function of the relative intensity of (200) plane. Anode chemistry appears to affect electrolyte chemistry and the deposit structure of starter sheets. The detrimental ductility efforts associated with using low molar anodes can be mitigated to some extent by the addition of thiourea. However, the use of anodes with a molar ratio greater than two is highly recommended based on the literature.

To further understand the root causes of the physical characterization reported in this report, detailed microstructural and textural analysis should be conducted.

Acknowledgements The first author would like to thank Dr. Franco C. Bubani (UFMG) for the valuable help during the bending tests and the Laboratory of XRD (DEMET/UFMG) for the analysis.

References

1. M. Moats, S. Wang, A. Filzwieser, A. Siegmund, W. Davenport, T. Robinson, Survey of copper electrorefining operations, in *Copper 2016*, Kobe, Japan (2016)
2. B. Wesstrom, The effects of high antimony in electrolyte, in *COM 2014-conference of metallurgists proceedings* (2014). ISBN 978-1-926872-24-7
3. T.J. O'Keefe, L.R. Hurst, The effect of antimony, chloride ion, and glue on copper electrorefining. J. Appl. Electrochem. **8**, 109–119 (1978)
4. P.I. Laforest, M.S. Moats, The effect of anode composition on electrorefined starter sheet ductility and electrolyte composition, in *IMPC 2016*: *XXVIII international mineral processing congress proceedings* (2016). ISBN 978-1-926872-29-2
5. P. Laforest, M. Moats, The effect of anode composition and thiourea on electrorefined starter sheet ductility and electrolyte composition, in *Copper 2016*, Kobe, Japan (2016, accepted)
6. X. Ye, M. De Bonte, J.-P. Celis, J.R. Roos, Role of overpotential on texture, morphology and ductility of electrodeposited copper foils for printed circuit board applications. J. Electrochem. Soc. **139**(6), 1592–1600 (1992)
7. V.A. Lamb, C.E. Johnson, D.R. Valentine, Physical and mechanical properties of electrodeposited copper III. Deposits from sulfate, fluoborate, pyrophosphate, cyanide, and amine baths. J. Electrochem. Soc. **117**(9), 291C–318C (1970)
8. D. Anderson, R. Haak, C. Ogden, D. Tench, J. White, Tensile properties of acid copper electrodeposits. J. Appl. Electrochem. **15**(5), 631–637 (1985)
9. P.I. Laforest, Understanding impurities in copper electrometallurgical processes. M.S. thesis. Missouri University of Science and Technology, 2015

Extraction of Copper from Sulfate-Chloride Solutions by Using Hydroxyoxime Extractants

M.C. Ruiz, I. Gonzalez, J. Salgado and R. Padilla

Abstract The mining companies in Chile are increasingly using chloride salts and/or sea water in the sulfuric acid leaching of copper minerals. For this reason, the effect of chloride ions on the solvent extraction equilibria was investigated. Copper extraction isotherms were determined at 25 and 35 °C for aqueous solutions containing 6 g/l of Cu, 7 g/l of Fe, chloride ion concentrations of 0, 60 and 110 g/l, and a pH of 2.0. The organic phases were 20% v/v solutions of a ketoxime (LIX 84-IC) or a salicylaldoxime (LIX 860 N-IC) . It was determined that the presence of chloride in the aqueous solution had a large negative effect on the extraction equilibria for both extractants, particularly the ketoxime. Therefore, as the chloride ion concentration increases, an efficient use of the extractant would become more difficult. An increase in the temperature from 25 to 35 °C positively affected the extraction equilibria for these extractants.

Keywords Solvent extraction · Copper extraction · Hydroxyoximes · LIX 84-IC · LIX 860 N-IC

Introduction

For many years, the solvent extraction process (SX) has been used successfully by the copper mining industry for the purification and concentration of aqueous solutions obtained by sulfuric acid leaching of copper ores in dumps, heaps, and agitated tanks. The process can be separated into two steps: extraction and stripping. In the extraction step the copper in the leaching solution is transferred to an organic phase that contains a highly selective copper extractant leaving an aqueous solution with low copper content (raffinate) that can be recycled back to the leaching operation. In the stripping step the copper in the loaded organic phase is

M.C. Ruiz (✉) · I. Gonzalez · J. Salgado · R. Padilla
Department of Metallurgical Engineering, University of Concepcion, Edmundo Larenas 285, Concepcion, Chile
e-mail: maruiz@udec.cl

S. Wang et al. (eds.), *Applications of Process Engineering Principles in Materials Processing, Energy and Environmental Technologies*,
The Minerals, Metals & Materials Series, DOI 10.1007/978-3-319-51091-0_14

transferred back into a concentrated electrolyte solution from where the copper can be recovered as high purity copper cathodes by electrowinning. The reaction that occurs during the extraction of copper can be written as:

$$Cu^{2+} + 2\overline{HR} = \overline{CuR_2} + 2H^+$$

This is a reversible reaction; therefore, in the stripping stage the reaction is reversed by contacting the loaded organic solution with a highly acidic aqueous solution (spent electrolyte) and the depleted organic phase can be recycled back to the extraction step.

Currently, several industrial SX plants of various sizes in Chile are treating sulfuric acid leaching solutions that also contain high concentrations of chloride [1, 2]. There are three sources of chloride in the solutions:

(i) The presence of the copper chloride mineral atacamite ($Cu_2(OH)_3Cl$) in the ore.
(ii) The use of sea water in the heap leaching of oxide copper minerals, an increasingly common practice due to the limited availability of fresh water for the mining projects.
(iii) The addition of chloride salts (NaCl or $CaCl_2$) in order to heap leach efficiently the secondary sulfide copper ores, a process developed as an alternative to bioleaching [3, 4].

Thus, the levels of chloride in the leaching solutions to be treated by SX can be higher than 100 g/L in some cases.

The industrial SX plants that treat sulfate-chloride leaching solutions in operation today use the same hydroxyoxime extractants that have been used traditionally for copper extraction from pure sulfate solutions. The transfer of chloride ions to the electrolyte is controlled by adding a washing stage of the loaded organic solution before stripping [2, 5, 6].

The presence of chloride ions changes the speciation of copper and iron in the pregnant leaching solutions due to the formation of Cu and Fe chloro-complexes. In particular, the concentration of the copper cation Cu^{2+} (the extractable copper species) would decrease, affecting the copper extraction equilibria. The chloride ions in the solution could also affect the extraction kinetics, the Cu/Fe selectivity, and the rate of phase separation. Basic knowledge on the extent of the effect of chloride ions on the copper solvent extraction would contribute significantly to the optimization of the solvent extraction process when using the traditional hydroxyoxime reagents.

Considering the above, in the present work experimental data are presented on the effect of chloride ions in the copper extraction equilibrium when using two hydroxyoxime extractants: a salicylaldoxime (LIX 860 N-IC) and a ketoxime (LIX 84-IC). These two reagents alone or mixed are used in many industrial copper solvent extraction plants [2, 6].

Experimental

The aqueous solutions used in the present study contained 6 g/l of copper and 7 g/l of iron, with 4 g/l of Fe(II) and 3 g/l of Fe(III), which were added as sulfate salts. Chloride ions were added to some solutions as sodium chloride. The pH of the solutions was regulated with sulfuric acid to a value of 2.0. The organic phase used in most tests was a 20% v/v solution of either LIX 84-IC or LIX 860 N-IC using Escaid 110 as diluent.

The experimental work included the determination of the maximum loading capacity of both extractants at ambient temperature and the determination of extraction isotherms at 25 and 35 °C for 0, 60 and 110 g/l of chloride ions in the aqueous phase.

The procedure for the determination of the maximum loading capacity consisted of mixing 50 ml of the organic phase (containing 5–35% v/v of extractant) with an equal volume of the aqueous phase in a separation funnel. After 5 min of energetic mixing, the two phases were allowed to separate and the aqueous phase was removed. Fresh aqueous phase was added to the funnel and the phases were mixed again. This procedure was repeated 4–5 times, until the pH of the aqueous phase did not show any change before and after contact. The complete procedure was carried out at room temperature and with no control of this variable. The copper loaded into the organic phase was determined by repeated contacts with a sulfuric acid solution and the resulting solutions were mixed and analyzed by atomic absorption spectroscopy.

The extraction isotherms experiments were carried out in a batch stirred vessel. The reactor was a cylindrical glass reactor of 1 L of capacity with four stainless steel baffles. During the experiments, the reactor was immersed in a constant temperature water bath. The stirring was carried out at 600 rpm for 10 min. The organic phase, 20% v/v of LIX 84-IC or LIX 860 N-IC, was contacted with the aqueous solution in various organic/aqueous (O/A) ratios in the range 3/1–1/6. After the specified mixing period the two phases were allowed to separate in the reactor. After complete separation, samples of the organic and aqueous phases were treated by centrifugation to separate any entrainments, before chemical analysis.

Results

Maximum Loading for LIX 84-IC and LIX 860 N-IC

The copper concentration values in the organic solution obtained in the maximum loading experiments for different concentrations of each of the extractants are shown in Fig. 1. The maximum loading capacity of each extractant (copper concentration loaded per 1% v/v) was determined from the slopes of the lines in this figure. The values obtained were 0.62 g/l for LIX 84-IC and 0.75 g/l for LIX

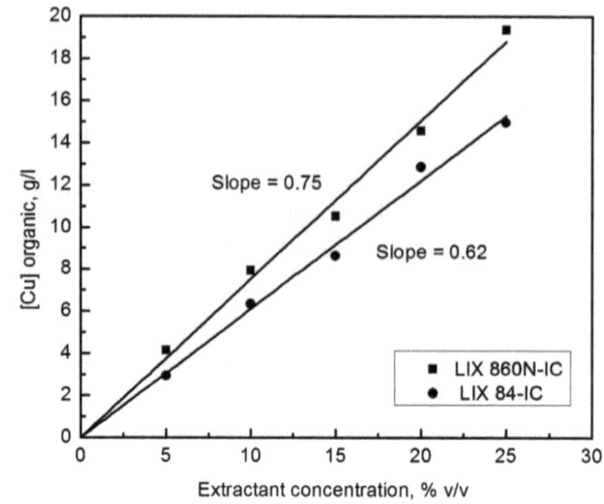

Fig. 1 Determination of the maximum loading capacity of the extractants. Aqueous phase: 6 g/l of Cu, 4 g/l of Fe^{2+}, 3 g/l of Fe^{3+} and pH 2.0

860 N-IC. These results are as expected since the salicylaldoxime (LIX 860 N-IC) is a stronger extractant and thus should load more copper at pH = 2 than the ketoxime (LIX 84-IC) which is a weaker extractant [7].

Extraction Isotherms

Extraction isotherms were determined for both extractants using aqueous solution with chloride concentrations of 0, 60 and 110 g/l. The results obtained for LIX 84-IC at 25 °C is presented in Fig. 2.

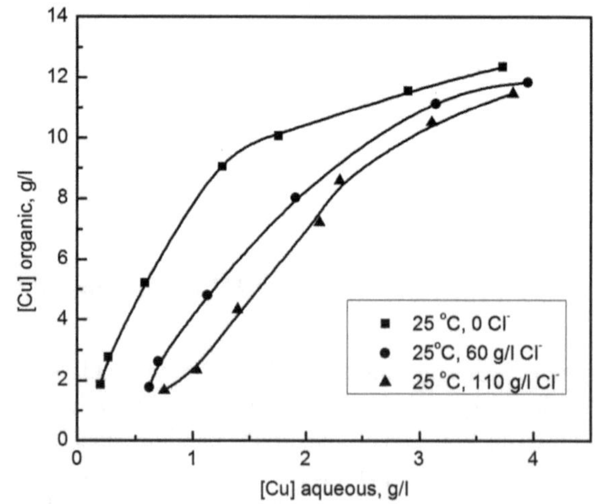

Fig. 2 Extraction isotherms for LIX 84-IC at 25 °C using aqueous solutions with different concentrations of chloride ions

As seen in this figure the presence of chloride in the aqueous solution has a large negative effect on the extraction equilibrium, as shown by the shifting of the equilibrium isotherms to the right in the presence of chloride in the aqueous solution. Therefore, as the chloride concentration in the aqueous phase increases, it will become increasingly difficult to obtain a raffinate with low copper concentration in a counter current extraction circuit.

To better visualize the situation; in Fig. 3 we present McCabe-Thiele diagrams for a counter-current extraction of copper. In this figure, we combined the extraction isotherms for the case of 0 and 110 g/l of chloride ions with the operation lines, which represent the mass balance in a continuous counter-current solvent extraction circuit.

It should be pointed out that to draw these McCabe-Thiele diagrams in Fig. 3, several assumptions were made; therefore, this figure is not applicable for other conditions and it is for illustration purposes only. We assumed that the loaded extractant can be discharged to a copper content of 2 g/l in the stripping operation. We also consider ideal extraction stages, and an O/A ratio of 0.8. As we can see in the figure, when working with the aqueous phase without chloride ions we can extract 95% of the copper in the solution by using a 2-stage extraction circuit, and the organic phase could be loaded to more than 9 g/l of copper. On the other hand, when the aqueous phase contained 110 g/l of chloride a 3-stage extraction could only extract about 83% of the copper in the solution. In addition, the organic phase would be loaded to about 8 g/l of copper. Therefore, this circuit will also have a less efficient use of the extractant.

The extraction isotherms for LIX 84-IC for 0, 60 and 110 g/l of chloride at a higher temperature of 35 °C are presented in Fig. 4.

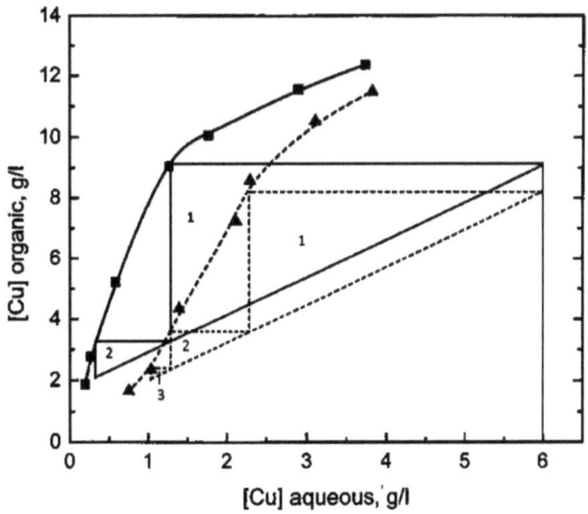

Fig. 3 McCabe—Thiele diagrams for the copper extraction with LIX 84-IC from solutions containing 0 and 110 g/l of chloride

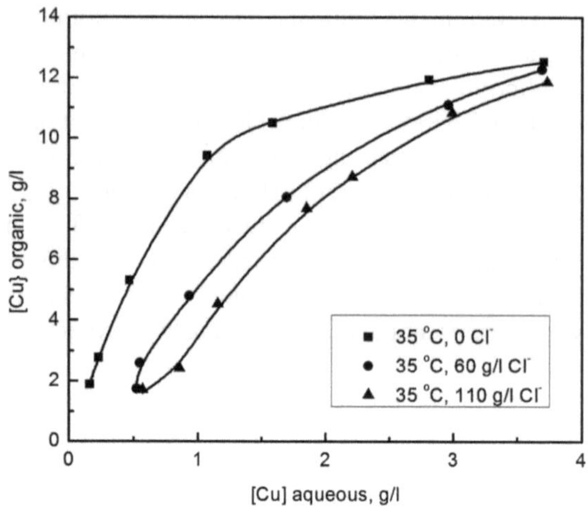

Fig. 4 Extraction isotherms for LIX 84-IC at 35 °C using aqueous phases with different concentrations of chloride ions

A comparison of these isotherms to the ones shown in Fig. 2 indicates that an increase in the temperature to 35 °C positively affected the extraction equilibria, raising the isotherms somewhat. However, the increase in the copper extraction when operating at 35 °C instead of 25 °C would be small. A larger temperature increase was not studied since operating at higher temperatures is not advisable due to an increase in the decomposition rate of the extractants.

For the case of the salicylaldoxime extractant, LIX 860 N-IC, the extraction isotherms at 25 °C are presented in Fig. 5. As seen in this figure, the presence of chloride in the solution also affects significantly the extraction equilibrium although the effect is not as large as observed with the ketoxime LIX- 84-IC. Therefore, this extractant will perform better in the counter-current copper extraction from chloride containing solutions than the ketoxime.

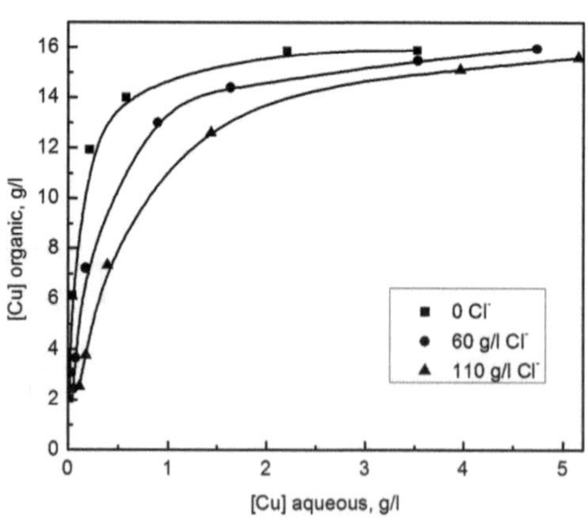

Fig. 5 Extraction isotherms for LIX 860 N-IC at 35 °C using aqueous phases with different concentrations of chloride ions

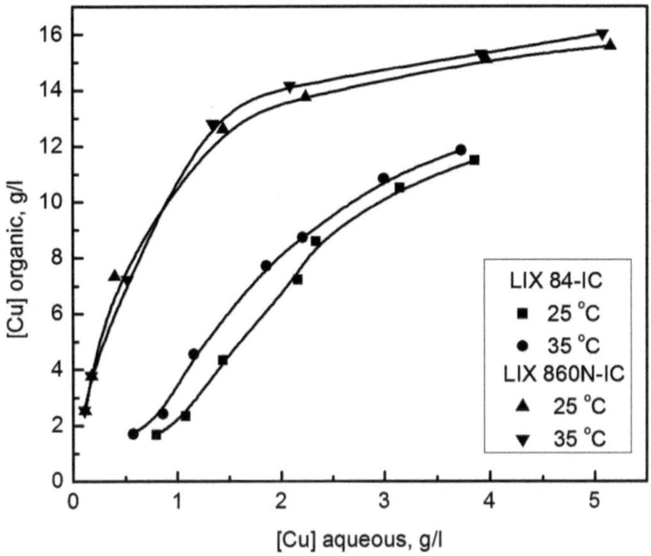

Fig. 6 Extraction isotherms for LIX 860 N-IC and LIX 84-IC at 25 and 35 °C using an aqueous phase with 110 g/l of chloride

On the other hand, an increase in the temperature to 35 °C affected very little the extraction equilibria for this extractant. This can be observed in Fig. 6, which compares the extraction isotherms at 25 and 35 °C for a solution with 110 g/l of chloride using both extractants.

Conclusions

- The presence of chloride ions in the aqueous solution has a large negative effect on the extraction equilibria of LIX 84-IC at 25 and 35 °C.
- As the chloride concentration in the aqueous phase increases, it becomes more difficult to obtain a raffinate with low copper concentration even when using more extraction stages.
- The effect of the chloride ions on the extraction of copper with LIX 860 N-IC is small as compared to the effect on the LIX 84-IC.
- An increase in the temperature from 25 to 35 °C positively affects the extraction equilibria of the hydroxyoxime extractants but the effect is not large, particularly for the LIX 860 N-IC.

Acknowledgements The National Fund for Scientific and Technological Development, FONDECYT, of Chile is acknowledged for the financial support of this study through Project No. 1150343.

References

1. P. Schmidt, Usos de Agua de Mar en Plantas Hidrometalúrgica. Paper presented at the 5th international seminar on process hydrometallurgy, hydroprocess 2013, Santiago, Chile, 10 July 2013
2. L. Ardiles, P. Jolly, Experiencia industrial con Alta Concentración de Cloruro. Paper presented at the workshop COPPERCLOR 2014, Santiago, Chile, 9 May 2014
3. J. Rauld, et al., Non-biochemical method to heap leach copper concentrates, US Patent, No. 6,926,753 B2, 9 Aug 2005
4. P. Schmidt, A. Backit, J. Jacob, Cuprochlor®: hydrometallurgical technology for leaching sulfide ores in chloride media, in *Proceedings copper 2013, IIMCH, 2013*, vol. IV, pp. 43–53
5. K. Sarangi et al., Separation of Iron(III), Copper(II) and Zinc(II) from a mixed sulphate-chloride solution using TBP, LIX 84I and Cyanex 923. Sep. Purif. Technol. **55**, 44–49 (2007)
6. M. Regel-Rosocka, F. Alguacil, Recent trends in metal extraction. Rev. Metal. (Madrid) **49**(4), 293–315 (2013)
7. H. Heinz, Selecting between weak and strong extractants on acid solutions in copper SX, in *Proceedings copper 2010, GDMB, 2010,* vol. 5, pp. 1925–1939

Hydrometallurgical Processes for the Recovery of Rare Earths, Nickel and Cobalt in Chloride Medium

V.I. Lakshmanan, R. Sridhar, D. Tait and M.A. Halim

Abstract Rare earth elements (REEs), nickel and cobalt find their use in developing strategic materials and their demand is consistently increasing in the world while the extractable ores of these metals are gradually depleting. Process Research ORTECH Inc. (PRO) has developed innovative mixed chloride leaching processes for the recovery of REEs from alumino-silicate ores, and nickel and cobalt from laterite ores. The addition of magnesium chloride in hydrochloric acid enhances the activity of the hydrogen ion by orders of magnitude, which permit rapid leaching rates at moderate temperature and atmospheric pressure, along with high metal recovery. Innovative solvent extraction process steps are used for the separation of these metals from pregnant leach solutions. PRO's processes also produce value added by-product of high purity iron oxide and recycle chloride lixiviant to the leaching stage. This paper describes the leaching and separation reaction mechanisms and potential process flowsheets.

Keywords Rare earths · Nickel · Cobalt · Laterite · Mixed chloride · Atmospheric leaching · Solvent extraction

Introduction

Rare earth elements (La to Lu plus Y) are being used in the making of hybrid electric cars, catalytic converters, wind power generators, LEDs, hard disc drives, flat panel displays and portable electronics due to their unique magnetic, electronic, catalytic and optical properties. Rare earth elements (REEs) are commonly divided in two groups: light rare earth elements (LREEs) and heavy rare earth elements (HREEs). Light rare earth elements include elements from lanthanum (La) to neodymium (Nd). Although promethium (Pm) is considered an LREE it is unstable

V.I. Lakshmanan · R. Sridhar · D. Tait · M.A. Halim (✉)
Process Research ORTECH Inc., 2350 Sheridan Park Drive,
L5K 2T4 Mississauga, ON, Canada
e-mail: halim@processortech.com

S. Wang et al. (eds.), *Applications of Process Engineering Principles in Materials Processing, Energy and Environmental Technologies*,
The Minerals, Metals & Materials Series, DOI 10.1007/978-3-319-51091-0_15

and scarcely occurs in nature and is generally excluded. Heavy rare earth elements include elements from samarium (Sm) to lutetium (Lu) plus yttrium. Although rare earth elements are not rare and the rare earth oxides are scattered around the globe, over 95% of REEs+Y are currently produced in China [1]. On one hand the Chinese government is restricting the export of rare earth elements and on the other hand the demand for REEs+Y is increasing in the rest of the world. This has created an imbalance in the supply and demand of rare earth elements and has led to a rush for developing new rare earths process flowsheets.

Demand for nickel and cobalt is consistently increasing in the production of strategic alloys in the world while the supplies of economically and environmentally recoverable resources are gradually depleting. Nickel is produced from laterites which have a number of layers including overburden (about <0.8% nickel) which is usually discarded, limonite ore zone (about 1.2–1.7% nickel, about 40% iron and 1–4% magnesium oxide), a saprolite ore zone (about 1.6–2.3% nickel and about 7–25% iron) and then a layer of rock [2, 3]. Laterite ore also contains other metals including cobalt, manganese and chromium.

High temperature and pressure based hydrometallurgical processes are also very capital intensive and require the use of autoclaves, high pressure pumping machines and costly materials of construction [4]. Chloride-based hydrometallurgical processes can be a potential alternative for the production of these value metals by meeting the more stringent environmental requirements, and reducing capital and operating costs compared to current process routes. Recycling of chloride lixiviant and minimizing of unwanted waste production are apparent advantages of chloride-based hydrometallurgical processes. Considering these advantages, innovative mixed-chloride processes have been developed by Process Research ORTECH Inc. [2, 3, 5–16]. In this paper, the recovery of REEs from alumino-silicate ores, and nickel and cobalt from laterite ores is described. Atmospheric pressure leaching of ores in lixiviant containing hydrochloric acid and magnesium chloride followed by separation and purification, and recovery of these metals are presented.

Mixed-Chloride Leaching Process for the Recovery of Rare Earth Elements

Process Research ORTECH Inc. (PRO) has developed an innovative mixed-chloride leaching process for the recovery of REEs from alumino-silicate ores that is cost effective and environmentally friendly [9]. Mixed-chloride containing hydrochloric acid and magnesium chloride is an effective leachant, which permit rapid leaching rates at modest temperatures and atmospheric pressures, with high metal recovery [2, 3, 9–18]. Based on PRO's proprietary mixed chloride leaching technology [9], a rare earths containing ore was subjected to grinding followed by leaching with a lixiviant of hydrochloric acid and magnesium chloride at 95 °C for 4 h. The leach slurry was subjected to solid/liquid separation.

Table 1 Results on extraction of REEs+Y by PRO's proprietary mixed chloride leaching technology [9]

100 mesh (%)	HCl (N)	$MgCl_2$ (g/L)	Solids (% w/w)	Temp (°C)	Time (h)	REEs (%)	REEs+Y (%)
85	5.8	225	10.4	95	4	86.1	83.3

The ore sample and the solid residue were digested and analyzed by Inductively Coupled Plasma (ICP). The elemental concentrations in the pregnant leach solution (PLS) were also determined by ICP. Leaching conditions and recovery of rare earth elements are shown in Table 1. Extractions of REEs and REEs plus yttrium (REEs+Y) were 86.1 and 83.3%, respectively.

The PLS undergoes multiple stage solvent extraction to selectively remove Fe and other impurities (including U, Th, Cu, and Zn) followed by the recovery of Zr. A ketone and a tertiary amine (Alamine 336) in kerosene were used for the removal of Fe and impurities, respectively, while 2-ethylhexyl phosphonic acid mono-2-ethylhexylester (PC-88A) in kerosene was used for separation of Zr. Free hydrochloric acid was extracted from Zr raffinate with EXXAL^{TM} 13, which can be recycled to the leaching stage. REEs+Y were precipitated from acid and Zr extracted raffinate with 10% oxalic acid at pH 2.0. Around 99% of the REEs+Y were precipitated at room temperature. Results are shown in Table 2. The rare earth oxalate obtained was washed with water, dried, calcined, and dissolved in

Table 2 Results on REEs+Y precipitation from Zr raffinate with 10% oxalic acid

	Fe	Zr	La	Ce	Pr	Nd	Sm	Eu	Gd	Tb
Zirconium raffinate, mg/L	0.5	5	768	1550	174	612	146	8	152	30
REEs+Y ppt.* filtrate, mg/L	0.5	<0.5	46.2	32.0	2.4	6.6	1.2	0.1	1.5	0.3
REEs+Y precipitation, %			94	98	99	99	99	99	99	99
	Dy	Ho	Er	Tm	Yb	Lu	Y	Cu	Zn	U
Zirconium raffinate, mg/L	208	41	115	15	71	8	772	1	2	<0.05
REEs+Y ppt.* filtrate	1.2	0.3	0.7	0.1	0.1	0.5	10	1	4	<0.05
REEs+Y precipitation (%)	99	99	99	99	100	94	99			
	Hf	Al	Ca	K	Mg	Mn	Na	S	Si	Sr
Zirconium raffinate, mg/L	1	1170	8750	915	58700	172	1590	123	9	125
REEs+Y ppt.* filtrate	<0.02	1710	7860	1320	66700	160	2140	121	17	155

*Precipitate

hydrochloric acid to give a REEs+Y concentrate for individual separation by solvent extraction.

Alternatively, PRO has tested the use of D2EHPA with EXXAL™ 13 in kerosene for direct extraction of REEs+Y from Zr raffinate after acid extraction (Table 3). Pregnant strip solution of REEs+Y could potentially be used as a feed for the individual separation of REEs+Y via solvent extraction process steps. The raffinate from the final solvent extraction stage is recycled to the leaching stage as shown in Fig. 1.

Table 3 Results on REEs+Y extraction from acid-extracted Zr raffinate (mg/L)

	La	Ce	Pr	Nd	Sm	Eu	Gd	Tb	Dy	Ho	Er	Tm
Zr Raffinate after acid extraction	50	70	6	54	13	<5	19	7	28	7	19	<5
REEs+Y Raffinate	<5	5	<5	<5	<5	<5	<5	<5	<5	<5	<5	<5
	Yb	Lu	Y	Zr	Fe	Ti	Al	Ca	Cu	K	Na	Zn
Zr Raffinate after acid extraction	16	5	184	<5	<5	<5	7180	5710	<5	1610	4400	<5
REEs+Y Raffinate	<5	<5	<5	<5	<5	<5	6680	5630	<5	1470	4410	<5

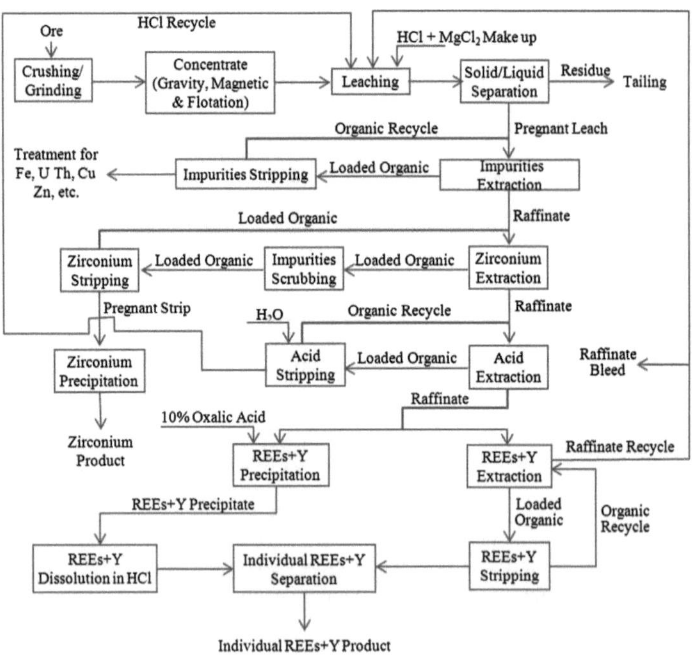

Fig. 1 Flowsheet for REEs+Y recovery by PRO's mixed chloride leaching process [9]

Mixed-Chloride Leaching Process for the Recovery of Nickel and Cobalt from Laterite

Process Research ORTECH Inc. (PRO) has developed an innovative atmospheric mixed-chloride leaching process for the recovery of nickel (Ni) and cobalt (Co) from laterite ores [2, 3]. As discussed in the previous section, addition of $MgCl_2$ to HCl solution increases the activity of the hydrogen ion (H^+) by orders of magnitude which help to increase Ni and Co extraction efficiency of the lixiviant with a lower concentration of HCl. Atmospheric leaching tests were conducted at 90–95 °C using a lixiviant containing HCl and $MgCl_2$ with a laterite ore having a composition of 0.74% Ni, 0.10% Co and 37.6% Fe; and with a concentrate which contains 4.02% Ni, 0.10% Co and 43.44% Fe. Under the optimal conditions, leach recoveries of Ni, Co and Fe were 98–99%, 86–93% and 98–99%, respectively (Table 4). After solid–liquid separation, pregnant leach liquor (PLS) went through a series of solvent extraction (SX) process steps to subsequently separate Fe, Co and Ni.

Under the oxidizing condition, the predominant oxidation state of iron is ferric ion (Fe^{3+}), which has strong affinity to form complexes with the chloride ion, such as $FeCl^{2+}$, $FeCl_2^+$, $FeCl_3{}^\circ$ and $FeCl_4^-$. Lee et al. [19] suggest that the mole fraction of $FeCl_4^-$ increases with increasing HCl concentration, while those of $FeCl^{2+}$ and Fe^{3+} decrease with the HCl concentration. Cobalt and Ni exist as divalent hexahydrated ions in aqueous solutions. The formation of ion complex often proceeds much more readily with divalent Co than with Ni due to the rate of water exchange on the cobalt ion is much higher than for nickel [20]. The trivalent Co ion is much less labile and forms in preference to Ni even though the redox potentials for the Co^{2+}–Co^{3+} and Ni^{2+}–Ni^{3+} couples are nearly identical. The divalent Co has a strong tendency to form tetrahedral anionic chloro-complex such as $CoCl_3^-$ and $CoCl_4^{2-}$ in a strong chloride ion solution, whereas divalent Ni does not form such complexes and Ni^{2+} retains a hexa-coordinated complex. This difference in chemical behavior allows for selective separation of Co^{2+} from Ni^{2+} with an anion exchange extractant in mixed chloride medium [2]. Understanding this chloro-chemistry of Fe, Co and Ni species has greatly helped to separate them by SX process steps (Table 5).

Iron was selectively extracted over Co and Ni by contacting the pregnant leach solution (PLS) with an organic phase (*S*) containing a solvating extractant, more specifically a high molecular weight of ketone, a modifier (e.g. Exxal™ 13 tridecyl alcohol) and a diluent (e.g. Exxsol™ D80) by the Reaction 1. Stripping of Fe^{3+}

Table 4 Leach recoveries of Ni, Co and Fe with atmospheric mixed chloride leaching process

	Ni	Co	Fe
	%		
Ore (grade)	0.74	0.1	37.6
Recovery	98	93	98
Concentrate (grade)	4.02	0.10	43.44
Recovery	99	86	99

Table 5 Separations of Fe, Co and Ni from PLS by subsequent SX process steps

	Fe	Co	Ni
	mg/L		
Pregnant leach liquor	33,150	58	3331
Fe raffinate at the O/A ratio of 1:2	<5	63	3376
Co raffinate at the O/A ratio of 1	<5	<5	3356
Ni raffinate at the O/A ratio of 1	<5	<5	28

A three stage Fe extraction was conducted

from the loaded organic phase was carried out with a dilute HCl solution. Iron raffinate was subjected to subsequent selective SX of Co^{2+} and Ni^{2+} with organic phases containing a tertiary amine (e.g. Alamine™ 336 (R_3N)) and oxime (e.g. LIX 63 (RH)), respectively, following the reactions of 2–4 [2, 3, 21]. In both organic phases, Exxal™ 13 tridecyl alcohol and Exxsol™ D80 were used as a modifier and a diluent, respectively. Stripping of both Co^{2+} and Ni^{2+} from loaded organic phases was conducted with a dilute HCl solution. The McCabe–Thiele extraction and stripping isotherms for Co^{2+} and Ni^{2+} are shown in Figs. 2 and 3, respectively.

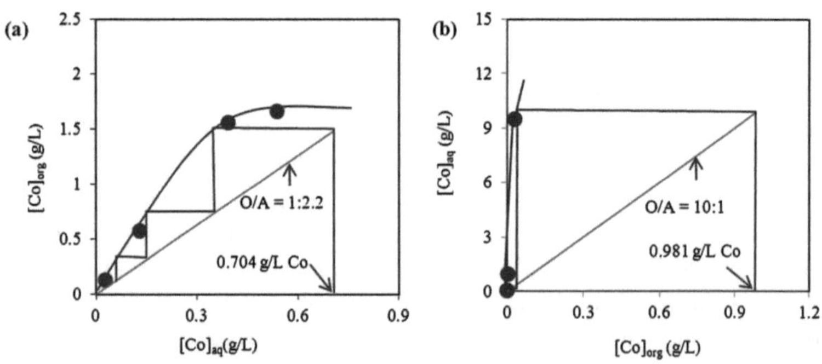

Fig. 2 The McCabe–Thiele isotherms for Co^{2+} **a** extraction isotherm and **b** stripping isotherm [2, 3]

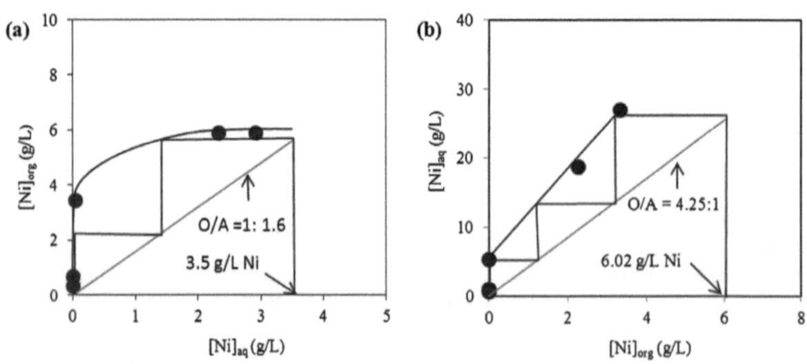

Fig. 3 The McCabe–Thiele isotherms for Ni^{2+} **a** extraction isotherm and **b** stripping isotherm [2]

$$H^+FeCl_4^- + 2S = H^+FeCl_4^-.2S \tag{1}$$

$$R_3N + HCl = R_3NHCl \tag{2}$$

$$2\,R_3NHCl + CoCl_4^{2-} = (R_3NH)_2CoCl_4^{2-} + 2\,Cl^- \tag{3}$$

$$2\,RH + NiCl_2 = R_2Ni + 2HCl \tag{4}$$

The extractants used in this study are highly selective for Ni^{2+}, Co^{2+} and Fe^{3+} (Table 5), which results in high purity pregnant strip solutions. Depending on the market value of the product, Ni and Co can be recovered from their respective pregnant strip solutions in different forms such as carbonate and oxide. Hydrochloric acid can be regenerated from the pregnant strip solution of Fe by pyrohydrolysis, which has been commercially practiced in the pickling steel industry [22–24]. This process also produces a value added by-product of high purity iron oxide (Fe_2O_3). The developed flowsheet for the recovery of Ni, Co and Fe from laterite is shown in Fig. 4 [2, 3]. The use of $MgCl_2$ in lixiviant and LIX 63 permits the recovery of Ni^{2+} at a lower pH of 0.5–2.5, compared to other processes that operate at a pH of 5–6.

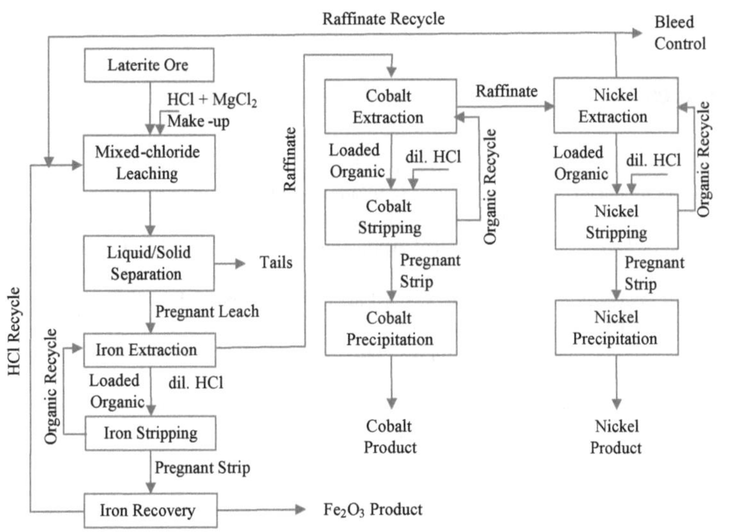

Fig. 4 PRO's mixed-chloride process flowsheet for the recovery of Ni and Co from laterite [2, 3]

Conclusions

Process Research ORTECH Inc. has developed hydrometallurgical processes for the recovery of rare earths from alumino-silicate ores; and nickel and cobalt from low grade laterite. Leaching of these ores was conducted with mixed chloride lixiviant containing HCl and $MgCl_2$ at atmospheric pressure and varying temperature between 90 and 95 °C. After leaching, the leach slurries were subjected to liquid/solid separation. By understanding the chloro-chemistry and separation technology, these metals were selectively separated from pregnant leach liquors by solvent extraction process steps. Around 99% of REEs+Y was precipitated from acid-extracted Zr raffinate with 10% oxalic acid. Alternatively, REEs+Y were directly extracted from acid-extracted Zr raffinate. PRO's mixed-chloride processes have several advantages over the conventional processes such as (i) addition of $MgCl_2$ to HCl in the leaching stage increased H^+ activity in the lixiviant by order of magnitude which helps to increase metal recovery with a lower concentration of HCl, (ii) regeneration and recycling of HCl and raffinate to the leaching stage reduces reagent costs, (iii) the use of $MgCl_2$ in lixiviant and LIX 63 permits the separation of Ni^{2+} at a lower pH of 0.5–2.5 and (iv) PRO process can be a potential alternative to the conventional hydrometallurgical and smelting processes by meeting the more stringent environmental requirements, and reducing capital and operating costs.

References

1. M. Pitts, Endangered elements. The Chem. Eng. 48–51 (2011). http://www.tcetoday.com/~/media/Documents/TCE/Articles/2011/844/844elements.pdf
2. V.I. Lakshmanan, R. Sridhar, M.A. Halim, J. Chen, R. DeLaat, *Recovery of Nickel in Leaching of Laterite Ores*. US Patent Application, No. 294702A1, 2 October 2014
3. V.I. Lakshmanan, R. Sridhar, J. Chen, M.A. Halim, R. DeLaat, *Separation of Iron from Value Metals in Leaching of Laterite Ores*. US Patent, No. 8916116 B2, 23 December 2014
4. G.O. Onyedika, C. Achusim-udenko, C.I.A. Nwoko, M.O.C. Ogwuegbu, Chemistry, processes and problems of complex ores utilization: hydrometallurgical options. Int. J. Chem. Sci. **10**, 112–130 (2012)
5. P.G. Christie, V.I. Lakshmanan, G.J. Lawson, The behaviour of Lix 63 in the extraction of Cu(II) and Fe(III) from Chloride Media. Hydrometallurgy **2**, 105–115 (1976)
6. G.B. Harris, V.I. Lakshmanan, T.J. Magee, R. Sridhar, *Atmospheric Chloride Leaching of Base Metal Sulfides*. eds. by Hydro-Sulfides. International Colloquium on Hydrometallurgical Processing of Copper Sulfides (Santiago, Chile, 2004), pp. 384–398
7. G.B. Harris, V.I. Lakshmanan, R. Sridhar, G. Puvvada, *A Process for the Recovery of Value Metals from Base Metal Sulfide Ores*. Canadian Patent, No. 2478516, 11 December 2011
8. G.B. Harris, V.I. Lakshmanan, R. Sridhar, G. Puvvada, *Process for the Recovery of Value Metals from Base Metal Sulfide Ores*. US Patent, No. 7736606 B2, 15 June 2010
9. V.I. Lakshmanan, R. Sridhar, M.A. Halim, *Process for extraction of Rare Earth Elements*. US Patent, No. 9115419, 25 August 2015

10. V.I. Lakshmanan, M.A. Halim, S. Vijayan, *Innovative Process for the Production of Titanium Dioxide*. eds. by V.I. Lakshmanan R. Roy, V. Ramachandran. Innovative Process Development in Metallurgical Industry- Concept to Commission (Springer International Publishing, Switzerland, 2016), pp. 359–83
11. V.I. Lakshmanan, R. Sridhar, M.A. Halim, *Chloride Process for the Leaching of Gold*. US Patent Application, No. 0283976 A1, 31 October 2013
12. V.I. Lakshmanan, R. Sridhar, G.B. Harris, G. Puvvada, *Process for the Recovery of Titanium in Mixed Chloride Media*. US Patent, No. 7803336, 28 September 2010
13. V.I. Lakshmanan, R. Sridhar, R. Roy, V. Ramachandran, *Recovery of Value Metals from Ores by Mixed Chloride Extraction*. Proceedings of Sixth International Symposium, Hydrometallurgy 2008, Phoenix, Arizona, USA, 2008, pp. 895–902
14. V.I. Lakshmanan, R. Sridhar, M. Rishea, R. DeLaat, *Methods for Separation of Titanium from Ore*. US Patent, No. 6699446 B2, 2 March 2004
15. V.I. Lakshmanan, R. Sridhar, M. Rishea, R. DeLaat, *Separation of Titanium Halides from Aqueous Solutions*. US Patent, No. 6500396 B1, 31 December 2002
16. V.I. Lakshmanan, G.B. Harris, *Hydrometallurgical Flowsheets for Tomorrow: The Role of Chemistry in Solution Purification in Impurity Control and Disposal in Hydrometallurgical Process*, eds. by G.B. Harris and E. Krause. Proceedings of the International Symposium on Impurity Control and Disposal in Hydrometallurgical Processes, 24th Annual Hydrometallurgical Conference of CIM (Toronto, Ontario, Canada, 1994), pp. 3–16
17. E. Königsberger, P. May, B. Harris, Properties of electrolyte solutions relevant to high concentration chloride leaching. II. density, viscosity and heat capacity of mixed aqueous solutions of magnesium chloride and nickel chloride measured to 90 °C. Hydrometallurgy **90**, 168–176 (2008)
18. J.J.C. Jansz, Estimation of Ionic activities in chloride systems at ambient and elevated temperatures. Hydrometallurgy **11**, 13–31 (1983)
19. M.-S. Lee, G.-S. Lee, K.Y. Sohn, Solvent extraction equilibrium of $FeCl_3$ with TPB. Mater. Trans. **45**, 1859–1863 (2004)
20. M. Cox, *Solvent Extraction in Hydrometallurgy*, 2nd edn. eds. by J. Rydberg, C. Musikas, G.R. Choppin. Principles and Applications of Solvent Extraction (Marcel Dekker Inc., New York, USA, 2004), pp. 455–505
21. Y.F. Shen, W.Y. Xue, W.Y. Niu, Recovery of Co(II) and Ni(II) from hydrochloric acid solution of alloy scrap. Trans. Nonferrous Met. Soc. China **18**, 1262–1268 (2008)
22. K. Adham, C. Lee, D. Small, *Energy Consumption for Iron Chloride Pyrohydrolysis: A Comparison Between Fluid Beds And Spray Roasters*. eds. by J.E. Dutrizac, P.A. Riveros. Iron Control Technologies, Proceedings of the Third International Symposium on Iron Control in Hydrometallurgy, 36th Annual CIM Hydrometallurgical Meeting (Montreal, Canada, 2006), pp. 815–30
23. E.M. Peek, O.F. Goedhart, G.V. Weert, *Process Evaluation of Steel Pickle Liquor Pyrohydrolysis in a Commercial Keramchemie Fluid Bed Reactor*. eds. by J.E. Dutrizac, G.B. Harris. Iron Control and Disposal, Second International Symposium on Iron Control in Hydrometallurgy (Ottawa, Canada, 1996), pp. 483–503
24. F. Baerhold, A. Lebl, J. Statrcevic, *Recycling of Spent Acids and Iron via Pyrohydrolysis*. eds. by J.E. Dutrizac, P.A. Riveros. Iron Control Technologies, Proceedings of the Third International Symposium on Iron Control in Hydrometallurgy, 36th Annual CIM Hydrometallurgical Meeting (Montreal, Canada, 2006), pp. 789–804

A Cr^{6+}-Free Extraction of Chromium Oxide from Chromite Ores Using Carbothermic Reduction in the Presence of Alkali

L. Escudero-Castejon, S. Sanchez-Segado, S. Parirenyatwa, Y. Hara and A. Jha

Abstract Oxidative alkali roasting of chromite is the state-of-the-art process for manufacturing chromium-containing chemicals, which involves dealing with serious environmental problems arising from handling Cr^{6+}-containing wastes generated in this process. In this article a new method for the extraction of Cr_2O_3 from chromite ores is explained, based on the carbothermic reduction of concentrates in the presence of alkali investigated in the temperature range of 950–1050 °C. Under these conditions, the iron oxides present in the ore body are reduced to metallic iron and the resulting separation of chromium occurs by forming sodium chromite ($NaCrO_2$). The reduced samples are magnetically separated for the recovery of an iron-rich fraction, and a non-magnetic fraction containing $NaCrO_2$, MgO and other impurities. The further treatment of the non-magnetic fraction by leaching yields a Cr_2O_3-rich product of approximately 85% purity, with remaining alumina, alkali and magnesia. The main advantage of the process is that, under reducing atmosphere and subsequent leaching conditions, the oxidation of Cr^{3+} to Cr^{6+} is completely avoided; thereby decreasing the risk of land, air and water pollution.

Keywords Chromite ore · Carbothermic reduction · Alkali · Chromium(VI) · Chromium extraction · Chromium oxide

Introduction

The future manufacturing of nonferrous metals is dependent on significant improvement of the traditional beneficiation processes by minimizing the generation of hazardous wastes and the energy consumption. Although the chromium chemicals industry has been reducing the environmental impact by designing safer disposal procedures for process waste and handling, the landfill sites require

L. Escudero-Castejon (✉) · S. Sanchez-Segado · S. Parirenyatwa · Y. Hara · A. Jha
The Institute for Materials Research, School of Chemical and Process Engineering, University of Leeds, LS2 9JT Leeds, UK
e-mail: pmlec@leeds.ac.uk

S. Wang et al. (eds.), *Applications of Process Engineering Principles in Materials Processing, Energy and Environmental Technologies*,
The Minerals, Metals & Materials Series, DOI 10.1007/978-3-319-51091-0_16

ongoing maintenance [1]. Only a fraction of chromium containing wastes are recycled in the industry, which implies that there is no long-term solution for the management of hexavalent chromium.

Reactive metal oxides, namely Cr_2O_3, because of their multiple valences determined by the 3d-electronic structure, form a range of complex oxides in nature; which in case of chromium occurs as chromite spinel $(Fe,Mg)(Cr,Al,Fe)_2O_4$ [2]. Six different spinels ($FeCr_2O_4$, $MgCr_2O_4$, $FeAl_2O_4$, $MgAl_2O_4$, $MgFe_2O_4$ and Fe_3O_4) coexist in the complex chromite spinel forming a solid solution.

Worldwide manufacturing of sodium chromate, sodium dichromate, chromic acid and many other chromium chemicals relies on the process of alkali roasting of chromite in oxidising conditions, which yields water soluble sodium chromate (Na_2CrO_4). Silica is often present in chromite and it can vary between 2 wt% in S. African ores and more than 8 wt% in lower grades of chromites used for chemicals manufacturing. Although the S. African ores are most widely used, there is a general tendency in the sector to use indigenous resources for chromium chemicals, which has then implications on using more energy for handling and managing chromium (6+) wastes. As a result of increasing silica, which forms a viscous silicate liquid, the oxygen transport during oxidation is reduced and there is higher consumption of alkali as it is required to neutralize silica [3, 4]. Hence, the extraction efficiency of chromium decreases with increasing silica content [5], and consequently, the generation of hazardous chromite ore processing residue (COPR) containing toxic Cr^{6+} have not been eliminated and it stands for the main limitation of the alkali roasting process.

Besides the use of chromite ores for chromium chemicals manufacture, the most important application of chromite is in the production of high-grade ferro-chromium, which is used in the metallurgical industry for manufacturing of stainless steels and chromium-containing metal alloys [6]. Ferro-chromium is produced by smelting of chromite ore in reducing atmosphere in an arc submerged furnace with coke at high temperatures (above 1600 °C), with partial solid state pre-reduction of chromite. The reduction of chromium oxide takes place as shown in reaction (1):

$$Cr_2O_3 + 3[C]_{Fe} \leftrightarrow 2[Cr]_{Fe} + 3\,CO(g) \tag{1}$$

where $[C]_{Fe}$ and $[Cr]_{Fe}$ represent the carbon and chromium in the alloy. The equilibrium constant for reaction (1) may be expressed as:

$$k = \frac{(P_{CO})^3 \cdot (a_{[Cr]_{Fe}})^2}{(a_{[C]_{Fe}})^3 \cdot a_{Cr_2O_3}} \tag{2}$$

where a_i is the activity of species i; and P_{CO} is the equilibrium partial pressure of CO gas. As it is evident from Eq. (2), the activity of carbon must be kept high so that higher Cr can be retained in the alloy phase because $P_{CO} \propto a_{[C]Fe}$ and $P_{CO}a_{[C]Fe} \propto (a_{[C]Fe})^{\frac{3}{2}}$ if $a_{Cr_2O_3}$ and P_{CO} are considered fixed at a given temperature. However, the presence of high amounts of carbon may lead to the formation of chromium carbides which are undesirable for the formability of stainless steel products. The production of Fe–Cr needs expensive metallurgical coke, cannot use chromite fines and requires a great supply of electrical energy for operating the electric arc furnace [7].

In this study, the production of chromium(III) oxide based on the reduction of chromite ores in the presence of alkali with carbon at moderate temperatures (950–1050 °C) was investigated. The reaction of chromium oxide with Na_2CO_3 to form sodium chromite ($NaCrO_2$) occurs in reducing conditions with simultaneous reduction of iron oxides to metallic iron, following reaction (3).

$$FeCr_2O_4 + Na_2CO_3 + 2\,C \rightarrow Fe + 2\,NaCrO_2 + 3\,CO(g) \tag{3}$$

The reduction of chromite under these conditions does not suffer from the complex balance between high carbon activity and partial pressure of CO gas, since the chromium oxide activity is lowered by forming the alkali chromite. Furthermore, the generation of Cr^{6+}-species is eliminated in this process due to the presence of a reducing atmosphere throughout the alkali complexation process at elevated temperature. The purpose of this investigation is therefore to demonstrate a new process methodology that does not yield Cr^{6+} containing waste in solid, liquid or gaseous forms, and therefore excludes the dependency on waste handling. The chromium(III) oxide derived from this process may then be used as raw material for the manufacturing of chromium chemicals, stainless steels, chrome metal or other chromium-containing alloys. The proposed flowsheet and the fundamentals of the different stages of the process are discussed in this manuscript.

Experimental

The ore used in this study was S. African chromite ore of the following composition 48.8% Cr_2O_3, 31.3% Fe_2O_3, 7.03% MgO, 7.15% Al_2O_3, 3.45% SiO_2 and 0.54% CaO. Sodium carbonate (Na_2CO_3) and activated charcoal of analytical grade were also employed in alkali reduction experiments. For the reduction experiments, the as-received mineral was mixed with alkali (Na_2CO_3) and carbon (activated charcoal) using a certain mineral:alkali:carbon weight ratio. Mixtures were heated in a tube furnace under isothermal reducing conditions for 2.5 h, in argon atmosphere and at temperatures ranging between 900 and 1050 °C.

Reduced samples were ground and magnetically separated. During magnetic separation the sample was sprayed with water to wash-off the non-magnetic fraction and dissolve the water-soluble components. The non-magnetic fraction, rich in $NaCrO_2$, was leached first in water and subsequently in acid media. Either oxalic acid or H_2SO_4 solutions were used as acids in leaching experiments. Both water and acid leaching were carried out in a leaching vessel with continuous stirring, at 50 °C and controlling pH.

Results and Discussion

Carbothermic Reduction of Chromite Ore in the Presence of Alkali

During reduction in the presence of alkali, the iron present in the spinel is reduced to metallic form (Fe), and the chromium oxide present in chromite solid solution reacts with Na_2O produced from the decomposition of sodium carbonate forming sodium chromite ($NaCrO_2$), as previously shown in Eq. (3). The carbon dioxide, which is a waste otherwise, combines with excess carbon and yields CO gas.

The equilibrium reduction condition of chromite in the presence of alkali may be explained on the basis of the Na–Fe–Cr–O–C predominance area diagram shown in Fig. 1. From this figure, it is apparent that metallic Fe co-exists with $NaCrO_2$ (area 4) at 1050 °C (1323 K) for a range of partial pressures of CO and CO_2. Reduction of iron oxides takes place by shifting the equilibrium condition corresponding to area 1 to area 4, resulting in the decreasing value of $P(CO)_2$. $NaCrO_2$ is stable at these temperatures and reductive pressure conditions, thereby ensuring that the Cr^{6+} state remains unstable.

The aim of the alkali reduction stage is to form a magnetic metallic iron-rich fraction and a non-magnetic fraction containing sodium chromite, which can be magnetically separated. Images from scanning electron microscopic (SEM) analysis and the energy dispersive X-ray elemental mapping of reduced chromite samples are shown in Fig. 2. The elemental mapping illustrates that the brighter phase corresponds to metallic Fe formed during reduction, while the darker grey phase is

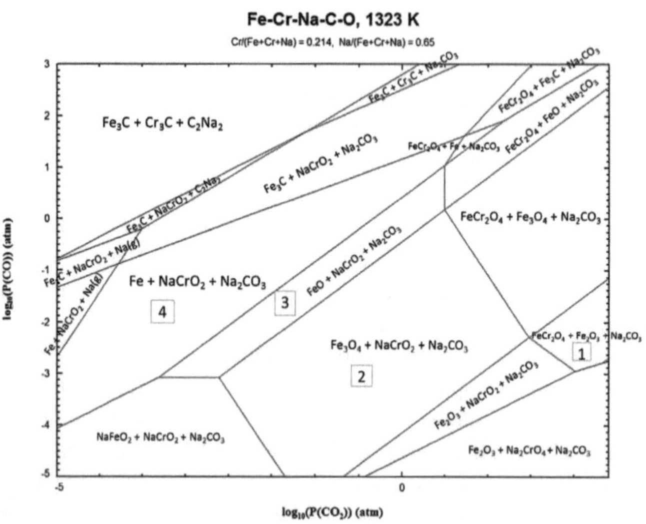

Fig. 1 Predominance diagram of the Na–Fe–Cr–O–C system at 1323 K (1050 °C) computed by Fact-Sage 6.4 software [8]

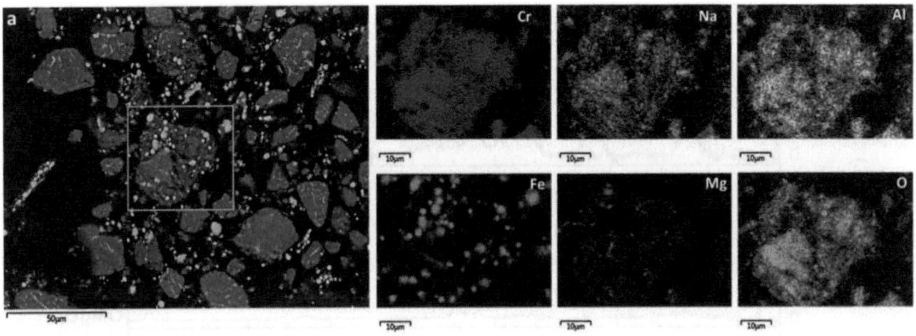

Fig. 2 Scanning electron microscopy image (**a**) of a chromite reduced sample with Na_2CO_3 and activated charcoal in argon atmosphere at 1050 °C for 2.5 h; and energy dispersive X-ray elemental maps of Cr, Na, Al, Fe, Mg and O of the selected area

rich in $NaCrO_2$. Sodium aluminate is also present in the reduced product as a result of the chemical reaction of aluminium oxide with excess Na_2CO_3, following reaction (4).

$$Al_2O_3 + Na_2CO_3 + C \leftrightarrow 2\,NaAlO_2 + 2\,CO(g) \tag{4}$$

$NaAlO_2$ is water soluble and it can be separated readily from the reaction product. A part of sodium carbonate may react with MgO and silica to form complex sodium magnesium silicates. Magnesium oxide (MgO) and calcium carbonate ($CaCO_3$) are also present as products of reaction.

Wet Magnetic Separation

The X-ray powder diffraction patterns of a reduced sample, magnetic fraction and non-magnetic fraction are compared in Fig. 3 in which $NaCrO_2$, Fe, $NaAlO_2$, MgO and $CaCO_3$ were the main phases identified in the reduced sample pattern (a). This is in good agreement with the elemental mapping results shown in Fig. 2.

For an efficient magnetic separation, it is necessary to optimize the formation of metallic Fe by adjusting the operation parameters of the reduction process. In Fig. 3, it can be seen that the magnetic fractions still contain residual sodium chromite after the separation, as low-intensity peaks for $NaCrO_2$ were identified in pattern (b). This is also confirmed by the comparative analysis of the microstructures of the magnetic fraction (left) and the non-magnetic fraction (right) shown in Fig. 4, in which extensive entrapment of metallic iron within $NaCrO_2$ particles and sintering between both phases are evident. The compositions of the magnetic and non-magnetic samples analysed by XRF are shown in Table 1, and these data are compared with the chemical composition of the initial chromite ore. The magnetic

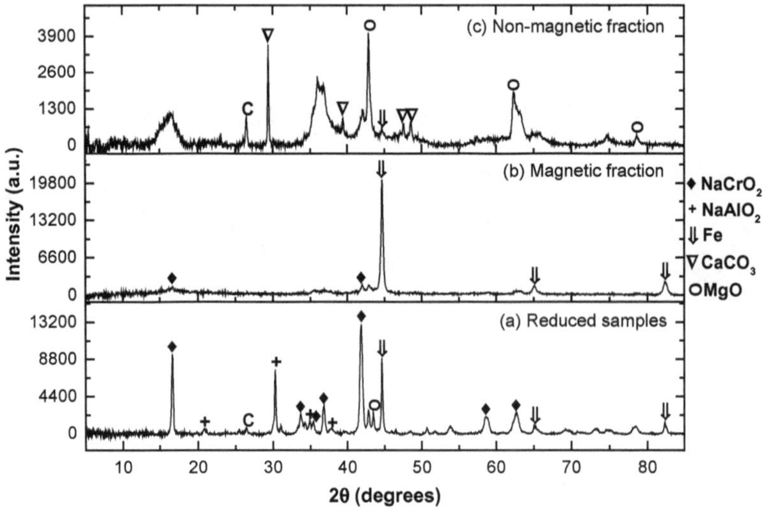

Fig. 3 X-ray powder diffraction (XRPD) patterns of reduced sample (**a**), magnetic fraction (**b**) and non-magnetic fraction (**c**) obtained from the reduction of chromite ore with sodium carbonate and activated charcoal (1050 °C, 2.5 h), followed by wet magnetic separation

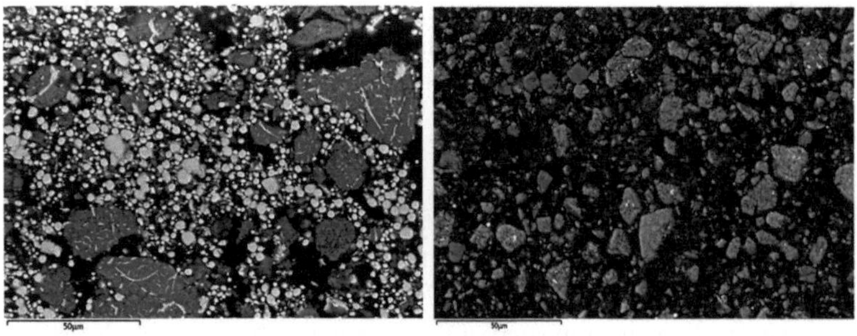

Fig. 4 Scanning electron microscopy images of magnetic fraction (*left*) and non-magnetic fraction (*right*) samples obtained after two steps magnetic separation of chromite samples reduced with Na_2CO_3 and activated charcoal at 1050 °C for 2.5 h

Table 1 XRF analysis of as-received chromite ore, magnetic and non-magnetic fractions after reduction of chromite ore with Na_2CO_3 and charcoal at 1050 °C for 2.5 h followed by magnetic separation

wt%	Cr_2O_3	Fe_2O_3	Na_2O	MgO	Al_2O_3	SiO_2	CaO
Chromite	48.80	31.30	–	7.03	7.15	3.45	0.54
Magnetic fraction	21.10	71.20	1.60	1.99	1.85	0.45	0.59
Non-magnetic fraction	50.20	6.16	7.42	13.30	7.01	4.34	6.63

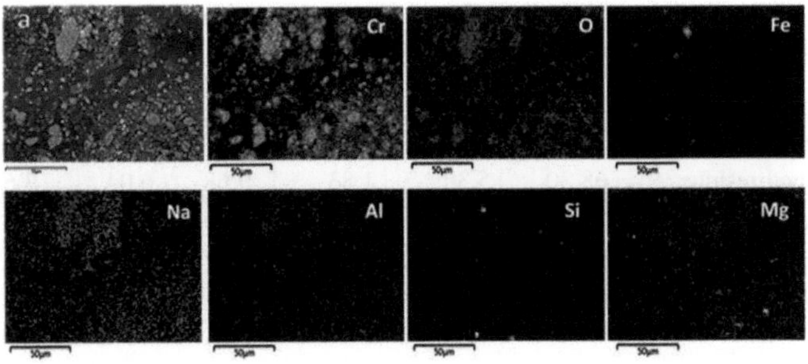

Fig. 5 Scanning electron microscopy image (**a**) and energy dispersive X-ray elemental maps of Cr, O, Fe, Na, Al, Si and Mg, of a non-magnetic fraction sample obtained after chromite reduction with Na_2CO_3 and activated charcoal at 1050 °C for 2.5 h followed by magnetic separation

fraction contains 21.1% Cr_2O_3, being most of it in the form of trapped $NaCrO_2$ with a minor quantity of chromium in the metallic iron matrix.

This tendency of entrapment and sintering apparently impedes complete recovery of $NaCrO_2$ into the non-magnetic fraction. The efficiency of the magnetic separation improves with increasing particle size of metallic iron, and thus the particle size dependence of Cr-entrapment as sodium chromite in the iron matrix is a key aspect for achieving high chromium recovery.

In Fig. 3, it could be seen that the XRPD pattern (c) of the non-magnetic fraction is not completely crystalline. Diffraction peaks of sodium chromite, which is the main product expected in this fraction, could not been identified possibly because of the non-crystallinity of $NaCrO_2$. However, evidence of the presence of $NaCrO_2$ in the non-magnetic fraction is given in the SEM image and elemental mapping of a non-magnetic fraction sample presented in Fig. 5. X-ray fluorescence results in Table 1 showed that the non-magnetic fraction contains 50.2% Cr_2O_3; and the remaining impurities are, from higher to lower wt%, MgO, Na_2O, Al_2O_3, CaO, Fe_2O_3 (which may be in form of metallic iron) and SiO_2.

Since magnetic separation is carried out with water, part of the unreacted Na_2CO_3 and the water soluble compounds formed during reduction, namely $NaAlO_2$ and water soluble silicates, will solubilise at this stage yielding an alkaline solution. The treatment of this solution in order to recover the alkali will be explained below in this text.

Treatment of the Non-magnetic Fraction

A Cr_2O_3-rich product can be extracted from the non-magnetic fraction by different leaching steps. The first leaching step was performed with water by removing as much sodium-bearing compounds as possible from the non-magnetic fraction. This is essential for the process since, if most of the sodium is extracted at this point,

Table 2 Chemical compositions of the non-magnetic fraction, water leached and acid leached Cr_2O_3 residues using either sulfuric or oxalic acid

wt%	Cr_2O_3	Fe_2O_3	Na_2O	MgO	Al_2O_3	SiO_2	CaO
Non-mag fraction	50.20	6.16	7.42	13.30	7.01	4.34	6.63
Water leached residue	68.30	5.30	1.83	9.68	6.04	3.64	3.08
Sulfuric leached residue	83.60	2.66	1.65	1.49	4.79	0.53	2.25
Oxalic leached residue	81.50	2.17	1.41	1.82	6.07	0.27	4.77

Samples were analysed by XRF

the alkali can be recovered and recycled back into the process, which will then help decreasing the consumption of acid during acid leaching.

The residue after water leaching is subsequently leached with acid with the aim of removing the remaining sodium ions from the partially leached sodium chromite ($Na_{1-x}CrO_2$). The remaining iron and any water insoluble silicates present in the non-magnetic fraction may be removed at this stage for obtaining an enriched Cr_2O_3 precipitate. Sulfuric acid and oxalic acid were tested for acid leaching, achieving 83–85% and 80–82% Cr_2O_3-rich products, respectively. In Table 2 the chemical compositions of the non-magnetic fraction and the water and acid leached residues obtained are compared, showing a significant enrichment of the residue after the leaching stages. The main impurities remaining on the enriched products after leaching with H_2SO_4 are Al_2O_3 and Fe_2O_3, whereas when leaching with oxalic acid the content of CaO is also significantly high.

Proposed Process Flowsheet

Based on the experimental results presented, a novel process for extraction of chromium oxide from chromite ores is proposed, and its flowsheet is presented in Fig. 6.

The first step of the new process design consists on the reduction of chromite in the presence of Na_2CO_3 and activated charcoal (T = 950–1050 °C), as explained before in this text. Reduced samples were cooled down to room temperature and ground using a mortar and pestle to liberate Fe particles. On the next step, reduced samples are magnetically separated using water which yields a Fe-rich magnetic fraction, a $NaCrO_2$-rich non-magnetic fraction and an alkaline solution.

It was shown in Table 1 that the magnetic fraction contains ≈20% Cr_2O_3, and therefore, it is worthwhile considering the recovery of chromium from this fraction, which needs to be further investigated. An option would be to treat this magnetic fraction at high temperature in order to separate the remaining $NaCrO_2$ (into slag) from the metallic Fe. The $NaCrO_2$ slag could be recycled back into the process for extraction of chromium, or fed into a ferrochromium alloy making plant.

The non-magnetic fraction is subjected to water leaching, generating a solid precipitate rich in Cr_2O_3 and a solution which is highly alkaline. The leachate

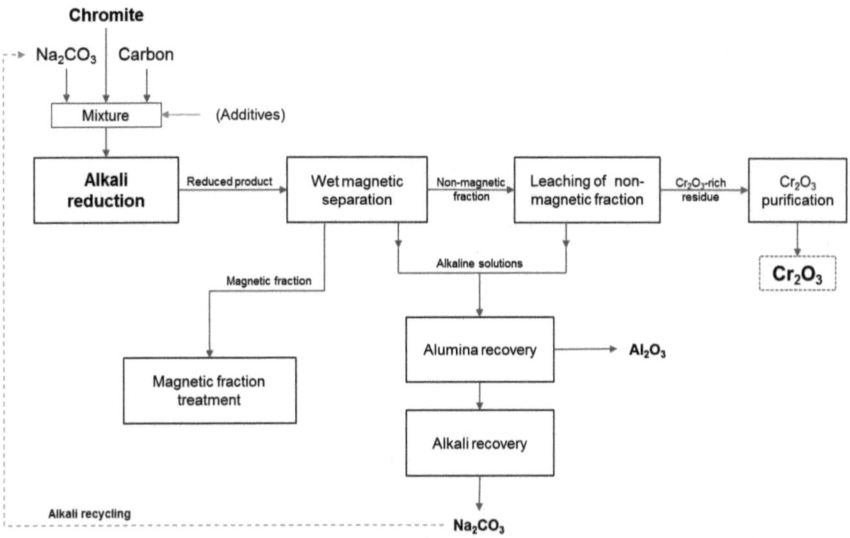

Fig. 6 Process flowsheet including the main stages of the alkali reduction process

solution obtained is rich in alumina, alkali and contains some silica, and may be combined and treated with the highly alkaline solution obtained from the wet magnetic separation stage for potential recovery of alkali. Alumina is precipitated from the solutions by CO_2 bubbling. Sodium species as $NaHCO_3$, which can be then recycled back into the reduction stage, are recovered from the remaining solution by evaporation. Maximization of the alkali recovery is a key aspect of the process from an economical point of view.

Water leaching is followed by leaching in acid media, giving as a result a Cr_2O_3-rich product of up to 85% purity. Further purification is required in order to achieve high purity chromium oxide.

Conclusions

A novel process for extraction of chrome oxides from chromite ores has been investigated. The process is based on the carbothermic reduction of chromite ore with solid carbon (activated charcoal) in the presence of an alkali in the temperature range of 950–1050 °C. The magnetic separation of reduced samples with water leaching yields a non-magnetic fraction residue which is the main source of Cr_2O_3. The leaching of the Cr_2O_3 residue with oxalic and sulphuric acid in one step produces better than 80% pure Cr_2O_3. So far it is evident from the analysis that there was no evidence for the presence of Cr^{6+} in the process from the leaching or reduction stage.

Optimization of the different steps of the process is now required, with special emphasis on the separation of iron from $NaCrO_2$ during magnetic separation and further purification of the Cr_2O_3-rich product.

Acknowledgements The authors acknowledge the financial support from a consortium of UK industry and the EPSRC standard grants (GR/T19889/01 and GR/L95977/01) and PhD studentships for research which were initiated in 1997 at the University of Leeds. AJ also acknowledges the support from the European Union's Marie Curie Fellowship grant number 331385 for Dr. Sanchez-Segado.

References

1. K. Freese et al., Review of chromite ore processing residue (COPR): past practices, environmental impact and potential remediation methods. Curr. Environ. Eng. **1**(2), 82–90 (2014)
2. S. Sanchez-Segado et al., Reclamation of reactive metal oxides from complex minerals using alkali roasting and leaching—an improved approach to process engineering. Green Chem. **17** (4), 2059–2080 (2015)
3. S. Parirenyatwa et al., Comparative study of alkali roasting and leaching of chromite ores and titaniferous minerals. Hydrometallurgy, 2015
4. V.D. Tathavadkar, A. Jha, M. Antony, The effect of salt-phase composition on the rate of soda-ash roasting of chromite ores. Metall. Mater. Trans. B **34**(5), 555–563 (2003)
5. M. Antony, A. Jha, V. Tathavadkar, Alkali roasting of Indian chromite ores: thermodynamic and kinetic considerations. Min. Process. Extract. Metall. **115**(2), 71–79 (2006)
6. J.F. Papp, *Chromium life cycle study* (US Department of the Interior, Bureau of Mines, 1994)
7. Y. Ding, N. Warner, Catalytic reduction of carbon-chromite composite pellets by lime. Thermochim. Acta **292**(1), 85–94 (1997)
8. C. Bale, A. Pelton, W. Thompson, *FactSage 6.4, Factsage thermochemical software and databases*

Part IV
Pyrometallurgy I

Market Dynamics, Recycling and Recovery of Magnesium from Aluminum Alloy Scrap

A.J. Gesing and S.K. Das

Abstract This paper provides an overview of North American markets for magnesium and aluminum alloys along with life cycle material flow diagrams. Additionally, we review the production processes of magnesium and its alloys from various magnesium-bearing raw materials, including scrap. The main focus of the paper is on the collection, recycling and processing of post- and pre-consumer metal scrap containing magnesium. We discuss the latest status and economic analysis of RE-12™, a scrap-based process for recovering magnesium from aluminum alloys and its potential application to can sheet, automotive (body and closure sheets) and aluminum foundry markets.

Keywords Magnesium recycling · Aluminum recycling · Magnesium electrorefining · Market economics · RE-12™ electrorefining process

Introduction

Aluminum and Magnesium Production and the Scrap Industry

Primary aluminum is produced exclusively from bauxite ore through refining to alumina (Bayer process) and electrowinning to primary Al metal melt (Hall-Héroult process). The energy cost of the electrowinning step is 13–15 kWh/kg, and a total mine-to-metal melt energy cost of nearly 56 kWh/kg. Primary magnesium metal can be produced by several methods. The two leading methods are silicothermic reduction of calcined magnesite (the Pidgeon process), and electrowinning of $MgCl_2$ that can be produced from seawater, lake salt deposits, or by carbo-chlorination of MgO-containing oxide ores. The Pidgeon process, which is practiced mostly in China using highly environmentally and energy inefficient

A.J. Gesing · S.K. Das (✉)
Phinix, LLC, P.O. Box 11668, Lexington, KY 40577, USA
e-mail: skdas@phinix.net

S. Wang et al. (eds.), *Applications of Process Engineering Principles in Materials Processing, Energy and Environmental Technologies*,
The Minerals, Metals & Materials Series, DOI 10.1007/978-3-319-51091-0_17

technologies, currently dominates the prime Mg market producing Mg at the lowest dollar cost with a total mine-to-metal melt energy cost of over 100 kWh/kg.

Aluminum metal recycling is an integral part of the global material recovery and recycling system that supplies the material needs of the industry and society from the growing urban mine at a small fraction of the capital and energy cost that it takes to mine and process raw materials. Recycled Al alloys contain Mg as one of the alloying elements. Hence Al recycling also manages to recycle a large proportion of Mg metal units contained in these Al alloys. This recycling system is critically important to achieving a sustainable society and economy. A magnesium-based alloy recycling system is still in its infancy with only new Mg die-casting scrap being thoroughly collected and closed-loop recycled. The small proportion of magnesium-alloy scrap in machine shops makes it inconvenient to source-segregate Mg cut-offs and machining chips by alloy. The small proportion of magnesium alloys in consumer products makes it more profitable to leave it in the Al scrap then to separate it as a Mg-scrap product. In cases where old scrap Mg alloy mix is separated from light metal scrap, it is typically consumed as a desulfurizing agent for steel production.

New scrap is produced during material production and product fabrication, while old scrap is a post-consumer material. While most new scrap can be segregated at source by alloy for recycling to a composition-compatible alloy, it is often not practical to separate clippings, machining chips and turnings in large machine shops. As a result, these items constitute a major mixed-new-scrap category that is of low value and is downgraded to low-grade Al foundry alloys, where it is used after chlorinating the magnesium content out of the aluminum alloy scrap melt. The magnesium content of aluminum scrap destined for Al foundry alloys ends up as Mg-chloride contaminant of chlorination dross to be disposed of as salt-contaminated waste.

Old aluminum scrap is gathered through well-established collection systems for end-of-life vehicles, building demolition waste, food packaging, electrical and electronic waste and other metals collected by scrap yards. In any collection system, Al is not collected separately from other materials. Aluminum cans and rigid foil are recovered from food packaging by sorting in municipal recycling facilities (MRFs). At scrap yards there is a certain amount of manual sorting and grouping by International Scrap Recycling Institute (ISRI) categories, but these separate scrap by parent metal and generic source and shape, but not by alloy.

Old Al Sheet, a mixed-alloy category that groups old wrought Al alloys, is among the largest of the Al categories. The remaining scrap (from vehicles, machinery and building demolition residue) is shredded and then mechanically separated into ferromagnetic metals, non-magnetic metals and non-metals. Aluminum and magnesium contents are recovered together from non-magnetic metals by density separation and eddy current rotor cleanup and are almost exclusively used in secondary aluminum foundry alloys.

Pre-sorting and pre-cleaning is an integral part of the global material recycling system. With new scrap the best practice is to maintain and track alloy segregation at source. With old scrap, currently, density-based physical separation methods produce a Mg- and Al-containing light alloy mixture as a single shredded fraction sold to secondary Al smelters. In the USA over 35,000 tons annually of Mg ends up as a contaminant in the Al foundry alloy melt to be chlorinated out to a specification of <0.1% Mg. This amounts to ~50% of primary Mg production in the USA. There is, therefore, a need for Mg recovery. A practical way of recovering Mg from Al melt is to electrorefine. A Mg electrorefiner uses Mg-contaminated Al alloy melt as an anode electrolytically oxidizing the Mg and transporting it as a Mg halide through a halide electrolyte to deposit Mg in a cathode metal melt. The resulting Mg product is pure enough to be used as a Mg alloying element in batching prime Al alloys.

Commercial Significance—Mg Recovery from Al Scrap Is Important Now

The processing of scrap has gained importance during the last 15 years.

With escalating energy costs for these very energy-intensive light metals, recycling saves energy costs as it requires only <5% of energy compared to Al or Mg production from ore.

As the primary aluminum and magnesium industry has largely left the USA for the cheaper electrical energy and environmental cost regions of the world (such as the Middle East in case of aluminum, and China for both aluminum and magnesium), US fabrication plants are paying a much higher price for imported primary aluminum and magnesium. These imports also lead to large trade deficits and may even have national security implications.

Public perception has moved towards sustainability and using recycle-friendly products and against landfills.

At the same time the demand for structural corrosion-resistant prime alloys that use Mg as an alloying element is increasing in the transportation marketplace for the construction of cars, trucks, busses, trains, boats, ships and aircraft. In the past, the supply of Mg in light-metal mixed-alloy scrap was insufficient to justify process development and operating economics. Going forward, Mg supply from scrap is steadily increasing, providing economic justification for Mg recovery and recycling.

Magnesium Recovery Potential

There is almost three million tons of Al scrap recycled annually in North America, which can produce about 45,000 tons of Mg content in primary-quality recycled Al–Mg and Mg–Al products.

The amount of 45 kt of recovered Mg is transformational when compared to the total US primary Mg production capacity of only 60 kt. It is estimated that 28 kt of

Mg, contained in the 2140 kt of domestically generated Al scrap, leaves US shores —only to return as more value-added products.

The lower cost of magnesium recovery from scrap will reduce imports of Mg and Al to the USA. Commercial implementation of a magnesium recovery process will help reduce the export of Al and Mg scrap and develop an aluminum-scrap-based Al and Mg industry in the USA. Subsequently, successful commercialization of a magnesium recovery process will also lower costs, energy and emissions of Al–Mg alloy production, thereby expanding the use of both Al and Mg in their major usage sectors of aerospace and automotive industries.

Magnesium and Aluminum Scrap North American Materials Flow

A detail discussion of aluminum and magnesium material flow including scrap is described in a recent TMS presentation [1]. Figure 1 illustrating Mg unit flows the importance of metal scrap to the Mg production system. Metal scrap supplies ~100 kt of the Mg content. Most of the Mg units are in the form of Al scrap. The opportunity for a new RE-12™ Mg electrorefining process (described below) is the 45 kt of Mg chlorinated out of Al foundry melts and Mg in consumable products that are currently lost to the system.

Figure 2, which illustrates Al unit flows in North America, underlines that even in the well-established Al recycling system, North America still exports and

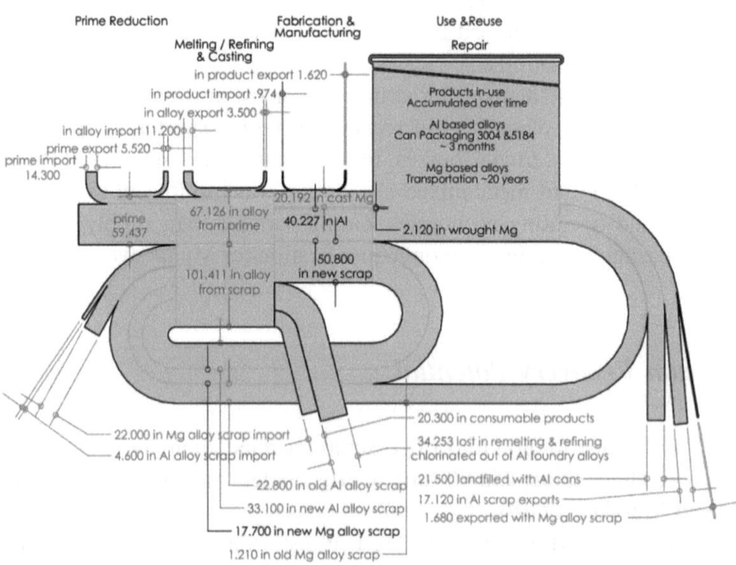

Fig. 1 North American Mg unit flow diagram based on 2012 USGS and USEPA data [1]

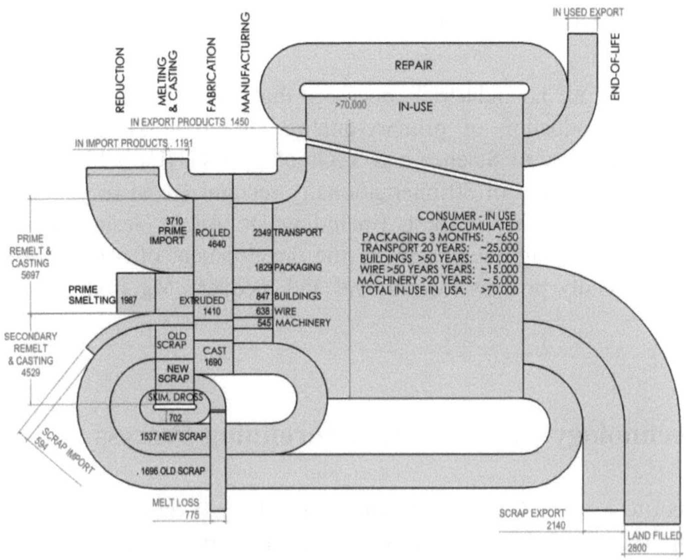

Fig. 2 North American Al unit flow diagram based on 2012 USGS and USEPA data [1]

landfills more Al scrap (nearly 5000 kt) than it imports as Al prime (3700 kt). An amount of 1537 kt of new Al alloy scrap and 1696 kt of old Al scrap provide potential raw material for the RE-12™ electrorefining process feed.

Current State-of-the-Art and Available Technology

Current aluminum refining uses chlorination with a gas mixture or reactive salts and transfers reactive alloying elements (Mg, Li, Na, Ca, Sr, and Rare Earths (RE)) from secondary Al foundry alloy into dross for disposal typically as salt-contaminated wastes. This practice is both costly and hazardous as it takes extensive time, leads to melt losses of both Mg and Al, and uses highly toxic chlorine gas. Chloride byproducts in the dross again go into a difficult-to-dispose salt-contaminated waste. At present there are no industrial processes that recover Mg and other reactive elements from Al or Mg alloys for production of primary-quality alloys.

Hoopes [2, 3] invented the electrolytic three-layer refining process for Al in 1925. Bowman [4] and Dewing [5] proposed such a three-layer refining process for recovery of Li from Al-Li melt in 1989. Other literature, by Das, Gesing and Fray [6–9] suggests improvements to the molten salt electrolyte Al refining cells.

Need for a New Technology

At present there are no industrial processes that recover Mg from Al–Mg alloy scrap for the production of primary-quality alloys. In 2013 SINTEF at the Norwegian University of Science and Technology (NTNU) held an International Al Recycling Workshop of 30 internationally recognized Al recycling experts in Trondheim, Norway. These experts (including Dr. Adam Gesing, a co-author of this paper) assigned highest priority to the development of a non-chlorine environmentally friendly process for removal and recovery Mg from Mg-bearing Al scrap [10].

New Technology—Re-12™ Electrorefining Process

The project to recover Mg from Al alloy melted scrap for recycling by Mg electrorefining was funded by the United States Department of Energy Advanced Research Project Agency (ARPA-E) [11].

This RE-12™ technology, which builds on the electrochemistry of the industrially established Hoopes process for the production of five-nine (99.999%) pure Al, has been designed to transfer >90% of the Mg from secondary Al melt to the refined product. Electrorefining selectively transfers Mg to the Mg–Al alloy floating top layer. The cathodic product would be alloyed to the desired Mg–Al or Al–Mg alloy composition by diluting it with primary-grade Al and adding other specified alloying elements. The anode product is the starting Al alloy with Mg content reduced to the specificaction level.

Calculations of metal-electrolyte melt phase diagrams, predicted product compositions, electrolyte chemical stability and physical properties (such as density, viscosity, electrical conductivity, cell construction materials, vapor pressure and superheat) were made using commerciality available thermodynamic modelling FactSage software and databases [12] leading to electrolyte selection for the RE-12™ process.

This project experimentally demonstrated that the Mg recovery electrolyzer can produce refined product with 99.9% of Mg plus Al. There is no commercial need to produce Al-free Mg product since the target market for Mg is alloying of Al. Even for use in Mg-based alloys, Al content of the refined product is not a problem since most commercial Mg-based alloys contain >4% Al. A schematic flow sheet of the process along with a depicted commercial set up are described in Figs. 3 and 4.

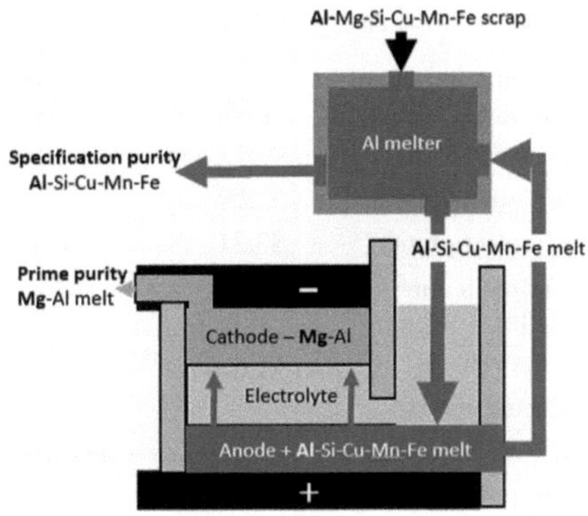

Fig. 3 Process schematic of Mg electrorefiner linked to an Al scrap melting furnace

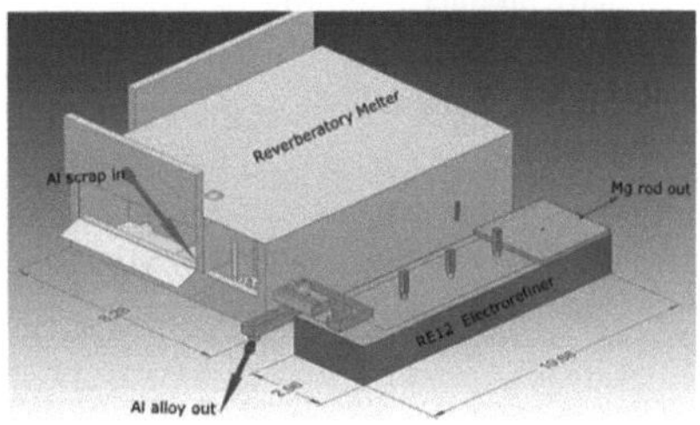

Fig. 4 RE-12™ process commercial concept; integration with reverberatory melter

Experimental and Analytical Results

Three recently published papers [13–15] describe the process details, laboratory-scale experimental results and product marketability, and economic analysis of the RE-12™ process. These results and analyses combine to give an experimental proof-of-concept. This RE-12™ process is a patent-pending [16] and trademark-pending [17] route for a Mg recovery for recycling from Al alloy melts.

The theoretical minimum energy requirements along with production costs, emissions and processing energy needed for the current magnesium producing (from Pidgeon and electrolysis) technologies have been extracted from the ARPA-e FOA document [18]. Comparable numbers for the RE-12™ process have been experimentally demonstrated in laboratory electrorefining cells as shown in Table 1.

Markets for RE-12™ Process Implementation

RE-12™ electrorefiner is typically integrated with Al remelting furnace in a secondary smelter remelting Al scrap. There are three major market opportunities.

Table 1 Cost, energy and emission analyses—current Mg versus RE-12™ process [13–15]

Current and target Mg metal reduction processes	Domestic production cost ($/kg Mg)	Emissions (kg CO_2/ kg Mg)	Processing energy ore to metal (kWh/kg Mg)	Theoretical minimum energy (kWh/kg Mg)
Mg-silicothermic	$2.50	37	102	5.8
Mg-electrolytic	$3.31	6.9	43.6	5.8
Mg reduction—ARPA-e target	≤$2.00	≤10	≤27	
RE-12™ electrorefining	$1.6–$1.7	2	<4	~0
RE-12™ Mg electrorefining process laboratory experimental results				Units
Mg product purity	Al		<1	wt%
	Fe + Si + Mn + Cu + Zn		<0.05	wt%
Al alloy product purity	Mg		<0.15	wt%
Electrolytic current efficiency			~96	%
Electrorefining step energy consumption			2.5	kWh/kg of Mg
	With 1% Mg electrorefined		~0.025	kWh/kg of Al alloy

Production of AA 6111 Automotive Closure Sheet from New Stamping Plant Scrap

Automotive sheet is currently the fastest growing market for Al alloys. Existing stamping plants tend to have a single scrap chopping and handling line. While the steel scrap is separated magnetically, new Al stamping scrap often is a comingled mixture of 6xxx and 5xxx alloys. Piece-by-piece Al alloy sorting has been considered but not used. The RE-12™ process provides an alternative where the high-Mg AA 5754 and lower-Mg AA 6111 mixed new stamping scrap can be melted together and the Mg content of the mix can be electrorefined down to the AA 6111 alloy target concentration.

Low Mg–Al Foundry Alloys 356, 380, 319

There are over two million tonnes of secondary foundry alloys produced in the USA. The typical mixed-alloy old scrap mix feeding this market has ~1.1 wt % Mg content. The Mg is chlorinated out producing hygroscopic Mg salt contamination of Al dross and leading to the loss of Mg units. The added value of the RE-12™ process comes from avoidance of chlorination costs, the value of the recovered Mg product, and reduction of Al losses.

Table 2 Magnesium removal potentials from selected North American market sectors

Market sector	Alloys	Expected Al alloy production volume	Magnesium removal potential			
		t/y	Start (wt%)	End (wt%)	Recovered Mg (%)	t/y
Automotive-sheet	6111/5754	750,000	2.25	0.75	1.5	11,250
Automotive-cast	319/356/380	2,000,000	1.1	0.1	1.0	20,000
Beverage cans	3104/5182	1,500,000	1.5	1.0	0.5	7500

Production of Can Body Sheet from Used Beverage Can (UBC) Scrap

There is over 1.5 million tons of AA3104 alloy can body sheet produced in North America, much of it supplied by UBC and new can sheet production scrap. Can alloys consist of ~1.1 wt% Mg AA3104 alloy body and 5.1 wt% Mg AA5184 alloy lid. The Mg concentration is reduced by deliberate preferential Mg oxidation melt, and by dilution with prime Al. Consequently, the RE-12™ electrorefiningprocess, which enables batching of the AA3104 alloy can body melt for 100% UBC feed, has the potential for substantial reduction of scrap feed costs, and for allowing melt practices that prevent excessive melt oxidation. These combine for the potential to actually reduce can body alloy production cost, and provide recovered Mg at the same unit price. Some of these numbers are summarized in Table 2.

Results and Conclusions

Based on extensive experimental and techno-economical modelling using commercially available data, inputs and experience, we have established the technical, economic and environmental viability of this market needed, driven and disruptive RE-12™ Mg electrorefining process for recovery of Mg from melted Al scrap for recycling.

Three large market segments in Al recycling system were identified as promising areas for profitable implementation of RE-12™ process. These total to ~4.25 million annual tons of potential Al alloy feed to the RE-12™ process and 45,000 annual tons of potential Mg recovery these three markets alone in North America.

Path Forward

Efforts are underway to scale up the RE-12™ process leading to commercialization and broader market adaptations to recover magnesium from aluminum-magnesium-based scrap.

Acknowledgements The authors gratefully acknowledge the financial and the technical assistance provided by the United States Department of Energy Advanced Research Project Agency (US DOE ARPA-e, Contract Number DE-AR0000413). James Klausner, Bahman Abbasi, Thomas Bucher and Daniel Matuszak were very helpful. Ray Peterson of Real Alloy provided industrial inputs regarding commercialization. Raouf Loutfy at MER Corporation and Mark Gesing of GCI. contributed significantly. Real Alloy and Alcoa provided scrap samples.

References

1. S.K. Das, A.J. Gesing, Presentation at energy technologies and carbon dioxide management conference, TMS 2014
2. W. Hoopes, Electrolitically refined Al and articles made therefrom. US Patent No. 1,534,315, 1925
3. W. Hoopes, Electrolitic refining of Al. US Patent No. 1,534,318, 1925
4. K.A. Bowman, Electrolytic process for recovering Li from Al-Li alloy scrap. US Patent No. 4,849,072, 1992
5. E.W. Dewing, Refining of Li containing scrap. US Patent No. 4,790,917, 1988S.
6. S.K. Das et al., Aluminum purification. US Patent No. 4,115,215, 1978
7. E.W. Dewing, A.J. Gesing, Electrolytic refining of molten metal. US Patent No. 4,405,415, 1983
8. D.J. Fray et al., Packed bed electrorefining and electrolysis. US Patent No. 4,118,292, 1978
9. D.J. Fray et al., Electrode for electrorefining. US Patent No. 4,904,356, 1990
10. Sintef, International Aluminium Recycling Workshop, Trondheim 9–11 June 2013
11. Advanced Research Project Agency (ARPA-e) of the United States Department of Energy
12. FactSage—http://www.crct.polymtl.ca/fact/
13. A.J. Gesing, S.K. Das, R.O. Loutfy, JOM **68**(2), 585–593 (2016)
14. A.J. Gesing, S.K. Das, R.O. Loutfy, REWAS, TMS 2016, pp. 253–254
15. A.J. Gesing, S.K. Das, Met. Trans. B (2016)
16. A.J. Gesing, S.K. Das, M.A. Gesing, US Patent Application US 2015/0225864 A1, 2015 (Phinix, LLC)
17. Phinix, LLC—Trade Mark: RE-12 TM Recycled Magnesium, 2014
18. Energy—Funding Opportunity No. DE-FOA- 0000882, 20 Mar 2013
19. R.A. Sharma, Fused fluoride electrolytes for MgO electrolysis in the production of Mg metal. US Patent No. 5,427,657, 1995
20. P.R. Kruesi, Methods of recovering and purifying secondary Al. US Patent No. 8,002,872, 2011

Alternative Ways of Using Nonferrous Slags as Feed Material in the Ferrous Production Industry

Mario Sánchez, Fernando Parada and José Palacios

Abstract As an alternative method of reducing the large amount of slag generated during metallurgical processes in the primary copper production, it is to use the remaining iron in the slag as feed material for the iron and steel industry. The slag is generated during copper concentrates smelting in ratios ranging from 2 to 5 ton per ton of produced copper, and it contains 42–45% of iron. However, copper, sulfur and other detrimental elements affecting iron properties need to be reduced prior to use the slag in steel production. Therefore, some laboratory experimental run has been conducted to reduce copper by slag oxidation followed by leaching and recovering the dissolved copper by SX-EW, and smelting the solid with coke to produce a metallic phase containing 98% iron and 0.9% molybdenum. In further reducing steps, copper content is lower down and a molybdenum rich iron phase is obtained. Recovery of iron from copper slags considering reduction and control of harmful elements for iron properties was studied.

Keywords Copper slag · Iron recovery

M. Sánchez (✉)
Universidad Andrés Bello, Autopista Concepción - Talcahuano,
7100 Concepción, Chile
e-mail: mario.sanchez@unab.cl

F. Parada
Universidad de Concepción, Barrio Universitario, Concepción, Chile
e-mail: fparada@udec.cl

J. Palacios
Universidad de Playa Ancha, Av. González de Hontaneda 855,
Valparaíso, Chile
e-mail: jose.palacios@upla.cl

S. Wang et al. (eds.), *Applications of Process Engineering Principles in Materials Processing, Energy and Environmental Technologies*,
The Minerals, Metals & Materials Series, DOI 10.1007/978-3-319-51091-0_18

Introduction

Slags produced during the pyrometallurgical treatment of copper concentrates considered as wastes are disposed close to smelter plants, and the amount ranges from 2.0 to 5.5 ton of slag per ton of copper produced, and they are usually subject to a cleaning operation [1] commonly involving holding the molten slag at temperature in a furnace under mildly reducing conditions allowing time for reduction of dissolved copper oxide and some magnetite and for coalescence and settling of metal and sulfide droplets.

However, they contain important concentrations of iron and other metals to be recovered, given their prizes in the international markets. The main components in the copper slags are: Fe (32–52%) and SiO_2 (17–37%) and, also contain amounts of Cu (0.5–3.7%), Co and/or Ni, Mo and some amount of As and heavy metals depending on the origins of the slags [2]. A typical copper smelter slag is chemical and mineralogical characterized in Table 1.

Smelting furnaces producing high matte grades, a high copper content in the slag is also obtained. Then, they are cleaned in an electric furnace using coke as a reductant [3, 4]. Therefore, Iron is a major element in the final slag and can be recovered as metal in an additional reducing step.

Thermodynamics Considerations

In Fig. 1, the stability of different components is shown as function of the oxygen potential and temperature. Also, the copper converting and slag cleaning processes are represented in the stability area, and the oxidation roasting condition and the reduction condition are also shown according to the oxygen potential and temperature conditions required for both processes.

In Fig. 1, it can be observed that the copper oxide content in the fayalite slag could be reduced with a lower energy requirement than the iron oxides therefore the obtained iron is going to be saturated with copper and carbon. Also, some silicon can be present in the metal phase in a small amount because the process takes place very close to the equilibrium condition.

Table 1 The slag chemical and mineralogical characterization

Compound	Cu_2O	Cu_2S	FeS	FeO	Fe_3O_4	SiO_2	Al_2O_3	CaO	Other
Wt%	0.76	2.00	1.17	40.61	12.38	33.0	2.12	0.69	7.27
Element	Cu	Fe	S	Si	Al	Ca	Mo		As mg/kg
Wt%	2.27	41.3	0.83	15.4	1.60	0.49	0.3		74

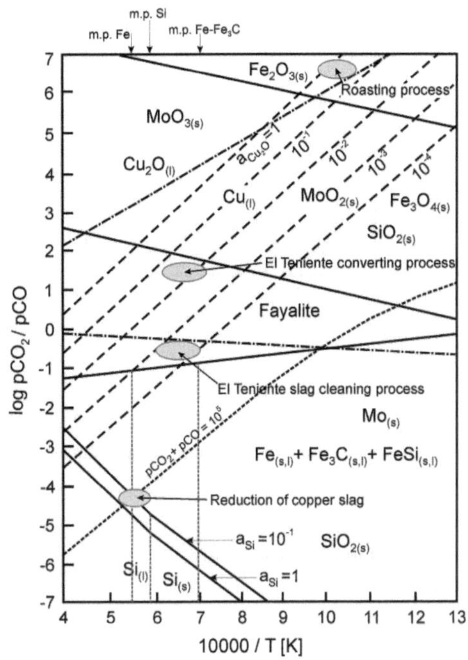

Fig. 1 Stability diagram for the Fe–Si–O, Cu–O and Mo–O systems [5]

Iron Recovery

The metals oxides presented in copper slags such as iron copper and molybdenum can be reduced by solid carbon in solid or liquid states in a wide temperature range as shown in Fig. 1. However, slag making oxides, such as silica, alumina and lime, are stable to reduction with carbon up to 1650 °C. Therefore, at temperature lower than 1650 °C, the direct reduction and separation of useful metals from oxides into separate liquid phase and Fe–Cu–C alloy is expected.

The behavior of the Fe–Cu–C system have been reported [6], and it is shown as function of copper content at 1723 K (a) and copper content at given temperature (b) in Fig. 2. In Fig. 2a, it can be observed the miscibility gap between copper rich (L_2) and iron rich (L_1) melts, and a very low solubility of carbon in copper melt while the melt rich in iron dissolves copper and carbon. The effect of temperature on the copper solubility of carbon saturated iron melts is shown in Fig. 2b, and it can be observed that copper concentration increase with temperature up to about 9% at 1823 K.

Therefore, under carbon saturation condition at a given temperature, a Fe–C–Cu alloy is produced as result of slag reduction process.

The kinetic of the reduction of slag process was studied by using a graphite rod to keep the carbon saturation condition at 1450 °C and a CaO/SiO_2 ratio of 60/40 in the $CaO–SiO_2–FeO$ system. The results are summarized in Fig. 3 where the concentration of copper and iron in function of time are described.

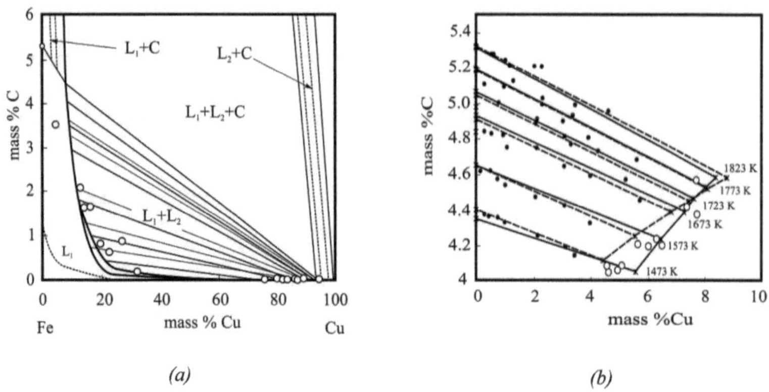

Fig. 2 Solubility of carbon in the Fe–Cu–C system as function of: (**a**) copper content at 1723 K and (**b**) copper content at several temperatures [6]

Fig. 3 Concentration of copper and iron as function of time

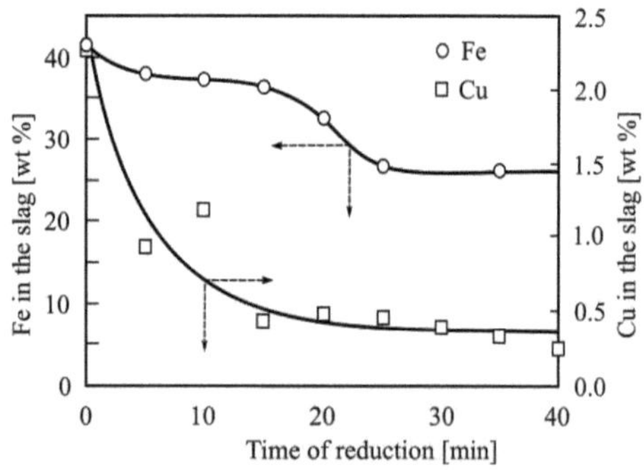

In this figure, it can be seen the rapid decreasing of copper content in the slag during the first 15 min while the iron concentration remain almost constant over same period of time.

Therefore, a selective reduction according to the thermodynamic stability of both compounds is possible, and during the first 15 min copper content decreases from 2.3 to 0.5% and to 0.3% in the final slag. The iron oxide reduction start after 15 min, when most of the copper oxide has been reduced, and it reaches the lowest value of 27% after 40 min of reduction time under these conditions. Therefore, the experimental time was stablished in 50 min.

The result of the slag reduction process is shown in Table 2, and in it can be observed that copper and iron content decreases in final slag and the metallic phase brake into an iron rich and a copper rich melts. Additionally, molybdenum contents in the slag decrease and concentrate in the iron rich melt.

The copper content in the final slag decreases to a minimum value of 0.04% and the iron decreases from 43.4 to 1.13% suggesting that is possible recover most of the copper content in the slag in the metal phase. However, the copper content in

Table 2 Compositions of slag and metallic melt at different steps of the process

Component (wt%)	Cu	Mo	Fe	S	Fe_3O_4	SiO_2	CaO	Al_2O_3
Slag before reduction	1.1	0.30	43.4	0.5	5.7	30.0	1.1	3.4
Modified slag	0.9	0.23	32.6	0.4	4.3	22.7	27.2	6.8
Slag after reduction	0.036	15 ppm	1.13	–	–	40.0	48.0	12.0
Metallic phases								
Iron alloy	4.79	0.89	91.33	–	–	–	–	–
Copper alloy	94.52	0.0	4.23	–	–	–	–	–

the iron alloy is high for steel application, and it is necessary to reduce the copper concentration in the iron alloy.

The final slag and the metallic alloy obtained in the experiment run at 1450 °C were analyzed by using a SEM with EDS, and the results are shown in Fig. 4. In this figure, several parts of the metal and the slag phases are analyzed, and they are indicated in the crucible placed at the bottom of each picture by a white circle.

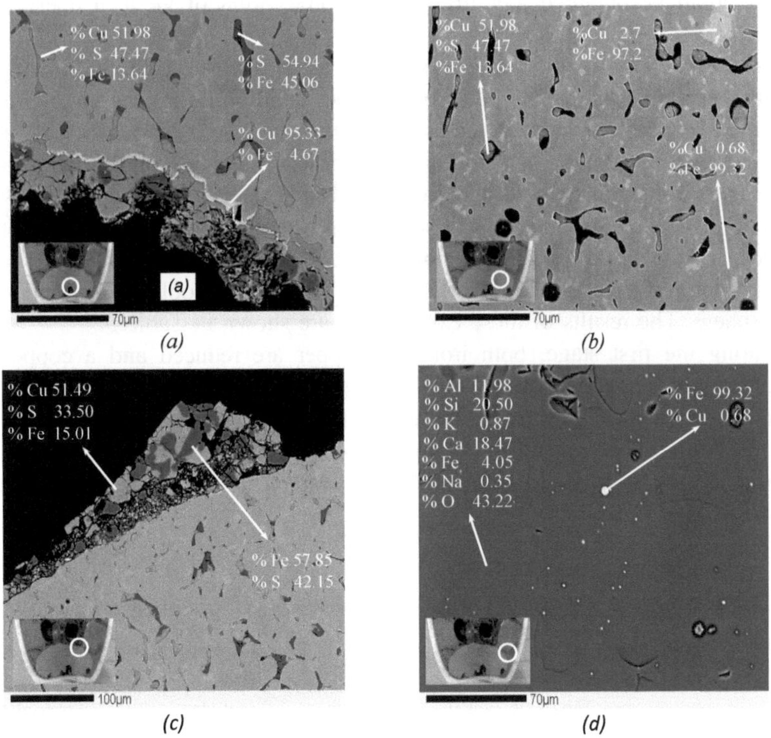

Fig. 4 Microstructures of metal and final slag phases: (**a**) lower part of the metallic alloy; (**b**) central part of the metallic alloy; (**c**) upper part of the metallic alloy in contact with the slag phase and (**d**) final slag phase

In Fig. 4a, a copper-iron alloy in the lower part of the metallic phase containing 95% of copper is observed, and there are also particles of matte and iron sulfide distributed around the copper alloy. In Fig. 4b, the central part of the iron alloy containing over 97% of iron is shown, and some particles of matte entrained in the metal phase are also present. The upper part of the metallic phase is shown in Fig. 4c, and it can be seen a major phase containing mainly iron with minor phases including iron and copper sulfides. The final slag phase contains mainly Si, Ca, Al, O and minor elements commonly present in copper slags as shown in Fig. 4d. Also, entrained particles of iron-copper alloy are seen in this figure.

Copper Removal

Slag roasting under oxidative condition followed by an acid leaching is an alternative to remove copper by leaching the copper oxide produced. This is an efficient way to recover copper and iron could be recovered by reduction with carbon from the leach residue forming an alloy with molybdenum. A good recovery of the three metals could be obtained, but from an economic point of view, it doesn't look attractive.

However, copper could be as a separated phase due to the limited mutual solubility between iron and copper under these conditions, as shown in Fig. 2a. Therefore, copper can be removed by conducting the slag reduction process in two stages; a decopperization step followed for an iron reduction under the following conditions: 100% of the coke required to reducing copper oxide and a CaO/SiO_2 ratio of 60/40 were used in the first step which was run during 30 min at 1450 °C; the second step was run during 50 min using coke addition of 100 and 150% of the carbon required to reduce iron oxide, and the sample was cooled down under a N_2 atmosphere. The results of these experiments are shown in Table 3.

During the first stage, both iron and copper are reduced and a copper rich metallic alloy is obtained, and on the second stage, a rich iron alloy is obtained. The copper content decreases to 0.24% in the final slag, and to 0.84% in the final metallic alloy. The iron alloy obtained by using 150% of carbon addition was micro-analyzed in order to observe the copper-iron interaction, and the result is shown in Fig. 5.

Table 3 Composition of the phases obtained during copper and iron selective reduction

Carbon addition wt%	1st. step: selective copper reduction					2nd. step: reduction to metallic iron				
	First slag rich in iron			Metallic phase		Final slag			Iron alloy	
	%Cu	%Fe	%S	%Cu	%Fe	%Cu	%Fe	%S	%Cu	%Fe
100	0.32	24.6	0.5	51.0	48.2	0.02	3.8	0.26	1.36	98.0
150	0.20	30.1	0.45	12.2	87.0	0.24	4.2	0.25	0.84	98.9

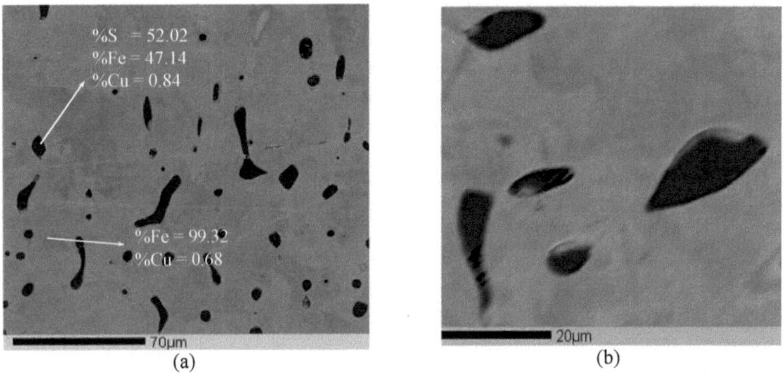

Fig. 5 Micrograph of the iron alloy contained 98.9% of iron: (**a**) 500× and (**b**) 1600×

In this figure, it can be seen a metallic phase containing 99.32% of iron and 0.68% of copper, which is in agreed with the chemical analysis shown in Table 3, and inclusions of iron sulfides containing 0.84% of copper.

Conclusions

Slag processed by direct reduction with carbon produces a carbon saturated iron alloy which contains copper and molybdenum, and a copper phase.

The copper content in copper smelting slag is selectively reduced with carbon at 1450 °C and its content decreases rapidly in the slag to values in order to 0.5% while iron content remains almost constant.

The coke addition is effective for decreasing the copper content in the slag reaching a value of 0.06% of copper in the final slag.

To decrease copper content in iron-copper alloy, more than one step is required. The first step to reduce copper content in the slag and produce a copper rich melt and a second step for reducing the iron oxide to a metallic iron rich alloy containing 0.84% of copper, which is one of the lowest obtained values.

References

1. B. Gorai, R.K. Jana, Premchand, Characteristics and utilization of copper slag—a review. Resour. Conserv. Recycl., 1–15 (2003)
2. Huiting Shen, E. Forssberg, Waste Manag. **23**, 933–949 (2003)
3. D. Busolic, F. Parada, R. Parra, E. Urday, J. Palacios, M. Hino, F. Cox, A. Sanchez, M. Sanchez, Recovery of Iron from copper flash smelting slags, in *VIII International Conference Molten2009*, eds. By M. Sanchez, R. Parra, G. Riveros, C. Díaz (Santiago, Chile, 2009)

4. T.J. Kim, Flash smelting in Korea, in *Copper Smelting—An Update*, eds. D.B. George, J.C. Taylor (AIME, New York, 1981), pp. 33–39
5. M. Sanchez, F. Parada, R. Parra, F. Marquez, R. Jara, J.C. Carrasco, J. Palacios, Management of copper pyrometallurgical slags: giving additional value to copper mining industry. in *VII International Conference on Molten Slags, Fluxes and Salts*. The South African Institute of Mining and Metallurgy (2004) pp. 543–550
6. M. Hino, Lecture on pyrometallurgical process for Chilean slags, in *FONDEF Project D021159*, Universidad de Concepción, August 2004

Insulating or Conductive Lining Designs for Electric Furnace Smelting?

J.D. Steenkamp, G.M. Denton and D.A. Hayman

Abstract Design of furnace containment systems can be based on one of two design philosophies: Insulating or conductive lining designs. In insulating lining designs, management of the compatibility between refractory materials and process materials (typically liquid metal and/or slag) is an important criterion for furnace design, and an important process parameter to control during furnace operation. In conductive lining designs, management of the integrity of the layer of frozen material (freeze lining) on the interface between the liquid process material, and the refractory lining, is important. For example in ferrochromium production, both lining design philosophies are applied. The work presented here discusses the significance of lining design philosophy in the context of ferrochromium production.

Keywords Furnace containment · Lining design · Ferrochromium · FeCr

Introduction

When designing a new industrial-scale furnace for the production of ferrochromium (FeCr), furnace operators have a choice between two operating philosophies: Submerged arc furnace (SAF) or open-arc furnace operations. The former typically applies an electrical system based on alternating current (AC) and the latter, direct current (DC). Choice in operating philosophy is primarily driven by the type of ore available [1]: SAF operation is dependent on lumpy raw materials (>6 mm) whilst DC open arc furnace operation was specifically developed for the reduction of fines (<6 mm) [2]. Another advantage of DC open arc furnace operation over conventional AC-SAF is the reduced dependence of process temperature on the electrical resistivity of the process material [3, 4]. Slag chemistry and process temperature can therefore be controlled independently to optimize activity of slag components

J.D. Steenkamp (✉) · G.M. Denton · D.A. Hayman
MINTEK, 200 Malibongwe Road, Randburg 2125, South Africa
e-mail: joalets@mintek.co.za

S. Wang et al. (eds.), *Applications of Process Engineering Principles in Materials Processing, Energy and Environmental Technologies*,
The Minerals, Metals & Materials Series, DOI 10.1007/978-3-319-51091-0_19

participating in reduction reactions (improved recovery of Cr, low levels of S and P), manage slag viscosity for tappability, and in the case of an insulating lining design, manage activity of refractory components in the slag to minimize potential for refractory dissolution or chemical reaction between slag and refractory. Other advantages of DC open arc furnace operations over AC-SAF for FeCr production have been discussed extensively by others [2, 3, 5, 6].

Similarly, furnace operators have a choice between two furnace containment philosophies [7, 8]: Insulating or conductive lining designs (defined in the next section in terms of the steady-state heat-transfer principles applicable [9]). During the conceptual design phase the furnace containment philosophy needs to be selected with care as it will have a significant impact on future furnace operability. It is the intention of this chapter to demonstrate the use of desktop tools available to aid in the decision-making process during the conceptual design phase, and to discuss some of the consequences of the philosophy selected on plant design, commissioning, and operations.

Background

In pyrometallurgical processes, the input energy requirement (E_{In}) depends on two factors (see Eq. 1): Process energy requirement ($E_{Process}$) and energy losses from the furnace (E_{Losses}).

$$E_{In} = E_{Process} + E_{Losses} \tag{1}$$

The energy input, supplied in electric arc furnaces as electrical energy, is typically one of the operational cost drivers. The role of the refractory containment system therefore is not only to contain the process but also to minimize energy losses to the environment.

In an insulating lining, heat is transferred from the inside of the furnace to the environment through convective and conductive heat transfer mechanisms. Using an electrical analogy [8]—see Fig. 1a—for a one-dimensional heat transfer problem applied to a circular furnace (Eq. 2), the heat flux (Q, in Wm^{-2}) is a function of the difference in temperature between the process material inside the furnace and the cooling medium acting as the external environment (T_{Liquid} and $T_{Coolant}$ respectively, measured in K), and the thermal resistances of the different components of the containment system as defined by the heat transfer mechanism applicable.

As an example, when heat is transferred from liquid slag to the hot face of the refractory layer through convection, the thermal resistance (R_{Liquid}) is dependent on the convective heat transfer coefficient (h, in Wm^{-2}) of the slag and the radius of the hot face (r_i, in m) of the refractory, as defined in Eq. 3. Another example would be conduction of heat through the refractory: The thermal resistance ($R_{Refractory}$) is dependent on the thermal conductivity (k, in $Wm^{-1}K^{-1}$) of the refractory and radius of the hot face and cold face (r_o, in m) of the refractory, as defined in Eq. 4.

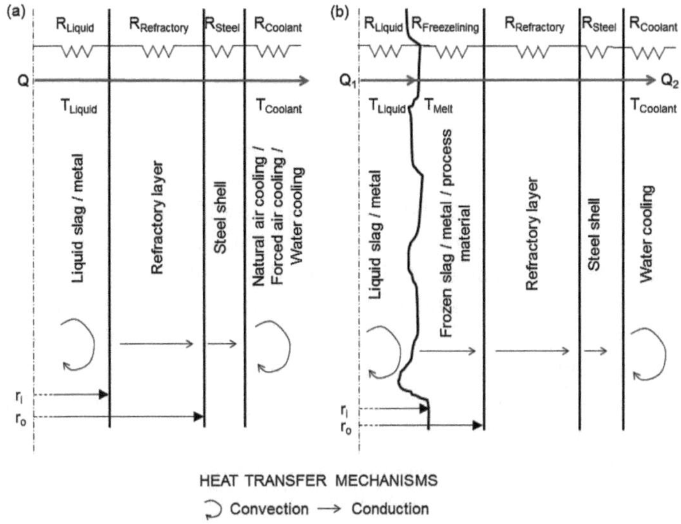

Fig. 1 Steady state heat transfer principles on which **a** insulating and **b** conductive lining designs are based

Descriptions of insulating lining designs, applied in FeCr production, were found for AC-SAF [10, 11] and DC-open arc furnace designs [12].

$$Q = \frac{T_{Liquid} - T_{Coolant}}{R_{Liquid} + R_{Refractory} + R_{Steel} + R_{Coolant}} \tag{2}$$

$$R_{Convection} = \frac{1}{2\pi r_i h} \tag{3}$$

$$R_{Conduction} = \frac{\ln \frac{r_o}{r_i}}{2\pi k} \tag{4}$$

In a conductive lining (Fig. 1b), furnace conditions are manipulated to ensure that a layer of slag, metal, or raw materials, or combinations thereof, are frozen on the hot face of the refractory [8]. A boundary condition is introduced to the heat transfer calculation: The temperature at which the frozen layer of process material starts to melt (T_{Melt}). Increasing the energy input to the process will result in melting the solid process material, reducing the thickness of the frozen layer without increasing the heat flux to the steel shell, therefore $Q_1 = Q_2$ (with Q_1 and Q_2 defined in Eqs. 5 and 6 respectively). Descriptions of conductive lining designs applied in FeCr production (albeit for significantly different refractory configurations), were found for AC-SAF [13, 14] as well as DC open arc furnace designs [6, 15–18].

$$Q_1 = \frac{T_{Liquid} - T_{Melt}}{R_{Liquid}} \tag{5}$$

$$Q_2 = \frac{T_{Melt} - T_{Coolant}}{R_{Freezelining} + R_{Refractory} + R_{Steel} + R_{Coolant}} \tag{6}$$

Although the intention with both design philosophies is to minimize energy losses to the environment, physical containment of the process is the most important consideration. In an insulating lining design, the hot face of the refractory material is exposed to liquid metal and/or slag. Chemical compatibility between the hot face refractory material and the liquid process materials is therefore important, thus adding an additional constraint in terms of slag conditioning. The refractory material should neither dissolve into, nor participate in, chemical reactions with process materials. In a conductive lining design the need for chemical compatibility between the hot face refractory material and the liquid process materials is greatly reduced. The potential for the process material to solidify on and attach mechanically to the surface of the refractory hot face is, however, important. The use of one-dimensional heat transfer, and thermodynamic and thermo-physical property calculations (utilising FactSage 7.0 software) [19] in selecting a suitable refractory containment system is illustrated in the next section.

Tools Available to Conduct Desktop Study

For the purpose of the discussion, a hypothetical case study was developed in which a producer of FeCr required the design of a furnace containment system for a new DC-open arc furnace. The calculations focus on the slag-line of the furnace—typically one of the high wear areas in a DC-open arc furnace [20]. Three slag compositions were evaluated (Table 1) selected from a range of compositions reported by Geldenhuys [1]. The selection was based on diversity in location of the ore body, and variation in bulk chemical composition of the slag.

The assumptions made for the one-dimensional steady-state heat transfer calculations are summarised in Fig. 2. T_{Melt} was estimated from the calculated percentage slag phase formed as a function of temperature, under equilibrium conditions. The Equilib module was applied, and FToxid and FactPS databases selected. All compound species were selected (gas, pure liquids, and pure solids),

Table 1 Three slag compositions (mass percent) evaluated in case study—after [1]

	Cr_2O_3	FeO	MgO	Al_2O_3	SiO_2	CaO	Total
Slag #1	3.8	2.1	31.3	33.7	28.0	1.1	100
Slag #2	4.3	3.4	16.7	37.2	22.9	15.6	100
Slag #3	6.2	1.4	43.8	19.1	28.8	0.7	100

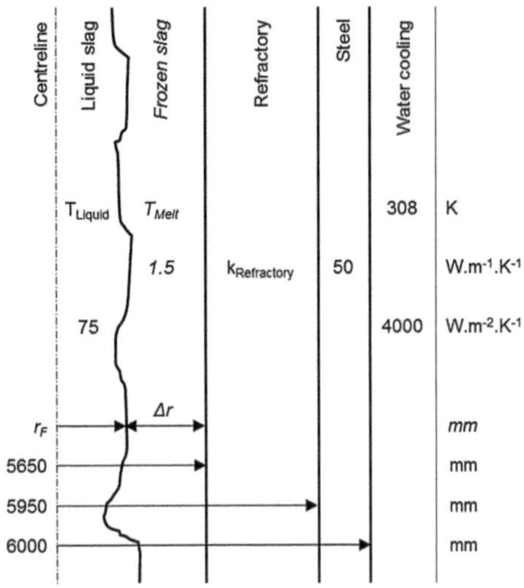

Fig. 2 Assumptions made for one-dimensional, steady-state heat transfer calculations where text in italics applied to conductive lining design calculations only

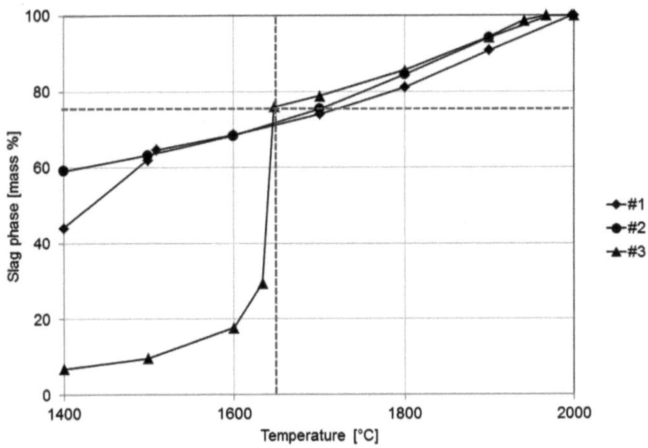

Fig. 3 Calculated equilibrium liquid phase formation in slags with compositions in Table 1 (*solid lines*) where the estimation of T_{Melt} estimation was based on the temperatures at which 80% slag phase formed (*broken lines*)

and duplicates were suppressed, with FToxid having preference over FactPS. All solution phases were selected except for B-olivine. The results are presented in Fig. 3. Wettability tests conducted on synthetic and industrial silicomanganese slag demonstrated that balling temperatures (considered to be the melting point of the slag) typically occurred in the temperature range where 80–95% of the slag phase is calculated to form [21]. Therefore, T_{Melt} was selected as the temperature where <80% slag phase formed: 1923 K (1650 °C). At lower temperatures, note the sensitivity of the slag phase formed for Slag #3, compared to Slag #1 and Slag #2 with changes in temperature.

Jones and Erwee [4] discussed ways in which slag compositions are designed for ferrochromium production, based on slag/metal separability (bulk liquidus) and tappability (viscosity). They stated that for ferrochromium slag to be tappable, the liquid viscosity—η_o—(calculated in the viscosity module of FactSage), should ideally be 0.2 Pa.s (2 poise) or less. Slag with a liquid viscosity of more than 0.45 Pa.s (4.5 poise) was considered not tappable. The Roscoe relationship [22] illustrates the effect of the volume fraction of solids (*f*) on the apparent viscosity (η_{app})—see Eq. 7.

$$\eta_{app} = \eta_o(1 - af)^{-n} \tag{7}$$

As illustrated in Fig. 4a, the apparent viscosity changes exponentially with the increase in the solids fraction. Therefore, assuming that a liquid viscosity of 0.2 Pa.s (2 poise) and solids mass fraction less than 10% are required for a tappable slag (see Fig. 4b, T_{liq} was assumed to be 2173 K (1900 °C). The typical tap temperatures for slag produced in a DC-open arc furnace, smelting South African ore—Slag #2 in Table 1—, is reported to be 1923 K (1650 °C) with bath temperatures of up to 1973 K (1700 °C) [15].

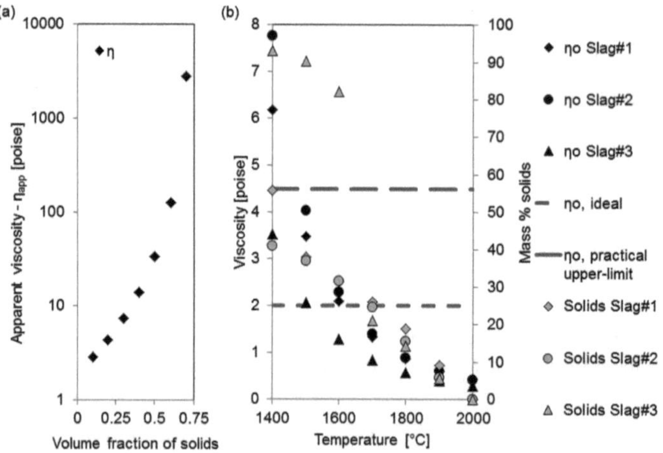

Fig. 4 **a** The effect of the volume fraction of solids on the apparent viscosity (η_{app})—estimated with the Roscoe relationship—as a function of temperature for the case where $\eta_o = 2$, $a = 1.35$, and $n = 2.5$. **b** On primary Y-axis: Calculated viscosity of the liquid portion of the slag (η_o) as a function of temperature, with ideal viscosity (similar to that of maple-syrup at 298 K (25 °C)) and the practical upper-limit for tappable slag [4] superimposed as *dash* and *long lines* respectively. On secondary Y-axis: Calculated mass per cent solids. Slag compositions were stated in Table 1

Table 2 Thermal conductivity ($k_{Refractory}$) of selection of refractory materials evaluated in case study as well as calculated radius of freeze lining (r_F), thickness of freeze lining (Δr), and heat flux (Q)

		$k_{Refractory}$ @ 1273 K (1000 °C) ($W.m^{-1}.K^{-1}$)	r_F (m)	Δr (mm)	Q (kW)
#A	Magnesia	4	5.631	18	663
#B	Sintered corundum	2.5	–	–	505
#C	Fireclay	1.4	–	–	297
#D	Spinel	2.5	–	–	505
#E	Carbon	10	5.565	85	656
#F	Graphite	80	5.527	123	651

Table 3 Composition (mass percent) of selection of refractory materials evaluated in case study—after [21, 23]

		C	FeO	MgO	Al_2O_3	SiO_2	CaO	Total
#A	Magnesia	–	0.4	97.9	0.4	0.3	1.0	100
#B	Sintered corundum	–	0.5	–	95.4	4.0	–	100
#C	Fireclay	–	2.7	–	22.1	75.2	–	100
#D	Spinel	–	0.9	29.9	67.7	1.0	0.5	100
#E	Carbon	90.2	–	–	9.8	–	–	100
#F	Graphite	100.0	–	–	–	–	–	100

For the insulating lining design, refractory material properties of interest are low thermal conductivity (Table 2) and chemical compatibility with the slag (Table 3), while for the conductive lining design high thermal conductivity is desirable. Magnesia-based [12, 15–18], and carbon-based [10, 13, 14] lining designs were found in literature, but Al_2O_3-, and SiO_2-based refractory were also evaluated for their lower thermal conductivity.

The results for the heat transfer calculations are summarised in Table 2: From the results of the one-dimensional steady-state heat transfer calculations (based on the assumptions discussed earlier), a solid layer of process material will only form for a magnesia or carbon-based lining design, not for the Al_2O_3-, and SiO_2-based refractory. Under these conditions, heat losses for the insulating lining designs are significantly (1½ to 2 times) lower than for the conductive designs i.e. an argument therefore exists for the selection of an insulating lining design.

To determine the chemical compatibility between the slag and oxide-based refractory material—as required for an insulating lining design—the equilibrium liquid phase formation when reacting 50 g of slag (compositions in Table 1) with 50 g of refractory (compositions in Table 3) was calculated. Although the calculation only applies to insulating lining designs, magnesia refractory was also included. The Equilib module was applied, and FToxid and FactPS databases selected. All compound species were selected (gas, pure liquids, and pure solids),

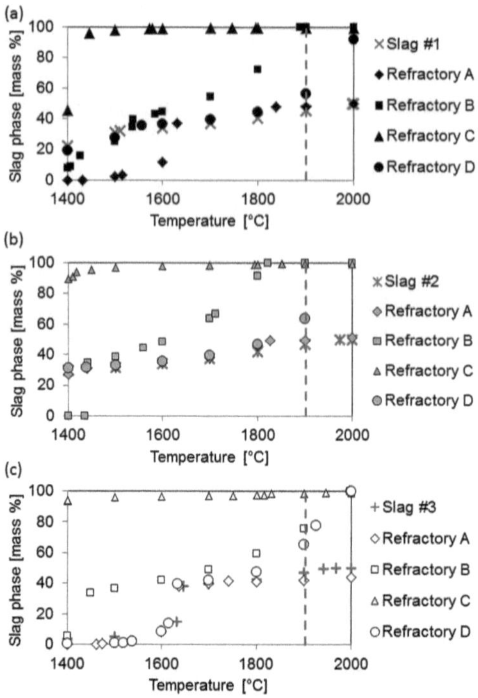

Fig. 5 Calculated equilibrium slag phase formation when reacting 50 g of slag (compositions in Table 1) with 50 g of refractory (compositions in Table 3) with equilibrium liquid phase formation in 50 g slag superimposed, for **a** Slag #1, **b** Slag #2, and **c** Slag #3, and the broken vertical *red-lines* indicate the typical tap temperature

and duplicates were suppressed, with FToxid having preference over FactPS. All solution phases were selected except for B-olivine, and mullite (FToxid-Mull). The results were presented in Fig. 5. Should the slag phase formation be less than or equal to the slag phase formation calculated for slag only, the refractory is considered compatible with slag. When more slag is formed it is not compatible with slag. From Fig. 5, refractory C is immediately excluded as the refractory will react with the slag. At the tapping temperature selected (2173 K (1900 °C)); refractory B and refractory D will also react with the slag. The only refractory suitable is refractory A. Therefore, even for an insulating lining design with slag in contact with refractory, magnesia-based refractory material at the hot face will be more suitable.

Consequences of Philosphy Selected

Irrespective of the furnace containment philosophy selected, slag composition and temperature need to be managed: For insulating lining designs, to ensure that the slag is saturated in the hot face refractory components; for conductive lining designs, to ensure that the slag freezes and remain frozen on the surface of the refractory. Managing the mass and the energy balance for the process is therefore important.

Referring back to Eq. 1: $E_{process}$ is recipe dependent (the enthalpy values of different raw materials differ), and is calculated by furnace operators in theoretical mass and energy balance calculations. The properties of importance in these calculations are masses, chemical compositions, temperatures, and enthalpies, of all inputs to and outputs from the furnace. Therefore, the following measurements on the plant are critical for process control:

1. Frequent, representative sampling of all input and output streams.
2. Accurate analyses, reported within a short time period from sampling (ideally within 30 min, but typically within 2 h for solid materials, and online analyses of off-gas), of all input and output streams.
3. Mass measurements (or volumetric flow measurements in the case of off-gas) of all input and output streams that are accurate, reliable, and validated.
4. Temperature measurements (especially for the output streams) that are accurate, reliable, and validated.
5. Enthalpy values of all input and output streams that are accurate and validated.

E_{losses} is calculated by furnace operators in energy loss calculations, based on water temperatures and flow rates for sections of the furnace that are water-cooled, and temperature differences in refractory material with known thermal conductivity for sections with no water-cooling. Therefore, the following accurate, reliable, and validated measurements on the plant are critical for process control:

1. Temperature measurements of water, at the inlet and outlet of cooling circuits.
2. Flow measurements of water, to or from, cooling circuits.
3. Temperature measurements in a single refractory brick, using dual thermocouples, where the exact positions of the two hot junctions are known.
4. Thermal conductivity of the single refractory brick, as a function of temperature.

These measurements should be included in the detailed design phase, taking into account not only how the systems will be managed during normal operation, but also how the systems will be checked and measurements validated during cold and hot commissioning of the plant.

A typical process control measure applied in DC-open arc furnace is the power-to-feed ratio, where power is the electrical energy input to the furnace and feed the total feed [16–18]. Typically, the furnace operator selects a specific power setting to operate at, and with known energy losses, adjusts the feed rate of the material fed to the furnace, in order to control the power-to-feed ratio for a specific recipe. At a constant power setting with constant energy losses, the power requirement will change when the recipe changes due to the differences in enthalpies of the input materials. The differences can occur when the ratios of feed materials change, e.g. ore/flux, or when the mineralogy of a specific feed material changes, e.g. changes in mineralogy within chromite ore.

If the energy requirement of the process changes without the awareness of the furnace operator, more or less energy can be added to the furnace than what is required by the process and energy losses—in layman's terms referred to as 'over-powering or under-powering the furnace'. In principle, the consequences of over-powering of the furnace are:

1. In open arc furnace operation, with an insulating lining design, the excess energy will increase the temperature of the slag saturated in refractory components (and metal, depending on the heat transfer mechanisms at play). An increase in slag temperature could potentially influence saturation of the slag in refractory components as illustrated in Fig. 5—an analogy would be when more sugar can be dissolved in boiled water compared to water at room temperature.
2. In open arc furnace operation, with a conductive lining design, the excess energy will melt away the freeze lining, but the metal and slag tap temperatures will remain essentially the same, until the freeze lining disappears. Thereafter, the heat transfer mechanisms essentially become similar to that of an insulating lining design with the risks of chemical incompatibility between refractory and process material, and temperature increases beyond the service temperature of the refractory material.
3. In SAF operation, the excess energy will allow for the burden to be consumed at a faster rate i.e. the production rate will increase (within the constraints of heat and mass transfer between the furnace slag bath and burden) and therefore the rate at which the furnace needs to be fed, but the metal and slag tap temperatures will remain essentially the same. Should the heat and mass transfer between the furnace slag bath and the burden become rate-limiting, the excess thermal energy will increase metal and slag temperatures resulting in the potential wear mechanisms described for open arc furnace operation.

Conclusion

In ferrochromium production, both insulating and conductive lining design philosophies are applied. Tools available for the evaluation of lining design philosophies at desktop level include the use of one-dimensional, steady-state heat transfer and thermodynamic and thermo-physical property calculations (utilising FactSage 7.0 software). Steady-state models tell only part of the story as insulating vs conductive linings have very different transient responses to disturbances in process conditions, which can affect the operability of the furnace significantly. Therefore the importance of further heat transfer modelling combined with laboratory and pilot-scale test work to validate the selection of lining philosophy as well as materials selected, cannot be over-emphasized. By not doing these tests on laboratory and pilot-scale, the industrial-scale furnace essentially becomes a pilot furnace. Once in operation, the power-to-feed ratio is one of the most important parameters to utilise for process control. It is therefore important to design for measurements that are accurate and reliable, and to validate these measurements during cold commissioning and hot commissioning of the industrial-scale furnace.

Acknowledgments This chapter is published with permission from MINTEK. Helpful discussions with Chris Pistorius, Rodney Jones, Quinn Reynolds, and Markus Erwee are gratefully acknowledged.

References

1. I.J. Geldenhuys, *Aspects of DC chromite smelting at MINTEK—an overview* (INFACON XIII, Almaty, 2013), pp. 31–48
2. T.R. Curr, N.A. Barcza, The production and treatment of ferrochromium, South African patent 82/7401, 1982
3. R.T. Jones, DC arc furnaces—past, present, and future, Celebrating the Megascale: the symposium of the Extraction and Processing Division of TMS 2014 in honour of D.G.C. Robertson, San Diego, California, 2014, 129–139
4. R.T. Jones, M.W. Erwee, Simulation of ferro-alloy smelting in DC arc furnaces using Pyrosim and FactSage. Calphad (2016). doi:10.1016/j.calphad.2016.05.003
5. R.T. Jones, N.A. Barcza, T.R. Curr, *Plasma developments in Africa, second international plasma symposium: world progress in plasma applications* (Palo Alto, California, 1993)
6. D. Slatter et al., *Technology for the production of new grades and types of ferroalloys using thermal plasma* (INFACON IV, Rio de Janeiro, 1986), pp. 191–204
7. P. Hloben, *Refractory materials—Major industrial applications* (Rexxon Corporation, Bryanston, 2000)
8. P.L. Duncanson, J.D. Toth, The truths and myths of freeze lining technology for submerged arc furnaces, INFACON X, Cape Town, South Africa, 2004, 488–499
9. J.D. Steenkamp, Chemical wear of carbon-based refractory materials in a silicomanganese furnace tap-hole, Ph.D. thesis, University of Pretoria, 2014
10. J. Chirasha, N.R. Shoko, *Refractory design effect on submerged arc furnace life and production performance—the Zimbabwe alloys experience* (INFACON XI, New Delhi, 2007), pp. 728–738
11. A.V.L. Narasimham, *Refractory lining failures in FeCr furnaces an overview* (INFACON XI, New Delhi, 2007), pp. 715–721
12. S.A.C. Hockaday, K. Bisaka, *Some aspects of the production of ferrochrome alloys in pilot DC arc furnaces at MINTEK* (INFACON XII, Helsinki, 2010), pp. 367–376
13. C. Coetzee, P.L. Duncanson, P. Sylven, *Campaign extensions for ferroalloy furnaces with improved tap hole repair system* (INFACON XI, New Delhi, 2007), pp. 857–866
14. C. Coetzee et al., *New refractory lining direction at Jindal stainless FeCr #1 and #2 furnaces* (INFACON XI, New Delhi, 2007), pp. 891–898
15. D. Sager et al., *Low cost ferroalloy extraction in DC-arc furnace at Middleburg Ferrochrome* (INFACON XII, Helsinki, 2010), pp. 803–814
16. H.J. Kammeyer et al., *Open-bath production of ferrochromium in a DC plasma furnace* (INFACON V, New Orleans, 1989), pp. 95–102
17. M. Ford, J.F. Oosthuizen, *The production of ferrochromium in a 40MVA DC plasma furnace* (INFACON VII, Trondheim, 1995), pp. 263–272
18. F.P. Greyling, W. Greyling, F.I. De Waal, *Developments in the design and construction of DC arc smelting furnaces* (INFACON XII, Helsinki, 2010), pp. 815–824
19. C. Wenzl, A. Spanring, *RHI refractory solutions—a reliable partner for the ferroalloys industry* (INFACON XIV, Kyiv, 2015), pp. 264–269

20. R.C. Nunnington et al., *Developments in refractories for plasma technology* (Institute of Refractory Engineers Bi-ennial Colloquium, Pretoria, 1985), pp. 38–78
21. J.D. Steenkamp, P.C. Pistorius, M. Tangstad, Wear mechanisms of carbon-based refractory materials in SiMn tap-holes—part 2: in-situ observation of chemical reactions. Metall. Mater. Trans. B. **46B**(2), 668–679 (2015)
22. R. Roscoe, The viscosity of suspensions of rigid spheres. Br. J. Appl. Phys. **3**, 267–269 (1952)
23. Gerald Routschka (ed.), *Pocket manual—refractory materials*, 2nd edn. (Vulkan-Verlag, Essen, 2004)

The Influence of Phosphorous Additions on Phase Evolution in Molten Synthetic Coal Slag

Hani Abu El Hawa, Jinichiro Nakano, Anna Nakano and James P. Bennett

Abstract Carbon feedstocks used in power, chemical and metallurgical industries typically contain numerous minerals or impurities that liquefy during high temperature processing to form slag as a byproduct. Slags have many industrial applications, for example, as construction materials and for element recovery. In addition to other variables, the extent of slag utilization is dependent on its composition and post treatments. In this work, the influence of temperature and phosphorus additions on phase evolution in coal slag was investigated. Synthetic slag mixtures with varying $AlPO_4$ contents (0–14 wt%) were heated from 1380–1470 °C in air, then water-quenched to preserve the phases formed. SEM/EDS, XRF and XRD analyses were performed before/after testing to gain insight about slag chemistry and phases present. It was found that coal slag exhibited a large solubility of phosphorous, affecting morphology of mullite formed under the conditions studied. Results obtained and their significance are discussed.

Keywords Coal slag · Mullite · Aluminum phosphate

Introduction

Coal gasifiers are commercially used to produce syngas (CO and H_2), an important gas mixture that can be converted into useful products ranging from chemicals to power. Gasifier feedstock materials typically contain impurities that can melt and oxidize during gasification to form slags [1, 2]. Slag properties (e.g. chemistry, solid morphology and phases present) are strongly dependent upon the operating conditions of the gasifier. Mullite, for instance, is one of the phases that may be found

H. Abu El Hawa (✉) · J. Nakano · A. Nakano · J.P. Bennett
U.S. Department of Energy, National Energy Technology Laboratory,
1450 Queen Ave SW, Albany, OR 97321, USA
e-mail: hanihawa@yahoo.com

J. Nakano · A. Nakano
AECOM, P.O. Box 1959, Albany, OR 97321, USA

S. Wang et al. (eds.), *Applications of Process Engineering Principles in Materials Processing, Energy and Environmental Technologies*,
The Minerals, Metals & Materials Series, DOI 10.1007/978-3-319-51091-0_20

in coal slags. Previous research [3] has shown that it has the ability to form a solid solution in a wide Al_2O_3/SiO_2 range and to incorporate a variety of foreign cations into the structure. Slag equilibria and mullite chemistry in coal slag were studied at 1500 °C and an oxygen partial pressure of 10^{-8} atm by Nakano et al. [1] where Al/Si ratios in mullite were affected by bulk slag chemistry.

This work investigates the influence of temperature and phosphorus additions on phase evolution in synthetic mixtures of coal slag. Phosphorous content was varied in samples by changing the quantities of $AlPO_4$ additions. It was changed to evaluate the impact of phosphate added to gasifier liner brick and how slag properties would be impacted. Samples were heated from 1380–1470 °C in air, then water-quenched to preserve the phases formed. Scanning electron microscopy/energy dispersive X-ray spectroscopy (SEM/EDS), X-ray fluorescence (XRF) and X-ray diffraction (XRD) analyses were performed before/after testing to gain insight about slag chemistry and phases present.

Materials and Methods

Three synthetic coal slag and slag/aluminum phosphate mixtures were prepared by blending the chemical reagents silica (99.50% pure), alumina (99.00% pure), hematite (99.80% pure) and calcia (99.95% pure) with varying ortho-aluminum phosphate (97.00% pure) additions (0–14.1 wt%). Each mixture was tumble-mixed (with the aid of dense alumina balls) for 8 h. Portions of these mixtures were pelletized at 2000 psi to create ~2.5 cm (1″) diameter pellet. This was necessary to allow for compositional analysis by XRF which was conducted using Rigaku ZSX Primus II wavelength-dispersive X-ray fluorescence (WDXRF) spectrometer. Table 1 shows the mixtures prepared along with their IDs and compositions as determined by XRF.

A total of twelve samples were tested in air in the temperature range (1380–1470 °C) as shown in Table 2. Typically, a ~2 g sample of the raw powder was placed in an alumina crucible and heated in a tube resistance furnace in air at 150 °C/h ramp rate and held at a designated temperature for 3 h. The samples were never heated above the designated temperature. A protected Pt-thermocouple (type B) was placed in the center of the tube furnace just underneath the alumina boat holding the samples and was used to monitor temperature. The temperature was maintained within 3 °C of the desired set point. After being held at temperature in

Table 1 Compositions as determined by XRF for the synthetic slag and slag/$AlPO_4$ mixtures

Mixture Id/composition	SiO_2 (wt%)	Al_2O_3 (wt%)	Fe_2O_3 (wt%)	CaO (wt%)	$AlPO_4$ (wt%)
0	44.9	24.4	23.6	7.1	0.0
7	41.7	22.7	21.9	6.6	7.1
14	38.6	20.9	20.3	6.1	14.1

Table 2 Sample ID versus $AlPO_4$ content and testing temperature

Temperature/composition (°C)	0.0 wt% $AlPO_4$	7.1 wt% $AlPO_4$	14.1 wt% $AlPO_4$
1380	A0	A7	A14
1400	B0	B7	B14
1440	C0	C7	C14
1470	D0	D7	D14

All samples were held at temperature in air for 3 h then water-quenched to room temperature

air for 3 h, the samples were water-quenched to room temperature. At each temperature, three samples of varying $AlPO_4$ composition were tested at the same time.

Once quenching was completed, slag samples were mounted in epoxy and polished for SEM/EDS (JSM-7000F and FEI Inspect F field performed at 20 kV) analyses to study phase composition and morphology. A portion of each sample was ground below 45 μm using a ceramic mortar and pestle. X-ray diffraction crystalline phase analysis was then performed using a Rigaku Ultima III XRD instrument.

Thermodynamic phase equilibria calculations were performed at 1 atm ($P_{O_2} = 0.2089$ atm) using FactSage^{TM} 7.0 with FactPS and FToxid databases in order to estimate the phases, which may be present in each mixture as function of $AlPO_4$ composition and testing temperature. For these calculations, a coal slag mixture containing 42.5% SiO_2, 42.0% Al_2O_3, 8.4% Fe_2O_3 and 7.1 wt% CaO (all in wt%) was used.

Results and Discussion

Figure 1 shows SEM micrographs of samples A0–A14 (top row), B0–14 (second row from top), C0–14 (third row from top) and D0–14 (bottom row). As mentioned before, these samples were tested in air for 3 h at 1380 °C (A samples), 1400 °C (B samples), 1440 °C (C samples) and 1470 °C (D samples) then water-quenched to room temperature. EDS scans were performed on selected points to determine specific microstructure chemistry. As can be seen from Fig. 1, distinct regions containing slag and another crystalline phase exist. The elemental compositions of the phases are presented in Tables 3, 4, 5 and 6. Based on these compositions, this phase has a chemistry similar to mullite as will be discussed below.

As can be inferred from the tables, the liquid slag has a very consistent composition, with (Al + Fe) to Si ratios of 0.6–0.8, 0.7–0.9, 0.8–0.9 and 0.7–0.8 for samples A, B, C and D, respectively. At a given temperature, the ratio tended to increase with increasing $AlPO_4$ content, showing high solubility in molten slag.

In addition to the slag, mullite with (Al + Fe) to Si ratios of 1.7–2.2, 2.2–2.5, 2.1–2.7 and 2.4–2.5 was also identified for samples A, B, C and D, respectively. For phase identification, it is worth noting that using (Fe + Al)/Si ratio is more rational than using the Al/Si ratio as Fe can be incorporated in the mullite phase

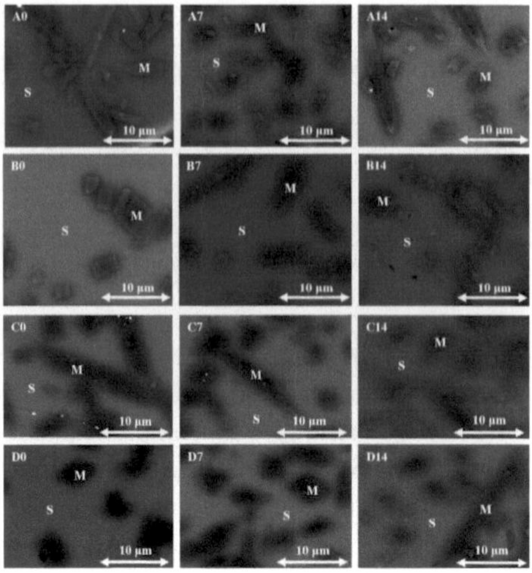

Fig. 1 SEM micrographs (backscatter mode) for samples A (*1st row from top*), B (*2nd row from top*), C (*3rd row from top*) and D (*bottom*) after quenching from the test temperature. "S" stands for slag while "M" stands for mullite

Table 3 Samples A average elemental composition (atom %) as determined by EDS along with phases identified at 1380 °C

Sample	Phase	O	Al	Si	Ca	Fe	P	Al/Si	(Al + Fe)/Si
A0	Slag	74.3 (0.1)	6.1 (0.2)	14.6 (0.1)	2.1 (0.0)	2.9 (0.0)	–	0.4 (0.0)	0.6 (0.0)
	Mullite	74.2 (0.4)	13.4 (0.9)	9.2 (0.4)	0.8 (0.1)	2.4 (0.0)	–	1.5 (0.2)	1.7 (0.2)
A7	Slag	74.5 (0.1)	6.5 (0.4)	12.0 (0.1)	1.9 (0.1)	3.5 (0.1)	1.7 (0.0)	0.5 (0.0)	0.8 (0.0)
	Mullite	74.0 (0.1)	14.6 (0.6)	8.1 (0.3)	0.6 (0.1)	2.3 (0.1)	0.6 (0.1)	1.8 (0.1)	2.1 (0.1)
A14	Slag	74.6 (0.3)	6.5 (0.3)	11.6 (0.0)	1.8 (0.0)	3.2 (0.1)	2.3 (0.0)	0.6 (0.0)	0.8 (0.0)
	Mullite	73.8 (0.1)	15.0 (0.4)	7.8 (0.1)	0.5 (0.1)	2.2 (0.1)	0.7 (0.1)	1.9 (0.1)	2.2 (0.1)

The numbers in brackets represent the standard deviation (Refer to Fig. 1)

Table 4 Samples B average elemental composition (atom %) as determined by EDS along with phases identified at 1400 °C

Sample	Phase	O	Al	Si	Ca	Fe	P	Al/Si	(Al + Fe)/Si
B0	Slag	74.3 (0.1)	6.1 (0.1)	14.0 (0.1)	2.1 (0.1)	3.4 (0.1)	–	0.4 (0.0)	0.7 (0.0)
	Mullite	73.4 (0.3)	16.2 (1.2)	7.7 (0.8)	0.3 (0.1)	2.5 (0.1)	–	2.1 (0.4)	2.5 (0.4)
B7	Slag	74.6 (0.1)	6.5 (0.5)	12.4 (0.1)	1.7 (0.1)	3.4 (0.1)	1.4 (0.1)	0.5 (0.0)	0.8 (0.0)
	Mullite	74.9 (1.4)	14.7 (1.5)	7.5 (0.3)	0.5 (0.2)	2.0 (0.1)	0.4 (0.1)	1.9 (0.2)	2.2 (0.2)
B14	Slag	74.6 (0.2)	6.5 (0.1)	11.4 (0.2)	1.9 (0.0)	3.3 (0.1)	2.3 (0.0)	0.6 (0.0)	0.9 (0.0)
	Mullite	73.5 (0.5)	15.0 (0.6)	7.9 (0.1)	0.5 (0.1)	2.3 (0.1)	0.8 (0.0)	1.9 (0.1)	2.2 (0.1)

The numbers in brackets represent the standard deviation (Refer to Fig. 1)

upon heating. Thus, adding Fe makes the stoichiometric ratio of mullite closer to theoretical one of 3.

Those ratios correspond to the mullite phase that was present at this temperature as outlined before. For samples A, these observations suggest that the addition of $AlPO_4$ to the synthetic coal slag seems to have some influence on the stability of the

Table 5 Samples C average elemental composition (atom %) as determined by EDS along with phases identified at 1440 °C

Sample	Phase	O	Al	Si	Ca	Fe	P	Al/Si	(Al + Fe)/Si
C0	Slag	74.2 (0.2)	6.6 (0.3)	13.4 (0.1)	2.0 (0.1)	3.8 (0.1)	–	0.5 (0.0)	0.8 (0.0)
	Mullite	73.2 (0.1)	17.2 (0.4)	7.1 (0.4)	0.2 (0.0)	2.2 (0.0)	–	2.4 (0.2)	2.7 (0.2)
C7	Slag	74.4 (0.2)	6.5 (0.1)	12.0 (0.1)	1.8 (0.0)	3.7 (0.1)	1.5 (0.0)	0.5 (0.0)	0.9 (0.0)
	Mullite	73.5 (0.3)	15.6 (0.1)	7.8 (0.1)	0.4 (0.0)	2.3 (0.1)	0.4 (0.2)	2.0 (0.0)	2.3 (0.0)
C14	Slag	75.0 (0.3)	7.1 (0.8)	11.1 (0.1)	1.7 (0.2)	3.1 (0.2)	2.1 (0.1)	0.6 (0.1)	0.9 (0.1)
	Mullite	74.5 (0.3)	14.5 (0.6)	7.9 (0.3)	0.5 (0.1)	2.0 (0.1)	0.7 (0.1)	1.8 (0.1)	2.1 (0.1)

The numbers in brackets represent the standard deviation (Refer to Fig. 1)

Table 6 Samples D average elemental composition (atom %) as determined by EDS along with phases identified at 1470 °C

Sample	Phase	O	Al	Si	Ca	Fe	P	Al/Si	(Al + Fe)/Si
D0	Slag	74.2 (0.2)	6.9 (0.1)	13.1 (0.1)	2.0 (0.0)	3.9 (0.1)	–	0.5 (0.0)	0.8 (0.0)
	Mullite	73.4 (0.3)	16.6 (0.7)	7.6 (0.5)	0.3 (0.2)	2.1 (0.2)	–	2.2 (0.2)	2.5 (0.2)
D7	Slag	74.9 (0.1)	6.9 (0.2)	11.9 (0.2)	1.6 (0.0)	3.2 (0.1)	1.4 (0.0)	0.6 (0.0)	0.7 (0.0)
	Mullite	73.9 (0.1)	17.1 (0.7)	7.0 (0.5)	0.2 (0.1)	1.6 (0.1)	0.2 (0.1)	2.5 (0.3)	2.5 (0.3)
D14	Slag	74.0 (0.2)	7.0 (0.1)	11.6 (0.1)	1.8 (0.0)	3.3 (0.1)	2.3 (0.1)	0.6 (0.0)	0.8 (0.0)
	Mullite	73.7 (0.3)	16.7 (1.0)	7.3 (0.4)	0.4 (0.1)	1.5 (0.2)	0.5 (0.1)	2.3 (0.3)	2.4 (0.3)

The numbers in brackets represent the standard deviation (refer to Fig. 1)

mullite phase at this temperature as evidenced by the increase in the (Al + Fe)/Si seen in samples A7 and A14 compared to what was seen in sample A0. While Fe content in mullite decreased, (Al + Fe)/Si tended to increase with $AlPO_4$ addition to the original slag, which may be caused by lower activity of Fe and higher activity of Al in slag. In general, the presence of smaller mullite particles were noted in samples with $AlPO_4$ additions (A7 and A14) as can be inferred from Fig. 1.

For samples B, mullite phase with (Al + Fe) to Si ratio of 2.2 for samples B7 and B14 was identified. Sample B0 shows slightly higher (Al + Fe) to Si ratio of 2.5 compared to 1.7 for sample A0. This increase is consistent with the finding of Schneider et al. [3] where higher Al content in mullite was reported at higher temperatures. From the B samples observations, it appears that the addition of $AlPO_4$ to the synthetic coal slag along with the higher treatment temperature affected the composition of the mullite phase (i.e. (Al + Fe)/Si); (Al + Fe)/Si tended to decrease with $AlPO_4$ addition to the original slag at this temperature. The presence of smaller mullite particles were also noted in samples with $AlPO_4$ addition (B14), which may be attributed to slag polymerization due to phosphate in addition to silicate in molten slag, which caused viscosity to increase (slower diffusion).

For samples C, however, mullite phase with (Al + Fe) to Si ratios of 2.7 for sample C0, 2.3 for sample C7 and 2.1 for sample C14 was identified. A different observation compared to what was observed for A and B samples. It appears that the addition of $AlPO_4$ to the synthetic coal slag along with the higher testing temperature affected composition of the mullite phase (i.e. (Al + Fe)/Si). At this temperature, the (Al + Fe)/Si ratio rapidly decreased with $AlPO_4$ addition to the original slag. The presence of smaller and thinner mullite particles were noted in samples with $AlPO_4$ addition as well (C14).

For the final test (Samples D), the mullite phase with (Al + Fe) to Si ratios of 2.4 for sample D14 and 2.5 for samples D0 and D7 was identified. From these results, it seems that the addition of $AlPO_4$ to the synthetic coal slag along with the higher testing temperature exhibited little effect on the composition of the mullite phase (i.e. (Al + Fe)/Si) compared to the A, B and C samples. (Al + Fe)/Si slightly decreased with $AlPO_4$ addition to the original slag at this temperature. The presence of smaller and thinner mullite particles was also noted in samples with $AlPO_4$ addition (D14), an observation that is similar to what was seen with the samples that were tested at lower temperature.

In order to investigate solid phases present in the synthetic coal slag samples that were tested in this study, XRD scans were performed as previously described. For samples A0–14, XRD scans were conducted while they were mounted with the alumina crucible using epoxy, which introduced XRD peaks corresponding to the alumina crucible. The overall goal was to identify peaks that correspond to crystalline phases present as a function of testing temperature and $AlPO_4$ contents. Figure 2 shows the XRD spectra for all A, B, C and D samples, grouped together based on testing temperature.

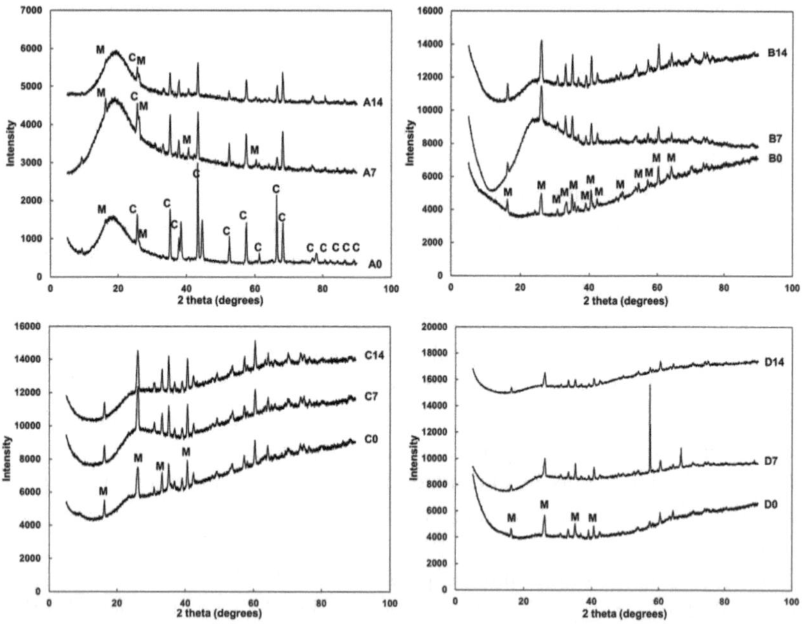

Fig. 2 XRD spectra along with major peak identification for samples tested at 1380 °C (*top left*), 1400 °C (*top right*), 1440 °C (*bottom left*) and 1470 °C (*bottom right*). M = Mullite and C = Al_2O_3

For the A samples, alumina peaks can be clearly seen; most likely from the alumina crucible as mentioned before. Major mullite peaks, however, were detected for samples A7 and A14 which is consistent with the EDS results.

For samples tested at temperatures higher than 1380 °C (B, C and D samples), all XRD spectra suggest the formation of mullite only, no other crystalline phases were noted which is consistent with EDS results reported previously for those samples.

Finally, another interesting observation from the XRD spectra for B and C samples is that the mullite area under its peaks is increasing with additions of $AlPO_4$. This suggests that $AlPO_4$ stabilized mullite in the synthetic coal slags at 1400 and 1440 °C. For samples D, on the other hand, the area under the mullite peaks is decreasing with increasing $AlPO_4$ additions. This implies $AlPO_4$ additions decreased the stability of mullite in the synthetic coal slags at 1470 °C. $AlPO_4$ peaks were difficult to distinguish owing to the overlap of some peaks with mullite.

Figure 3 shows the thermodynamic phase equilibria diagram calculated for synthetic coal slag-$AlPO_4$ as a function of temperature. Individual equilibrium calculations at 100% coal slag composition without $AlPO_4$ (A0, B0, C0, and D0) indicated equilibrium phase regions containing slag + mullite + Fe_2O_3 (containing 7 wt% Al_2O_3) at 1380 °C, slag + mullite at 1400 °C and only slag at 1440 and 1470 °C. That is, a slight addition of $AlPO_4$ to coal slag would change phase equilibria, and anorthite may form with $AlPO_4$ additions at or below 1400 °C, according to the calculations. In the present experiment with 3 h hold, anorthite was

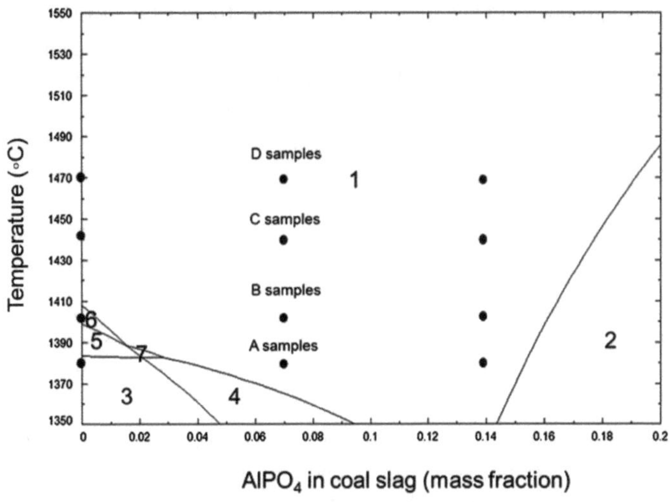

Fig. 3 Thermodynamic phase equilibria diagram (1 atm) for the synthetic coal slag as function of $AlPO_4$ content and temperature. Region *1* has slag only, *2* slag + mullite + $AlPO_4$, *3* slag + mullite + corundum + anorthite, *4* slag + mullite + corundum, *5* slag + spinel + mullite + anorthite, *6* slag + mullite + anorthite and *7* slag + spinel + mullite. The *solid dots* represent the experimental points

not found. Mullite present in samples at 1440 and 1470 °C may have formed at lower temperatures where a holding length used in this work was insufficient to attain equilibrium.

While those theoretical calculations suggest the formation of Fe_2O_3 at 1380 °C for the 100% coal slag, it was not found in the present sample at that temperature, following XRD results; presumably due to slow kinetics at this temperature. At testing temperatures 1440 and 1470 °C, formation of mullite may have required more time than the soaking time used in this study (3 h); that might explain the disagreement between the EDS/XRD results and the theoretical calculations.

Overall, the reported study gives a thermodynamic equilibrium knowledge which would aid predictions of the crystallization behavior in slags to optimize industrial operations. Phosphorous, which may come out form the P-containing gasifier liner brick, will affect the molten slag properties including its viscosity. The $AlPO_4$ addition into coal slag stabilized the mullite phase precipitation, with the finer crystals formed at higher $AlPO_4$ concentrations. In general, fine crystals or no crystals is beneficial in order to allow the slag flowing toward the gasifier exit.

Conclusions

Coal slag properties can be strongly affected by the high temperature treatment as well as the presence of impurities. This work evaluated the influence of temperature and phosphorus additions on phase evolution in synthetic coal slags. Synthetic slag mixtures with varying $AlPO_4$ contents were heated from 1380–1470 °C in air, then water-quenched to preserve the phases formed. SEM/EDS and XRD results showed that coal slag exhibited a large solubility for phosphorous, affecting morphology and chemistry of mullite formed under the conditions studied. Mullite was the only crystalline phase identified by XRD for all samples at all temperatures. Mullite's chemistry appears to be affected by temperature as well as $AlPO_4$ content.

Acknowledgements This technical effort was performed in support of the National Energy Technology Laboratory's ongoing research under the RES contract DE-FE0004000. This research was also supported in part by an appointment to the National Energy Technology Laboratory Research Participation Program, sponsored by the U.S. Department of Energy and administered by the Oak Ridge Institute for Science and Education. The authors would like to thank Eric Krabbe, Richard Chinn and Keith Collins of the National Energy Technology laboratory, Albany, OR for their assistance with testing and SEM, XRD and XRF runs.

liability or responsibility for the accuracy, completeness, or usefulness of any information, apparatus, product, or process disclosed, or represents that its use would not infringe privately owned rights. Reference herein to any specific commercial product, process, or service by trade name, trademark, manufacturer, or otherwise, does not necessarily constitute or imply its endorsement, recommendation, or favoring by the United States Government or any agency thereof. The views and opinions of authors expressed herein do not necessarily state or reflect those of the United States Government or any agency thereof.

References

1. J. Nakano et al., Phase equilibria in synthetic coal–petcoke slags (Al_2O_3–CaO–FeO–SiO_2–V_2O_3) under simulated gasification conditions. Energy Fuels **25**(7), 3298–3306 (2011)
2. J. Bennett et al., Slag characterization for the development of new and improved service life materials in gasifiers using flexible carbon feedstock, in *Advances in Materials Science for Environmental and Nuclear Technology II* ed. by S.K. Sundaram et al. (2011), pp. 3–16
3. H. Schneider, J. Schreuer, B. Hildmann, Structure and properties of mullite—a review. J. Eur. Ceram. Soc. **28**(2), 329–344 (2008)

The Recovery of Copper from Smelting Slag by Flotation Process

Jiaqi Fan, Hongxu Li, Liangtian Wei, Chao Li and Shi Sun

Abstract Aiming at the recovery of copper from smelting slag, a flotation approach was studied. It was found that this slag composed of fine particles with complex association and distribution, in which bornite was the main copper-bearing mineral after a detailed mineralogy analysis via polarizing microscope, SEM and XRD. Consequently, flotation was attempted to recycle Cu in slag containing copper sulphide components. Three key factors affecting flotation were ascertained, namely, grinding fineness, the collector and the pH value, while the recovery of Cu estimated by ICP and XRF. Copper concentrate grading at 14.47 with 79.66% Cu recovery was obtained, in the condition of the grinding fineness of −0.074 mm and the proportion of 80%, reagent dosages of 50 g/t for butyl xanthate and a pH value of 10 adjusted using Na_2CO_3.

Keywords Smelting slag · Copper recovery · Mineralogy analysis · Flotation

Introduction

It is well-known that copper is very important metal to the economic development, the copper concentrate gradually becomes a scarce resource. But, the world's copper production is 80% using pyrometallurgical method. More than 97% of China's copper output is generated by pyrometallurgy, which is a dominant production method for copper metal [1]. Accompanying with the pyrometallurgical processes of copper, a large number of copper smelting slag is generated, in which

J. Fan · H. Li (✉) · L. Wei · C. Li · S. Sun
School of Metallurgical and Ecological Engineering, University of Science and Technology Beijing, 30# Xueyuan Road, Beijing 100083, China
e-mail: lihongxu@ustb.edu.cn

S. Wang et al. (eds.), *Applications of Process Engineering Principles in Materials Processing, Energy and Environmental Technologies*, The Minerals, Metals & Materials Series, DOI 10.1007/978-3-319-51091-0_21

the average content of copper reaches more than 1% (reaching or even exceeding China's copper ore grade for smelting). China's output of copper slag is up to 15 million tons per year [2]. Such huge quantities of slag will be a tremendous waste of resources if dumped or disposed directly, therefore it has important practical significance to recover sufficient amount or quantity copper from smelting slag. Besides, it is also important for eventual sustainable development of copper smelting enterprises.

So far, many studies have been carried out on the recovery of copper from slag, in which the methods used are leaching, flotation, direct reduction and bio-leaching [3–6]. Among those methods, flotation procedure, a method for mineral separation, is widely applied in smelters because of its great processing capacity as well as relatively small investment cost for equipment and reagents [7]. Most importantly, for the recovery of copper in the form of sulfide, the flotation usually possesses better effects. But, because of the difference in the composition of the slag, each process may have different effects. Therefore, a flotation approach was studied, aiming at the recovery of copper from smelting slag.

Experimental

Materials

The smelting slag used in this study was obtained from a copper plant in China. Mineralogical analysis, performed using scanning electron microscope (SEM) and energy dispersive X-ray spectroscope (EDS), polarizing microscope and X-ray diffraction (XRD), indicated that fayalite (Fe_2SiO_4) and magnetite (Fe_3O_4) phases were major components while bornite (Cu_5FeS_4) was the main copper-bearing mineral in the slag. In addition, chemical composition of the slag was analyzed by inductively coupled plasma-atomic emission spectrum (ICP-AES), atomic absorption spectrometry (AAS) and X-ray fluorescence (XRF), which was given in Table 1. The results showed that the contents (wt%) of the main elements were Cu 1.22, Co 0.43 and Fe 40.72%.

Table 1 Chemical composition of slag

Component (%)								
Cu	Co	Pb	Zn	Fe	Si	Al	Ca	S
1.22	0.43	0.12	0.49	40.72	15.48	2.21	3.13	0.40

Flotation Tests

The slag was crushed in a jaw crusher and ground in a ball mill and then sieved. N-ethyl-O-isopropyl thiocarbamate (Z-200), ethyl xanthate, butyl xanthate, isopentyl xanthate and butylamine dithiophosphate were used as collectors; terpineol oil as frother; sulphuric acid, lime, sodium hydroxide and sodium carbonate as pH regulator. Only the direct one-stage of single-stage flotation (rougher) was used in all experiments, except the last flotation test which was in the optimal condition.

Results and Discussion

Characterization of the Slag

The chemical analysis shows that the main phases of slag are fayalite (Fe_2SiO_4) and magnetite (Fe_3O_4), while the main copper-bearing minerals are bornite, copper, chalcopyrite, covellite and chalcocite from the X-ray diffraction pattern and polarizing microscope images (Fig. 1). Besides, copper-bearing minerals, fayalite, magnetite and other amorphous minerals were mutually wrapped and embedded. Further more, contents and sizes of major minerals in the slag were determined via polarizing microscope, which could be seen from Tables 2 and 3.

It could be seen from Table 3 and Fig. 1, the vast majority of copper-bearing phases showed complex association and distribution among slag with the particle size less than 0.02 mm. To recover copper in such fine minerals, grinding was essential in flotation test.

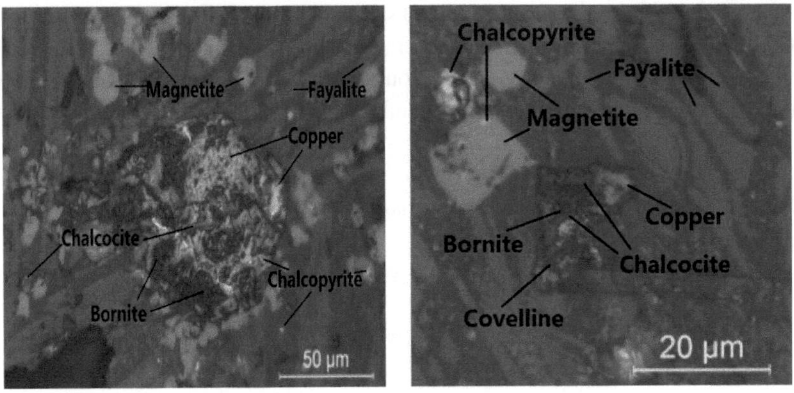

Fig. 1 Representative polarizing microscope image of slag showing fine copper-bearing particles embedded intricately with each other

Table 2 Mineral composition of slag

Minerals	Wt./%	Minerals	Wt./%
Bornite	0.41	Fayalite	52.06
Copper	0.39	Amorphous minerals	39.69
Covelline	0.33	Magnetite	6.68
Chalcopyrite	0.20	Others	0.15
Chalcocite	0.09		

Table 3 Size distribution of main minerals in slag

Minerals	Size			
	>0.05 mm	0.02–0.05 mm (%)	0.005–0.02 mm (%)	<0.005 mm (%)
Magnetite	0	20	60	20
Bornite	25%	35	30	10
Copper	8%	45	44	3
Covelline	0	3	25	72
Chalcopyrite	0	2	53	45

Flotation Results

Figure 2 shows the effect of grinding fineness on recovery rate of copper. It is observed that a rise of the proportion of the grinding fineness of −0.074 mm from 60 to 80% obviously increased the copper recovery, then decreased when it exceeded 80%. This is because the initial appearance was dissociation fragmentation, namely separation among different minerals. With the continuation of grinding time, penetrating fragmentation occurred, which meant that the copper-bearing minerals began to separate. This is evident in the SEM images from Fig. 3. Dissociation fragmentation is beneficial for the flotation, while penetrating fragmentation may interfere with it which is undesired.

A series of experiments was carried out to determine the best flotation collector among Z-200, ethyl xanthate, butyl xanthate, isopentyl xanthate and butylamine

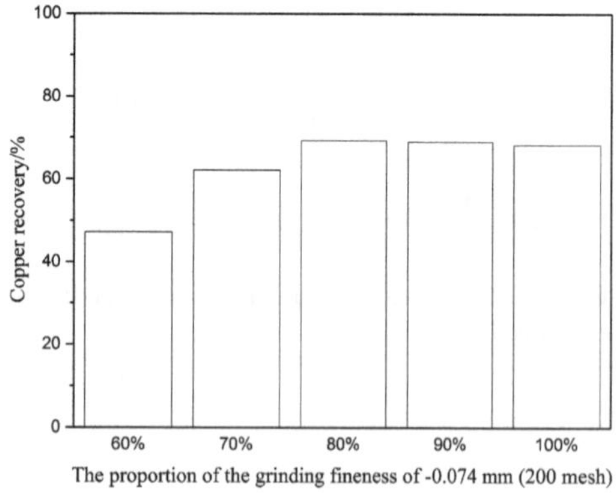

Fig. 2 Influence of grinding fineness on copper recovery. Experimental conditions: pulp density = 33%, collector (butyl xanthate) 50 g/t, terpineol oil 50 g/t, no pH regulator

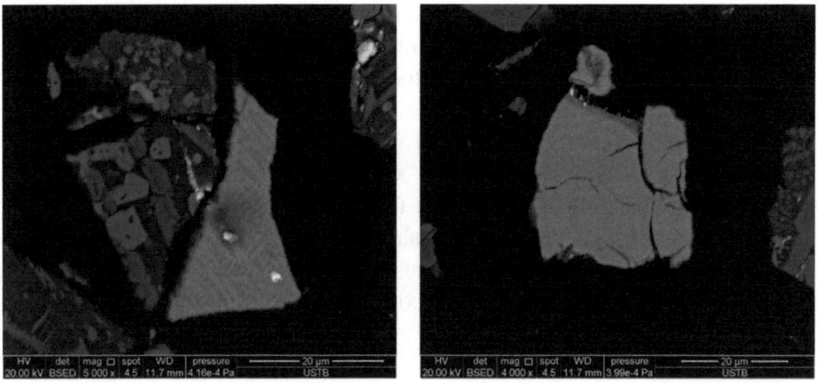

(a) Dissociation fragmentation (b) Penetrating fragmentation

Fig. 3 Images displaying the separation among mineral particles taken under SEM, in which the *light gray* parts were copper-bearing minerals

Table 4 Results of flotation tests with different collector

Collectors	Copper in concentrate/%	Copper in tail/%	Copper recovery (%)
Z-200	5.05	0.687	50.53
Ethyl xanthate	6.00	0.547	61.02
Butyl xanthate	6.42	0.484	69.27
Isopentyl xanthate	4.41	0.552	62.60
Butylamine dithiophosphate	8.10	0.561	58.08

Table 5 Results of flotation tests with different pH value

pH value	Copper in concentrate/%	Copper in tail/%	Copper recovery (%)
pH = 4	4.49	0.601	58.27
pH = 7	6.42	0.484	69.27
pH = 10	7.44	0.431	73.01
pH = 12	5.46	0.618	55.67

Table 6 Results of flotation using different pH regulator

pH regulator	Copper in concentrate/%	Copper in tail/%	Copper recovery (%)
CaO	10.22	0.500	70.67
Na_2CO_3	7.44	0.431	73.01
NaOH	6.97	0.480	75.23

dithiophosphate. The experimental conditions were pulp density = 33%, collector 50 g/t, terpineol oil 50 g/t without pH regulator. It is shown in Table 4 that the application of butyl xanthate as a collector achieved 69.27% copper recovery rate, which is the best flotation result.

In order to determine the impact of pH value and pH regulators on flotation, some experiments were conducted using butyl xanthate as a collector. The experimental conditions were same as before, but with the addition of pH regulator. It is inferred from Tables 5 and 6 that the best pH value to recover copper was 10, which was adjusted using sodium carbonate. The cost also needs to be considered, although sodium hydroxide showed slightly more copper recovery than sodium carbonate.

Conclusions

Taking into account that the slag contained more copper than ore, a series of flotation tests was conducted on copper smelting slag produced in an electric furnace of a copper plant in China. Based on the mineralogical analyses of slag, the tested copper smelting slag was composed of fine particles with complex association and dissemination, in which bornite was the main copper-bearing mineral. Under the condition of the grinding fineness of −0.074 mm and the proportion of 80%, flotation exhibited highest efficiency. The results also prove that the copper smelting slag particles with finer size do not imply that they show or exhibits better copper recovery. Addition of butyl xanthate at the concentration of 50 g/t gave the best recovery effect. By changing the pH of the pulp solution to 10 with sodium carbonate, the recovery significantly increased. In summary, copper concentrate grading at 14.47% with 79.66% Cu recovery was obtained, under the optimal condition of the grinding fineness of −0.074 mm and the proportion of 80%, with the reagent dosages of 50 g/t of butyl xanthate and a pH value of 10 adjusted using Na_2CO_3.

Acknowledgements The authors gratefully acknowledge the financial support of the National Science Foundation of PRC for the Research Project (2012-51234008).

References

1. K. Zhao, X. Cheng, Y. Qi, J. Gao, X. Shi, Analysis of copper slag treatment process and new technology of comprehensive utilization. China Nonferrous Metall **41**(1), 56–60 (2012)
2. H. Yang, Y. Yuan, L. Zhang, C. Wang, Q. Tang, Present situation and proposed method of recycling iron and copper from copper slag. Metal Mine (5), 65–168 (2012)
3. H.S. Altundogan, M. Boyrazli, F. Tumen, A study on the sulphuric acid leaching of copper converter slag in the presence of dichromate. Miner. Eng. **17**(3), 465–467 (2004)

4. W.J. Bruckard, M. Somerville, F. Hao, The recovery of copper, by flotation, from calcium-ferrite-based slags made in continuous pilot plant smelting trials. Miner. Eng. **17**(4), 495–504 (2004)
5. K. Maweja, T. Mukongo, I. Mutombo, Cleaning of a copper matte smelting slag from a water-jacket furnace by direct reduction of heavy metals. J. Hazard. Mater. **164**(2–3), 856–862 (2009)
6. F. Carranza, R. Romero, A. Mazuelos, N. Iglesias, O. Forcat, Biorecovery of copper from converter slags: slags characterization and exploratory ferric leaching tests. Hydrometallurgy **97**(1–2), 39–45 (2009)
7. B. Gorai, R.K. Jana, Premchand, Characteristics and utilisation of copper slag—a review. Resour Conserv Recycl **39**(4), 299–313 (2003)

Reaction Mechanisms in the Silicothermic Production of Magnesium

M. Chen, B.J. Zhao, Y.H. Chen, F.L. Han and L.E. Wu

Abstract Magnesium is the lightest structural metal with a density only two-thirds of aluminium and one-quarter of steel. It can be produced by electrolysis of fused magnesium chloride or high temperature reduction of the oxide. The predominant method for magnesium production is the Pidgeon process which is a batch process starting from calcined dolomite and ferrosilicon. CaF_2 is commonly used in the process as the catalyst. Analyses of the slag samples and thermodynamic calculations indicated that liquid phase plays important role in the reduction process. Effects of chemistry, temperature and gas pressure on the production have been discussed and compared with the operating data. Understanding of the reaction mechanisms will provide a fundamental base for alternative processes of the magnesium production.

Keywords Pidgeon process · Mg slag · Factsage calculation · Mechanisms

Introduction

Magnesium, as a new star of the structural alloy materials, has attracted more and more attentions from the researchers [1]. Due to its low density and outstanding mechanical strength, the Mg-containing alloy can be used to make lighter cars, airplanes and etc., which will also beneficial for energy saving. World production of primary magnesium is increased significantly from 737 kt in 2010 to 973 kt in 2014 [2]. In 2014 China's output was 874 kt which accounted for 90% of the global total. Although several new magnesium production technologies have been developed recent years [3, 4], the Pidgeon process is still the major process for Mg production, particularly in China.

M. Chen · B.J. Zhao (✉)
The University of Queensland, Brisbane, Australia
e-mail: baojun@uq.edu.au

Y.H. Chen · F.L. Han · L.E. Wu
Beifang University of Nationalities, Yinchuan, China

S. Wang et al. (eds.), *Applications of Process Engineering Principles in Materials Processing, Energy and Environmental Technologies*,
The Minerals, Metals & Materials Series, DOI 10.1007/978-3-319-51091-0_22

In Pidgeon process ferrosilicon is used as the reductant to reduce MgO to produce Mg under low pressures. It is usually claimed to be a solid-solid reaction. CaF_2 has been used as the catalyst in the process to accelerate the chemical reaction since the technology started. However, CaF_2 in the slag is harmful to the environment and operators' health and needs to be replaced or limited. To replace CaF_2 in the Pidgeon process, the reaction mechanisms involving CaF_2 have to be clearly understood. In the present study, the role of CaF_2 in the Pidgeon process will be studied and the possible substitution of CaF_2 by B_2O_3 as the catalyst is also discussed.

Experimental and Thermodynamic Calculation Conditions

The experiments to investigate the Pidgeon process were undertaken in pilot-scale ($\sim$10 kg) in Beifang University of Nationalities, China [5, 6]. The experiments were carried out in the heat resistant steel retorts (ϕ 159 mm * 760 mm) heated by muffle furnace with the roots-type vacuum pump to simulate the Pidgeon process Mg production. Kilos of industrial grade dolomite, ferrosilicon $FeSi_3$ and catalysts (CaF_2 or B_2O_3) were ground, mixed and pelletized to be briquettes. The briquettes were charged into the retorts at temperature 1130 °C, and then the vacuum pump started to work. The experimental conditions were controlled to be 1200 °C and 10 Pa at steady state. The bulk compositions of the starting materials in two experiments are given in Table 1. According to these receipts, the maximum Mg can be produced is 15 g per 100 g feed. After the experiments, the Mg ingots and Mg slags were collected and weighted. The Mg slags were then sent to the University of Queensland, Australia for EPMA analysis.

The thermodynamic calculations were conducted using FactSage thermodynamic software [7]. It should be noted that the predictions cannot be performed by FactSage 7.0, and the FactSage 6.2 is used in the present study. The FactSage calculations give predictions of the reactions achieving the thermodynamic equilibrium under various conditions. The optimum conditions were obtained and discussed based on the predictions. In the FactSage thermodynamic software, the solution species chosen for calculations were: (1) for the calculations of CaF_2-containing system: FToxid-SLAGH, FToxid-MeO_A, FToxid-bC2S, FToxid-aC2S and FACT-FeLQ; (2) for the calculations of B_2O-$_3$-containing system: FToxid-SLAGA, FToxid-MeO_A, FToxid-bC2S, FToxid-aC2S and FACT-FeLQ.

Table 1 The compositions of the starting materials in Pidgeon process experiments using CaF_2 and B_2O_3

Experimental no.	Si	Fe	MgO	CaO	Al_2O_3	Na_2O	CaF_2
1	12.0	4.0	24.8	53.7	2.1	1.0	2.5
	Si	Fe	MgO	CaO	Al_2O_3	Na_2O	B_2O_3
2	12.0	4.0	24.8	53.7	2.1	1.0	2.5

Use of CaF_2 as the Catalyst

As the conventional catalyst in Pidgeon process, the experiment with the addition of CaF_2 is used as a comparison. The Mg ingot collected after the experiment were analysed, and the Mg content in the Mg ingots were around 93%. The yield of the production reached the targeted requirement. The slag samples were analysed by EPMA. The typical microstructure of the Mg slag using CaF_2 as the catalyst is shown in Fig. 1. It can be seen that Ca_2SiO_4 solid phase is the major phase, and Si–Fe alloy distributes in the sample. The glass phase was identified beside one of the large ferrosilicon alloys. The compositions of the three phases are listed in Table 2.

It should be kept in mind that the sample was slow-cooled after the experiment. The proportions and compositions of the phases are different from those at operating temperature. The presence of the glass phase indicates that liquid slag was present at the reaction temperature. The shape of the Si–Fe alloy in the sample indicates that the ferrosilicon was liquid at the reaction temperature. From the Fe–Si binary phase diagram between 30 and 75 wt% Si, the liquid formation temperature can be as low as 1203 °C, which may be even lower if C and/or Al is present. Therefore, the high temperature reactions could be liquid (Si–Fe)—solid (MgO) or even liquid (Si–Fe)—liquid (MgO-containing slag). FactSage calculations below will be used to explore the possible reactions further.

The Experiment 1 shown in Table 1 is simulated thermodynamically using FactSage 6.2 [7] as the starting composition. The calculations show that in gas phases, the Mg gas, Ca gas, Na gas and fluoride gases will be present in all the

Fig. 1 Typical microstructure of the SEM microstructure of the Mg slag using CaF_2

Table 2 Compositions of three phases: glass, Ca_2SiO_4 and ferrosilicon (wt%)

Phases	CaO	SiO_2	MgO	Al_2O_3	CaF_2
Glass	58.9	24.5	0.5	14.4	1.7
Ca_2SiO_4	64.6	34.6	0.2	0.6	0
		Si	Fe		
Si–Fe alloy		31.1	68.9		

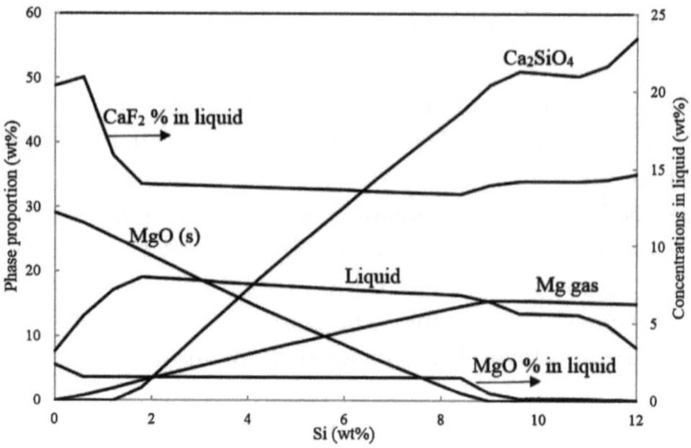

Fig. 2 Proportions of the main phases and concentrations of CaF_2 and MgO in liquid as a function of Si at 1200 °C, 10 Pa predicted by FactSage 6.2 [7]

conditions, and the amounts of Na and fluoride in the gas are less than 1 wt%. Ca_2SiO_4 is the major solid phase and small amounts of CaF_2, CaO and CaSi may also be formed during the reactions. For the presentation purpose only, only the major phases are shown in the figures.

The core reaction in the Pidgeon process is the silicothermic reduction to produce Mg gas. The reduction mechanism can be analysed by the gradual reaction of Si with the oxides. The equilibrium states of the present systems with Si additions from 0 to 12 were calculated using FactSage 6.2 [7] at 1200 °C and 10 Pa. The results are shown in Fig. 2. It can be seen that at the beginning of the reaction, the liquid is formed with nearly 20% CaF_2 and 4–5% MgO. In addition, Al_2O_3, Na_2O, CaO and SiO_2 are also present in the liquid phase. With addition of Si, amounts of liquid and MgO in the liquid are almost constant. Solid MgO is reducing and Ca_2SiO_4 is increasing as a result of CaO–SiO_2 reaction. The solid MgO phase disappears when 8.4 wt% Si is added and the MgO concentration in liquid starts to decrease until zero at 9.6 wt% Si. Between 0–9.6 wt% Si additions, the Mg gas is generated continuously. Throughout the reduction process, CaF_2 concentration in the liquid is kept between 15 and 20 wt% which confirms that CaF_2 is the major liquid-forming component. Based on the present calculations the mechanisms of the silicothermic reduction of MgO can be proposed as the follows:

1. Liquid formation: Si–Fe–Al–C (s) → Si–Fe–Al–C (l)

$$MgO(s) + CaF_2(s) + Al_2O_3(s) + Na_2O(s) + CaO(s) + SiO_2(s) \rightarrow \text{liquid slag}$$

2. Reduction of MgO in liquid slag by Si in liquid alloy:

$$MgO(l) + Si\ (l) \rightarrow Mg(g) + SiO_2(s)$$

3. SiO_2 reacts with liquid slag to form solid Ca_2SiO_4:

$$SiO_2(s) + \text{liquid slag} \rightarrow Ca_2SiO_4(s) + \text{liquid slag}$$

4. Solid CaO and MgO dissolve into liquid slag to continue Reaction 2:

$$CaO(s) + MgO(s) \rightarrow \text{liquid slag}$$

Figure 3 shows the changes of main phases in the Pidgeon process as a function of temperature with fixed pressure 10 Pa. It can be seen that theoretically Mg gas can be formed at a temperature as low as 1000 °C. However, at low temperatures, the reaction rate is very slow due to the solid—solid reactions. The formation of liquid phase will be able to increase the reaction rate of the reduction. It can be seen from Fig. 3 that, at the temperature above 1060 °C the solid CaF_2 will completely dissolve into the liquid and nearly 10% liquid phase is present in the system. This is the minimum temperature in which a reasonable reaction rate can be obtained. On the other hand, it can be seen that at temperatures above 1260 °C, the amount of liquid phase starts to decrease due to vaporisation of CaF_2. Reduced liquid will decrease the contact area between the reactants and slow down the reactions. It also can be seen that, Ca gas will be increased at high temperature which will consume the reductant and introduce impurities into the Mg product. Therefore, the maximum reaction temperature should be limited to 1260 °C.

Effect of pressure on the reduction process is presented in Fig. 4 at fixed temperature 1200 °C. It is noted that amount of liquid decreases with decreasing gas pressure. When the gas pressure is lower than 6 Pa the amount of liquid decreases significantly due to the rapid vaporisation of F. In contrast, Ca gas will increase

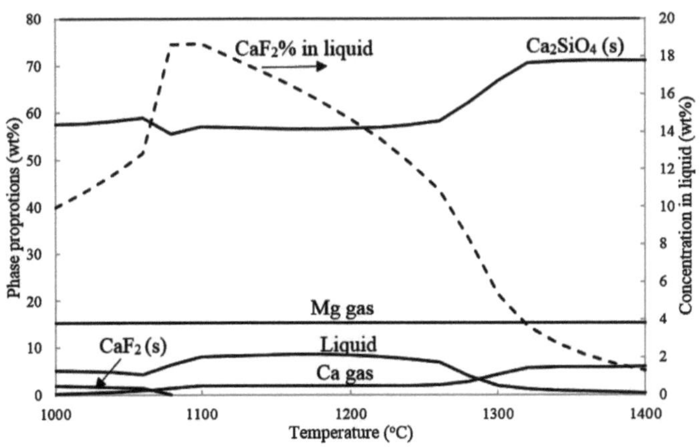

Fig. 3 The proportions of the main phases in bulk and concentration of CaF_2 in liquid as the function of temperatures at 10 Pa predicted by FactSage 6.2 [7]

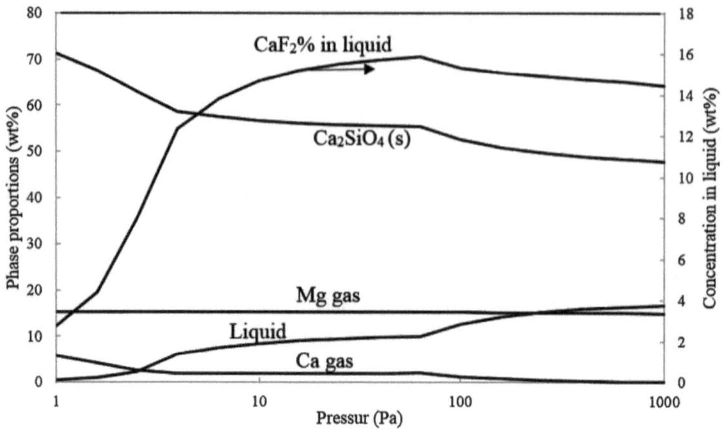

Fig. 4 The proportions of the main phases in bulk and concentration of CaF_2 in liquid as the function of pressures at 1200 °C predicted by FactSage 6.2 [7]

with decreasing gas pressure. 100–1000 Pa seems to be an optimum range for the gas pressure to have the required liquid and little Ca gas.

As a catalyst, it is important to control the amount of CaF_2 used in the process. CaF_2 is the key to form liquid to accelerate the reduction of MgO. On the other hand, extra CaF_2 will not only increases the direct cost, but also causes difficulties in the slag removal from the furnace. The effect of CaF_2 addition on phase equilibria of the system is shown in Fig. 5 at 1200°C and 10 Pa. It can be seen that the Mg formation is not affected by the addition of CaF_2. The proportion of liquid formed and CaF_2 in the liquid increase with increasing CaF_2 added. In contrast,

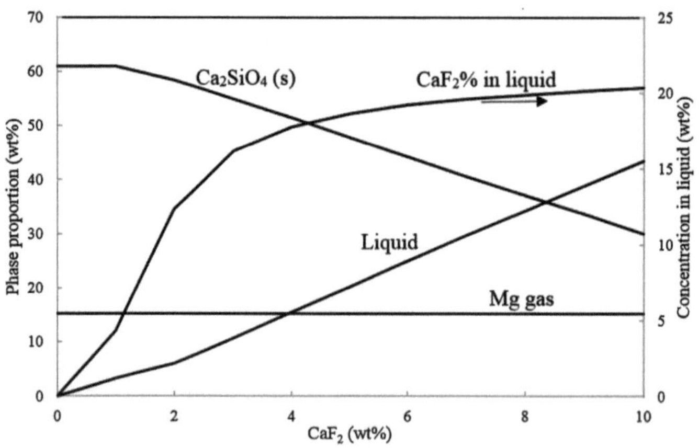

Fig. 5 Proportions of the main phases and concentration of CaF_2 in liquid as a function of CaF_2 addition at 1200 °C and 10 Pa predicted by FactSage 6.2 [7]

solid Ca_2SiO_4 decreases with increasing CaF_2 addition. Industry practices have shown that 2–3 wt% CaF_2 is the optimum range to ensure the smooth operation. This indicates that 10 wt% liquid will be enough to maintain the reduction reaction to continue. At this CaF_2 addition, approximately 15 wt% CaF_2 is present in the liquid phase which also ensures the liquid slag to have a low viscosity.

Use of B_2O_3 as the Catalyst

If the assumption for the catalytic mechanism of CaF_2 is confirmed, any chemicals that can maintain the required liquid proportion and viscosity will be able to be used as catalyst in the silicothermic reduction of MgO. B_2O_3 has the similar functions in reducing formation temperature and viscosity of liquid phase. On the other hand, magnesium borate (ascharite) is present naturally as a clean and cheap sources of both magnesium and boron. It could be considered as a replacement of CaF_2 in the production of Mg to avoid the fluoride pollutions. As shown in Table 1, 2.5 wt% B_2O_3 was used in the experiment at the same conditions as the experiment with CaF_2 addition. It was confirmed that the same quality and quantity of Mg can be produced with the addition of B_2O_3. The typical microstructure of the slag after the experiment is shown in Fig. 6 and the compositions of the phases measured by EPMA are listed in Table 3. It can be seen from the figure that Ca_2SiO_4 is the dominate phase and glass, Si–Fe alloy and $MgO–B_2O_3$ phase are also present. Surprisingly it is found that nearly 20 wt% Al_2O_3 is present in the glass phase. However, the composition of the glass should be different from that of the liquid at operating temperature as the sample was slow-cooled.

Thermodynamical simulation of the Pidgeon process Mg production using B_2O_3 is also conducted using FactSage 6.2 [7]. Similarly, Ca gas and Na gas are found to be present in all the calculations together with the major Mg gas. In the solid phases, Ca_2SiO_4 is not always the major phase in the solid phases. At high B_2O_3 addition, the boric compounds are also formed according to the predictions. Limited

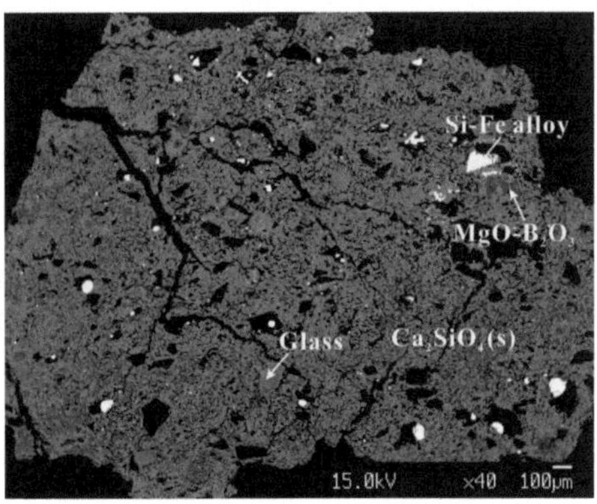

Fig. 6 Typical microstructure of the SEM microstructure of the Mg slag using B_2O_3

Table 3 Compositions of liquid, Ca_2SiO_4, $MgO–B_2O_3$ and ferrosilicon (wt%)

Phases	CaO	SiO_2	MgO	Al_2O_3	B_2O_3
Glass	51.7	24.7	0.2	19.4	4
Ca_2SiO_4	64.5	34.0	0.3	1.0	0.2
$MgO–B_2O_3$	0	0	90.9	0	9.1
		Si	Fe		
Si–Fe alloy		20.2	79.8		

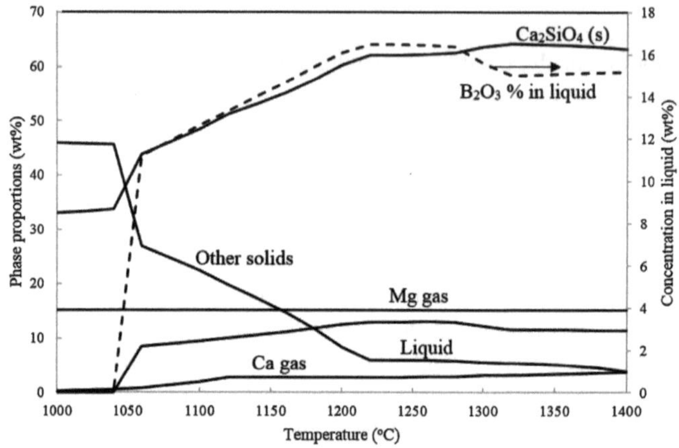

Fig. 7 The proportions of the main phases in bulk and concentration of B_2O_3 in liquid as the function of temperatures at 10 Pa predicted by FactSage 6.2

amount of CaSi may also be formed. For the presentation purpose, only the major phases are shown in the following figures.

Figure 7 shows the changes of main phases as a function of temperature at fixed pressure 10 Pa. It can be seen from the figure that, thermodynamically MgO can be fully reduced to Mg gas all over the temperatures calculated. The liquid will start to form from 1040 °C and then maintain approximately 10 wt% at temperatures above 1060 °C. The B_2O_3 concentration in the liquid is approximately 15 wt% at 1200 °C which is similar to that with CaF_2 addition. Ca_2SiO_4 is always the major solid phase. However, $Ca_3Al_2O_6$ and $Ca_{11}B_2Si_4O_{22}$ may also be present at low temperatures. Ca gas can also form at high temperature indicating that operating temperature needs to be limited to avoid impurities.

The effect of pressure on the Pidgeon process at 1200 °C is shown in Fig. 8 for the B_2O_3-containing system. It seems that the changes of pressure do not have significant effects on the proportions of Mg gas and liquid phase at the pressure range calculated. B_2O_3 concentration in the liquid is maintained the same level 16.5 wt% at the pressure range calculated. However, the predictions indicate that, when the gas pressure is lower than 60 Pa, Ca gas formation will be significant and the amount of CaO solid phase is reduced.

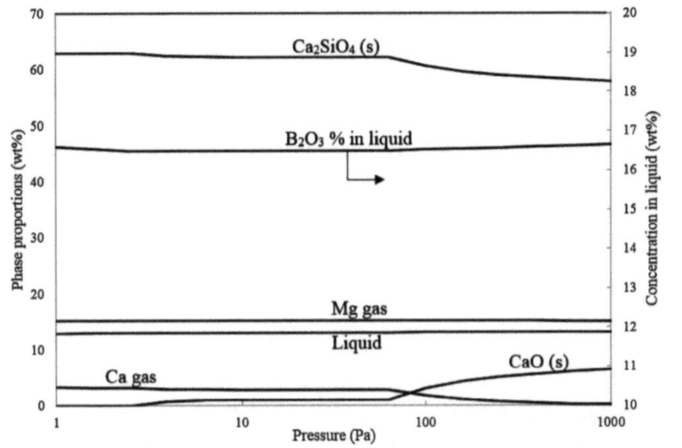

Fig. 8 The proportions of the main phases in bulk and concentration of B_2O_3 in liquid as the function of temperatures at 1220 °C predicted by FactSage 6.2 [7]

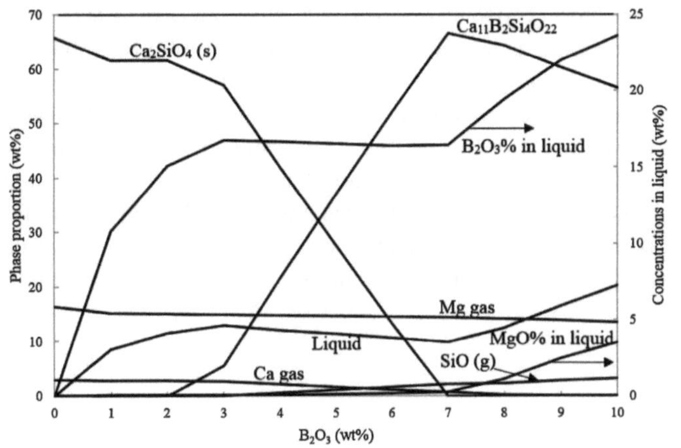

Fig. 9 The proportions of the main phases and concentrations of B_2O_3 and MgO in liquid as a function of B_2O_3 addition at 1220 °C and 10 Pa predicted by FactSage 6.2 [7]

The effect of B_2O_3 addition on the phase equilibria of the system is calculated at 1200 °C and pressure 10 Pa as shown in Fig. 9. The absolute amount of Mg gas formed is not affected by the addition of B_2O_3. The liquid phase proportion will reach 10 wt% when 1.5 wt% B_2O_3 is added into the system. A further increase of B_2O_3 addition will cause the formation of $Ca_{11}B_2Si_4O_{22}$ and decrease of Ca_2SiO_4. With the increase of B_2O_3 in the system, Ca gas decreases but SiO gas will form and increase.

It seems that B_2O_3 can be used in the Pidgeon process Mg production to replace CaF_2 as demonstrated by experiments and thermodynamic calculations. The comparisons show that B_2O_3 has a similar role as CaF_2 in the Pidgeon process. Similar levels of liquid phase can be generated in the system to accelerate the reduction reaction of MgO by Si. Potentially the fluoride pollutions will be significantly reduced.

Conclusions

Experimental studies and thermodynamic calculations have been conducted for silicothermic reduction of MgO using CaF_2 and B_2O_3 as catalysts. The catalytic mechanisms of CaF_2 and B_2O_3 in the Pidgeon process have been proposed. The following conclusions may be drawn: (1) ferrosilicon with impurities of aluminium and carbon could be liquid during the reduction process; (2) it is the MgO dissolved in the liquid slag to be reduced by Si; (3) the role of the CaF_2 as a catalyst is to form liquid and maintain low viscosity of the liquid; (4) extra liquid formed by higher temperature or more CaF_2 causes difficult slag removal after the production; (5) B_2O_3 may be used as a catalyst to replace CaF_2 in the Pidgeon process.

The present study proposed new concepts for the Pidgeon process and further experiments are required to obtain the quantitative results. The samples in the present study were not quenched from high temperature after the experiments. Analyses of the slags did not give accurate information at the operating temperatures. CaF_2- and B_2O_3-containing systems are not well optimised in the FactSage database due to limited experimental data available. Systematic and accurate experimental data are required to confirm and improve the thermodynamic predictions.

References

1. L.Y. Chen, J.Q. Xu, H. Choi, M. Pozuelo, X.L. Ma, S. Bhowmick, J.M. Yang, S. Mathaudhu, X.C. Li, Processing and properties of magnesium containing a dense uniform dispersion of nanoparticles. Nature **528**, 539–543 (2015)
2. E.L. Bray, *2014 minerals yearbook, magnesium* (USGS, US, 2016)
3. F. Habashi, *Handbook of Extractive Metallurgy* (Wiley-VCH Verlag GmbH, Weinheim, Germany, 1997)
4. M.A. Schoukens, M. Freeman, Technological breakthrough of the mintek thermal magnesium process. J. S. Afr. Inst. Min. Metall. **106**(1), 25–29 (2006)
5. L.E. Wu, F.L. Han, Q.X. Yang, S.W. Guo, Fluoride emissions from Pidgeon process for magnesium production, in *Proceedings of the International Conference on Waste Technology, Philadelphia, US* (2012), pp. 150–161

6. F.L. Han, Q.X. Yang, L.E. Wu, C. Du, Innovative utilization of a borate additive in magnesium production to decrease environmental impact of fluorides from Pidgeon process. Adv. Mater. Res. **690–693**, 378–389 (2013)
7. C. Bale, E. Bélisle, P. Chartrand, S. Degterov, G. Eriksson, K. Hack, I.H. Jung, Y.B. Kang, J. Melançon, A. Pelton, FactSage thermochemical software and databases–recent developments. Calphad **33**, 295–311 (2009)

Influences of $CaO/SiO_2/MgO/Al_2O_3$ on the Formation Behavior of FeO-Bearing Primary-Slags in Blast Furnace

Dongdong Wang, Kaihui Ma, Yang Xu, Jian Xu and Liangying Wen

Abstract Slags play an important role in blast furnace operation, and their compositions are based on the $CaO–SiO_2–MgO–Al_2O_3$ quaternary system. However, there is not a clear understanding of the effects of gangue on of the formation of FeO-bearing primary-slags process in blast furnace. In this work, the softening and dripping experiments under the blast furnace conditions are designed to explore the influences of CaO-/SiO_2-/MgO-/Al_2O_3 on the softening and melting properties of FeO respectively. The results indicate that additions of CaO or Al_2O_3 decrease the starting softening temperature and no dripping behaviors are found in comparison with the base case results when only FeO is used. On the contrary, the addition of SiO_2 or MgO rises the starting and ending softening temperature, as well as the dripping temperature. The lowest maximum pressure drop is obtained in the case with addition of SiO_2. According to XRD analysis results, the initial phase with CaO addition in the primary-slags should be $CaO·Fe_3O_4$ and that with SiO_2, MgO, Al_2O_3 additions are fayalite ($2FeO·SiO_2$), magnesioferrite (MgO·FeO), hercynite ($FeO·2Al_2O_3$), respectively.

Keywords Primary-slags · Softening and dripping · FeO-bearing

Introduction

The cohesive zone inside blast furnace, where the ferrous burden softens and melts, significantly affects the productivity of the blast furnace [1]. A Reduction Degree (RD) of a ferrous burden is typically above 50% in the cohesive zone area of a BF [2, 3]. At this point the iron bearing burden consists of metallic iron (Fe), wustite (FeO) and other oxides (CaO, SiO_2, MgO, Al_2O_3 etc.) which form a slag phase. The phase compositions are dependent on the chemistry of the iron bearing burden [4–8].

D. Wang · K. Ma · Y. Xu · J. Xu (✉) · L. Wen
College of Materials Science and Engineering,
Chongqing University, Chongqing 400044, China
e-mail: jxu@cqu.edu.cn

S. Wang et al. (eds.), *Applications of Process Engineering Principles in Materials Processing, Energy and Environmental Technologies*,
The Minerals, Metals & Materials Series, DOI 10.1007/978-3-319-51091-0_23

Many researchers have found that the softening and melting properties of ferrous materials are controlled by reduction degree, basicity, slag viscosity, gangue content and their distribution in the microstructure [9–12].

However, though various studies are conducted to research the properties of blast furnace slags, no satisfactory knowledge has been acquired about gangue on of the formation of FeO-bearing primary-slags process in blast furnace. Because the mixture between sinter and lump ores in blast furnace was not simply physical mixing, it is of great significance to study the influences of CaO-/SiO_2-/MgO-/Al_2O_3 on the softening and melting properties of FeO respectively [13–15]. In the present work, the softening and dripping experiments under the blast furnace conditions are designed to explore the influences of respective CaO-/SiO_2-/MgO-/Al_2O_3 on the softening and melting properties of FeO. The chemical compositions, liquidus temperature of primary-slags were also studied.

Experimental

Experimental Method

The samples were prepared from $CaCO_3$ ($\geq$99.9%), SiO_2($\geq$99.9%), MgO ($\geq$99.9%), Al_2O_3($\geq$99.9%) and $FeC_2O_4 \cdot 2H_2O$ ($\geq$99.9%) and the powdery $CaCO_3$/SiO_2/MgO/Al_2O_3 were uniformly mixed with $FeC_2O_4 \cdot 2H_2O$ respectively with a 70:10:10:5:5 (FeO:CaO:SiO_2:MgO:Al_2O_3) mass ratio and then pressed to cylindrical shape samples. The samples were finally roasted to 1173 K (900 °C) with a triple-fired furnace in the inert gas atmosphere (N_2, 99.9%). This temperature

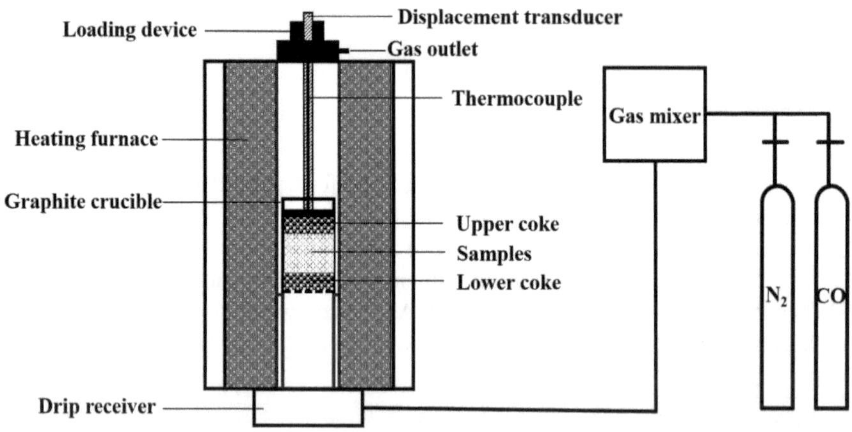

Fig. 1 Schematic diagram of softening-melting equipment

was held for 1 h, in order to make sure the complete decomposition of $FeC_2O_4{\cdot}2H_2O$. Finally, the samples were cooled to the room temperature in the inert gas atmosphere. Both the above mentioned stages were accomplished in the softening-melting equipment with the schematic diagram shown in Fig. 1.

Samples having a layer thickness of 50 mm were charged in a graphite crucible and cokes with a layer thickness of 15 mm were placed over and below the samples. The iron ore samples are 10–15 mm in diameter and the cokes are 6–10 mm in diameter. Subject to the condition of constant layer thickness, the charge of the iron ore samples weighted different due to the difference of density. The inner diameter of the graphite crucible is about 50 mm.

Experimental conditions for the softening and dripping behavior are shown in Fig. 2. The heating up rate is 10 °C/min below 900 °C and 5 °C/min over 900 °C to make sure the iron ore samples are adequately reduced after 900 °C. Gas flow is 5 L/min of N_2 below 900 °C and 10 L/min of reducing gas over 900 °C, the composition of reducing gas is $CO:N_2$ in mole fraction = 40:60(%). The load of 1 kg/cm^2 is added to the samples in the whole process. The experiment was stopped when pressure drop reached the maximum. The samples are cooled down to room temperature by N_2 with the flow rate 5 L/min. In this work, a series of five experiments to determine the impact of CaO-/SiO_2-/MgO-/Al_2O_3 to the softening and melting properties of samples, the chemical compositions of the samples for softening and melting experiment as listed in Table 1.

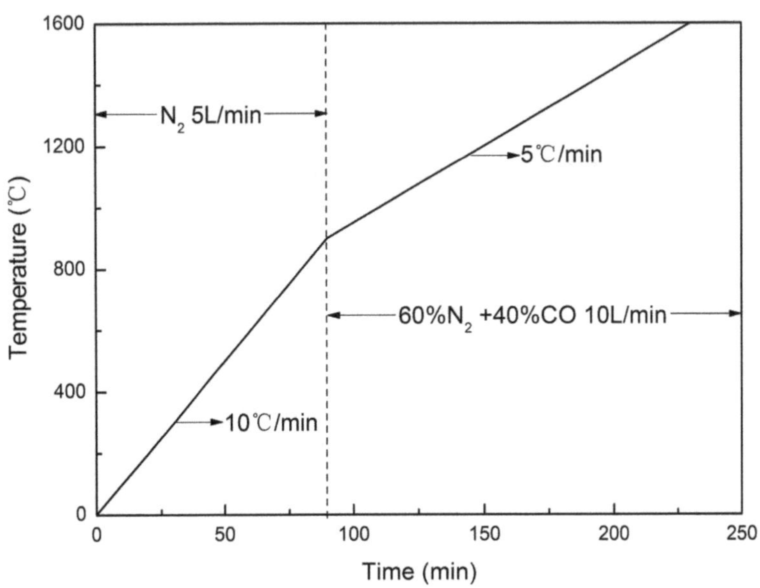

Fig. 2 Experimental conditions for softening and dripping experiment

Table 1 Chemical compositions of samples

Samples	FeO	CaO	SiO_2	MgO	Al_2O_3
Base case	✓	–	–	–	–
A	✓	✓	–	–	–
B	✓	–	✓	–	–
C	✓	–	–	✓	–
D	✓	–	–	–	✓

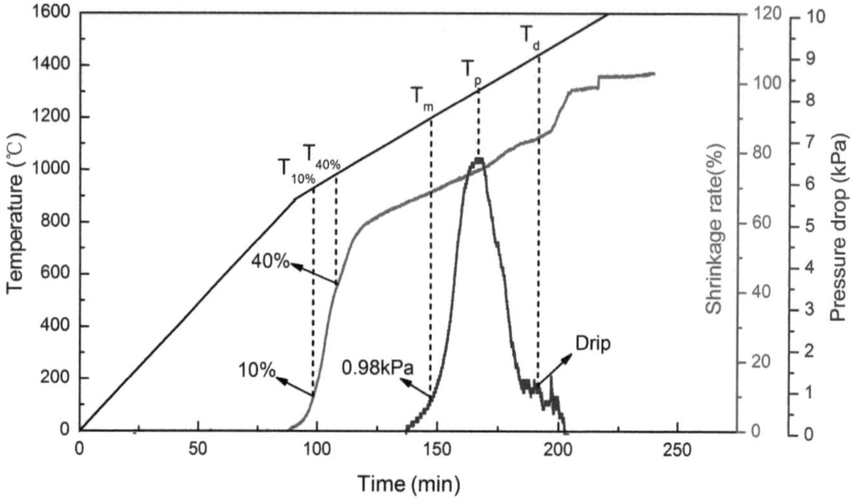

Fig. 3 A typical set of softening and dripping experiment

Results and Discussion

Softening and Melting Properties of Base Case

A typical set of softening and dripping test results is shown in Fig. 3, in which pressure drop, temperature and shrinkage rate are conducted to evaluate the primary slags formation behaviors of iron ore samples. It is difficult to observe the behavior directly as the internal of the softening-melting equipment is not visible. Then some indexes are conducted to evaluate the primary-slags formation behavior of iron ores. $T_{10\%}$ is the temperature when the shrinkage rate of the samples reaches 10%, indicating the samples start to soften, softening end temperature ($T_{40\%}$) is the temperature when the shrinkage of the samples reaches 40%. The temperature interval ($T_{40\%}$–$T_{10\%}$) represents the softening zone of samples. T_m is the temperature when the pressure drop of the samples reaches 0.98 kPa, meaning the samples begin to melt. T_p is the temperature when the pressure drop of the samples reaches the highest (ΔP_{max}), meaning the primary–slags are totally produced. T_d is the temperature at which dripping starts and the temperature interval (T_d–$T_{10\%}$) reflect the thickness of the cohesive zone.

Table 2 Softening and melting properties of base case

Samples	$T_{10\ \%}$/°C	$T_{40\ \%}$/°C	$T_{40\ \%}$-$T_{10\ \%}$/°C	T_d/°C	T_d-$T_{10\ \%}$/°C	ΔP_{max}/kPa
Base case	784	895	111	1436	652	6.6

Table 3 Softening and melting properties of samples

Samples	$T_{10\ \%}$/°C	$T_{40\ \%}$/°C	$T_{40\ \%}$-$T_{10\ \%}$/°C	T_d/°C	T_d-$T_{10\ \%}$/°C	T_p/°C	ΔP_{max}/kPa
Base case	784	895	111	1436	652	1174	6.6
A	692	823	132	–	–	–	–
B	918	1080	162	1525	607	1403	1.5
C	799	947	148	–	–	–	–
D	704	918	214	–	–	1158	3.8

To study the whole process of the high temperature behaviors of iron ores and gain the primary-slags at T_p, every test contains two steps. Firstly, the whole process of samples from room temperature to dripping finished is conducted. Secondly, stop heating the samples at T_p and cool down the samples to room temperature under the protection of pure N_2.

The results of the softening and melting properties of base case were shown in Table 2. The results show that when sample is pure FeO, the softening zone is 111 °C, thickness of the cohesive zone is 652 °C, and the dripping temperature is 1436 °C.

Effects of the Samples Composition on Softening and Melting Properties

In order to study the impact of $CaO/SiO_2/MgO/Al_2O_3$ on softening and melting properties of iron ores, the samples composition was changed as listed in Table 1.

Table 3 shows the softening and melting properties of samples. For the respective samples, the starting and finishing softening temperature of sample B and C higher than base case and the the softening temperature interval are also wider than base case, of which the sample B temperature is the highest. Figure 4 shows the starting softening temperature of sample D is lower than pure FeO, but the finishing softening temperature is higher than base case, and softening temperature interval is extremely wide. For the sample A, both the starting and finishing softening temperature are lower than base case. While the dripping start temperature of sample B is higher (1525 °C) than base case, indicated that addition of SiO_2 increase the softening temperature and dripping start temperature as well as the dripping temperature. Although the maximum pressure drop of the case with the addition of SiO_2 decreases to 1.5 kPa, the Tp is the highest among all the cases. The addition of CaO, MgO and Al_2O_3 have no dripping temperature.

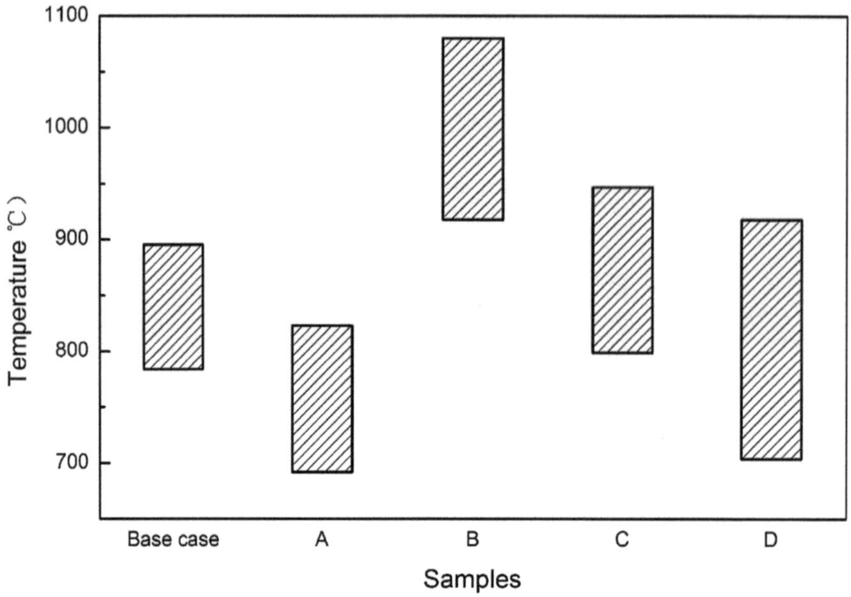

Fig. 4 Softening temperature intervals of samples

To further study the influence of $CaO–SiO_2–MgO–Al_2O_3$ on the formation behavior of FeO-bearing primary-slags, the primary-slags samples of A, B, C and D were analyzed through X-Ray diffraction. Figure 5 shows the diffraction patterns of primary-slags of samples. The main primary-slags phases of CaO additions are $CaO{\cdot}Fe_3O_4$ (liquidus temperature 1104 °C) and FeO, and that with SiO_2, MgO, Al_2O_3 additions are fayalite ($2FeO{\cdot}SiO_2$, 1889 °C), magnesioferrite (MgO·FeO, 1791 °C), hercynite ($FeO{\cdot}2Al_2O_3$, 1780 °C), respectively. The results of the liquidus temperature of the slags were obtained from FactSage6.2. The changes in the softening temperature of the slags are owing to the new compounds generated when compositions changed. Therefore, the decreasing softening temperature of the primary slag is mainly caused by its high CaO. While the slag has a higher SiO_2 and higher MgO content, the softening temperature is increased, because it will produce high liquidus temperature substances.

On the other hand, there was no doubt that FeO-bearing primary-slags was circumscribed, because the environment of blast furnace is complexed. The above analysis results inspired to design more experiments to study the impact of slag phase on the softening and melting properties.

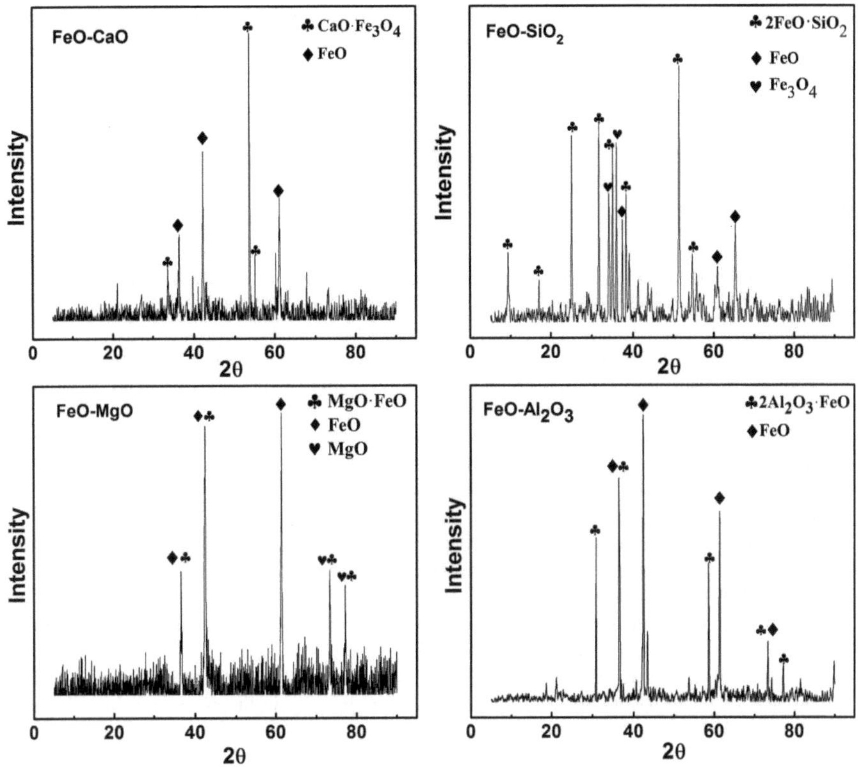

Fig. 5 X-ray diffraction analysis of samples

Conclusions

The influences of $CaO/SiO_2/MgO/Al_2O_3$ on the formation behavior of FeO-bearing primary-slags is experimentally demonstrated, and the following conclusions are obtained.

1. Additions of CaO or Al_2O_3 decrease the starting softening temperature and no dripping behaviors are found in comparison with the base case results when only FeO is used.
2. The addition of SiO_2 or MgO rises the starting and ending softening temperature, as well as the dripping temperature. In addition, the addition of SiO_2 decreases the maximum pressure drop to as low as 1.5 kPa while increases Tp to as high as 1403 °C.
3. According to the phase diagram and XRD analysis results, the initial phase with CaO addition in the primary-slags should be $CaO{\cdot}Fe_3O_4$ and that with SiO_2, MgO, Al_2O_3 additions are fayalite ($2FeO{\cdot}SiO_2$), magnesioferrite ($MgO{\cdot}FeO$), hercynite ($FeO{\cdot}2Al_2O_3$), respectively.

Acknowledgements The authors would like to thank the project supported by the National Natural Science Foundation of China (51304257), and the financial support from the Chongqing Research Program of Basic Research and Frontier Technology (cstc2015jcyjA50014) and National Natural Science Foundation of China (51374263) is also acknowledged.

References

1. T. Bakker, et al., *Softening in the Blast Furnace Process: Local Melt Formation as the Trigger for Softening of Iron Bearing Burden Materials*. Delft University of Technology (1999)
2. P. Kaushik et al., Mixed burden softening and melting phenomena in blast furnace operation Part 2—Mechanism of softening and melting and impact on cohesive zone. Ironmaking Steelmaking **33**, 520–528 (2006)
3. P.F. Nogueira et al., Blast furnace burden softening and melting phenomena: Part I. Pellet bulk interaction observation. Metall. Mater. Trans. B **35B**, 829–838 (2004)
4. P.F. Nogueira et al., Blast furnace burden softening and melting phenomena: Part II. Evolution of the structure of the pellets. Metall. Mater. Trans. B **36B**, 583–590 (2005)
5. P.F. Nogueira et al., Blast furnace burden softening and melting phenomena: Part III. melt onset and initial microstructural transformations in pellets and R. J. Fruehan. Metall. Mater. Trans. B **37B**, 551–558 (2006)
6. J.R. Kim et al., Influence of basicity and FeO content on viscosity of blast furnace type slags containing FeO. ISIJ Int. **44**, 1283–1290 (2004)
7. M. Hino et al., Simulation of primary-slag melting behavior in the cohesive zone of a blast furnace, considering the effect of Al_2O_3, Fe_tO, and basicity in the sinter ore. Metall. Mater. Trans. B **30B**, 671–683 (1999)
8. Y.S. Lee et al., Influence of basicity and FeO content on viscosity of blast furnace type slags containing FeO. ISIJ Int. **44**, 1283–1290 (2004)
9. J.R. Kim et al., Effect of Al_2O_3 and CaO/SiO_2 on the viscosity of calcium-silicate–based slags containing 10 mass Pct MgO. Metall. Mater. Trans. B **44**, 5–12 (2013)
10. M. Matsunura et al., Improvement of sinter softening property and reducibility by controlling chemical compositions. ISIJ Int. **45**, 594–602 (2005)
11. T. Nishimura et al., Evaluation of softening, shrinking and melting reduction behavior of raw materials for blast furnace. ISIJ Int. **51**, 1316–1321 (2011)
12. S.L. Wu et al., Increasing lump ores proportion in blast furnace based on the high-temperature interactivity of iron bearing materials. ISIJ Int. **50**, 686–694 (2010)
13. A. Shankar et al., Experimental investigation of the viscosities in $CaO–SiO_2–MgO–Al_2O_3$ and $CaO–SiO_2–MgO–Al_2O_3–TiO_2$ slags. Metall. Trans. B **38B**, 911–915 (2007)
14. S.L. Wu et al., Evaluation of lump ores for use in modern blast furnaces as part of mixed burden practice. Ironmaking Steelmaking **36**, 19–23 (2009)
15. A. Kemppainen et al., Effect of H_2 and H_2O on the reduction of olivine pellets in CO and CO_2 gas. ISIJ Int. **52**, 1973–1978 (2012)

Desulfurization of High Sulfur Coal Leached with H_2O_2 and NaOH by Microwave Irradiation

Pengqi Zhang, Shengfu Zhang, Lixiong Shao, Mingcheng Bing, Shuxing Qiu and Qingyun Zhang

Abstract Coal desulfurization has been regarded as an effective pretreatment for clean fuel and less environmental pollution. The microwave irradiation was used to desulfurize the Nantong coal leached with H_2O_2 and NaOH. The results showed the desulfurization rate of the coal is very low by directly microwave irradiation, while the coal samples leached with H_2O_2 and NaOH by microwave irradiation could effectively improve the desulfurization rate. The optimum desulfurization conditions were obtained as 10-min irradiation time, 800 W irradiation power, particle size below 74 μm, 1:1 concentration ratio of H_2O_2:NaOH, in which the desulfurization rate reached 51.14% of the total sulfur content. Attenuated total reflectance Fourier transform infrared (ATR-FTIR) spectra of coals structure before and after treatment have shown the bands in both coals changed slightly except the relative intensity of some absorption ones. It can be found that most of FeS_2 and –S–S– functional groups transferred after treatment, while –SH and –S=O– had almost no changing by ATR-FTIR spectra analysis. The structures of coal matrix would not be broken by microwave irradiation in reasonable the reaction conditions.

Keywords Coal · Desulfurization · Microwave irradiation · Sulfur

Introduction

Coal plays a critical role in the development of industry and economy. However, high sulfur coal accounts for 30% of the total coal reserves in China. Using the high sulfur coal directly would generate plenty of hazardous gas, such as SO_2 and H_2S, which will pollute the environment [1]. In iron and steel industry, if coke with a lot of sulfur can make steel more fragile and decrease its mechanical property. Therefore, it is necessary to remove sulfur from coal prior to its utilization [2, 3].

P. Zhang · S. Zhang (✉) · L. Shao · M. Bing · S. Qiu · Q. Zhang
College of Materials Science and Engineering, Chongqing University, Chongqing 400044, China
e-mail: zhangsf@cqu.edu.cn

S. Wang et al. (eds.), *Applications of Process Engineering Principles in Materials Processing, Energy and Environmental Technologies*,
The Minerals, Metals & Materials Series, DOI 10.1007/978-3-319-51091-0_24

Sulfur in coal contains inorganic sulfur and organic sulfur. The inorganic sulfur exists mostly in pyrites and sulfate. The organic sulfur includes mercaptan (R–SH), disulfides (R–S–S–R′), sulfides (R–S–R′), sulphone (R–SO–R), sulfoxide (R–SO_2–R) and thiophene (C_4H_4S). Generally, the organic sulfur is fixed in the chemical structure of coal matrix and is more difficult to remove it [4, 5].

At present, many methods have been used for coal desulfurization before combustion by physical, chemical and biological processes. Although physical desulfurization methods are in an effective cost and simple operation, it can only remove inorganic sulfur from coal [5, 6]. Chemical desulfurization methods can remove inorganic sulfur and organic sulfur, but they need harsh conditions and may break the coal properties. Biological desulfurization methods still stay in the laboratory stage due to a lack of stability and controllability [7].

Microwave is a kind of electromagnetic wave with frequencies ranging from 300 MHz to 300 GHz. Microwave irradiation can heat substance quickly, uniformly and selectively. It is a new desulfurization method in the field of coal clean technology. Moisture, some minerals and chemical reagents can increase the microwave absorption of coal [2, 3], which will avoid significantly changing the fundamental properties of the coal. Peter Wardle studied the selective heating of pyrite in coal using microwave energy. Using a multimode microwave cavity it was found that significant amounts of pyrrhotite were formed in 20s when using 15 kW of microwave power in coal. Treatment at lower power levels in this type of cavity was found to be less effective [7]. Desulfurization of Tabas coal with microwave irradiation and peroxyacetic acid washing has been reported by E. Jorjani. The results showed that microwave irradiation had a positive effect on desulfurization with peroxyacetic acid. The desulfurization rates enhanced from 36.1 to 61.9%. Peroxyacetic acid was believed to produce the hydroxyl cations that are strong electrophiles and react with sulfur atoms [8, 9]. Xu studied the sulfur forms change during coal desulfurization with NaOH by microwave irradiation. After the desulfurization, only a small amount of thiophene sulfur was removed, some sulfur-containing components were oxidized to sulfates sulfur. Under the condition of microwave irradiation, NaOH not only reacted with pyrites but also with partial organic sulfur [10]. Oxidative desulfurization techniques is another effective method to remove sulfur from the high sulfur coal. H_2O_2 has become a popular oxidant due to environment-friendly. It is convenient and effective, forming no toxic by-products which only is water [2, 5]. It is worth trying to remove sulfur from the coal using cooperation of microwave irradiation and leach of H_2O_2 and NaOH.

In this study, the microwave irradiation was used to desulfurize the raw Nantong coal and the coal leached with H_2O_2 and NaOH. The effects of microwave irradiation time, irradiation power, particle size and concentration ratio (H_2O_2:NaOH) on desulfurization were investigated by single factor analysis. Furthermore, the chemical structure change of coal matrix before and after treated was analyzed by Attenuated total reflectance Fourier transform infrared (ATR-FTIR).

Table 1 Characterization of Nantong coal representative sample

Proximate analysis (wt%)		Ultimate analysis (wt%)		Forms of sulfur (wt%)	
M_{ad}	1.24	C_{daf}	87.62	$S_{t,d}$	2.83
A_d	10.37	H_{daf}	4.48	$S_{s,d}$	0.04
V_d	17.92	N_{daf}	1.39	$S_{p,d}$	0.30
FC_d	71.71	O_{daf}	3.68	$S_{o,d}$	2.49

M moisture, *A* ash, *V* volatiles, *FC* fixed carbon, *ad* air dry *d* dry, *daf* dry ash free, *t* total, *s* sulfate, *p* pyrite, *o* organic

Experimental

Coal Sample

The coal samples to be studied was prepared from Nantong, Chongqing municipality. It is a kind of high organic sulfur coal. The sample was ground to a particle size of <300 μm after drying under ambient atmosphere and stored in a glass desiccator until further experiments. The proximate and ultimate analysis of the coal sample were performed with a WS-G410 elemental analyser. The GB/T 214-2007 and GB/T 215-2003 were applied to determine the total, pyrite and sulfate forms of sulfur in coal. Organic sulfur was calculated by difference [7]. The results were shown in Table 1.

Desulfurization Experiments by Microwave Irradiation

In microwave irradiation experiments, about 5 g coal was added into a 100 ml beaker. Microwave irradiation was carried out in a microwave oven with 2.45 GHz frequency and 100–1000 W energy. The aim in this work was to study the effects of irradiation time, irradiation power and coal particle size on the desulfurization rates of Nantong coal. The irradiation time of 1, 3, 5, 10 and 15 min were selected under 800 W and <74 μm. The irradiation powers of 100, 300, 500, 800 and 1000 W were employed under 10 min and <74 μm. The coal particle sizes of <74, 74 ~ 150, 150 ~ 300 μm were chosen under 800 W and 10 min.

Desulfurization Experiments Leached with H_2O_2 and NaOH by Microwave Irradiation

In leached with H_2O_2 and NaOH by microwave irradiation experiments, about 3 g coal,1 ml C_2H_5OH, 25 ml H_2O_2 and 25 ml NaOH were added into a 100 ml flask. The experiments were performed in the flask and microwave heating (Fig. 1).

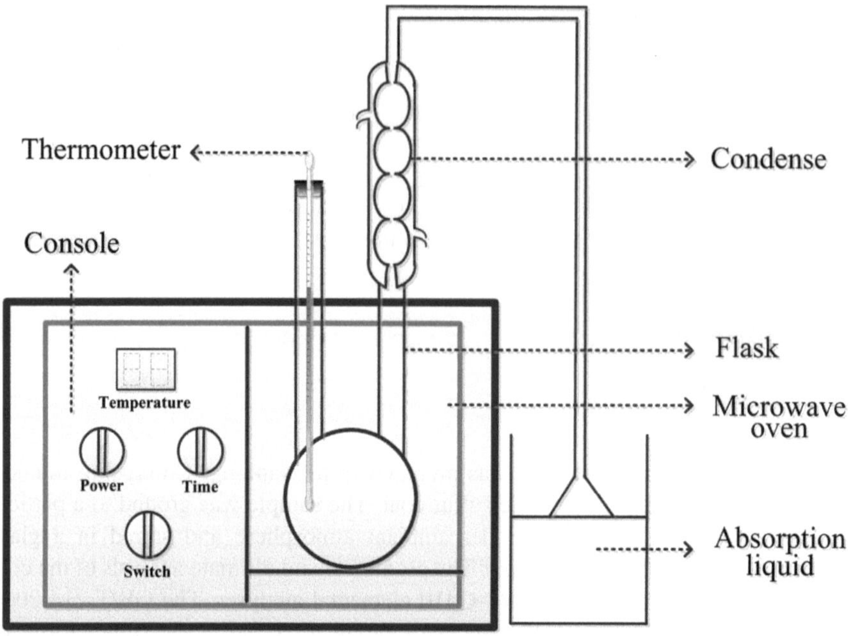

Fig. 1 The schematic of the experimental equipment

The aim in this work was to study the effects of irradiation time, irradiation power, coal particle size and the concentration ratio (H_2O_2:NaOH) on the desulfurization rates of Nantong coal. The irradiation time of 1, 3, 5, 10 and 15 min were selected under 800 W, <74 μm and H_2O_2:NaOH = 1:1. The irradiation powers of 100, 300, 500, 800 and 1000 W were employed under 10 min, <74 μm and H_2O_2:NaOH = 1:1. The coal particle sizes of <74, 74 ~ 150, 150 ~ 300 μm were chosen under 800 W, 10 min and H_2O_2:NaOH = 1:1. The concentration ratios (H_2O_2:NaOH) of 1:1 were selected under 10 min, 800 W and <74 μm. The coal after treated was filtered with distilled water many times until the PH of the filtrate became neutral and then dried for 2 h at 105 °C.

ATR-FTIR Analysis

Attenuated total reflectance Fourier transform infrared spectra (ATR-FTIR) of raw and treated coal were measured with a Nicolet Is5 FTIR in the range of 4000–400 cm^{-1}. The samples were formed into a KBr pellet for spectroscopic examination. Integrated area (A) of different absorption peaks obtained from the selected ATR-FTIR data were employed for the semi-quantitative analysis [6].

Results and Discussion

The results of coal desulfurization rates of two methods were shown in Figs. 2, 3, 4 and 5. From these results, it can be seen that the desulfurization rates varied from 1.11 to 12.34% by microwave irradiation and from 34.28 to 51.14% leached with H_2O_2 and NaOH by microwave irradiation. The microwave is absorbed selectively. The pyrites' capacity to absorb microwaves is more than ten times higher than coal. The pyrites convert into pyrrhotite while the sulfide was released in the form of hydrogen sulfide. The reaction equation was as following [7, 11]:

$$FeS_2 \rightarrow Fe_{1-x}S + H_2S \uparrow \rightarrow FeS + H_2S \uparrow \quad (1)$$

H_2O_2 and NaOH were added to increase the desulfurization rates of the coal. The sulfur content of pyrite, sulfate and organic sulfur declined with a varying degrees. The reaction equations were as following [9, 12]:

$$H_2O_2 + NaOH + FeS_2 \rightarrow Fe(OH)_3\downarrow + Na_2SO_4 + H_2O + Fe_2(SO_4)_3 \quad (2)$$

$$H_2O_2 + NaOH + R_1 - S - R_2 \rightarrow R_1OH + R_2OH + Na_2S \quad (3)$$

$$O_2R - SH + NaOH \rightarrow R + Na_2S + H_2O \quad (4)$$

$$H_2O_2 + NaOH + R_1 - S - S - R_2 \rightarrow R_1OH + R_2OH + Na_2S + S \quad (5)$$

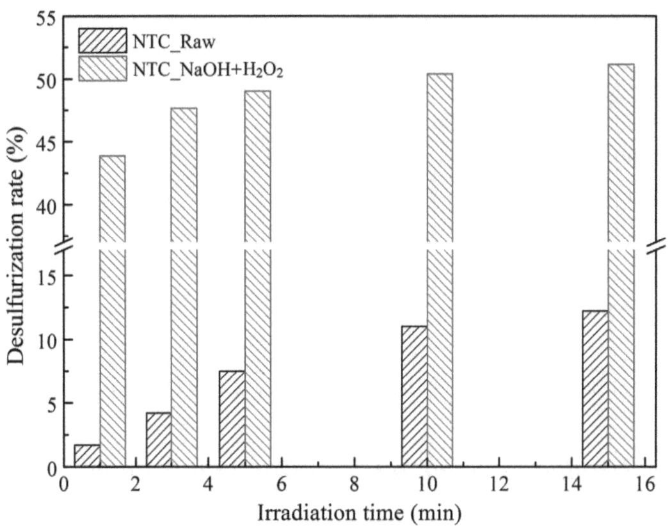

Fig. 2 Effects of irradiation time on desulfurization rates of raw and treated coals

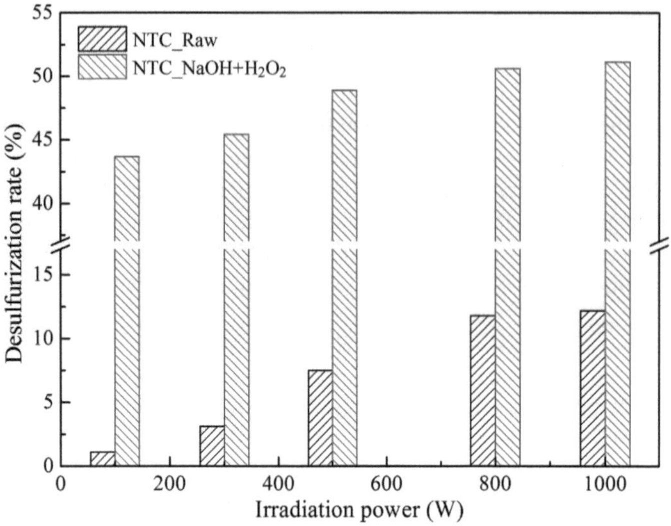

Fig. 3 Effects of irradiation power on desulfurization rates of raw and treated coals

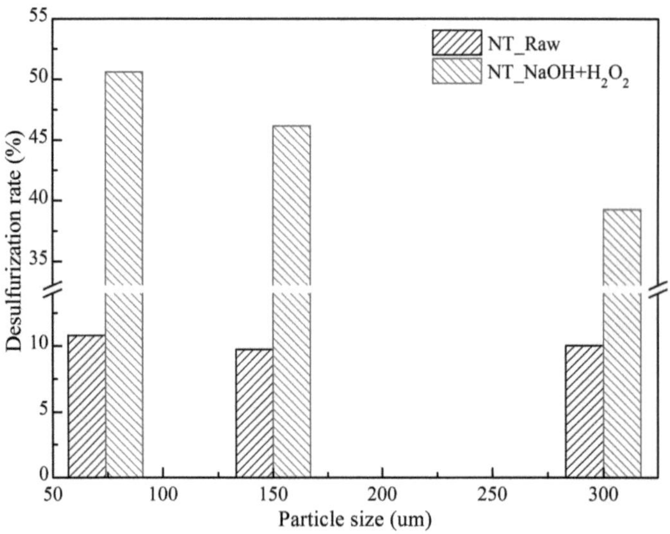

Fig. 4 Effects of particle size on desulfurization rates of raw and treated coals

Effects of Irradiation Time on Desulfurization Rates

The effects of irradiation time on desulfurization rates of raw and leached with H_2O_2 and NaOH were presented in Fig. 2. Generally, as the irradiation time increased, the desulfurization rates increased. Desulfurization rates increased from

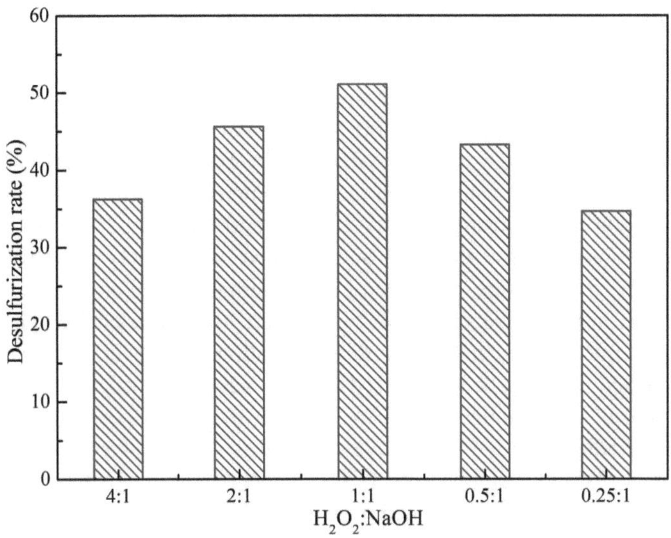

Fig. 5 Effects of H_2O_2:NaOH on desulfurization rates of treated coals

1.72 to 12.24% for the raw coal while the samples leached with H_2O_2 and NaOH were from 43.86 to 51.14%. The desulfurization rates increased sharply within 8 min while very slowly after 10 min. Considering the change of ranges, the following experiments were carried out using 10 min irradiation time. The sulfur removing processes were in two obvious stages based on the order and rates, in the first stage, the sulfur in FeS_2 and disulfides were easily removed, but the other forms of the sulfur belonged to the second stage for their hardly eliminating due to their solid structure [12]. Therefore, controlling the reaction time was responsible for enhancing the desulfurization rates.

Effects of Irradiation Power on Desulfurization Rates

The effects of irradiation power on desulfurization rates of raw and leached with H_2O_2 and NaOH coals were presented in Fig. 3. Generally, as the irradiation power increased, the desulfurization rates increased. It can be seen that the desulfurization rates were from 1.11 to 12.24% for the raw coal as the function of irradiation power while from 43.67 to 51.12% for samples. The desulfurization rates increased sharply within 800 W while exceedingly slow after 800 W. The irradiation power in the following experiments were 800 W taking into account the changes of desulfurization rates [13].

Effects of Particle Size on Desulfurization Rates

To evaluation of particle size effects on desulfurization process. The coal sample has divided to the fractions of 150–300, 74–150 and <74 μm, microwaved (800 W and 10 min) and leached with H_2O_2 and NaOH (H_2O_2:NaOH = 1:1). The results were shown in Fig. 4. It can be seen that with increasing of particle size the total sulfur reduction varied from 9.74 to 10.80% and from 39.25 to 50.61%, were respectively observed for the raw and leached with H_2O_2 and NaOH coals. Decreasing the particle size can increase the external surface area per unit mass of coal. For the particle size range of 74–300 μm, the particle size didn't clearly effected the microwave irradiation desulfurization. However, the particle size clearly effected the leached with H_2O_2 and NaOH by microwave irradiation desulfurization. This implied that there was no mass transport limitation during microwave irradiation [13, 14], however, there was mass transprot liminsatin in the process of H_2O_2 and NaOH leach. H_2O_2 and NaOH had not free access to the reactive sites on the surface and within particles [24]. Therefore, it is necessary to grind the coal to enhance the desulfurization rates when leached with H_2O_2 and NaOH.

Effects of the Concentration Ratio on Desulfurization Rates

To evaluation of the concentration ratio (H_2O_2:NaOH) effects on desulfurization process, five concentration ratios (H_2O_2:NaOH) were chosen from 4:1 to 0.25:1. The results were shown in Fig. 5. It can be seen that with the decreasing of concentration ratio the desulfurization rates first increased from 36.02 to 50.58% and then decreased from 50.58 to 34.58%. The optimum concentration ratio of H_2O_2: NaOH was 1:1. The concentration radio was found to be the most effective parameter on desulfurization process of the coal [15].

Analysis of the ATR-FTIR Spectra

A semi-quantitative approach has been made through ATR-FTIR spectra to figure out the sulfur contents of different type sulfur forms. From the comparison between the ATR-FTIR spectra of raw coal and samples leached with H_2O_2 and NaOH by microwave irradiation (Fig. 6). The bands in both coals showed slight changes except the relative intensity of some absorption ones. All the functional groups that were originally existed in raw coal were still present in the treated coal. The peaks at 3624 and 3688 cm^{-1} were mainly hydroxy of alcohol or phenol functional groups. The intensity of peaks at 2950 and 2850 cm^{-1} were aliphatic C–H_2 groups, which were also no significant difference between raw and treated coals. The main

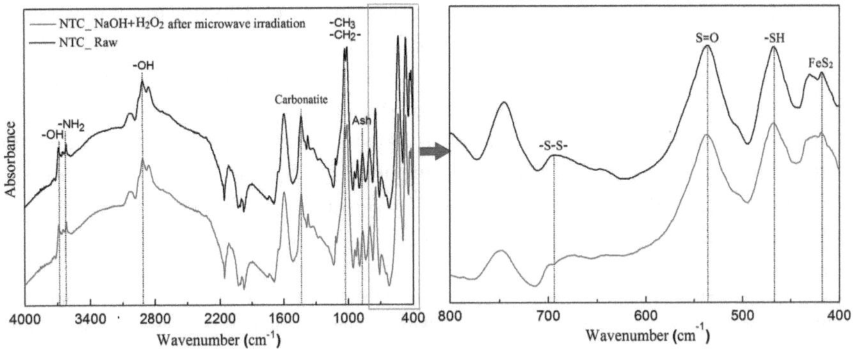

Fig. 6 ATR-FTIR spectra of raw and leached with H_2O_2 and NaOH by microwave irradiation coals

Table 2 Sulfur contents of different type sulfur forms raw and treated coals

Sulfur forms	FeS_2	–SH–	–S=O-	–S–S–
Wavenumbers (cm^{-1})	420	470	535	649
Raw area	0.181	0.153	3.912	0.311
Treated area	0.012	1.185	2.837	0.045
Raw S content (w/%)	0.300	0.600	1.643	0.166
Treated S content (w/%)	0.012	0.431	1.170	0.019

peaks between 4000 and 800 cm^{-1} were C, O, H functional groups, where these groups were not obviously effects after treated. The peaks between 800 and 400 cm^{-1} region were related to sulfur containing functional groups [12, 16]. All the peaks intensity of 694 cm^{-1} (–S–S–), 540 cm^{-1} (–S=S–), 470 cm^{-1} (–SH), 420 cm^{-1} (FeS_2) declined with a varying degrees [17]. The sulfur contents of these four types sulfur forms raw and treated coals were shown in Table 2. It can also be found that most of FeS_2 and –S–S– functional groups transfered after treatment, which may be caused by the oxidation of sulfur. The functional groups of –SH and –S=O– had almost no changing by ATR-FTIR spectra analysis.

Conclusions

Microwave irradiation of coal leaching with H_2O_2 and NaOH is an effective process for desulfurization. The desulfurization rate of the coal is very low by directly microwave irradiation, while it can be tremendously enhanced after leach of H_2O_2

and NaOH. The optimum conditions are obtained as 10-min irradiation time, 800 W irradiation power, particle size of <74 μm, 1:1 concentration ratio of H_2O_2: NaOH, which are collectively responsible for a 51.14% desulfurization rate among the total sulfur content. ATR-FTIR spectra shown that the desulfurization method of leaching with H_2O_2 and NaOH by microwave irradiation haven't broken the construction of raw coal. The functional groups of FeS_2 and –S–S– almost transfer while the functional groups of –SH and –S=O– are in a relative stability and less transformation.

Acknowledgements This word was supported by the National Natural Science Foundation of China (Grant No. 51474042). The authors acknowledge the 421 support provided by sharing fund of Chongqing University's large-scale equipment.

References

1. P. Prasassarakich, T. Thaweesri, Kinetics of coal desulfurization with sodium benzoate. Fuel **75**(7), 816–820 (1996)
2. W. Sihao, W. Jie, Tracking pyritic sulfur during the microwave-chemical desulfurization of coal using Mossbauer spectroscopy. Nucl. Sci. Tech. 121–125 (1992)
3. E. Jorjani, B. Rezai, M. Vossoughi et al., Desulfurization of Tabas coal with microwave irradiation/peroxyacetic acid washing at 25, 55 and 85 °C. Fuel **83**(7), 943–949 (2004)
4. H.G. Alam, A.Z. Moghaddam, M.R. Omidkhah, The influence of process parameters on desulfurization of Mezino coal by HNO_3/HCl leaching. Fuel Process. Technol. **90**(1), 1–7 (2009)
5. Z. Li, T. Sun, J. Jia, An extremely rapid, convenient and mild coal desulfurization new process: sodium borohydride reduction. Fuel Process. Technol. **91**(9), 1162–1167 (2010)
6. K. Charutawai, S. Ngamprasertsith, P. Prasassarakich, Supercritical desulfurization of low rank coal with ethanol/KOH. Fuel Process. Technol. **84**(1), 207–216 (2003)
7. P. Wardle, The selective heating of pyrite in coal using microwave energy (Ph.D. thesis, University of Nottingham, 2011)
8. G. Elsamak, N.A. Öztaş, Y. Yürüm, Chemical desulfurization of Turkish Cayirhan lignite with HI using microwave and thermal energy. Fuel **82**(5), 531–537 (2003)
9. S.C. Chelgani, E. Jorjani, Microwave irradiation pretreatment and peroxyacetic acid desulfurization of coal and application of GRNN simultaneous predictor. Fuel **90**(11), 3156–3163 (2011)
10. N. Xu, X. Tao, Changes in sulfur form during coal desulfurization with microwave: Effect on coal properties. Int. J. Min. Sci. Technol. **25**(3), 435–438 (2015)
11. T. Uslu, Ü. Atalay, Microwave heating of coal for enhanced magnetic removal of pyrite. Fuel Process. Technol. **85**(1), 21–29 (2004)
12. X. Tao, N. Xu, M. Xie et al., Progress of the technique of coal microwave desulfurization. Int. J. Coal Sci. Technol. **1**(1), 113–128 (2014)
13. S. Marland, A. Merchant, N. Rowson, Dielectric properties of coal. Fuel **80**(13), 1839–1849 (2001)
14. L. Xu, T. Peng, D. Zhang et al., Fractal desulfurization kinetics of high-sulfur coal. Int. J. Min. Sci. Technol. **22**(1), 111–114 (2012)

15. X. Ma, M. Zhang, Study of enhanced low-quality coal oxidative desulphurization by using HNO_3 and microwave pretreatment. Environ. Technol. **35**(1), 36–41 (2014)
16. X. Ma, M. Zhang, F. Mins et al., Fundamental study on removal of organic sulfur from coal by microwave irradiation. Int. J. Miner. Process. **139**, 31–35 (2015)
17. O. Folorunso, C. Dodds, G. Dimitrakis et al., Continuous energy efficient exfoliation of vermiculite through microwave heating. Int. J. Miner. Process. **114**, 69–79 (2012)

Part V
Pyrometallurgy II

Chloridizing Roasting of Bismuthinite with Sodium Chloride–Oxygen

R. Padilla, L. Salinas and M.C. Ruiz

Abstract Chloridizing roasting of bismuthinite (Bi_2S_3) with NaCl–O_2 has been studied to remove the impurity Bi by volatilization from copper concentrates that contain bismuthinite. The study was conducted using mixtures of Bi_2S_3 and NaCl in a horizontal furnace and TGA apparatus. The variables were temperature, oxygen partial pressure, and NaCl concentration. The chlorination was analyzed by weight loss measurement method. XRD results of calcines reacted for short times in 21% O_2, and 850–1000 °C showed the presence of Bi, BiOCl, and Na_2SO_4, while at longer times in the same conditions, Bi_2O_3 was identified as a stable phase. Thus, the chloridizing-volatilization of Bi_2S_3 proceeds through the formation of Bi and BiOCl. Temperature affected significantly the fraction of weight loss of samples. At 600 and 900 °C in 10.3% oxygen, the maximum weight loss was 4 and 20%, respectively. An increment in the partial pressure of oxygen in the range 1–21% affected negatively the Bi_2S_3 conversion. At normal roasting temperatures (550–700 °C), bismuth volatilization by NaCl roasting was found to be marginal.

Keywords Bismuthinite · Bismuth · Chloridizing roasting · Sodium chloride

Introduction

Copper concentrates contain bismuth as an impurity, which accompanies the copper in every step in the smelting/converting up to the copper cathode. This means that when the bismuth content in the concentrates is high, bismuth must be removed from the concentrates before smelting to avoid contamination of the copper cathodes. Bismuthinite (Bi_2S_3) is the bismuth sulfide that is present frequently in the copper concentrates. Roasting of the concentrates has been a common practice to remove the bismuth and other impurities from the copper concentrates that contain

R. Padilla (✉) · L. Salinas · M.C. Ruiz
Department of Metallurgical Engineering, University of Concepcion,
Edmundo Larenas 285, Concepcion, Chile
e-mail: rpadilla@udec.cl

S. Wang et al. (eds.), *Applications of Process Engineering Principles in Materials Processing, Energy and Environmental Technologies*,
The Minerals, Metals & Materials Series, DOI 10.1007/978-3-319-51091-0_25

high levels of impurities [1]. On this matter, a recent study [2] has shown that Bi_2S_3 decomposes readily to metallic liquid bismuth and sulfur gas at the temperature range of 500–700 °C in both inert and oxidizing atmospheres. The conclusion was that the volatilization of bismuth from bismuthinite in the presence of oxygen is negligible due to the slow rate of volatilization of liquid bismuth at roasting temperatures and also due to the rapid oxidation of liquid bismuth to Bi_2O_3. This bismuth trioxide is nonvolatile at roasting temperatures and thus, the removal of bismuth through the gas phase in the range 500–700 °C was not feasible. Therefore, the Bi removal from "dirty" concentrates by conventional roasting in inert or oxidizing atmospheres, is not feasible and thus, alternative routes must be seek to remove bismuth from impure concentrates before smelting.

Metal chlorides have normally low melting points, high vapor pressure at low temperatures, and thus, they can be vaporized more easily compared to the sulfides and oxides [3]. This means that bismuth chloride could also be volatilized easily at roasting temperatures. Thus, the present study was primarily concerned with the investigation of the behavior of bismuthinite in the temperature range of 600–1000 °C in the presence of sodium chloride and oxygen.

Experimental Work

This study was conducted using reagent grade Bi_2S_3 and NaCl as the source of chloride. The chloridizing reaction was studied conducting the experiments in a horizontal furnace setup and TGA apparatus. Homogenous mixtures of Bi_2S_3–NaCl were heated in the temperature range of 600–1000 °C and the extent of bismuth chloridation was analyzed by weight loss measurement. The variables studied were temperature, oxygen partial pressure, and NaCl concentration.

Results

For the chloridizing of bismuthinite with sodium chloride in oxidizing roasting, the following reactions (1)–(4) can be written a priori.

$$Bi_2S_3 + 6NaCl + 6O_2 = 2BiCl_3(g) + 3Na_2SO_4 \quad (1)$$

$$Bi_2S_3 + 2NaCl + 4.5O_2(g) = 2BiOCl + Na_2O + 3SO_2(g) \quad (2)$$

$$Bi_2S_3 + NaCl + 4.75O_2(g) = 0.5Bi_4O_5Cl_2 + 0.5Na_2SO_4 + 2.5SO_2(g) \quad (3)$$

$$Bi_2S_3 + 2NaCl + 5O_2(g) = 2BiOCl + Na_2SO_4 + 2SO_2(g) \quad (4)$$

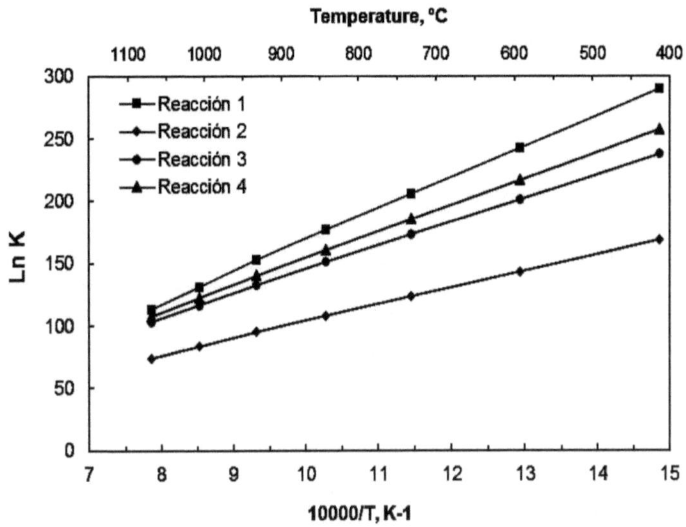

Fig. 1 Equilibrium diagram for the Bi_2S_3–NaCl–O_2 system considering reactions (1)–(4)

The equilibrium constants for these reactions were calculated by using the HSC data base [4], which were used to draw a diagram showing the equilibrium constant as a function of temperature for the reactions (1)–(4). The diagram is presented in Fig. 1.

Even though this diagram shows that reactions (1)–(4) are all likely to occur, clearly reaction (1) is the most favorable thermodynamically. Thus, in this preliminary study, the bismuth chloridization was analyzed assuming reaction (1) as the overall chloridizing reaction occurring on heating mixtures of Bi_2S_3–NaCl in the presence of oxygen. It was anticipated that the decomposition of bismuthinite according to reaction (5) should also occur in the system [2].

$$Bi_2S_3 = 2Bi(l) + 3S_2(g) \quad (5)$$

Previous research has shown that the kinetics of reaction (5) depends on the temperature and partial pressure of oxygen. Thus, reaction (5) could be an intermediate step of the overall chloridizing reaction (1).

Effect of Temperature

The effect of temperature was studied in the range 600–900 °C under 10.3% partial pressure of oxygen in the gas phase. The sample used was a mixture of molar ratio Bi_2S_3/NaCl equal to 1/6, which corresponds to the stoichiometry of reaction (1).

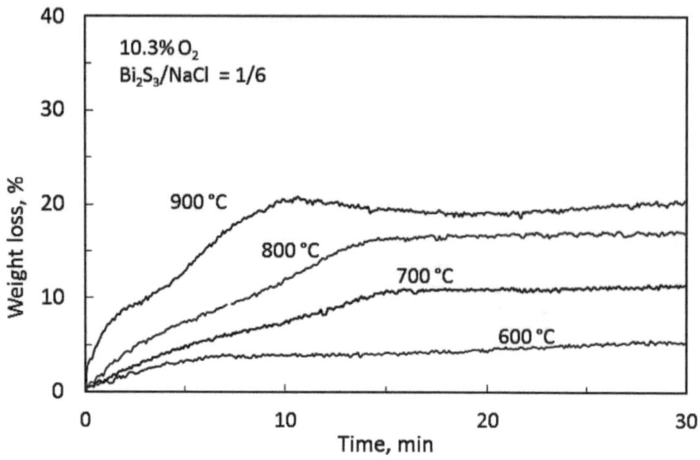

Fig. 2 Influence of temperature on the chloridizing of bismuthinite using sodium chloride. Molar ratio Bi_2S_3:NaCl = 1:6

The results are shown in Fig. 2, where it is plotted the weight loss as a function of temperature and time.

We can observe in Fig. 2 that all the weight loss curves attain a different maximum value at each temperature. The higher the temperature the higher the maximum, nevertheless, all the maximum values are much smaller than the theoretical weight loss for complete bismuthinite chloridization according to reaction (1), which is 50.7%. On the other hand, the decomposition of bismuthinite according to reaction (5) produces a maximum weight loss of 11.1% for the Bi_2S_3–NaCl mixture used in these experiments. Therefore, the low values of the experimental weight loss of the samples at the temperatures of 600 and 700 °C suggest that in these conditions it is unlikely that the formation of any volatile bismuth occurred in the system.

Effect of Partial Pressure of Oxygen

The effect of partial pressure of oxygen on the chloridizing reaction was studied at 850 and 1000 °C. The results are shown in Figs. 3 and 4. It is interesting to observe that at 850 °C, an increase in the oxygen partial pressure increases not only the maximum weight loss of the sample but also the weight loss rate as shown in Fig. (3). On the other hand, at the higher temperature of 1000 °C, increasing oxygen partial pressure affects negatively the weight loss; at this temperature, the higher weight loss occurred when using an atmosphere with 1% oxygen. The negative effect of a higher oxygen contents on the weight loss is most likely due to the formation of bismuth oxides, which in general are nonvolatile.

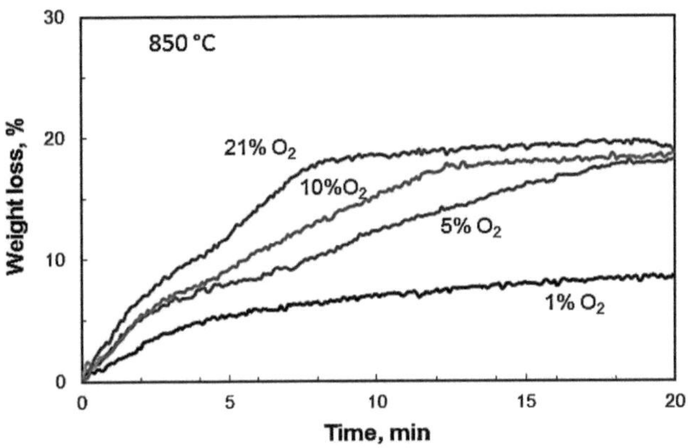

Fig. 3 Effect of oxygen partial pressure on the chloridizing of bismuthinite with sodium chloride at 850 °C. Molar ratio Bi_2S_3/NaCl = 1/6

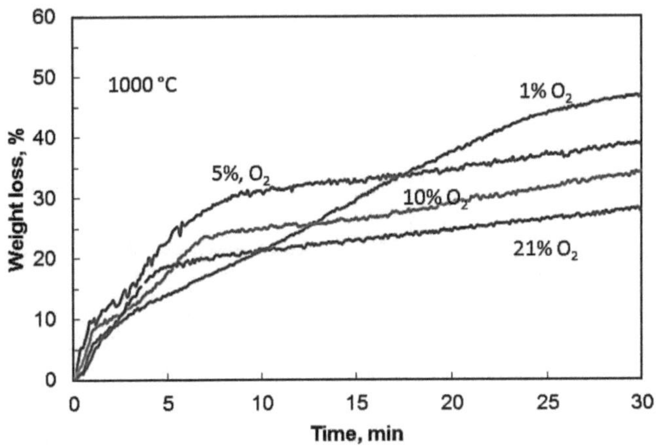

Fig. 4 Effect of oxygen partial pressure on the chloridizing of bismuthinite with sodium chloride at 1000 °C. Molar ratio Bi_2S_3/NaCl = 1/6

Characterization of the Reaction Products

In order to determine the phases formed by the chloridizing reaction in the various experimental conditions, batch experiments were conducted to obtain partially reacted samples in the horizontal furnace setup. The weight loss and the conditions of chloridizing of the samples are shown in Fig. 5. The partially reacted calcines thus obtained at the times 40, 120, and 420 min, labelled in the curve as A, B, and C, were analyzed by X-ray diffraction (XRD) spectroscopy.

The results of the XRD analysis of the three calcines in Fig. 5 are shown in Fig. 6. The XRD pattern for 40 min shows diffraction lines only for the unreacted

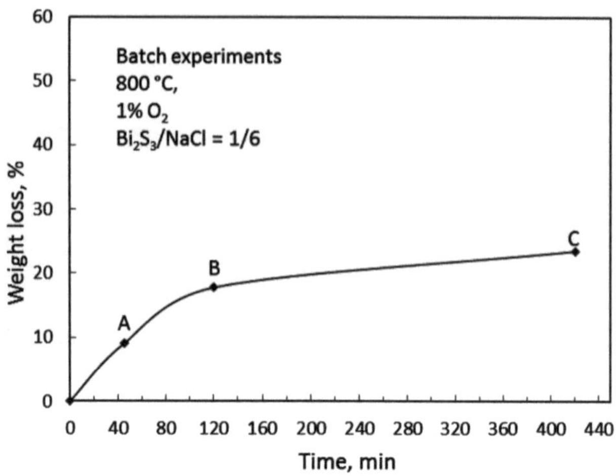

Fig. 5 Batch experiments for the production of samples for XRD analysis of the calcines

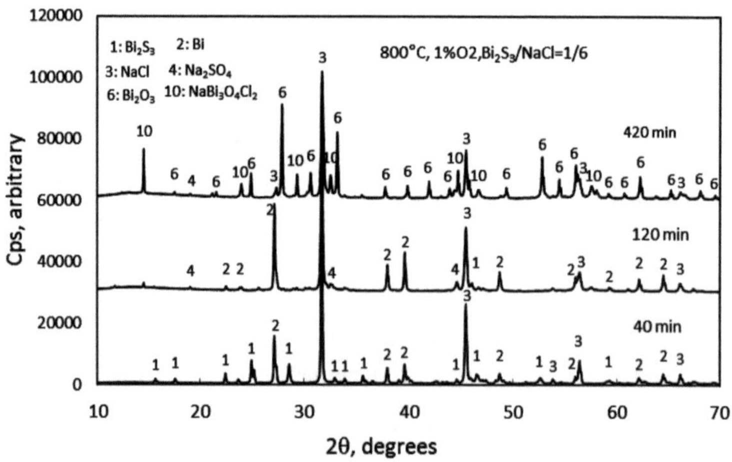

Fig. 6 XRD patterns for calcines obtained at 800 °C in 1% oxygen partial pressure

bismuthinite, metallic bismuth and sodium chloride. At 120 min in addition to the initial reactants, sodium sulfate is also present. Finally, at the long time of 420 min, sodium sulfate and the bismuth oxide Bi_2O_3 are the major identified compounds.

Additional XRD results, corresponding to calcines obtained at 1000 °C and 21% oxygen in the gas phase for various times of reaction, are shown in Fig. 7. We can see in this figure that in the sample reacted for 2.5 min at 1000 °C bismuthinite, bismuth, and sodium chloride were identified as the predominant compounds. However, at about 8 min, in addition to the latter compounds, BiOCl is present, and at 67 min, Bi_2O_3 and $Bi_{12}Cl_2O_{17}$ are the predominant compounds, while the BiOCl

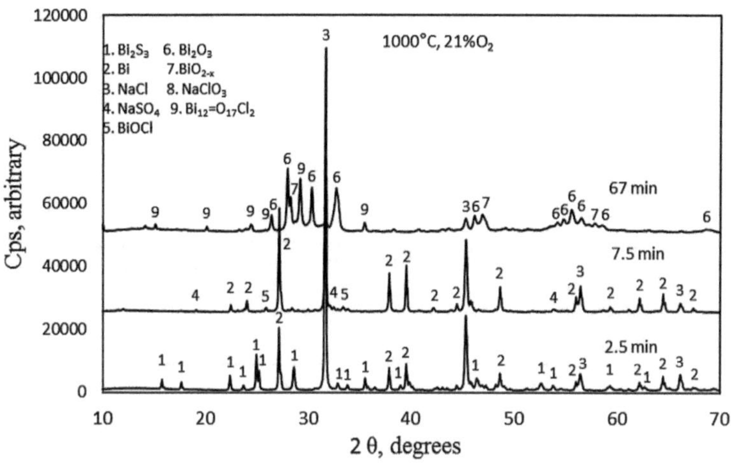

Fig. 7 XRD patterns for calcines obtained at 1000 °C in 21% oxygen

peaks are no longer present. These results indicate that bismuthinite decomposes initially to metallic Bi according to reaction (5).

Then, further reaction of the bismuth occurs with time to form volatile chlorides and nonvolatile bismuth oxides including Bi_2O_3, and $Bi_{12}O_{17}Cl_2$. The relative rates of these reactions would determine the degree of bismuth elimination by volatilization at each experimental condition. The compound BiOCl, appears to be an intermediate compound in the overall reaction.

Bismuthinite Oxidation with and without Sodium Chloride

Few additional experiments were conducted to compare the behavior of pure Bi_2S_3 to the behavior of Bi_2S_3 in the presence of NaCl in oxidizing atmosphere. The results are shown in Figs. 8 and 9. In these figures, the values of weight loss for the mixture (Bi_2S_3 + NaCl) were corrected to compare to the weight loss of the pure Bi_2S_3 curve. Thus, both curves show truly the weight loss of Bi_2S_3. As seen in Fig. 8, at the conditions of 700 °C and 10% oxygen, Bi_2S_3 behaves almost as if NaCl was not there. In other words, NaCl did not produce appreciable changes in the weight loss of the samples since the maximum weight loss detected corresponds closely to the theoretical weight loss for the decomposition of Bi_2S_3 according to reaction (5), which is 18.7% (corrected value). Therefore, at 700 °C and 10% oxygen, liquid bismuth remains as a stable phase thereafter. However, at 1000 °C and 21% oxygen, there is a large difference in the weight loss behavior of the two samples. The sample of pure Bi_2S_3 shows a rapid weight loss that can be attributed to reaction (5) followed by a weight gain due to oxidation of the liquid bismuth to Bi_2O_3. On the other hand, the weight loss of the Bi_2O_3–NaCl sample is significantly

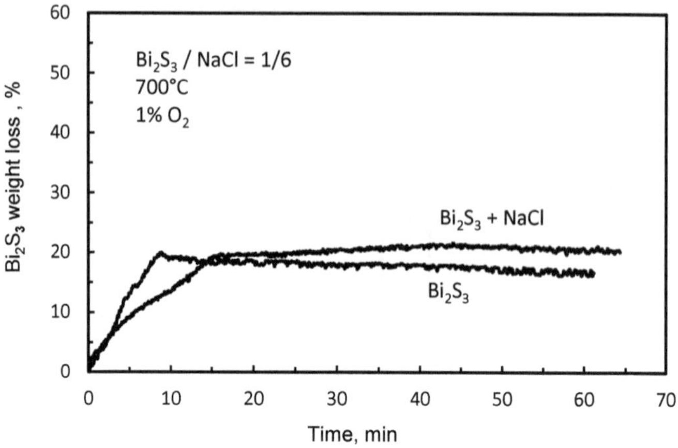

Fig. 8 Weight loss fraction of bismuthinite as compared to the behavior of bismuthinite in the presence of NaCl at 700 °C and 10% oxygen

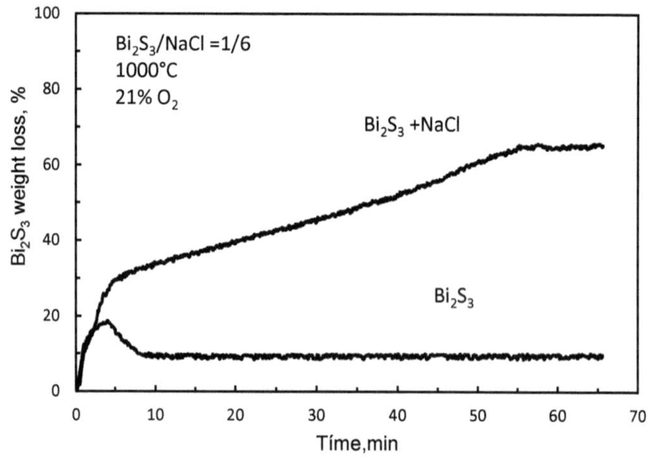

Fig. 9 Weight loss fraction of bismuthinite as compared to the behavior of bismuthinite in the presence of NaCl at 1000 °C, and 21% oxygen

higher than for the pure Bi_2S_3 sample. This difference in the weight loss is attributed to the formation of volatile bismuth compounds presumably chlorides.

Based on the weight loss and the XRD results, at temperatures lower than 700 °C in the presence of NaCl, bismuthinite decomposes readily to Bi and gaseous sulfur and then Bi is oxidized slowly to Bi_2O_3 at longer times. The formation of volatile bismuth chloride compounds is not evident at these temperatures. Therefore, at the roasting temperatures (550–700 °C) bismuth removal through the gas phase would be marginal.

Conclusions

From the experimental results, the following can be concluded:

- Temperature changes in the range 600–900 °C increase the maximum weight loss of Bi_2S_3–NaCl mixtures, while at 1000 °C the partial pressure of oxygen in the gas phase has a negative effect on the maximum weight loss of the samples.
- At temperatures lower than 700 °C in oxidizing atmosphere, Bi_2S_3 decomposes to Bi and sulfur in the presence of NaCl and oxygen.
- At 1000 °C and high oxygen concentration, bismuth oxychloride (BiOCl) was found to be an intermediate compound in the reaction of bismuthinite with sodium chloride and oxygen.
- The overall reaction of the chloridizing of bismuthinite with sodium chloride can be represented by: $Bi_2S_3 + 6NaCl + 6O_2 = 2BiCl_3(g) + 3Na_2SO_4$.
- Finally, at roasting temperatures in the range of 550–700 °C, bismuth volatilization by roasting with NaCl was found to be marginal.

Acknowledgements The authors acknowledge the National Fund for Scientific and Technological Development (FONDECYT) of Chile for the financial support of this research through project No. 1150339.

REFERENCES

1. L. Winkel et al., Decomposition of copper concentrates at high-temperatures: an efficient method to remove volatile impurities. Miner. Eng. **21**, 731–742 (2008)
2. R. Padilla, R. Villa, M.C. Ruiz, Decomposition and oxidation of bismuthinite in oxygen–nitrogen atmospheres. Second International Symposium on High Temperature Metallurgical Processing, ed. By J.-Y. Hwang et al (Wiley, Hoboken, NJ, 2011), pp. 221–228
3. E. Peek, Chloride metallurgy-process technology development. EPD Congress 2011, ed. By S. N. Monteiro et al. (Wiley, Hoboken, NJ, 2011), pp. 477–501
4. A. Roine, HSC chemistry 6.0, Outokumpu Research Oy, Pori, Finland (2006)

Natural Gas Utilization in Blast Furnace Ironmaking: Tuyère Injection, Shaft Injection and Prereduction

P. Chris Pistorius, Jorge Gibson and Megha Jampani

Abstract Increased utilization of natural gas in blast furnace ironmaking can decrease both the cost and the carbon intensity of ironmaking, given current US natural gas prices. In this paper, three ways to utilize natural gas are compared: tuyère injection, prereduction of iron ore, and shaft injection. The basis for comparison includes coke replacement ratios, carbon intensity and furnace productivity. These were calculated using relevant mass and energy balances, a blast furnace productivity correlation based on the bosh gas flow rate, and measured and modeled prereduction kinetics. Of the natural gas utilization methods, prereduction has the highest effective coke replacement ratio (and hence the largest advantage in raw material cost), but is likely to have the highest capital requirement.

Keywords Blast furnace ironmaking · Carbon intensity · Coke replacement · Natural gas

Introduction

Ironmaking is inherently energy intensive, because it involves breaking the chemical bonds between iron and oxygen atoms. Large-scale blast furnace ironmaking is also carbon intensive, since coke is the main fuel used in blast furnaces. One method to decrease the carbon intensity of blast furnace ironmaking is partial substitution of coke with less carbon-intensive fuels. In particular, natural gas (when injected through the blast furnace tuyères) can replace coke on an approximately equal mass basis (that is, for each kilogram of natural gas injected, approximately one kilogram of coke is saved—the "coke replacement ratio" is approximately 1 in this case [1]). Because the carbon content of natural gas (around 75% by mass) is lower than that of coke (which typically contains 85–90% carbon),

P.C. Pistorius (✉) · J. Gibson · M. Jampani
Center for Iron and Steelmaking Research, Department of Materials Science and Engineering, Carnegie Mellon University, 5000 Forbes Avenue, Pittsburgh 15213, PA, USA
e-mail: pistorius@cmu.edu

S. Wang et al. (eds.), *Applications of Process Engineering Principles in Materials Processing, Energy and Environmental Technologies*,
The Minerals, Metals & Materials Series, DOI 10.1007/978-3-319-51091-0_26

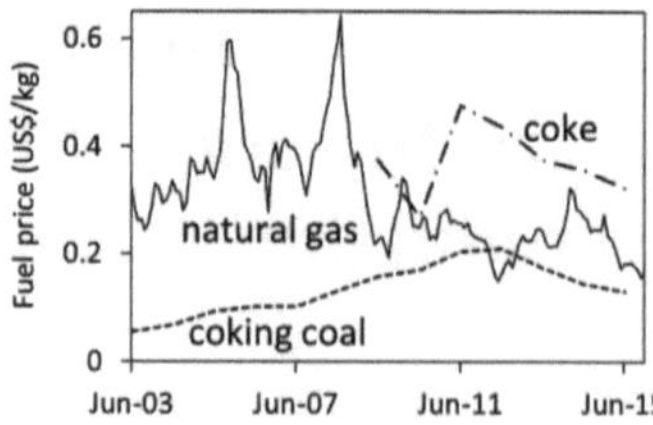

Fig. 1 Average US industrial prices of natural gas (converted assuming a density of 44.8 lb per thousand cubic feet; monthly averages), coal receipts at coking plants (annual averages), and open-market coke sales (annual averages). *Source* US Energy Information Administration [2]

partial replacement of coke with an equal mass of natural gas would yield a decrease in process carbon intensity. In addition, recent decreases in the US price of natural gas (Fig. 1) [2] means that partial replacement of coke by injected natural gas would both save raw-material cost and lower the carbon intensity.

Natural gas is increasingly used as injectant in North American blast furnaces: In 2015, natural gas was used as the sole injectant at 13 blast furnaces (with a total production of 16.6 million metric tons of hot metal [THM]) in North America, with a weighted average injection rate of 83 kg/THM; both natural gas and coal were injected at another 12 furnaces [3]. The furnace with the highest injection rate in 2015 was AK Steel Middletown, at 117 kg/THM [3].

Limits to Tuyère Injection Rates

Given the cost advantage of natural gas over coke, an obvious question is why the natural-gas injection rates are not even higher. One possible reason is that, as coke is replaced by natural gas as fuel, less coke is available to support the furnace burden and distribute the gas flow within the furnace. However, modern blast furnaces can operate at low coke rates: this has been best documented for furnaces operating at high rates of pulverized-coal injection (PCI) [4], with a low coke rate of approximately 260 kg/THM reported at a PCI rate of 240 kg/THM. This means that there would be sufficient coke in the furnace feed even at much higher natural-gas rates than the current North American maximum.

The second constraint is thermal: tuyère injection of natural gas tends to lower the flame temperature; while this can be counteracted by enriching the blast air with oxygen, oxygen enrichment in turn causes the top-gas temperature (at the furnace exit) to decrease, leading to condensation of water in the upper part of the furnace if the blast contains too high a percentage of oxygen. These two thermal constraints are conveniently illustrated with an "operating window", with examples shown in Fig. 2. In these diagrams, the top-gas temperature constraint is shown as one line (for a temperature of 125 °C in this case); oxygen concentrations (in the blast air) *below* this line would give higher top-gas temperatures, as required for stable

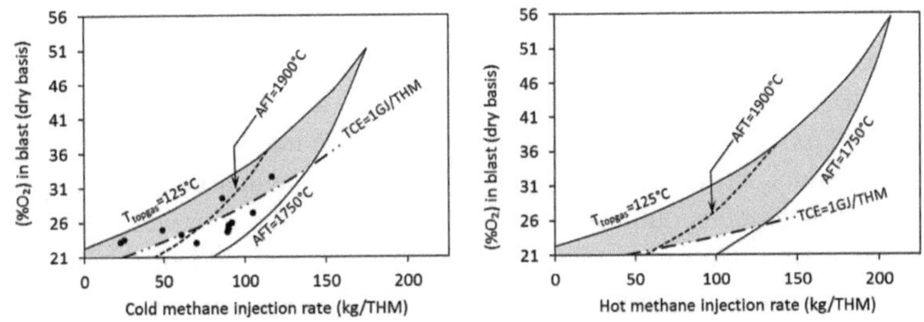

Fig. 2 Calculated operating windows for blast furnaces with different tuyère injectants: CH_4 at 298 K (*left*), and CH_4 preheated to 1200 K (*right*). "AFT" is the calculated adiabatic flame temperature, "T_{topgas}" is the gas temperature at the top of the furnace, and "TCE" is the total "thermal and chemical energy" of the "hearth gas". Shaded regions are proposed stable operating conditions. Data points are for North American blast furnaces which used only natural gas as injectant in 2015 [3]. See Table 1 for assumed furnace conditions

operation. The constraint value was chosen to be higher than what is absolutely necessary to prevent condensation, to provide a margin of safety, recognizing that the top-gas temperature in a blast furnace is not uniform (it tends to be higher at the center-line and lower near the walls of the shaft). The flame-temperature constraint is shown as another set of lines, each drawn for a separate (theoretical) adiabatic flame temperature (AFT) at the tuyères; the AFT would be higher (as required for stable operation) if the oxygen concentration in the blast is *above* the relevant line. Thus the region between the lines representing the top-gas temperature constraint and the flame-temperature constraint give conditions for stable furnace operation—the "operating window". It is not clear what the correct lower limit to the flame temperature is; Figure 2 shows lines for 1900 and 1750 °C, and in that figure all data (from 2015) for North American furnaces can be seen to lie above the 1750 °C line. It should be noted that for the present work the adiabatic flame temperature has been calculated with a mass and energy balance. There are other methods for estimating this value, such as the AISI method, and values obtained from various methods often differ when applied to the same operating conditions.

Figure 2 also shows a third constraint: the TCE (thermal and chemical energy) of the gases in the lower part of the furnace [5]. The principle is that the "hearth gases" produced by combustion at the tuyères must have sufficient enthalpy to transfer heat to and melt the iron and slag; an assumption of the calculation is that the presence of hydrogen in the hearth gases reduces the energy required for stable furnace operation, because utilizing hydrogen as reductant partially avoids the highly endothermic solution loss (carbon gasification) reaction. The TCE is calculated as the remaining enthalpy of the hearth gases after melting and heating the hot metal and slag to 1482 °C (2700 °F) (a reasonable tapping temperature), adding a credit of 46.5 kJ per mole of hydrogen in the hearth gases. The latter value represents the difference between the heats of reduction of FeO with C (CO as product) and H_2 (H_2O as product), multiplied by 35% utilization of hydrogen [5]. The heat of reaction was obtained from FactSage, and used instead of the values from the original TCE paper [5]. In the work presented here, the enthalpy of the "hearth gases" was calculated from an energy balance of the region of the furnace below the 1420 °C isotherm, assuming coke (including ash formers) and unmelted Fe and slag-forming oxides (at 1420 °C) as condensed-phase inputs, blast air and injectants as gaseous inputs, hot metal and slag (at 1482 °C) as condensed-phase outputs, and hearth gases as gaseous outputs. For cases where metallic iron was included in the feed, a credit of 146 kJ per mole of Fe° in the feed was added; this value corresponds to the heat of reaction of FeO with C (to produce solid Fe and CO) around 1600 K.

If a blast furnace is operated with a target of a constant TCE, the result is that the adiabatic flame temperature would be lower at higher rates of natural gas injection, as actually observed in blast furnace operations [5]. Operating data for North American furnaces (for the year 2015) in Fig. 2 cluster around a TCE of 1.0 GJ/THM, and do support the suggestion that the adiabatic flame temperature can be lower for higher natural-gas injection rates while opearating stably [5].

Using Pre-reduced Feed

Including metallic iron (as scrap or direct-reduced iron [DRI], or more typically hot-briquetted iron [HBI]) in the furnace feed greatly decreases the coke rate (because less reduction is needed per ton of hot metal) and increases furnace productivity (because of the smaller gas volume per ton of hot metal) [8]. The typical reduction in coke rate is around 0.37 kg coke saved per kg of metallic iron in the burden [9]. Because the typical rate of natural-gas use in production of DRI or HBI is around 0.2 kg natural gas per kg of metallic iron [9], the effective coke replacement rate by natural gas is $0.37/0.2 \approx 1.8$. This is much higher than the value for tuyère injection, which (as mentioned earlier) is around 1.0.

The larger increase in furnace productivity (typically 8% increase in hot metal production rate for every 10% of the ferrous feed which is in metallic form [10]) can have strategic advantages. For example, the increased productivity resulting from use of HBI at the AK Steel Middletown furnace allowed a less-efficient blast furnace to be closed [10].

Shaft Injection

One potential way to avoid the detrimental effect of tuyère injected natural gas on the flame temperature would be to inject gas in the furnace shaft instead [11, 12]. However, even in this case the endothermic effect of natural gas would have to be counteracted; this could be achieved by a combination of preheating and partially combusting the gas. Preheating the natural gas (and the oxygen for combustion) to 1200 K and partially combusting it to produce CO and H_2 would eliminate the endothermic effect [11]. (Partial combustion would have to be completed outside the furnace, to avoid introducing unreacted oxygen which would tend to reoxidize metallic iron in the furnace shaft.) If the gas is taken to be CH_4, this means that 0.5 mol of O_2 would be required per mol of CH_4. The combination of preheating and partial combustion is similar to what is used in the HYL Energiron ZRA process [13] for production of direct-reduced iron. In that process, the off-gas of the shaft furnace is cooled to condense out water, CO_2 is removed, then the gas (mainly CO + H_2, with unreacted CH_4) is enriched with natural gas, preheated to around 950 °C and partially combusted with oxygen before being returned to the shaft furnace inlet. Preheating temperatures higher than 950 °C are likely not feasible because of the danger of cracking of methane within the preheaters.

If a mixture of CH_4 and O_2 (2:1 molar ratio, initial temperature 1200 K) is allowed to react to form CO and H_2, the product gases would be at 1421 K under adiabatic conditions. This means that, in terms of the mass and energy balance of a blast furnace, injecting CH_4 and O_2 at 1200 K would have the same effect on the coke rate as injecting the equivalent amount of CO and H_2 at 1421 K. Given the relative amounts and masses, 1 kg of CH_4 would yield 2 kg of the partially combusted mixture, containing CO and H_2 in a 1:2 molar ratio. (The actual composition of the gas would be somewhat different because of the effect of reaction kinetics—it would contain some unreacted CH_4, CO_2 and H_2O in addition to CO and H_2 and would be at a higher temperature; however, the presence of these species does not change the overall mass and energy balance since the overall gas enthalpy and elemental molar amounts remain the same if partial combustion is assumed to be adiabatic.)

Calculation Approach

A mass and energy balance for that part of the furnace below the 1200 K isotherm was used. Similar to the approach of Rist and Meysson [1, 14], it was assumed that the furnace gas reaches thermal and chemical equilibrium with Fe/FeO at 1200 K as the gases rises through the furnace, and that coke remains unreacted until it is heated to above 1200 K (that is, coke passes unreacted through the 1200 K isotherm as it descends in the furnace). Detail of the (typical) compositions and conditions assumed in the calculation are given in Table 1. A hypothetical furnace with a hearth diameter of 12 m was considered. Changes in furnace productivity were calculated by assuming that the gas flow rate in the raceway region limits furnace productivity. Based on published data for operating furnaces, the raceway gas flow rate for a 12 m furnace was estimated to be approximately 7500 Nm^3/min [9]. The furnace production rate was found by dividing the raceway gas flow rate by the volume of raceway gases produced per ton of hot metal (as found from the mass and energy balance). In cases where shaft injection was considered, the total volume of injected gas plus that produced by combusted at the tuyères was used to estimate furnace productivity. Natural gas was approximated as methane in these

Table 1 Assumed furnace conditions for calculations

Iron ore/pellets	71.2% Fe_2O_3; 22.8% Fe_3O_4; 4.2% SiO_2; 0.4% Al_2O_3; 1.1% CaO; 0.3% MgO
Coke	12.5% ash yield (composition 65% SiO_2, 25% Al_2O_3, 6% CaO, 4% MgO); bal. carbon
Blast air	variable oxygen content; moisture content 18 g per Nm^3 dry blast; 1273 K
Direct-reduced iron	86.4% Fe, 8.4% FeO, 5.2% SiO_2
Flux	100% CaO
Hot metal	4.4% C; 0.5% Si; balance Fe; 1800 K
Slag	Basicity: (%CaO)/(%SiO_2) = 1; 1800 K
Hearth diameter	12 m
Bosh gas flow rate	7500 Nm^3/min
Heat loss	30 MW (from region of furnace below 1200 K isotherm) [1]
Slag enthalpy	Estimated using the model of Björkvall et al. [6]
WRZ[a] temperature	1200 K
AFT[a] calculation	Carbon (from coke) enters at 1800 K; blast air and injectant compositions and temperatures as specified; combustion products are N_2, CO and H_2
TCE[a]	Enthalpy of hearth gases after heating hot metal and slag to 1482 °C, plus 46.5 kJ per mole H_2 and 146 kJ per mole metallic Fe in the furnace feed
Oxygen plant	Energy requirement for oxygen production 0.3 kWh/Nm^3 [7]; CO_2 intensity of electricity 0.53 kg CO_2/kWh (US average [2])
Dust losses	Assumed to be zero

[a]*WRZ* wüstite reserve zone; *AFT* adiabatic flame temperature; *TCE* thermal and chemical energy of hearth gases

calculations. Reoxidation of the metallic feed in the upper parts of the furnace was assumed to be negligible.

Five cases were considered—a baseline with only coke as reductant, one with 180 kg DRI (containing 156 kg metallic iron) in the feed per THM, 180 kg/THM methane injected through the tuyères (while maintaining the same TCE as the baseline case, by adding oxygen enrichment to the blast air), 180 kg/THM preheated methane partially combusted with preheated O_2 (also 180 kg/THM) to yield at CO–2H_2 mixture at 1421 K, and 180 kg/THM preheated methane (at 1200 K) injected through the tuyères (adding oxygen enrichment to the blast air to achieve an adiabatic flame temperature of 1750 °C). The value of 180 kg/THM methane was chosen since this is approximately the maximum possible tuyère injection rate (of cold methane) within the constraints of TCE, top-gas temperature and adiabatic flame temperature (as indicated in the results, the AFT for this case was calculated to be 1711 °C).

The aim of the calculations was to confirm the general trends stated in the introduction, and to provide more detail of predicted changes in carbon intensity and coke replacement rates. In calculating the carbon intensity, each mole of carbon in coke and methane (whether used in the furnace, or in the production of DRI) was assumed to produce one mole of carbon dioxide at the ironmaking site (with no credit given for potential utilization of the blast furnace top gas). If oxygen enrichment was required, the associated CO_2 was estimated by using a typical energy requirement for oxygen production of 0.3 kWh/Nm^3 [7] and assuming all of this energy would be provided by electricity produced at the US average carbon dioxide intensity of 0.53 kg CO_2/kWh [2].

Results and Discussion

Coke Replacement Ratios and CO_2 Intensity

The main results are summarized in Fig. 3 and Table 2. For metallic iron in the feed, the results confirm the expected large increase in production rate, decrease in coke rate and carbon intensity, and high effective coke replacement rate (around 1.9 kg coke per kg methane used to produce DRI). Tuyère injection of cold methane does decrease the coke rate, with a replacement rate around 1 kg coke per kg methane; the reduction in CO_2 intensity is partially offset by the CO_2 associated with producing the oxygen to enrich the blast (to 48.5% O_2 on a dry basis, as required to match the TCE of the baseline case).

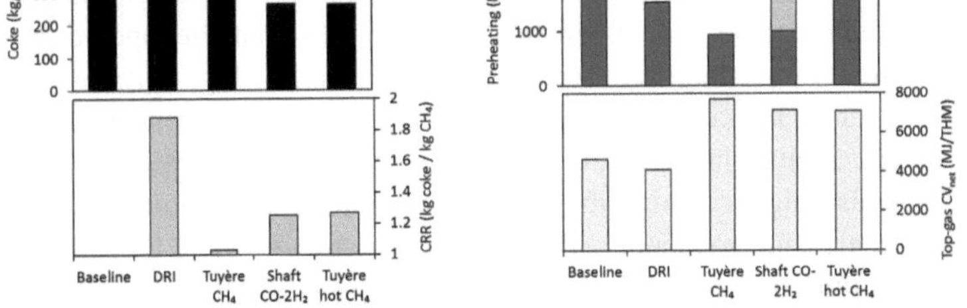

Fig. 3 *At left*, comparison of the coke rate, CO_2 intensity and coke replacement rate (CRR) relative to the baseline case, for 180 kg/THM direct-reduced iron in the furnace feed, 180 kg/THM cold CH_4 injected through the tuyères, 180 kg/THM CH_4 (preheated to 1200 K) partially combusted with O_2 (1200 K) to yield hot $CO + 2H_2$ and then injected in the furnace shaft, and 180 kg/THM CH_4 (preheated to 1200 K) injected through the tuyères; *At right*, comparison of the gas flow rate into the furnace, and the heat transfer required to preheat the blast air and injectants for these cases, together with the net calorific value of the blast furnace top gas

Table 2 Comparison of the blast furnace performance for the base case, with addition of 180 kg/THM of direct-reduced iron to the furnace feed, 180 kg/THM cold CH_4 injected through the tuyères, 180 kg/THM CH_4 (preheated to 1200 K) partially combusted with O_2 (1200 K) to yield hot $CO + 2H_2$ and then injected in the furnace shaft, and 180 kg/THM preheated (1200 K) CH_4 injected through the tuyères

Additive	$T_{additive}$ (K)	CH_4 used (kg/THM)	Coke rate (kg/THM)	O_2-blast (%)	Production (kTHM/day)	T_{topgas} (°C)	TCE (GJ/THM)	AFT (°C)
None	–	0	492	21.0	7.1	157	1.08	2134
DRI	298	31	433	21.0	8.0	127	1.15	2134
Tuyère CH_4 (cold)	298	180	305	48.5	7.2	152	1.08	1711
Shaft $CO + 2H_2$	1421	180	265	21.0	6.6	231	N/A	2134
Tuyère CH_4 (hot)	1200	180	263	39.2	7.2	178	1.19	1750

For the hot $CO–2H_2$ mixture (produced by partial combustion of preheated CH_4 with preheated oxygen, then injected into the shaft) the coke replacement ratio would be around 1.3; tuyère injection of hot methane would give a similar coke replacement ratio. The high coke replacement ratio for the cases involving hot $CO–2H_2$ and hot methane largely results from preheating; as the results show, injection of cold methane would give a replacement ratio of approximately 1.

For injection of the hot $CO–2H_2$ mixture, recovery of energy from the blast furnace top gas through preheating the injected gases would be similar to the current approach of using blast furnace stoves. Process complexity would be increased, though: in addition to stoves for the blast air, separate preheaters would need to be added for the natural gas and the oxygen (although a simpler "hot oxygen" heating system could perhaps be used for the oxygen [15]).

Tuyère injection of preheated methane (at 1200 K) appears to be an interesting alternative. Injection of hot methane would have a smaller effect on the theoretical flame temperature than cold methane. The result is that the operating window extends to higher injection rates; see the right-hand side of Fig. 1. As expected from the large effect on the coke rate, the energy requirement for preheating is large, as shown in Fig. 3 (at right). However, as the figure shows, the calorific value of the top gases is also higher at high methane injection rates, and should be sufficient to supply the energy for preheating: The values show that it would be feasible to meet the energy requirement if the stoves and preheater have a thermal efficiency of 40%; in comparison, modern furnace stoves typically have much higher thermal efficiencies, around 60% [16].

Costs

All of the process variants considered here would decrease the raw-material cost, since natural gas costs less than coke (per unit mass), and the coke replacement ratios are larger than 1 in all cases.

Of these variants, production of DRI would have the largest capital cost. For the hypothetical case considered here (a 12 m furnace, with 180 kg DRI per THM), the production rate of DRI would need to be approximately 500 kt per annum, with a total furnace production rate of 2.8 million metric tons per year. For comparison, the Nucor Louisiana plant has a capacity of 2500 kt DRI per annum, and cost US $750 million [17]. If it assumed that the plant cost would scale with $(\text{capacity})^n$, with $n = 0.5$, the estimated cost of a 500 kt/a plant would be approximately US $330 million. At a discount rate of 10% and 10-year project lifetime the capital cost would be approximately US$53 million per year, corresponding to US$19/THM.

Conclusion

Several US blast furnace operators have already taken advantage of the low natural-gas price in the US, increasing natural-gas injection rates to an average of some 80 kg/THM, with a maximum rate around 120 kg/THM. The analysis presented here indicates that higher tuyère injection rates (greater than 150 kg/THM) should be feasible, provided the level of oxygen enrichment is carefully increased to meet all the thermal requirements. Natural-gas utilization rates higher than this would require additional equipment—whether in the form of a plant to produce direct-reduced iron, or preheaters for natural gas, or both natural gas and oxygen.

Acknowledgements Support of this project by the industrial members of the Center for Iron and Steelmaking Research is gratefully acknowledged.

References

1. J.G. Peacey, W.G. Davenport, *The Iron Blast Furnace, Theory and Practice* (Pergamon, London, 1979)
2. US Energy Information Adminstration: http://www.eia.gov/electricity/state/unitedstates/; http://www.eia.gov/coal/production/quarterly/; https://www.eia.gov/naturalgas/. Last visited Sept 2016
3. 2016 AIST North American Blast Furnace Roundup. Iron Steel Technol. 13(3), 256–259 (2016)
4. P. Bennett, T. Fukushima, *Impact of PCI coal quality on blast furnace operations*. in Proceedings of the 12th International Conference on Coal Science. Cairns, Australia (2003). Available at http://www.coaltech.com.au/LinkedDocuments/Bennett_Fukushima.pdf. Last visited Sept 2016

5. J.C. Agarwal et al., The use of total thermal-plus-chemical energy above 2700 °F as a parameter to describe blast furnace operations. Ironmaking Conf. Proc. 473–479 (1991)
6. J. Björkvall, D. Sichen, S. Seetharaman, Thermodynamic model calculations in multicomponent liquid silicate systems. Ironmaking Steelmaking **28**, 250–257 (2001)
7. W.F. Castle, Air separation and liquefaction: recent developments and prospects for the beginning of the new millennium. Int. J. Refrig. **25**, 158–172 (2002)
8. M.M. Fine, N.B. Melcher, Prereduced Iron ore Pellets: Preparation, Properties, Utilization. (Bureau of Mines, United States Department of the Interior, Bulletin 651, Washington, D.C., 1970)
9. J. Gibson, P.C. Pistorius, *"Natural Gas in Ironmaking: On the Use of DRI and LRI in the Blast Furnace Process", AISTech2015 Proceedings* (Association for Iron and Steel Technology, Warrendale, Pa, 2015), pp. 657–671
10. R. Hunter, *Massive savings in CO_2 generation by use of HBI* (Third/Fourth Quarter, Direct from Midrex, 2009), pp. 9–11
11. M. Jampani, P.C. Pistorius, Increased use of natural gas in blast furnace ironmaking. Iron Steel Technol. **12**(3), 37–43 (2015)
12. M. Jampani, P.C. Pistorius, *"Increased use of natural gas in blast furnace ironmaking: methane reforming", AISTech2016 Proceedings* (Association for Iron and Steel Technology, Warrendale, Pa, 2016), pp. 573–580
13. J. Becerra, A. Martinis, Alternative energy sources, CO_2 recovery technology and clean environment compliance—integral components of Energiron technology. Iron Steel Rev. **51**(8), 107–110 (2008)
14. A. Rist, N. Meysson, A dual graphic representation of the blast-furnace mass and heat balances. J. Metals **19**(4), 50–59 (1967)
15. M.F. Riley, Hot oxygen blast injection system, United States Patent number 6,090,182. 18 July 2000 (2000)
16. K.R. Muske et al., *Advanced Control of Operations in the Blast Furnace Project. Phase One Effort Documentation: Optimal Operation and Control of the Blast Furnace Stoves* (Technical Report LA-UR-99-5051, Los Alamos National Laboratory, 1999)
17. *NUCOR 2013 annual report*. Available at http://www.nucor.com/investor/performance/annual/. Last visited Sept 2016

Selective Sulfation Roasting of Rare Earths from NdFeB Magnet Scrap

B.N. Carlson and P.R. Taylor

Abstract Rare earth magnets play an increasingly important role in high end technology, and the manufacture of rare earth magnets, such as the NdFeB type, consume a large amount of the rare earths produced. Recycling of this material could provide an important domestic source for these materials. A selective sulfation roast has been proposed; the central principle of which is to take advantage of the differing stabilities of the rare earth sulfates as compared to those of iron under an air/SO_2 atmosphere. Experiments show iron is found to be stable as an oxide, however, neodymium is found to form an oxy-sulfate phase.

Keywords Sulfation roasting · Rare earth magnet recycling

Introduction

NdFeB Magnets

Sintered NdFeB type magnets are by far the strongest commercially available variety, and are used where size and weight are important factors. The reason for this, as demonstrated in Table 1, is that for a given required field strength, a much smaller NdFeB magnet can be used in comparison to more traditional magnetic materials, such as ferrite or alnico [1]. As such, rare earth magnets play an integral

B.N. Carlson (✉) · P.R. Taylor
Kroll Institute for Extractive Metallurgy, Colorado School of Mines,
1500 Illinois Street, Golden, CO, USA
e-mail: BCarlson@mines.edu

S. Wang et al. (eds.), *Applications of Process Engineering Principles in Materials Processing, Energy and Environmental Technologies*,
The Minerals, Metals & Materials Series, DOI 10.1007/978-3-319-51091-0_27

Table 1 Required magnet size to achieve 1000 gauss field (after Darcy et al. [1])

Magnet composition	Required volume (cm^3)
NdFeB	0.22
$SmCo_5$	0.37
Ferrite	19.6
Alnico 9	14.3
Alnico 5–7	11.9

role in many high end technologies, such as hybrid and electric vehicles, computer hard drives, and high efficiency wind turbines [2].

During the manufacture of sintered rare earth magnets, up to 30% of the starting alloy is lost to machining waste. This waste, known as magnet "swarf" represents a significant waste stream in terms of value. This material cannot be re-used directly, however, as it is contaminated with cutting fluids and grinding media, as well as being partly oxidized during the machining process [3].

Sulfation Roasting

Sulfation roasting has been used as a selective method for the extraction of metals from primary sources such as the extraction of nickel from laterite ores [4]. The concept of this method is to form a soluble sulfate species of one component while leaving the other in an insoluble form. The desired sulfates species can then be selectively leached away from the insoluble phase with water. A recent study has successfully demonstrated the selective sulfation of rare earths from NdFeB magnet scrap via sulfuric acid sulfation and selective decomposition, showing that is possible to selectively form sulfate species of rare earths while leaving the iron as an oxide [5].

In this study, a method for selective, gas phase sulfation of neodymium from iron is proposed. The basis for this selectivity is the differing thermodynamic stability of iron and neodymium sulfate. Using a stability diagram, such as a Kellog diagram, it is possible to ascertain what species of certain metals may be stable at a set temperature and atmosphere. This can be seen in the Kellog diagram in Fig. 1, which shows, at 750 °C, there exists a region where it is possible to form ferric oxide and neodymium sulfate. Therefore, at this temperature, there exists a certain partial pressure SO_2 and O_2 which allows for the stability of iron oxide and neodymium sulfate.

As a method to increase the kinetics of sulfation roasting, alkali metal sulfates can be added. These pyro-sulfate forming compounds, such as potassium sulfate, allow for the penetration of reactant gases through the sulfate product layer. This is accomplished by formation of a liquid phase which acts to form a porous product layer [6].

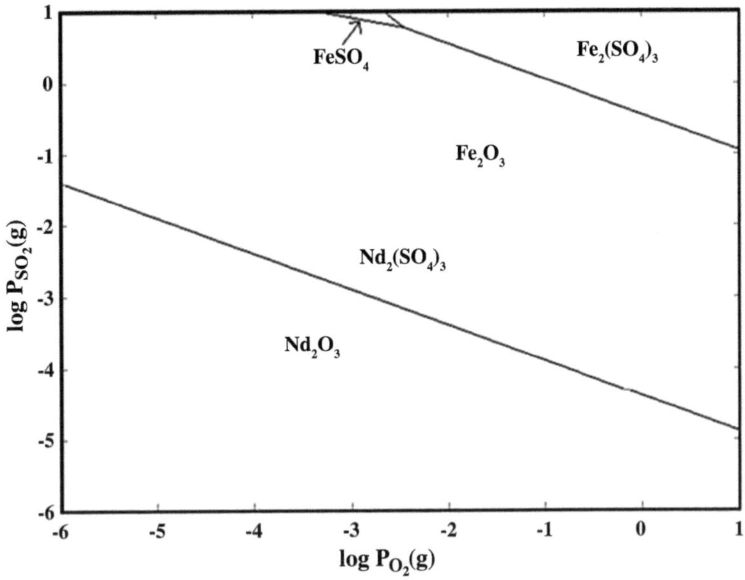

Fig. 1 Kellogg diagram composed of superimposed (Nd and Fe)–O_2–SO_2 diagrams at 750 °C

Experimental

Sample Preparation

In this study, porous pellets several compositions were prepared. Reagent grade components were utilized, all of which were sieved to −74 μm and +53 μm. In the samples containing Nd_2O_3 and Fe_2O_3, a 1:7 mol Fe:Nd ratio was used, corresponding to the $Nd_2Fe_{14}B$ phase utilized in NdFeB type magnets. Where potassium sulfate was added, a 0.1 mol per mol oxides ratio was used. Each pellet composition was mixed with ammonium bicarbonate so that this fraction took up 1/7 of the total volume. These components were pressed into 20 mm cylindrical die. These pellets were heated at 100 °C for one hour, sublimating the contained ammonium bicarbonate, leaving behind a porous structure, and then sintered at 600 °C for 6 h. They were then turned to rough spheres by rotating the pellet in a 12.7 mm hole drilled in 4.8 mm steel plate.

Experimental Procedure

Pellets were loaded in the thermal gravimetric analysis system (TGA), shown in Fig. 2, and heated to 750 °C under a flow of N_2 gas. When the desired temperature was reached, a flow of SO_2 and air was introduced into the furnace which provides

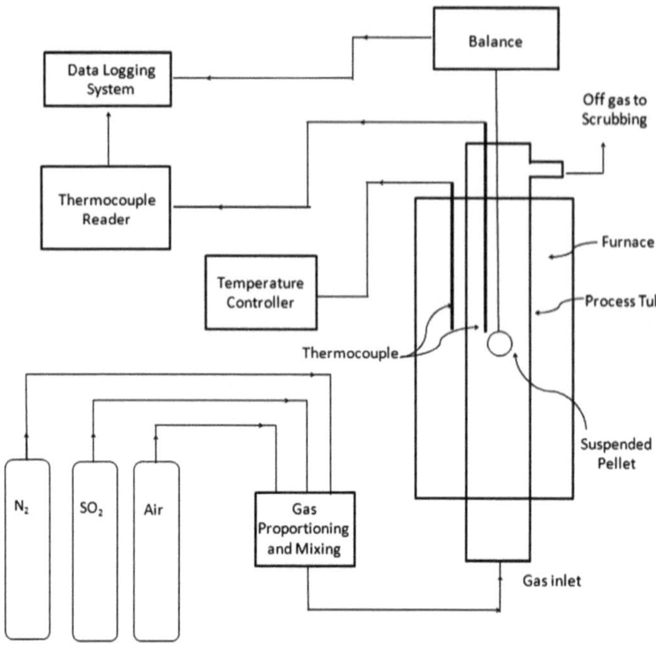

Fig. 2 TGA utilized for this study (after Anderson [7])

Table 2 Change in weight of sample after roasting

Sample	% Weight change
Fe_2O_3	−0.37
Nd_2O_3	12.27
Nd_2O_3–Fe_2O_3	7.03
Nd_2O_3–Fe_2O_3–K_2SO_4	7.96

an atmosphere which corresponded to a $-\log(P_{O_2},\ P_{SO_2})$ of 0.76. As seen in Fig. 1, this lies within a region which iron is stable as an oxide, while neodymium is stable as a sulfate. The pellet was reacted for 1 h. after which the furnace was flushed with N_2 and allowed to cool. The resulting pellet weighed was crushed to −295 μm for analysis by X-ray diffraction (XRD) to determine the phases present in the sample.

Results

Weight Change

The weight change of each pellet after sulfation roasting can be seen in Table 2.

XRD Spectra

The X ray diffraction pattern of the iron oxide pellet can be seen in Fig. 3.

The X ray diffraction pattern of the neodymium oxide pellet can be seen in Fig. 4.

The X ray diffraction pattern of the iron oxide and neodymium oxide pellet can be seen in Fig. 5.

The X ray diffraction pattern of the iron oxide and neodymium oxide pellet that included potassium sulfate can be seen in Fig. 6.

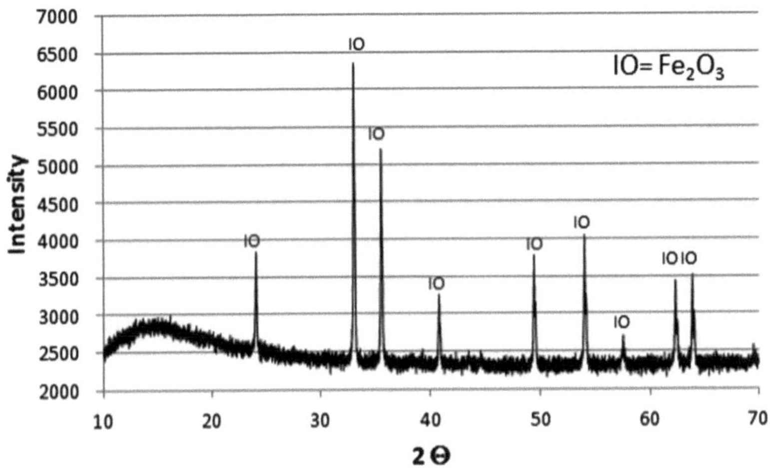

Fig. 3 XRD pattern of roasted Fe_2O_3 pellet

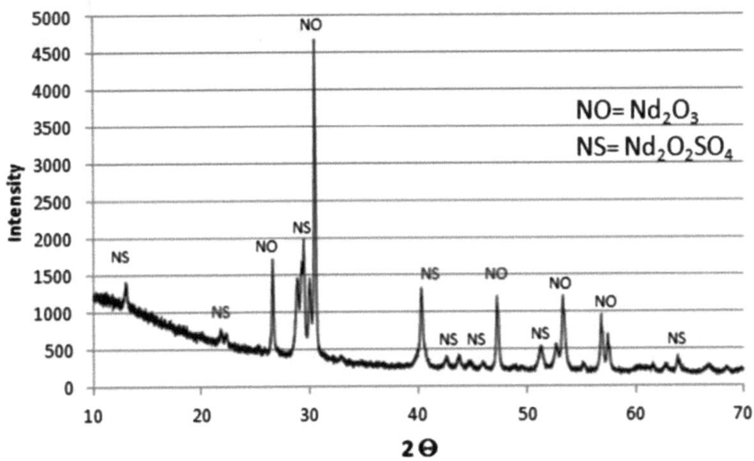

Fig. 4 XRD pattern of roasted Nd_2O_3 pellet

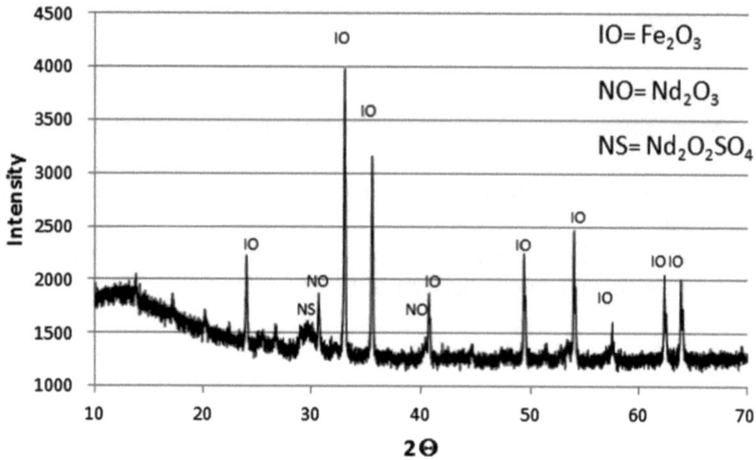

Fig. 5 XRD pattern of roasted Nd_2O_3–Fe_2O_3 pellet

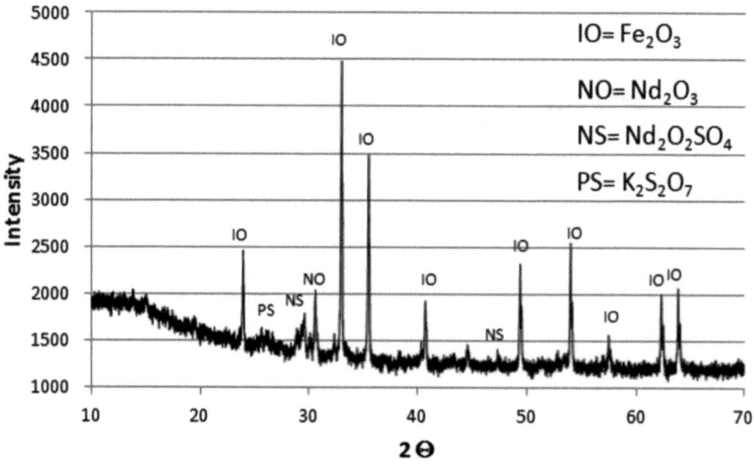

Fig. 6 XRD pattern of roasted Nd_2O_3–Fe_2O_3–K_2SO_4 pellet

Discussion

As seen in the XRD pattern in Fig. 3, the pellet of iron oxide appears to not have reacted at the experimental conditions used. This is in agreement with the Kellog diagram in Fig. 1, which shows that Fe_2O_3 is the thermodynamically stable phase at 750 °C and $-\log(P_{O_2}, P_{SO_2}) = 0.76$.

The samples containing neodymium oxide were shown to be reactive, as seen in Table 2, as they all exhibited a weight gain. Looking at the XRD patterns, however, it appears that in place of the sulfate species expected, a neodymium oxy-sulfate phase is dominant. It may be that this phase is an intermediate which forms in between the oxide and sulfate, and simply more reaction time is necessary.

Conclusions

Pellets of several compositions were created and reacted in an atmosphere and at a temperature which predicted iron to be stable as an oxide while neodymium is thermodynamically stable as a sulfate. By XRF analysis, iron oxide does not seem to react at the conditions used, as predicted by the thermodynamic stability diagram. However, neodymium was found to form an oxy-sulfate species. More study is merited to locate conditions where it may be possible to form neodymium sulfate species.

Acknowledgements The authors would like to acknowledge the Critical Materials Institute for their support of this research. Also, a thank you is due to the friends, colleagues, and faculty at the Colorado School of Mines for all of their help and guidance.

References

1. J.W. Darcy, H.M. Dhammika Bandara, B. Mishra, B. Blanplain, D. Apelian, M.H. Emmert, Challenges in recycling end-of-life rare earth magnets. JOM **65**(11) (2013)
2. M.J. Kramer, R.W. McCallum, I.A. Anderson, S. Constantinides, Prospects for non-rare earth permanent magnets for traction motors and generators. JOM **64**(7), 752–763 (2012)
3. C.O. Bounds, The Recycling of Sintered Magnet Swarf. in *Metals and Materials Waste Reduction, Recovery and Remediation* (1994), pp. 173–186
4. B.B. Kar, Y.V. Swamy, Extraction of nickel from Indian lateritic ores by gas-phase sulphation with SO_2—air mixtures. Miner. Process. Extr. Metall. **110**(2), 73–78 (2001)
5. M.A.R. Önal, C.R. Borra, M. Guo, B. Blanpain, T. Van Gerven, Recycling of NdFeB magnets using sulfation, selective roasting, and water leaching. J. Sustain. Metall. **1**(3), 199–215 (2015)
6. B. Kar, Y. Swamy, Some aspects of nickel extraction from chromitiferous overburden by sulphatization roasting, Pergamon (2000)
7. C.G. Anderson, *An Experimental and Mathematical Investicgation of Chemical Beneficiation of Low-Grade Chromite Ores, University of Idaho* (1987)

Gold Solubility in Smelting Slags for the Recycling of Industrial and Mining Wastes

Jun-Gil Yang, Joo Hyun Park and Hyun-Sik Park

Abstract Gold is one of the most valuable and precious metals. But, the extraction efficiency of gold from natural resources is very low and labor-intensive due to very low concentration of gold, i.e., 1–5 ppm, which means that gold extraction processes produce extensive amounts of tailings. The gold content in printed circuit board (PCB) of waste mobile phone is about 100–400 ppm, which is enormously greater than that found in natural ores. Consequently, a requirement for recycling waste PCB and gold mine tailings becomes increasing for maintaining sustainable society. Because the pyrometallurgical processes are issued due to economic reasons, it is needed to understand the thermodynamic behavior of gold in smelting slags under oxidizing and reducing atmosphere at high temperatures. Therefore, in this study, the effect of slag chemistry, oxygen potential and temperature on gold solubility in smelting slags are discussed.

Keywords Gold · Solubility · Mine tailings · PCB wastes · Pyrometallurgy · Smelting slags

Introduction

Gold has good physicochemical properties. Thus, it plays an increasingly important role in industrial applications. For example, over 300 tons of gold are used annually in electronic components [1]. However, 75% of all gold ever produced has been extracted since 1910 and gold reserves in the world are becoming rapidly depleted [2]. The cost of gold production and its price in the market have increased rapidly, and interest in the recycling of gold-containing materials has naturally increased.

J.-G. Yang · J.H. Park (✉)
Department of Materials Engineering, Hanyang University, Ansan, Korea
e-mail: basicity@hanyang.ac.kr

H.-S. Park
Korea Institute of Geoscience and Mineral Resources (KIGAM), Daejeon, Korea

S. Wang et al. (eds.), *Applications of Process Engineering Principles in Materials Processing, Energy and Environmental Technologies*,
The Minerals, Metals & Materials Series, DOI 10.1007/978-3-319-51091-0_28

The wastes of electronic equipments (e-wastes) contain large amounts of precious metals compared to their own respective ores, and therefore such wastes may be considered as a secondary source of valuable metals. For instance, the concentration of gold in natural ore is commonly 0.5–15 g/ton-ore (0.5–15 ppm), while its concentration in electronic circuit boards is over 10 times higher, typically being about 150 ppm in expansion cards and over 10,000 ppm in central processing units (CPUs) [3]. Also, gold mine tailings are discharged after floatation process as slurry form then stacked near the operating plants (approx. 100 ton per day in one of Korean gold mines). Gold mine tailings contain gold about 1-2 mg/kg. This is the reason why Au-containing materials such as e-wastes and mining wastes are highly important sources [4].

From the viewpoint of pyrometallurgical processing, the distribution ratio of gold between metal and slag and the solubility of gold in slags have been scarcely reported. Richardson and Billington reported that the solubility of gold in lead glass was 220 ppm but only 30 ppm in plate glass at 1673 K [5, 6]. Toguri and Santander measured the distribution ratio of copper between Cu–Au alloy and silica-saturated fayalite slags at temperatures ranging from 1523 to 1623 K as a function of the oxygen partial pressure from $p(O_2) = 10^{-8}$ to 10^{-7} atm [7]. Although the distribution of gold was plotted against the alloy composition for various conditions of temperature and oxygen partial pressure, there was a high degree of scatter.

The solubility of Au in fayalite slag was reported to be approx. 80 ppm at 1473 K by Altman and Kellogg [8]. Nagamori and Mackey [9] measured the distribution ratio of gold between copper/slag and matte/slag. However, the experimental results were scattered due to the poor reproducibility of gold analysis in the slag, segregation and sampling difficulty, and the very low level of Au in the slag. Recently, the dissolution mechanism of gold into metallurgical slags was systematically investigated by Swinbourne et al. [6] at extensive temperatures and oxygen partial pressures, viz. from 1373 to 1573 K, and $p(O_2) = 10^{-10}$ to 1.0 atm. The solubility of gold in iron silicate, calcium ferrite and lead oxide slags increased with increasing oxygen partial pressure, from which gold was found to be present probably as Au^+ ions in molten slags.

Based on the above background, not only the temperature dependency of gold solubility in slags but also the gold solubility in the calcium silicate based slags, which are more easily reused from iron- and steelmaking companies, has not been investigated yet. The operation temperature and slag composition are very important factors affecting the recovery of gold from electronic and mining wastes to design the cost-effective pyrometallurgical processing routes. Consequently, in the present study, gold solubility in the $CaO–SiO_2–Al_2O_3$ based slags was measured at 1673–1773 K to determine the dissolution mechanism of gold in the calcium silicate based slags.

Experimental Procedure

The thermochemical equilibration experiments were carried out using a super-kanthal electric furnace with a $MoSi_2$ heating element. The temperature was controlled within ±2 K using a B-type thermocouple and a PID controller. Pure gold (4 N purity) was used, and slags were prepared by mixing reagent-grade chemicals. Gold (0.38 g) and slag (3 g) were loaded into a fused alumina or magnesia crucible placed in a porous alumina holder. The oxygen partial pressure of the system was controlled by CO-CO_2 equilibrium, i.e., $p(O_2) = 10^{-10}$ to 10^{-8} atm at 1673 to 1773 K. After 12–24 h equilibration, the samples were quickly extracted from the furnace and quenched by plunging the crucible into brine. The slag and gold samples were carefully separated from the crucible. The gold content in the slag was analyzed by ICP-OES and the equilibrium composition of the slag was analyzed with an XRF spectroscopy.

Results and Discussion

Gold dissolution reaction into the slag can be described by the following general equation:

$$\mathrm{Au} + m\mathrm{O_2(g)} + n(\mathrm{O^{2-}}) = (\mathrm{AuO}^{2n-}_{2m+n}) \tag{1}$$

$$K_{[1]} = \frac{a_{\mathrm{AuO}^{2n-}_{(2m+n)}}}{a_{\mathrm{Au}} \cdot p^m_{\mathrm{O_2}} \cdot a^n_{\mathrm{O^{2-}}}} = \frac{f_{\mathrm{AuO}^{2n-}_{(2m+n)}} \cdot \left(\%\mathrm{AuO}^{2n-}_{(2m+n)}\right)}{p^m_{\mathrm{O_2}} \cdot a^n_{\mathrm{O^{2-}}}} \tag{2}$$

where a_i, f_i, and p_{O_2} are the activity and activity coefficient of component i, and the oxygen partial pressure, respectively. Because pure gold was used in the experiments, the activity of gold is unity. Therefore, the above Eq. (2) can be expressed as follows.

$$\log(\%\mathrm{AuO}^{2n-}_{(2m+n)}) = n\log a_{\mathrm{O^{2-}}} + m\log p_{\mathrm{O_2}} - \log f_{\mathrm{AuO}^{2n-}_{(2m+n)}} + C \tag{3}$$

From Eq. (3), the solubility of gold is expected to have a linear relationship with the oxygen partial pressure in logarithmic form (slope = m) at a fixed temperature and slag composition. The gold solubility in the 23%CaO–38%SiO_2–13%Al_2O_3–MgO_{sat} system at 1773 K is plotted against the oxygen partial pressure in Fig. 1. Gold solubility, log (%Au), linearly increases with increasing oxygen partial pressure, log $p(O_2)$, with a slope of 0.20(±0.02).

Gold has two oxidation states, Au^+ and Au^{3+}. If gold oxide is Au_2O, viz. gold is stable as Au^+, the stoichiometric coefficient m will be 0.25. Otherwise, if Au_2O_3 is the case, viz. as Au^{3+}, m will be 0.75. From the measured results in the present

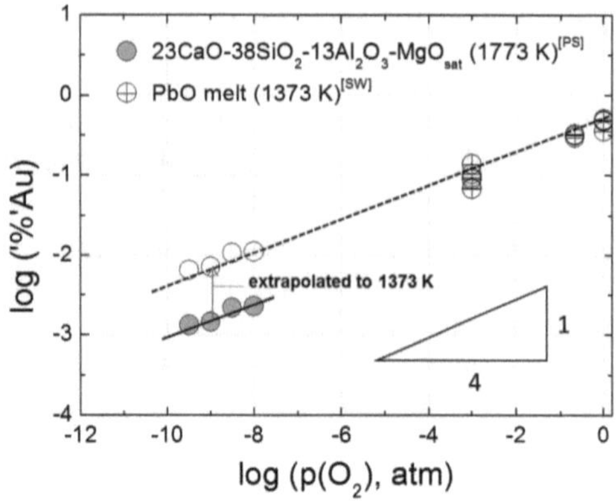

Fig. 1 Dependence of gold solubility in the $CaO–SiO_2–Al_2O_3–MgO_{sat}$ slag system on the oxygen partial pressure at 1773 K

study (Fig. 1), the theoretical slope of 0.25 is more reasonable than 0.75, indicating that the '$O_2(g)$' term should have the stoichiometric coefficient '$m = 1/4$'. It is significant that the oxygen potential dependence of gold solubility in the $CaO–SiO_2–Al_2O_3–MgO_{sat}$ slag system is very similar to that in the PbO slag at 1373 K, which was investigated over a more extensive $p(O_2)$ range, and supports the view that gold dissolves as the Au^+ oxidation state [6].

The measured results in the present slag system at 1773 K were extrapolated to 1373 K, at which temperature the system is assumed to be supercooled, based on the temperature dependence of gold solubility [10]. It is very interesting in Fig. 1 that the extension of gold solubility in the MgO-saturated calcium aluminosilicate melts at moderately reducing atmosphere, i.e. $p(O_2) = 10^{-10}$ to 10^{-8} atm, is surprisingly consistent with the solubility in the PbO slag under oxidizing atmospheres, i.e. $p(O_2) = 10^{-3}$ to 0.2 atm at 1373 K [6]. This means that the experimental results in the current study and Swinbourne et al.'s results are in good agreement and that the influence of oxygen partial pressure on the dissolution mechanism of gold in the $CaO–SiO_2–Al_2O_3–MgO_{sat}$ slag and PbO slag is identical through a very wide range of oxygen potentials, i.e. from $p(O_2) = 10^{-10}$ to 0.2 atm, irrespective of the slag system.

The solubility of gold in the $CaO–SiO_2–Al_2O_3$ based slags is plotted against the modified Vee ratio (MVR), i.e., $(\%CaO)/(\%SiO_2 + \%Al_2O_3)$, in Fig. 2. The solubility of gold in the slags increases with increasing MVR. The quantitative analysis for the influence of basicity on the dissolution behavior of gold can be given using the following equation, which is modified from Eq. (3).

$$\log\left(\%AuO^{2n-}_{(2m+n)}\right) = n\log a_{O^{2-}} + \frac{1}{4}\log p_{O_2} - \log f_{AuO^{2n-}_{(2m+n)}} + C \quad (4)$$

Because the basicity of slag, i.e. the activity of free oxygen, cannot be experimentally measured due to thermodynamic constraints, it can be replaced by the activity of CaO, assuming that the CaO activity is proportional to the O^{2-} ion activity at fixed oxygen partial pressure and temperature. In the current study, the

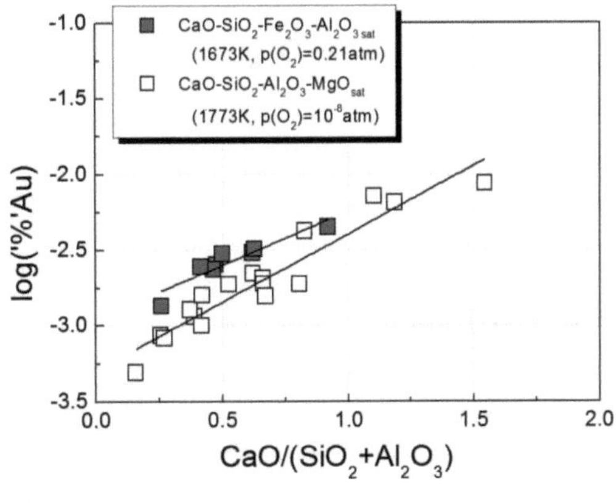

Fig. 2 Dependence of gold solubility in the CaO–SiO_2–Al_2O_3–MgO_{sat} and CaO–SiO_2–Fe_2O_3–Al_2O_3 $_{sat}$ slags on the modified Vee ratio

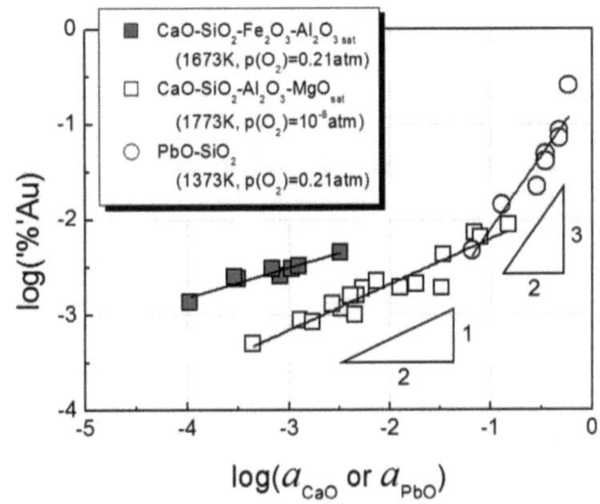

Fig. 3 Dependence of gold solubility in the CaO- and PbO-based slags on the activity of CaO and PbO

CaO activity in the slag was calculated by FactSage™7.0. The solubility of gold is plotted against the activity of CaO on logarithmic axes in Fig. 3, wherein a good linear relationship between log (%Au) and log a_{CaO} is found. Because the slopes are close to 0.5 within the experimental scatters, the stoichiometric coefficient n in Eq. (4) is 1/2 in the CaO-based slags.

In Fig. 3, the experimental results for the gold solubility in the PbO–SiO_2 slag measured at 1373 K under atmospheric condition are shown for the sake of comparison [8]. If the activity of PbO, which can be obtained using FactSage™, in the PbO–SiO_2 slag were employed as an indirect basicity index, the slope of the line obtained from a linear regression analysis is close to 1.5, viz. $n = 3/2$. Therefore, based on the above thermodynamic discussion, the dissolution mechanism of gold into the silicate melts can be summarized as follows according to the activity of basic oxides (BO).

$$Au + \frac{1}{4}O_2(g) + \frac{1}{2}(O^{2-}) = (AuO^{-}), \quad (a_{BO} < 0.1) \tag{5}$$

$$Au + \frac{1}{4}O_2(g) + \frac{3}{2}(O^{2-}) = (AuO_2^{3-}), \quad (a_{BO} > 0.1) \tag{6}$$

Conclusions

From the present results, it is concluded that the predominant species of gold is strongly dependent on the basicity of slag. Furthermore, the more basic slags provided the higher solubility of gold at a given oxygen partial pressure and the more oxidizing atmosphere enhanced the dissolution of gold to the molten slag phase. Consequently, it is necessitated to carefully control the slag composition during pyro-processing of gold-containing waste materials in order to improve the gold recovery.

Acknowledgements The authors express their appreciation to Prof. Douglas R. Swinbourne, RMIT University, Melbourne, Australia, for the fruitful discussion regarding the gold dissolution mechanism in the slags. Also, this research was partly supported by the Basic Research Project (GP2015-013) of the Korea Institute of Geoscience and Mineral Resources (KIGAM), funded by the Ministry of Science, ICT and Future Planning of Korea.

References

1. C. Hageluken, C.W. Corti, Gold Bull. **43**, 209–220 (2009)
2. C. Louis, O. Pluchery, *Gold Nanoparticles for Physics* (Chemistry and Biology, World Scientific, 2012), p. 3
3. L. Barbieri et al., Environ. Chem. Lett. **8**, 171–178 (2010)
4. E. Yamasue et al., Mater. Trans. **50**, 1536–1540 (2009)
5. F.D. Richardson, J.C. Billington, Miner. Process. Ext. Metall. (TIMM C) **65**, 273–297 (1956)
6. D.R. Swinbourne, S. Yan, S. Salim, Miner. Process. Ext. Metall. (TIMM C) **114**, 23–39 (2005)
7. J.M. Toguri, N.H. Santander, Metall. Trans. **3**, 586–588 (1972)
8. R. Altman, H. H. Kellogg, Miner. Process. Ext. Metall. (TIMM C) **81**, 163–75 (1972)
9. M. Nagamori, P.J. Mackey, Metall. Trans. B **9B**, 567–579 (1978)
10. Y.S. Han, D.R. Swinburne, J.H. Park, Metall. Mater. Trans. B **46B**, 2449–2457 (2015)

Solid State Reduction of Iron, Manganese and Chromium Oxide Ores with Methane

R.H. Eric, A. Bhalla, P. Halli and P. Taskinen

Abstract Sustainable development requires less energy consumption with lower carbon footprint. In this review the solid state reduction behaviour of three oxide ores using methane gas are briefly summarized based on experimental work conducted in our laboratories. The discussion concentrates mainly on mechanisms of the reduction studied through SEM-EDAX and X-ray diffraction techniques. Limited preliminary kinetic input is also mentioned. Iron ore containing titanium and vanadium could be reduced significantly in the solid state with methane-hydrogen mixture up to 400 °C lower than is needed in ordinary solid state carbothermic process. The reduction of chromite and manganese ores were possible similarly at lower temperatures with potential energy savings lower carbon footprint.

Keywords Metal oxides · Reduction · Methane

Introduction

Methane is widely known for its high energy potential and is already globally adopted as a source of energy and heat [1], but it is not as commonly studied yet as potential reductant for metallurgical processes. Currently most oxide ores are reduced through the use of solid carbon. This process is energy demanding due to the highly endothermic nature of reduction reactions conducted at temperatures well above 1200 °C and production of CO_2 which has to be converted to CO again by the highly endothermic Boudouard reaction to maintain high reaction rates and high recoveries. However, possibility of lowering both exists through the use of

R.H. Eric (✉) · A. Bhalla
School of Chemical and Metallurgical Engineering,
University of the Witwatersrand, Johannesburg, South Africa
e-mail: rauf.eric@wits.ac.za

R.H. Eric · P. Halli · P. Taskinen
Department of Materials Science and Engineering, Aalto University, Espoo, Finland

S. Wang et al. (eds.), *Applications of Process Engineering Principles in Materials Processing, Energy and Environmental Technologies*,
The Minerals, Metals & Materials Series, DOI 10.1007/978-3-319-51091-0_29

hydrocarbon gases such as methane as a reductant. The investigation of the potential alternative reductants has been carried out by some authors [2–9], but the literature is quite limited despite similar results were observed regardless of the raw materials used indicating the possible lowering of reduction temperatures compared to ordinary carbothermic reduction. The fundamental nature of hydrocarbons is that they are not thermodynamically stable at high temperatures and especially in the presence of a solid phase which may act as a catalyst for decomposition. In the case of methane this decomposition occurs approximately at 600 °C [10] as in Eq. (1).

$$CH_4(g) \rightarrow C(s) + 2H_2(g) \tag{1}$$

The overall reduction reaction of metal oxides, MeO, with methane is shown in Eq. (2).

$$4MeO(s) + CH_4(g) \rightarrow 4Me(s) + CO_2(g) + 2H_2O(g) \tag{2}$$

Moreover, as the system contains hydrogen and elemental carbon according to Eq. (1), they will act as the real reducing agents forming gaseous H_2O, CO and CO_2 along with metal. According to Randhawa and Rehmat, [11], hydrogen gas can react with carbon dioxide and according to Lin et al. [12], water vapour can also react with the carbon at high temperatures regenerating the reductant gases; CO and H_2. The decomposition of methane will produce potentially nanometre sized solid elemental carbon and hydrogen gas in the system. The thermodynamic activity of the nanometre sized carbon is greater than 1.0 in the system and the produced carbon will react with any CO_2 present producing carbon monoxide, which is the main reductant. In the Ellingham diagram for oxides, the standard line for the reaction of solid carbon (with respect to pure solid graphite as the standard state) with oxygen forming carbon monoxide gas at 1.0 atm pressure has a negative slope facilitating carbothermic reduction of different metal oxides the Ellingham lines of which have positive slopes. Thus the intersection point of the negatively sloped CO line and the positively sloped metal oxide line is thermodynamically the minimum temperature for the carbothermic reduction of the metal oxide. When the activity of carbon is greater than 1.0 the standard line for CO shifts in clockwise direction with more negative slope thus its intersection with a metal oxide line occurs at lower temperature. Moreover as the partial pressure of product CO gas will potentially be less than 1.0 atm due to presence of other gases the clockwise shift of the CO line will even be more severe. This is the fundamental principle of lowering the temperature of metal oxide reductions through the use of a hydrocarbon gases. Furthermore, hydrogen gas would also support the reduction of especially iron oxides and all of these would contribute to a lower temperature operation reducing energy requirements and less carbon consumption leading to lower the CO_2 emissions and lower energy requirements. The thermodynamic activity of carbon is illustrated in Fig. 1 as a function of CH_4/H_2 ratio in the gas phase and temperature calculated using HSC Chemistry 7.0 [13]. During the experimental work summarized below the CH_4/H_2 ratio of the gas was kept constant in certain values in order

Fig. 1 The behaviour of carbon activity in CH_4–H_2 gas mixtures as a function of temperature

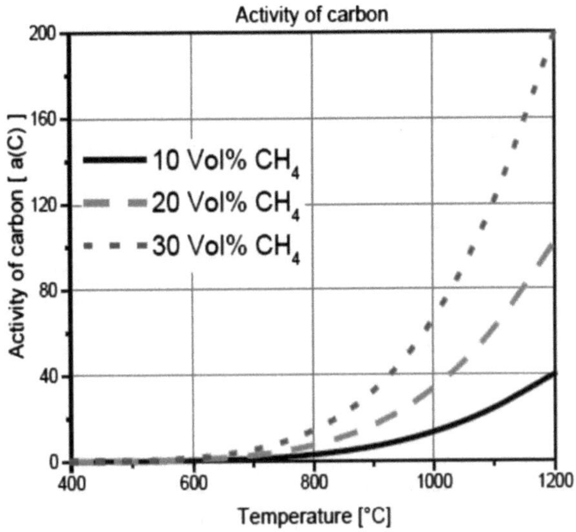

to fix the thermodynamic activity of carbon to create a thermodynamically controlled experimental reduction system. In an actual potential process the carrier gas, hydrogen, can be replaced with air and the oxygen of the air could be utilized to combust the excess part of methane to provide heat for the endothermic reduction reactions. In many cases the end product of the reduction reactions is not pure metallic phases but carbides due to the high activity and concentration of carbon resulting in carbon saturation and also due to the relevant stability of the carbides.

Experimental Procedure

Reduction experiments were carried out isothermally in batches of constant mass at temperatures from 800 to 1350 °C depending on the type of ore reduced. The treatment times were kept constant at each temperature to 10, 20, 30, 60, 90 and 120 min. The total gas flow rate in every experiment was 800 ml/min (STP). The contents of methane were 10, 20 and 30 vol.% and the balance in the mixture was hydrogen. Argon was employed as an inert gas to flush the furnace before and after every experiment. The argon gas flow rate was 500 ml/min. Experiments were carried out in a laboratory scale horizontal electrical furnace. The furnace is described in detail in a previous work [4]. The temperature variance through the hot zone was ± 2 °C. Both sample holder boats and the working tube were pure impervious alumina. Figure 2 present the schematic diagram of the experimental furnace employed. The temperature was measured with a calibrated thermocouple (Pt/Pt10Rh) and the thermoelectric force was converted into degrees with a

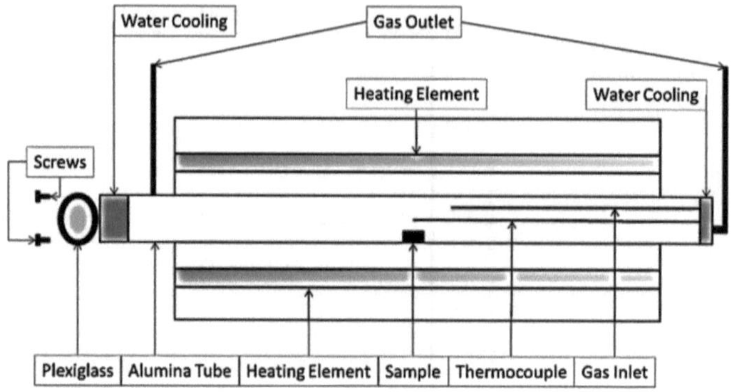

Fig. 2 Schematic diagram of the experimental furnace

Keithley measurement software and the data was collected with a National Instruments Labview temperature logging program.

XRD (X-ray diffraction), SEM-EDS (scanning electron microscopy, energy dispersive spectroscopy), optical microscopy and selective leaching with ICP-OES (inductively coupled plasma, optical emission spectroscopy) were employed to the experimental samples. The selective leaching with ICP-OES was employed to dissolve the reduced metallic elements, and the metallic carbides in order to define the metallization degree of (partially) reduced samples [14, 15]. XRD was the method used to determine the formed carbides. The degree of metallization is defined as:

$$M_i = (\%\,\text{of metallized i in the sample})/(\text{total}\,\%\,\text{of element i in the sample}) \quad (3)$$

where "% of metallized i in the sample" is obtained from chemical analysis ICP-OES mentioned above and total "% of element I" in the sample is known from the chemical composition of the ore.

The titanomagnetite ore used was produced by Mustavaaran Kaiso's Oy, Finland. The mean size was 19.63 μm. The chemical analyses of the ore used is presented in Table 1 and the XRD revealed two principal phases: magnetite and ilmenite.

The South African Mamatwan manganese ore from Kalahari district was used for this study. The composition of the ore is presented in Table 2. The mean particle size of the ore used in the tests was 95.78 μm. The manganese ore was calcined at 1000 °C prior to reduction tests to derive off its CO_2 and H_2O contents. The principal phases of the calcined ore determined by XRD were Hausmannite; Mn_3O_4, Braunite; Mn $(Mn, Fe)_6$ SiO_{12}, Manganosite; MnO, Jacobsite; $MnFe_2O_4$, and Lime; CaO.

The chromite ore used in the experiments was from the LG-6 seam of the Bushveld Complex of South Africa. The composition of the ore is shown in Table 3. The mean particle size of the ore used in the experiments was 104.62 μm.

Table 1 The chemical analysis of Mustavaara titanomagnetite ore

Compound	Average (wt%)	Element	Average (wt%)
SiO_2	0.805	Si	0.376
TiO_2	6.3	Ti	3.775
Cr_2O_3	0.007	Cr	0.004
V_2O_3	1.02	V	0.695
FeO	80.95	Fe	62.9
MnO	0.191	Mn	0.148
MgO	0.625	Mg	0.38
CaO	1.095	Ca	0.785

Table 2 Chemical composition of Mamatwan manganese ore

Component	Average wt%
Mn	37.69
Fe	4.71
SiO_2	4.51
CaO	16.51
MgO	3.29
Al_2O_3	0.231
P	0.0174
CO_2	17.80
H_2O	4.20

Table 3 Chemical composition of LG-6 chromite ore

Component	Average wt%
Total Cr	31.60
Cr_2O_3	46.20
Total Fe	19.60
Fe^{2+}	14.90
FeO	19.20
Fe_2O_3	6.71
MgO	9.98
Al_2O_3	13.70
SiO_2	1.46
TiO_2	0.54
Cr/Fe	1.61
Fe^{3+}/Fe^{2+}	0.315

The main phase of the ore is the spinel which is made up of divalent cations of Fe and Mg, three-valent cations of Fe, Cr and Al with O^{2-} anions. It can be represented in a very simplified way as (Fe, Mg) $[Fe, Al, Cr]_2O_4$. The only other phase was the silicate which actually was below the detection limit of XRD.

Results and Discussion

Within the limited scope and space of this presentation the results on each of the three ores tested will be given without going into details of the reduction mechanisms and kinetic aspects. The aim here is to reveal the possibility of using methane as a reductant and the potential benefits in terms of lower temperature operations in comparison to ordinary solid state carbothermic reductions.

Reduction of Mustavaara Titano Magnetite Ore

Two reference experiments were performed on the titanomagnetite ore with the same experimental setup as described earlier. The first reference experiment was carbothermic in argon atmosphere and the second reference experiment was carried out in hydrogen gas atmosphere. The treatment time in both cases was 120 min at 1200 °C. The carbothermic reduction achieved 18.2% metallization degree for titanium, 37.7% for vanadium and 97.2% for iron. When employing pure hydrogen, the final metallization degree of titanium, vanadium and iron were 0.40, 17.00 and 94.2% respectively.

The metallization of titanomagnetite ore as a function of temperature using 10 vol.% of CH_4 in the feed gas mixture can be seen in Fig. 3. The reduction of titanium under 1000 °C, as expected, did not take place due to high stability of rutile. The reduction of iron was practically complete after an hour at 1100 °C and in half an hour at 1200 °C and even at 800 °C the final metallization degree of iron achieved was 80%. The reduction rate of vanadium depended highly on the temperature and eventually the highest metallization degree was obtained at 1100 °C,

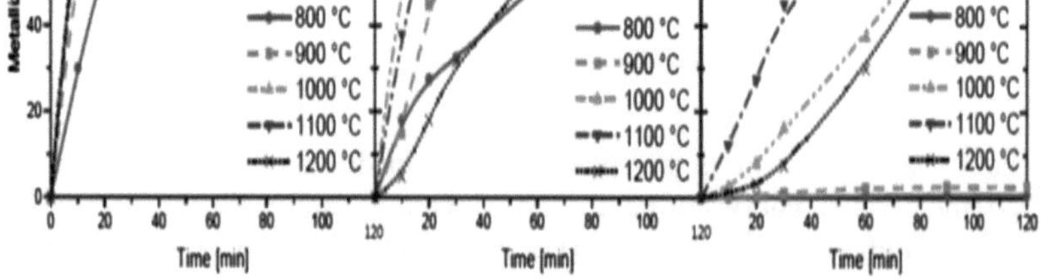

Fig. 3 Metallization curves of iron, vanadium and titanium of the titanomagnetite ore with 10 vol.% CH_4 containing methane-hydrogen gas mixture

Fig. 4 XRD and SEM of the reduction sequence of the titanomagnetite ore particles at 900 °C with 20 vol.% CH_4

Treatment time	Phases detected by XRD	SEM-micrograph
10 min	Magnetite, Iron, Rutile	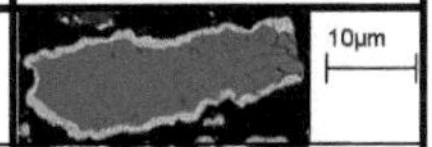
20 min	Iron, Wustite, Magnetite, Cementite, Rutile	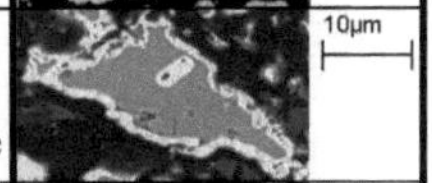
30 min	Iron, Cementite, Wustite, Rutile	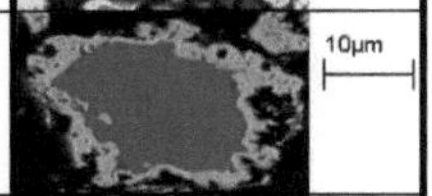
60 min	Iron, Cementite	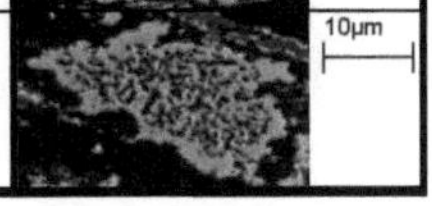

approximately 95%. The highest metallization degree of titanium achieved was around 80% at 1200 °C. The metallization values for vanadium and titanium when methane is used are spectacularly high in comparison to ordinary solid state carbothermic and hydrogen reductions conducted at the same temperatures and for the same reduction periods illustrating the potential benefits of methane/natural gas. Figure 4 presents the reduction mechanism of titano magnetite ore reduced at 900 °C with 20 vol.% of CH_4. The shrinking core kinetics is the main reduction mechanism at the earlier stages of the reduction but after 60 min, the inner core collapses completely regardless of the methane content employed in the feed gas mixture. Above 900 °C even with a treatment time of 10 min the inner oxide core disappears. Hence it can be concluded that at temperatures higher than 900 °C the shrinking core kinetics would be very fast. At 1000 °C and above the reduction rate is very high. Partial melting of the metallic iron phase started at 1100 °C just after 10 min and was more pronounced at 1200 °C. The methane content in the feed gas mixture did not affect to the partial melting of the metallic particles. Titanium carbide was first detected at 1200 °C after 1 h of reduction with every content of methane used. The SEM-EDS analyses revealed the formation of metallic vanadium potentially entering into solution with iron carbide.

Reduction of Mamatwan Manganese Ore

Reduction of calcined Mamatwan manganese ore was performed in the temperature range of 1000 to 1200 °C with 10, 20 and 30 vol.% of methane in the gas mixture. A typical metallization plot is illustrated in Fig. 5. Total metallization values achieved after 2 h ranged from 21 to 72% depending upon temperature and methane content of the gas mixture. The associated Mn/Fe metallization ratio were 0.19–0.69 all much higher than those achieved with ordinary solid state carbothermic reduction. The most common occurrence was rapid leveling of reduction rate after

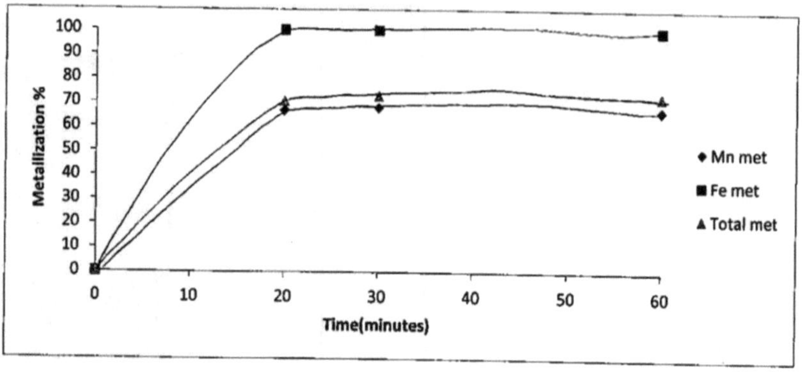

Fig. 5 Progress of metallization versus time curves for reduction of Mamatwan manganese ore for tests using 30% CH_4 in hydrogen at T = 1200 °C

first 20 or 30 min for all methane concentrations and temperatures revealing that the process will take 20–30 min to complete the reactions which is much quicker than ordinary carbothermic reduction. For example at 1100 °C reduction % achieved by carbothermic reduction with excess carbon was only around 35% [14] as compared to 70% metallization and above reached at same temperature of 1100 °C in this work. Manganese ore was reduced primarily to carbide Mn_7C_3 at lower temperature range of the experiments, but at 1200 °C the dominant reaction product was Mn_5C_2. The XRD analysis revealed that with the progress of metallization, the concentration of the original oxide phase decreased with reduction time and a new oxide phase; Ca_2FeMnO_5, began to appear. Fe_7TC_3 appeared as the predominant iron carbide phase early in the reduction test but with progress of metallization, this phase disappeared and was replaced in increasing amounts Fe_3C. Generally, the higher the concentration of methane in the reducing gas the sooner the Fe_3C appeared.

The particle morphologies were studied mainly by means of SEM and energy dispersive analysis of X-rays (EDAX). Two typical images are presented in Figs. 6 and 7 indicating the possibility of surface coverage type of kinetics where the metal appears almost everywhere on the surface of the particle and eventually covering it.

Reduction of LG-6 Chromite Ore

Methane reduction of LG-6 chromite ore was conducted in the temperature range of 1050–1250°C again with 10, 20 and 30 vol.% CH_4 in either methane-argon or methane-hydrogen gas mixtures. When argon was used as the carrier gas total metallization values achieved after 2 h of reduction ranged from 59 for 10% CH_4 in Ar at 1050 °C, to 79 for 30% CH_4 in Ar at 1250 °C. The associated Cr/Fe metallization ratios were 0.51 and 0.65 respectively, all much higher than those achieved with solid state carbothermic reduction [15, 16] conducted at these

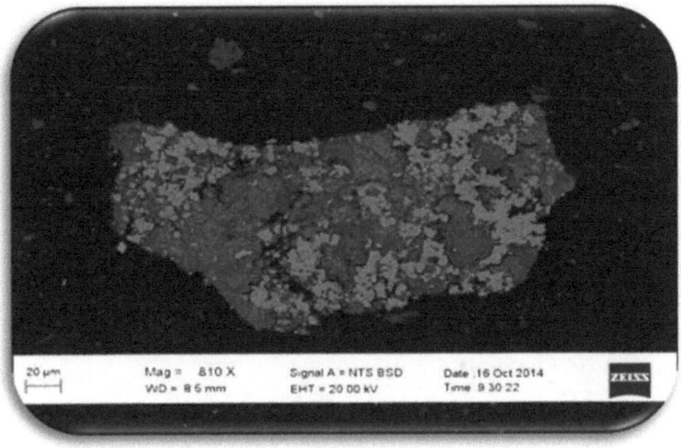

Fig. 6 Mamatwan manganese ore particle reduced at 1100 °C for 20 min using 10% CH_4 in the gas mixture

Fig. 7 Mamatwan manganese ore particle reduced at 1200 °C for 20 min using 10% CH_4 in the gas mixture

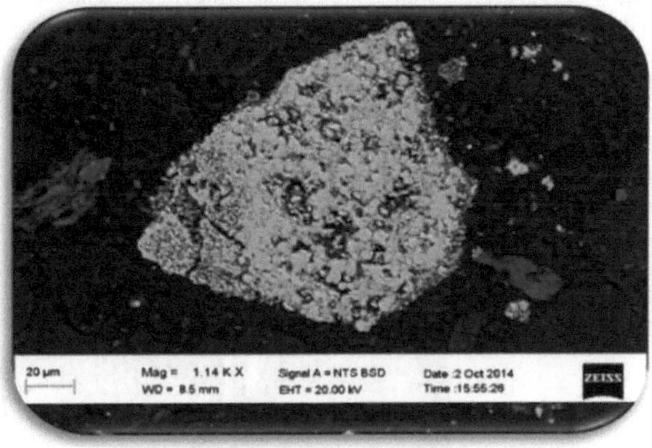

temperatures. When hydrogen was used as the carrier gas the total metallization values achieved after 2 h of reaction ranged from 44 for 10% CH_4 in H_2 at 1050 °C to 85 for 30% CH_4 in H_2 at 1250 °C. The associated metallic Cr/Fe ratios were 0.48 and 0.78. A typical metallization versus time curve for the reduction of LG-6 chromite ore is shown in Fig. 8.

The images presented in Fig. 9 illustrate the particle morphologies of partially reduced chromite ore. Pore formation appeared to initiate on the particle surfaces, at the crack tips and along the spinel crystallographic planes in the iron and chromium rich central core of the particle. This enabled the reducing gas to penetrate the inner core causing metallization there. In general the kinetics follow shrinking core characteristics. Changing CH_4 concentration had the effect of changing the kinetic rate and extent of reduction. Up to 1200 °C, product morphologies of particles

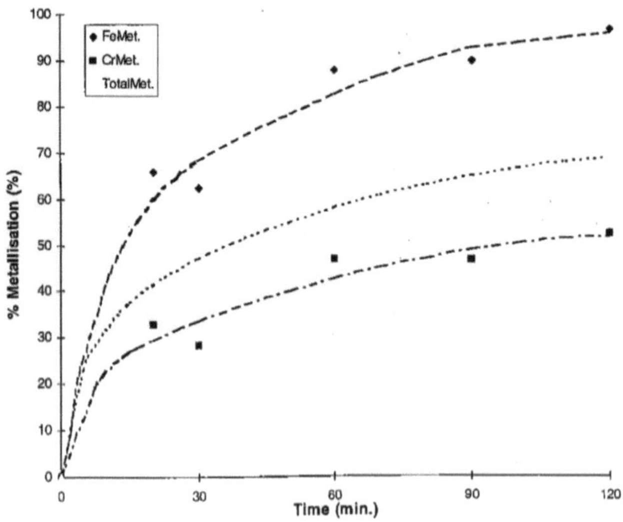

Fig. 8 Progress of metallization versus time for test using% 30 CH4 in Ar at T = 1100 °C for the reduction of LG-6 chromite ore

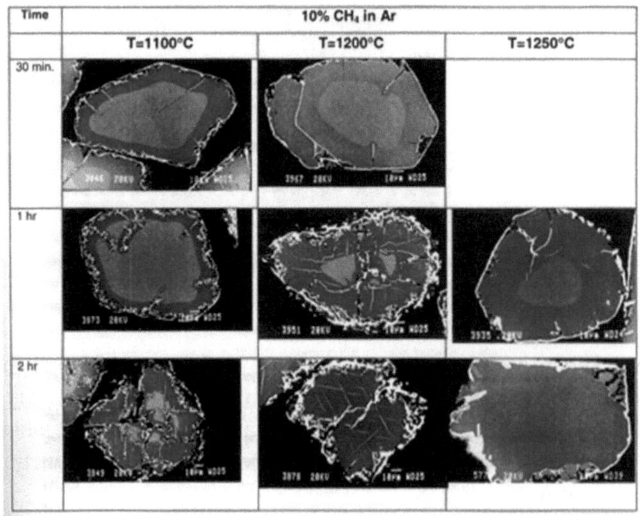

Fig. 9 Summary of particle morphologies for 10% CH_4 in Ar

reacted under 10% in CH_4 in Ar and H_2 looked nearly identical. At higher temperatures, however, some differences were observed indicating a change/shift in reaction mechanism and rate control. This aspect requires further study.

Conclusions

This study explored the use of different concentrations of CH_4 gas both in Ar and in H_2 as the reducing gas mixture for a better understanding of the reduction behavior of titanomagnetite, manganese and chromite fines at different temperatures. Generally, CH_4 was an effective reductant because it supplied both C and H_2 to the reduction site. H_2 is a kinetically rapid redundant. On site formed carbon from CH_4 cracking, in its atomic form with very high thermodynamic activity well above unity ensured the rapid regeneration of the reductants CO and H_2 from CO_2 and H_2O, ensuring the maintenance of a very low reduction potential shifting the minimum reaction temperature to lower values.

The experimental results for the reduction of Mustavaara titano magnetite showed high metallization degrees of iron and vanadium and at higher temperatures even titanium. The reduction mechanism of the Mustavaara titano magnetite ore clearly follows shrinking core kinetics which is clear at lower temperatures studied while at the higher temperatures the reduction kinetics was very fast and the shrinking core was not visible having collapsed at the very early stages of the reduction.

In the case of the reduction of Mamatwan manganese ore the reaction sites were mostly limited to the particle surface and there were no significant crack and pore formation. This resulted in a surface coverage type of mechanism and kinetics. The higher the concentration of the methane in the gas phase, the higher was the resulting reaction rate and extent reaching over 75% metallization at 1200 °C.

For the reduction of LG-6 chromite ore the reduction sites were not limited to the particle surface only and included cracks and pores. The rate of reduction was generally highest within the first 20 and 30 min of reaction. The higher the concentration of CH_4 in the gas mixture diffusing to the particle surfaces from the bulk phase, the higher were the resulting reaction rate and extent. Simultaneous to the surface reduction, the ionic solid-state diffusion caused the original solid solution to form a shrinking central core in the particle, with a secondary partially reduced spinel oxide forming the outer rim obviously following shrinking core mechanism and kinetics.

The solid state reduction of titanomagnetite, manganese and chromite ores into useful products with methane gas mixtures could potentially result in development of more sustainable process options which could reduce the energy requirements and carbon footprint due to being able to operate at lower temperatures to reach the metallization levels achieved by the existing processes operating at higher temperatures.

Acknowledgements The research summarized in this presentation has been supported and financed both: by the "Finland Distinguished Professor" project awarded to Professor R H Eric by Tekes; the Finnish Funding Agency for Innovation, and the Ferro Alloy Producers Association of South Africa. The authors are deeply grateful to both of these organizations.

References

1. Sang-Yong Lee, Gerald D. Holder, Methane hydrates potential as a future energy source. Fuel Process. Technol. **71**(1–3), 181–186 (2001)
2. Raymond James Longbottom, Oleg Ostrovski, Eungyeul Park, Formation of cementite from titanomagnetite ore. Iron Steel Inst. Jpn. Int. **46**(5), 641–646 (2006)
3. N.A. El-Hussiny, A. El-Amir, S.T. Abdel-Rahim, M.E.M.H. Shalabi, Kinetics of direct reduction titano-magnetite concentrate briquette produced from rosetta-ilmenite via hydrogen. Open Access Libr. J. **1**, e662 (2014). doi:10.4236/oalib.1100662
4. P. Halli, *Solid-State Reduction of Vanadium Containing Mustavaara Titano-Magnetite Concentrate, Master's Thesis* (Aalto University, Espoo, Finland, 2015)
5. E.R. Monazam, R.W. Breault, R. Siriwardane, G. Richards, S. Carpenter, Kinetics of the reduction of hematite (Fe_2O_3) by Methane (CH_4) 155 during chemical looping combustion: a global mechanism. Chem. Eng. J. **232**, 478–487 (2013). doi:10.1016/j.cej.2013.07.091
6. Somaye Nasr, Kevin P. Plucknett, Kinetics of iron ore reduction by methane for chemical looping combustion. Energy Fuels **28**(2), 1387–1395 (2014). doi:10.1021/ef402142q
7. G. Zhang, O. Ostrovski, Reduction of titania by methane-hydrogen-argon gas mixture. Metall. Mat. Transac. B **31**(1), 129–139 (2000)
8. G. Zhang, O. Ostrovski, Reduction of ilmenite concentrates by methane-containing gas: part I. Effects of ilmenite composition, temperature and gas composition. Can. Metall. Q. **40**(3), 317–326 (2001)
9. Guangqing Zhang, O. Ostrovski, Reduction of ilmenite concentrates by methane containing gas, part II: effects of preoxidation and sintering. Can. Metall. Q. **40**(4), 489–497 (2001)
10. B.J. Wood, H. Wise, The reaction kinetics of gaseous hydrogen atoms with graphite **73**(5), 1348–1351 (1964)
11. S.S. Randhawa, A. Rehmat, The hydrogenation of carbon dioxide in parts-per-million levels. ACS Fuels **14**(3), 1–9 (1970)
12. Shi-Ying Lin, Yoshizo Suzuki, Hiroyuki Hatano, Michiaki Harada, Hydrogen production from hydrocarbon by integration of water-carbon reaction and carbon dioxide removal (HyPr-RING Method). Energy Fuels **15**(2), 339–343 (2001)
13. HSC Chemistry 7.0, Outotec0
14. R.H. Eric, E. Burucu, The mechanism and kinetics of the carbothermic reduction of Mamatwan manganese ore fines. Miner. Eng. **5**(7), 795–815 (1992)
15. O. Soykan, R.H. Eric, R.P. King, The reduction mechanism of a natural chromite at 14,160 C. Metall Transac B, 22B, 53–63 (1991)
16. P. Weber, R.H. Eric, The reduction mechanism of chromite in the presence of silica flux. Metall Transac B, 24B, 987–995 (1993)

Stibnite Chloridizing with Calcium Chloride-Oxygen at Roasting Temperatures

R. Padilla, I. Moscoso and M.C. Ruiz

Abstract Antimony and arsenic are impurities in copper concentrates found normally as stibnite (Sb_2S_3), tetrahedrite ($Cu_{12}Sb_4S_{13}$) and enargite (Cu_3AsS_4). When copper concentrates contain appreciable amounts of As and Sb, they must be removed before smelting to avoid environmental pollution. Oxidizing volatilization roasting at 500–700 °C has been used to eliminate these impurities. However, only a fraction of Sb volatilizes in those conditions; thus, a more effective process to eliminate antimony from concentrates is needed. In this study, the chloridizing roasting of Sb_2S_3 using $CaCl_2$–O_2 was investigated to remove the antimony from concentrates. XRD analysis of calcines indicated that the overall reaction could be written as:

$$Sb_2S_3 + 3CaCl_2 + 6O_2 = 2SbCl_3 + 3CaSO_4$$

Temperature and oxygen partial pressure have a significant effect on the rate of reaction. An estimated conversion of about 90% was obtained at 750 °C, 5.4% oxygen in 20 min. High oxygen concentrations arrest the advance of the reaction forming various antimony oxides including Sb_6O_{13}, $Ca_5Sb_5O_{17}$ and Sb_2O_4.

Keywords Stibnite · Antimony · Volatilization · Chloridizing roasting · Calcium chloride

Introduction

Copper sulfide minerals are the major source for the production of metallic copper, and these sulfides usually contain variable amounts of non-desirable minor elements such as As, Sb and Bi, which are present in the concentrates as enargite (Cu_3AsS_4), antimonite (Sb_2S_3) and bismuthinite (Bi_2S_3), respectively. Antimony

R. Padilla (✉) · I. Moscoso · M.C. Ruiz
Department of Metallurgical Engineering, University of Concepcion,
Edmundo Larenas 285, Concepcion, Chile
e-mail: rpadilla@udec.cl

S. Wang et al. (eds.), *Applications of Process Engineering Principles in Materials Processing, Energy and Environmental Technologies*,
The Minerals, Metals & Materials Series, DOI 10.1007/978-3-319-51091-0_30

may also be present in some concentrates as tetrahedrite or famatinite. In the froth flotation processing of copper minerals, the majority of these impurities report to the copper concentrates. Since smelting and converting are the primary processes for the production of copper from copper sulfide concentrates, the ever increasing content of these noxious elements (As, Sb, and Bi) in the concentrates produce operational complications, leading to the pollution of the environment and also to the production of the final copper product with unacceptable high levels of impurities. Therefore, these contaminating elements must be separated or eliminated from the copper concentrates before smelting.

The conventional method for the pretreatment of dirty copper concentrates has been oxidizing roasting, where the minerals of the minor elements (As, Sb, and Bi) tend to decompose rather easily to volatile compounds, allowing separation of these elements by volatilization from the copper minerals [1]. Some alternative hydrometallurgical processes for cleaning the copper concentrates have also been proposed [2, 3]; however, they have not reached commercial application so far. Thus, roasting methods are still preferred over leaching methods for the elimination of As, Sb, and Bi from concentrates by the copper mining industry.

Regarding the high temperature behavior of stibnite, Padilla et al. [4] reported that in nitrogen atmosphere antimony could be volatilized efficiently as Sb_2S_3 only at temperatures higher than 900 °C. In the case of roasting in the presence of oxygen, antimony can be volatilized efficiently as Sb_2O_3 at temperatures higher than 900 °C when the gas phase contains less than 5% oxygen [4, 5]. At higher oxygen concentration in the gas phase, the volatilization of antimony is hindered by the formation of the nonvolatile oxide SbO_2. Thus, the elimination of antimony through the gas phase in conventional roasting processes using air or oxygen-deficient air in the roasting temperatures (500–700 °C) is not feasible and the antimony would remain mostly in the calcine as nonvolatile oxide. Therefore, the objective of this investigation was to determine the feasibility of volatilizing antimony and bismuth in the presence of calcium chloride in oxidizing roasting. In particular, in this paper we discuss the behavior of stibnite in the presence of calcium chloride at roasting temperatures.

Experimental Work

Materials and Procedure

The experimental work was carried out using mixtures of Sb_2S_3 and $CaCl_2*2H_2O$ (both Aldrich chemicals). The molar ratios of the sample mixtures used are shown in Table 1.

Table 1 Molar ratios of mixtures for the antimony chloridizing experiments

Sample	$Sb_2S_3/CaCl_2*2H_2O$
M9	1/4.5
M10	1/3

The methodology used for the determination of the chloridizing/volatilization of antimony was thermogravimetric. The experimental apparatus consisted of a horizontal furnace with a quartz reaction tube, and ancillary equipment including a temperature controller and a gas system to maintain a controlled atmosphere in the reaction tube. TGA setup for continuous monitoring of the chloridizing reaction was also used.

Results

The chloridizing reaction was followed by analyzing the weight loss fraction of the sample given as $\Delta W/W_o$, where W_o is the initial sample weight. Figure 1 shows the weight loss fraction as a function of time for experiments carried out to study the effect of temperature on the chloridizing of stibnite.

Clearly, the sample weight loss depends greatly on the temperature. In order to verify that the weight loss of the sample is associated to the volatilization of antimony species from the mixture sample, few experiments were run to capture condensed material from the gas phase over the surface of a cold finger set at the exit of the reaction tube. After a few runs, the precipitated solids on the surface of the cold finger were removed, and analyzed by X-ray diffraction spectroscopy. The results are shown in Fig. 2. The identified precipitated compound was Sb_2O_4 which confirms that the weight loss is related to antimony volatilization. Based on this result and the thermodynamic analysis of the Sb_2S_3–$CaCl_2$ system, the following reaction was considered a priori as the likely chloridizing reaction for the analysis of our experiments.

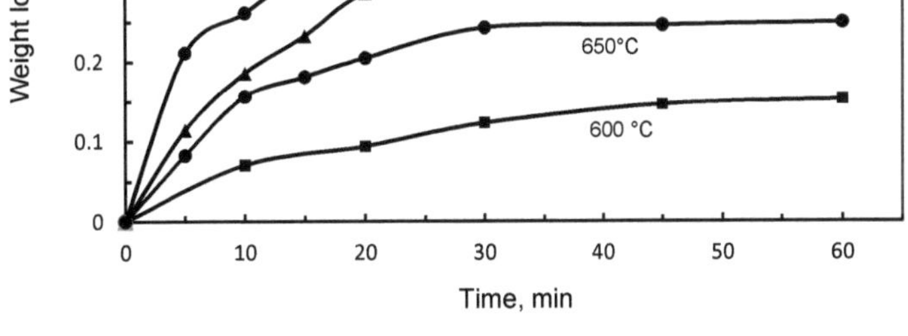

Fig. 1 Effect of temperature on the chloridizing/volatilization of stibnite sample M10 with calcium chloride in 5.4% oxygen in the gas phase

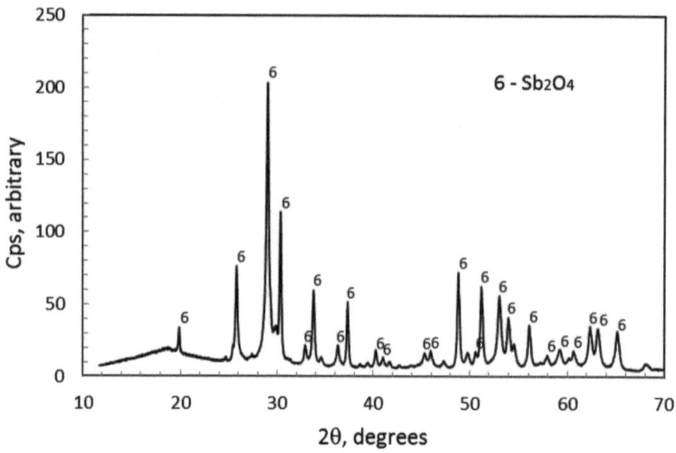

Fig. 2 Diffractogram of collected condensed material showing well crystallized Sb_2O_4

$$Sb_2S_3 + 3CaCl_2 + 6O_2(g) = 2SbCl_3(g) + 3CaSO_4 \quad (1)$$

Therefore, assuming that reaction (1) is the overall reaction for the antimony removal from the sample, the weight loss data can be used to estimate the conversion of this reaction. Therefore, the conversion of reaction (1) was calculated by $\Delta W/W_t$, where ΔW is the weight loss of the sample at any time and W_t is the theoretical final weight for complete reaction according to the stoichiometry of reaction (1). The resulting data are shown in Fig. 3.

Figure 3 shows that conversions over 90% can be obtained in a very short time, about 20 min at 750 °C. On the other hand, about 40% maximum conversion is attained at 600 °C in 30 min. Therefore, the maximum conversion attained in the experiments depends strongly on the temperature.

Effect of the Partial Pressure of Oxygen in the Reaction Atmosphere

The influence of the partial pressure of oxygen on the antimony removal from the sample is shown in Fig. 4. The batch experiments were conducted using mixture M10, which has the stoichiometric molar ratio of Sb_2S_3 and $CaCl_2$ for reaction (1). In these experiments, the antimony removed from the sample was determined by chemical analysis of the calcines for antimony.

We can observe in this figure that the maximum volatilization of antimony from the sample decreases with an increase in the oxygen content in the gas phase. This negative effect of the oxygen concentration on the antimony volatilization is probably due to the formation of nonvolatile compounds of antimony such as

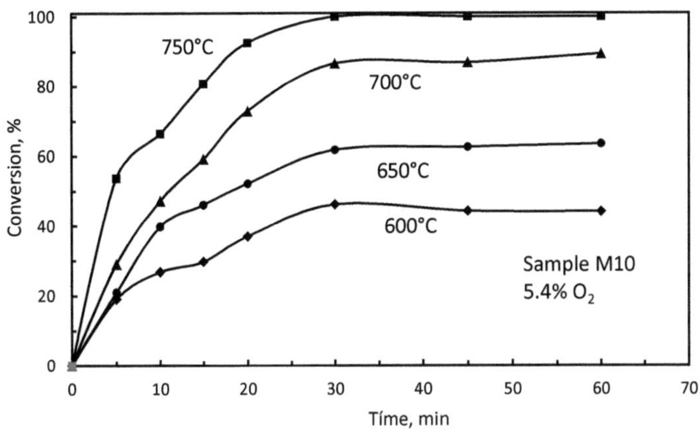

Fig. 3 Antimony chloridizing with calcium chloride as a function of time for samples with stoichiometric molar ratios of Sb_2S_3 and $CaCl_2$, (M10)

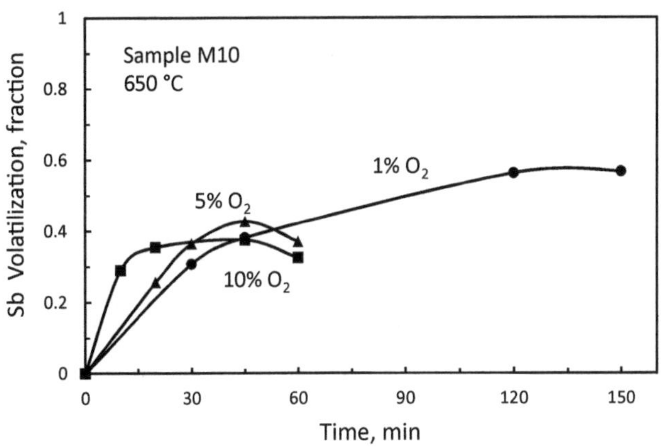

Fig. 4 Influence of the oxygen concentration on the antimony volatilization

Sb_2O_4, and Sb_2O_5 and other complex oxides as shown in the diffractograms of calcines shown in Figs. 6 and 7. Nevertheless, we can observe that the rate of the antimony volatilization increases with increasing the oxygen content in the gas phase.

Comparing the maximum fraction of arsenic volatilized at each oxygen concentration shown in Fig. 4 with the conversions values in Fig. 3 for the same temperature, it is evident that the conversion values calculated from the weight loss

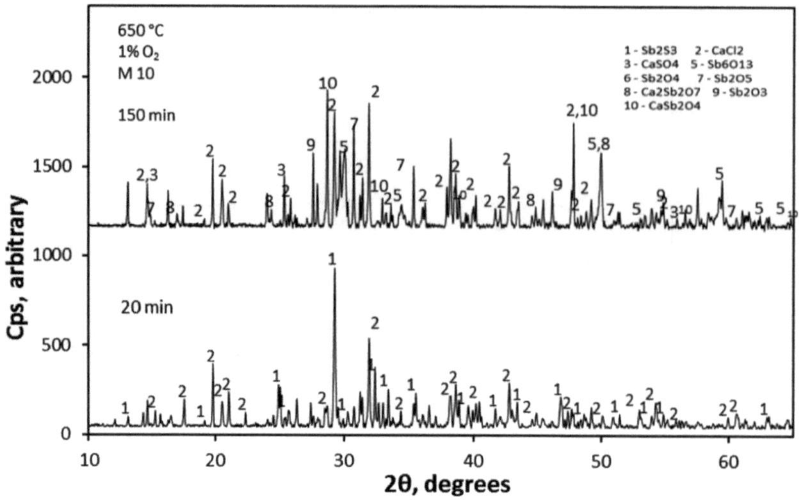

Fig. 5 XRD patterns of calcines obtained at 650 °C and 1% oxygen in the gas phase

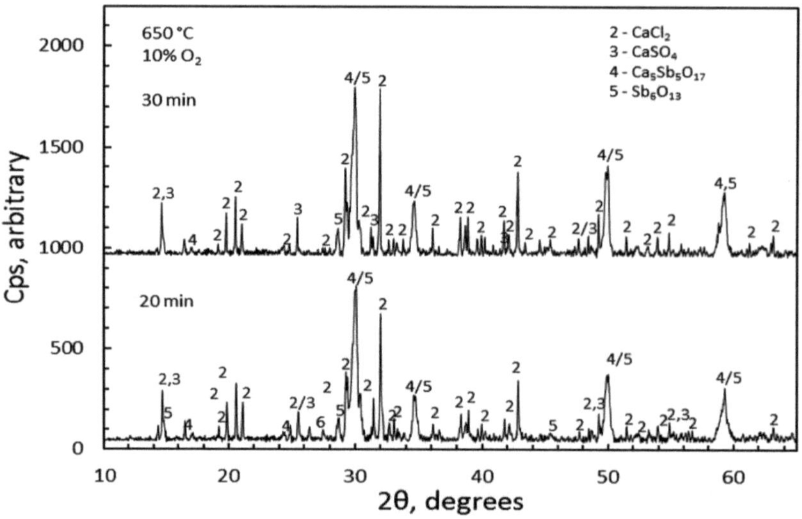

Fig. 6 XRD patterns of calcines obtained at 650 °C and 10% oxygen partial pressures

overestimate the fraction of arsenic volatilized from the samples. This again is a consequence of the fact that not all the antimony reacts according to reaction (1) but a fraction of the antimony oxidize to nonvolatile compounds and remains in the calcine. Therefore, estimated conversions based on weight loss would be better for lower oxygen partial pressures.

Characterization of the Reaction Products

The identification of the phases present in the calcines was done mainly by XRD spectroscopy. The results for samples obtained at 650 °C in 1% oxygen in the gas phase are shown in Fig. 5. In this figure, we can observe two XRD patterns; in the pattern corresponding to 20 min of reaction, the major compounds identified were the reactants Sb_2S_3 and $CaCl_2$. This means that antimony chloridizing under this low oxygen content and low temperature is a very slow process. However, at 150 min, $CaSO_4$ was identified together with several oxides of antimony including Sb_6O_{13}, $Ca_5Sb_5O_{17}$, and $CaSb_2O_4$.

In contrast, Fig. 6 shows the XRD patterns of samples obtained at the same temperature of 650 °C but higher oxygen content of 10%. We can observe in this figure that the major phases identified were $CaSO_4$, $Ca_5Sb_5O_{17}$, Sb_6O_{13} for both 20 and 30 min of reaction times. These results confirm that in the presence of high oxygen concentration the antimony removal through the gas phase is not complete due to the formation of $Ca_5Sb_5O_{17}$ and Sb_6O_{13} compounds.

At the higher temperature 675 °C, and 10% oxygen, the formation of $CaSO_4$, Sb_6O_{13}, and Ca–Sb oxides are again evident as shown in Fig. 7. The identification of $CaSO_4$ phase in the calcines is an indication that the reaction (1) does occur in the system for the chloridation of antimony. However, other reactions must also occur in parallel leading to the formation of the various antimony oxides identified.

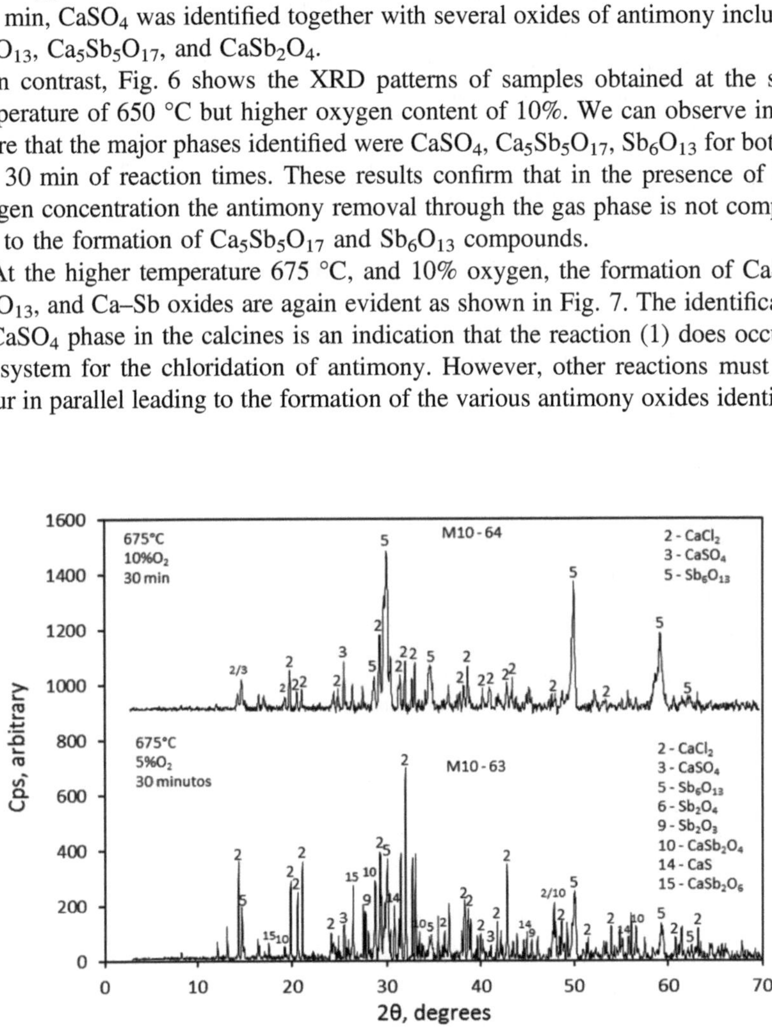

Fig. 7 XRD spectrums of calcines produced at 675 °C and 30 min of reaction time a function of oxygen partial pressures

Conclusions

From the data shown above, the following can be concluded.

- Temperature has a significant influence on the stibnite chloridization in oxidizing atmospheres of Sb_2S_3–$CaCl_2$ mixtures.
- An increase in the oxygen concentration in the gas phase has a detrimental effect on the antimony volatilization from the Sb_2S_3–$CaCl_2$ samples.
- An estimated conversion of about 90% was obtained at 750 °C in 5.4% oxygen and 20 min of reaction time.
- High oxygen concentrations in the gas phase arrest the advance of the chloridizing reaction forming various antimony oxides including Sb_6O_{13}, $Ca_5Sb_5O_{17}$ and Sb_2O_4.
- Finally, the XRD results suggest that the overall reaction for the chloridation of stibnite could be represented by:

$$Sb_2S_3 + 3CaCl_2 + 6O_2(g) = 2SbCl_3(g) + 3CaSO_4$$

Acknowledgements The authors acknowledge the National Fund for Scientific and Technological Development (FONDECYT) of Chile for the financial support of this research through project No. 1150339.

References

1. L. Winkel et al., Decomposition of copper concentrates at high-temperatures: an efficient method to remove volatile impurities. Miner. Eng. **21**, 731–742 (2008)
2. A. Roca. et al., A leaching process for removing arsenic from enargite-bearing copper concentrates. in *Proceedings of Copper 2003*, ed. by Riveros et al. (Canadian Institute of Mining, Metallurgy and Petroleum, Montreal, Quebec, 2003), pp. 631–644
3. M.C. Ruiz, R. Bello, R. Padilla, Removal of arsenic from enargite rich copper concentrates. *Materials Processing Fundamentals*, ed. by L. Zhang et al. (John Wiley &Sons, Hoboken NJ, 2013), pp. 217–223
4. R. Padilla, G. Ramirez, M.C. Ruiz, High-temperature volatilization mechanism of stibnite in nitrogen-oxygen atmospheres. Metall. Mater. Trans. B **41B**, 1284–1292 (2010)
5. R. Padilla, A. Aracena, M.C. Ruiz, Kinetics of stibnite (Sb_2S_3) oxidation at roasting temperatures. J. Min. Metall. Sect. B Metall. **50**(2)B, 127–132 (2014)

Investigations on Rotary Tool Near-Dry Electric Discharge Machining

Vineet Kumar Yadav, Pradeep Kumar and Akshay Dvivedi

Abstract Near-dry electric discharge machining (EDM) is a process variant of EDM, which uses two-phase flow of liquid and gas as dielectric medium. This article reports the results of an investigation pertaining to the drilling of holes on high speed steel (T2 grade) by a rotary tool near-dry EDM. The One-Factor-At-a-Time (OFAT) approach was used for experimentation. The effects of process parameters, viz. tool rotation speed, peak current, pulse duration, gas pressure and liquid flow rate on material removal rate (MRR) and overcut were investigated. The dielectric medium used was a mixture of glycerin-air. The experimental results reveal that rotation of the tool electrode along with high pressure of two-phase dielectric medium contribute in an effective flushing of inter electrode gap (IEG). Further, in rotary tool near-dry EDM process, higher values of current can be used. This improves process economics in terms of MRR. The MRR measured with rotary tool near-dry EDM was three to four times higher in comparison with conventional EDM. Additionally, appreciable surface quality was achieved.

Keywords Near-dry EDM · Material removal rate (MRR) · Surface roughness · Two-phase flow

Introduction

Electric Discharge Machining (EDM) is an advanced machining process. This process is primarily used to machine conductive materials irrespective of their hardness, into complex shapes with high precision. In EDM, material removal takes place due to erosion caused by a series of discrete discharges between the tool electrode and the workpiece. Both the electrodes are submerged in the dielectric medium such as hydrocarbon oil or deionized water. Commercially available EDM is also known as conventional EDM uses hydrocarbon oil as dielectric medium. The

V.K. Yadav (✉) · P. Kumar · A. Dvivedi
Indian Institute of Technology (IIT), Roorkee, India
e-mail: vineet437@gmail.com

S. Wang et al. (eds.), *Applications of Process Engineering Principles in Materials Processing, Energy and Environmental Technologies*,
The Minerals, Metals & Materials Series, DOI 10.1007/978-3-319-51091-0_31

burning of hydrocarbon oil produces harmful fumes, that pollute the environment and also harmful to the operator. Low material removal, high tool wear rate and issues related to environmental pollution are the main disadvantage of conventional EDM. In order to overcome these problems, researchers have developed several process variants such as dry EDM, powder mixed EDM and near dry EDM, etc.

In dry EDM, gaseous dielectric medium is used instead of hydrocarbon oil [1]. Kunieda et al. observed that MRR can be improved by using the oxygen gas as dielectric medium. Which resulted in higher MRR due to the exothermic reaction in IEG [2]. To further enhance the performance of dry EDM, various methods were proposed such as ultrasonic assisted dry EDM, hybrid dry EDM, quasi explosion mode dry EDM and dry EDM with the piezoelectric servo system, etc. The ultrasonic assisted dry EDM increased the material removal by two times while reducing surface quality [3]. The hybrid dry EDM process was performed in a pulsating magnetic field, which increased productivity by 130% with no tool wear [4]. The MRR was improved by performing dry EDM in quasi explosion mode assisted by oxygen and air at 15 psi [5]. The quasi explosion mode was very difficult to control during machining. Further, a gas suction dry EDM technique was proposed to overcome debris attachment to both electrodes and to achieve better surface quality. This technique resulted in better surface quality, but debris blocked the gas flow hole. Dry EDM performed with a piezoelectric servo system also improved MRR but repetitive oscillation was incompetent in this process [6]. However, low MRR with non-oxygen gas, short of stability, deposition of debris on the electrodes, and odor of burning are the disadvantage of dry-EDM [3, 7].

Near-dry EDM, another process variant of EDM was developed by Tanimura et al. [8] in 1989 to resolved the above mentioned problem. Near dry EDM uses two-phase flow of liquid and gas as a dielectric medium. The dielectric in this process is supplied through the hollow tool. Due to the presence of liquid phase in dielectric, debris reattachment to electrodes is minimal in near-dry EDM. The presence of liquid medium in the dielectric helps to solidify the debris particle and thus flush away from the IEG. Tao et al. identified that lower discharge current and lower pulse duration were significant process parameters of near-dry EDM for enhancing the surface quality. It was reported that this process produced better surface finish and also produced holes with nearly zero taper [9]. Higher MRR was found when near-dry EDM performed with oxygen as a gaseous medium in the dielectric and copper as tool electrode. Tao reported mirror-like surface finish with 0.09 μm R_a by near-dry EDM [10]. To predict the dielectric flow rate, Fujiki et al. [11] developed a model based on computational fluid dynamics (CFD). Further, the results obtained by developed model were compared with experimental values. The model showed that the dielectric flow rate was directly proportional to material removal and inversely proportional to tool wear. Thinner recast layer and negligible tool wear were found in near-dry EDM as compared to conventional EDM [12]. Dhakar et al. [13] compared the effect of glycerin-air, water-air, and EDM oil-air mixtures. They reported that glycerin-air dielectric medium produced higher MRR. This dielectric has a high Prandtl number and high viscosity that yields more thermal energy in IEG. Glycerin-air dielectric also resulted in fine surface integrity with no recast layer.

In this investigation, the rotary tool near-dry EDM drilling was investigated to understand the effect of process parameters. The effect of tool rotation speed, pulse duration, peak current, liquid flow rate and gas pressure on MRR and overcut were investigated. Further, the benefits and potentials of rotary tool near-dry EDM drilling process were identified and discussed.

Experimental Setup

Materials and Methods

High speed steel (T2 grade) was used as the workpiece material, which is widely used in tool and die making industries. The dimension of the workpiece was 100 mm × 60 mm × 6 mm. The surface of the specimen used for near dry EDM was smooth and polished. The workpiece was clamped in a vice attached to the machine table. A hollow copper tool electrode with 99.9% purity was used in this investigation. The tool of 6 mm outer diameter, 3 mm inner diameter and 45 mm length was used in this investigation.

Experiments were conducted on an EMS-5030 EDM die sinking machine. An additional attachment was mounted on EDM machine for mixing gaseous and liquid medium to form mist [12]. This mist was injected by a dielectric mixing unit through the hollow tool electrode in the inter electrode gap. A self-designed rotary system was attached to the EDM machine, to provide rotation to the tool electrode. Belt drive was used to transmit motion from an electric motor to the pulley attached to the hollow shaft as shown in Fig. 1. The rotation speed of the tool electrode was controlled by a voltage variator. The tool electrode was clamped in a chuck attached

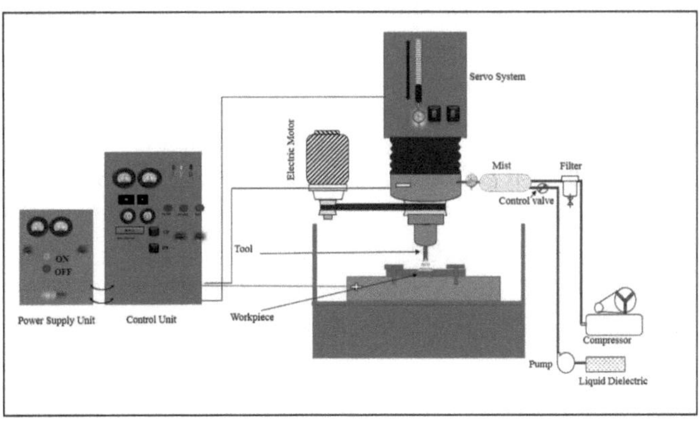

Fig. 1 Schematic of rotary tool near-dry EDM setup

to the hollow shaft. In this investigation, glycerin-air was used as a dielectric medium. The air was supplied by a compressor and glycerin was supplied by a precision pump.

Experimental Procedure

In the present investigation tool rotation speed, peak current, pulse duration, flushing pressure and liquid flow rate were used to perform the experiments on HSS workpiece.

The One-Factor-At-a-Time (OFAT) approach was used in this investigation. Running OFAT experiments is a sequential learning process and continuously receive information from each run. OFAT gives a more rapid effect of input parameters on responses [14]. The experimental settings are summarized in Table 1.

The weight of the workpiece before and after drilling was measured using a SHIMADZU AUW220D electronic balance with a least count of 0.01 mg. The MRR (mm^3/min) was calculated by (Eq. 1).

$$\text{MRR} = \frac{\text{Weight difference before and after machining (gm)}}{\text{Density of workpiece}\left(\frac{\text{gm}}{\text{mm}^3}\right) * \text{Machining time (min)}} \tag{1}$$

Overcut is the difference between the diameter of the machined hole and the diameter of the tool electrode before machining. These diameters were measured using NIKON SMZ745T optical microscope. The first set of experiments was conducted by varying the tool rotation speed by keeping all the other parameters at middle value. Highest MRR was found at 2500 rpm thus tool rotation speed of 2500 rpm was kept constant for remaining experiments. The next set of experiments was performed with the best setting of process parameters from the previous experiments. Current (12 A), pulse duration (390 μs), pressure (55 psi) and liquid

Table 1 Experimental settings

Variable parameters		Constant parameters	
N (rpm)	500, 1000, 1500, 2000 and 2500	Work material	HSS
I (A)	4, 8, 12, 16 and 20	Polarity	Straight (tool negative)
T_{on} (μs)	90, 190, 290, 390 and 490		
F (ml/min)	2, 3, 4, 5 and 6	Dielectric medium	Glycerin-air
P (psi)	25, 40, 55, 70 and 85	Tool material	Copper
		Lift	2
		Sensitivity	3
		Gap control	4

N: Tool rotation speed, I: Current, T_{on}: Pulse duration, F: Liquid flow rate, P: Gas pressure flow rate (4 ml/min) were found the best settings of input parameters. The effects of these process parameters on MRR and overcut were investigated.

Results and Discussions

It was observed during the experiment that the MRR increased with an increase in tool rotation speed as well as current. With an increase in pulse duration, gas pressure and liquid flow rate, MRR increased up to a certain value then start decreasing. The overcut first reduced and then enlarged as tool rotation speed increased. An increased in overcut was found with increase in electrical parameters. With an increase in the liquid flow rate and gas pressure, the diametral overcut increased up to a certain value and then deceases.

Effect of Process Parameters on MRR

The effect of input parameters on MRR is shown in Fig. 2. With an increase in tool rotation speed, MRR increased continuously. Rotation of the tool electrode increased tangential velocity of the two phase dielectric medium, which resulted in higher centrifugal force. Higher centrifugal force contributed in an effective flushing of inter electrode gap and stable machining.

An increase in current and pulse duration increased MRR. When current and pulse duration increased, discharge energy between the tool electrode and workpiece also increased for longer duration. Due to transfer of more energy from cathode to anode, more material melted and vaporized. Which resulted in increased MRR. Very high value of pulse duration resulted in expansion of plasma channel and low discharge energy. Hence, MRR decreased after 390 μs. Similar results were reported by Tao et al. [9] and Dhakar and Dvivedi [12]. MRR first increased and then decreased with an increase in gas pressure. This was because, an increase

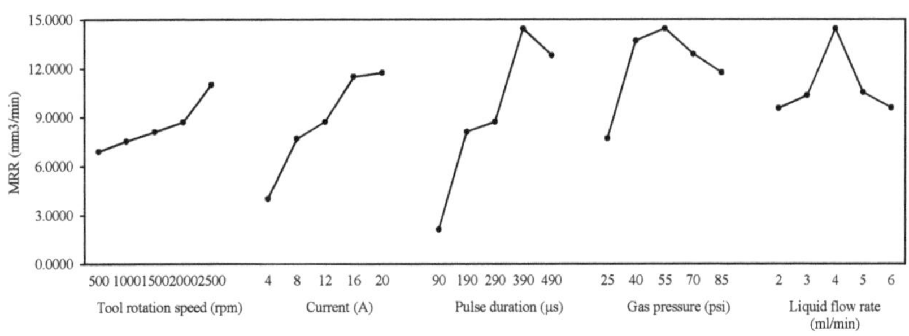

Fig. 2 Effect of process parameters on MRR

in gas pressure expelled debris particle from IEG more effectively. Improved flushing of IEG resulted in effective sparks, minimizes debris accumulation and concentrated thermal energy. At very high pressures, MRR decreased because the circulation of dielectric medium became very fast, which led to the process instability. As the liquid (glycerin) flow rate increased, the MRR increased up to 4 ml/min then start decreasing. The increase in flow rate provided more number of glycerin molecules at IEG. Higher amount of glycerin molecules increased the number of sparks in IEG.

Effect of Process Parameters on Overcut

The effect of input parameters on diametral overcut is shown in Fig. 3. With an increase in tool rotation speed, there was an increase in diametral overcut. As the tool rotation increased, centrifugal force and tangential velocity of the dielectric medium were increased. Which resulted in removing more debris from IEG and promoted discharge between the side surface of the copper tool and workpiece. Discharge between the side surface of the copper tool and workpiece resulted in higher overcut. An increase in current increased diametral overcut. The higher value of current resulted in more discharge energy in IEG. Higher thermal energy resulted in larger crater size, which resulted in higher overcut. Higher pulse duration led to the prolonged spark between the tool and workpiece. More dominant discharges led to higher overcut. With the increase in gas pressure diametral overcut increased up to 55 psi and as the pressure of the gas increased further the diametral overcut decreased. When high pressure combined with rotation, provide cyclic movement to the debris particle. More interaction time increased side spark between the side surface of the copper tool and workpiece, resulted in higher overcut. The overcut decreased above 55 psi. This was because at pressure above 55 psi improved debris flushing from inter electrode gap, resulted in effective sparks between tool electrode and workpiece.

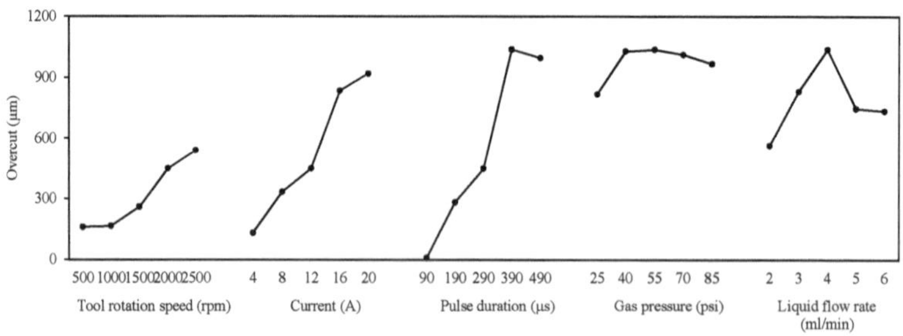

Fig. 3 Effect of process parameters on overcut

Conclusions

Effect of tool rotation speed, pulse duration, peak current, gas pressure and liquid flow rate were experimentally investigated and analyzed using OFAT approach. The main conclusions obtained in this investigation are as follows:

- Rotation of the tool electrode along with the high pressure of two-phase dielectric medium contribute in an effective flushing of inter electrode gap. MRR obtained by rotary tool near-dry EDM was three to four times higher than that of other EDM variants. Further, better quality holes were achieved.
- Higher current and higher pulse duration can be used in a rotary tool near-dry EDM. The MRR increased with an increase in tool rotation speed, peak current. MRR increased with pulse duration up to 390 μs, with gas pressure up to 55 psi and with the liquid flow rate up to 4 ml/min.
- Overcut reduced at higher values of gas pressure and liquid flow rate, whereas it increased with an increase in tool rotation speed and electrical parameters.
- The best settings for input parameters obtained by OFAT with rotary tool near-dry EDM were tool rotation speed (2500 rpm), gas pressure (55 psi), peak current (12 A), pulse duration (390 μs) and liquid flow rate (4 ml/min).

References

1. NASA, Inert-gas electrical-discharge machining. NASA Technical Brief No. NPO-15660 (1985)
2. M. Kunieda, M. Yoshida, Electrical discharge machining in gas. Ann. CIRP **46**, 143–146 (1997)
3. Q.H. Zhang, J.H. Zhang, S.F. Ren, J.X. Deng, X. Ai, Study on technology of ultrasonic vibration aided electrical discharge machining in gas. J. Mater. Process. Technol. **149**, 640–644 (2004)
4. S. Joshi, P. Govindan, A. Malshe, K. Rajurkar, Experimental characterization of dry EDM performed in a pulsating magnetic field. CIRP Ann. Manufact. Technol. **60**, 239–242 (2011)
5. M. Kunieda, Y. Mlyoshl, T. Tsutomu, High speed 3D milling by dry EDM. CIRP Ann. Manufact. Technol. **52**, 147–150 (2003)
6. M. Kunieda, T. Takaya, S. Nakano, Improvement of dry EDM characteristics using piezoelectric actuator. CIRP Ann. Manufact. Technol. **53**, 183–186 (2004)
7. C.C. Kao, J. Tao, S.W. Lee, A.J. Shih, Dry wire electrical discharge machining of thin workpiece. Trans. NAMRI/SME **34**, 253–260 (2006)
8. T. Tanimura, K. Isuzugawa, I. Fujita, A. Iwamoto, T. Kamitani, Development of EDM in the mist. in *Proceedings of Ninth International Symposium of Electro Machining (ISEM IX)* (Nagoya, Japan, 1989), pp. 313–316
9. J. Tao, A.J. Shih, J. Ni, Experimental study of the dry and near-dry electrical discharge milling processes. J. Manuf. Sci. Eng. **130**, 11002–11009 (2008)
10. J. Tao, Investigation of dry and near-dry electrical discharge milling processes. A dissertation, The University of Michigan, 2008

11. M. Fujiki, J. Ni, A.J. Shih, Investigation of the effects of electrode orientation and fluid flow rate in near-dry EDM milling. Int. J. Mach. Tools Manuf. **49**, 749–758 (2009)
12. K. Dhakar, A. Dvivedi, Parametric evaluation on near-dry electric discharge machining. Mater. Manuf. Process. **31**, 413–421 (2016)
13. K. Dhakar, A. Dvivedi, A. Dhiman, Experimental Investigation on effects of dielectric mediums in near-dry electric discharge machining. J. Mech. Sci. Technol. **30**, 2179–2185 (2016)
14. X. Qu, C.F. Wu, One-factor-at-a-time designs of resolution V. J. Stat. Plann. Infer **131**, 407–416 (2005)

Dependence of Ti_2O_3 and Temperature on Electrical Conductivity of TiO_2–FeO–Ti_2O_3 Slags

Shengping Li, Xuewei Lv, Gangqiang Fan, Wei Lv and Yingyi Zhang

Abstract The electrical conductivity of high titania slag, as being an important fundamental physical property utilized in the theoretical understanding of mass transfer phenomena and requirement of practical electric smelting processes, were measured by AC Impedance Spectroscopy using four-electrode method for ternary TiO_2–FeO–Ti_2O_3 slag system based on 85 type of titania slag in Panzhihua in this work. The experiment results show that the slags have a high magnitude of the order of 80–180 s/cm^{-1} in the whole experimental temperature range from 1523 to 2013 K. The Ti_2O_3 exhibits very strong effect on increasing the electrical conductivity of high-titania slag since the specific values averagely increase by 55 s/cm^{-1} with corresponding Ti_2O_3 content range from 10 to 16 wt%. the X-ray diffraction gram and phases calculated by *FactSage* are presented and interpreted to discuss the impact of phase change on the electrical conductivity.

Keywords High titania slag · Electrical conductivity · AC impedance spectroscopy · Four-electrode method · Phase changes

Introduction

The electrical conductivity is an important property in understanding the structure of molten slag and operation of electric smelting furnace, in practice, it is directly related to the temperature of the molten pool and the electric power consumption [1, 2]. The electrical conductivity of TiO_2–BaO [3], TiO_2–SiO_2–CaO–MgO–Al_2O_3 [4], and TiO_2–CaO–SiO_2 [5], had been measured and reported in the published literatures. However, the basic data of high titania slags from literatures are especially limited due to the high melting point, high electrical conductivity as well as high reactivity. Since 1990, the growing requirements for properties of high-titania

S. Li · X. Lv (✉) · G. Fan · W. Lv · Y. Zhang
College of Materials Science and Engineering, Chongqing University,
Chongqing 400044, China
e-mail: lvxuewei@163.com

S. Wang et al. (eds.), *Applications of Process Engineering Principles in Materials Processing, Energy and Environmental Technologies*,
The Minerals, Metals & Materials Series, DOI 10.1007/978-3-319-51091-0_32

slag at high temperature aroused great interest by researchers [6–9] who focused on the effects of various composition on the properties and smelting process of high-titania slag. this work was conducted to investigation the fundamental electrical conductivity on TiO_2–FeO–Ti_2O_3 slag system by means of AC Impedance Spectroscopy [10, 11].

Experimental

In the present study, the electrical conductivity of ternary TiO_2–FeO–Ti_2O_3 slag were experimentally measured. The typical compositions of industrial high-titania slag in Panzhihua area and slag investigated in this work are separately given in Tables 1 and 2, as showed in the Table 1, partial titanium dioxide TiO_2 is inevitably reduced to low valence titanium sesquioxide Ti_2O_3 under strong practical reducing conditions.

Apparatus for Electrical Conductivity

In the present work, the electrical conductivity measurements were determined by AC Impedance Spectroscopy (Electrochemical Workstation: Princeton 2273) using four-electrode method. The experimental arrangement including the dimensions of the crucibles and electrodes was schematically shown in Fig. 1, the Princeton 2273 device was connected with a computer to simultaneously record the correlation between impedance and frequency of slag. The graphite-tube resistance furnace was specially used to meet the ultra-high temperature requirement and it can also guarantee a long enough and uniform hot zone. The crucible and electrodes made of molybdenum were employed for electrical conductivity measurement and the slag sample was loaded in the molybdenum crucible that was further put into a graphite crucible. With a graphite crucible in the outer layer, the left oxygen in the furnace can be effectively eliminated to a large extent. Additionally, before the experiment,

Table 1 Compositions of Panzhihua 85 type of high-titania slag

TiO_2	Ti_2O_3	FeO	SiO_2	MgO	CaO	Al_2O_3	MnO	Cr_2O_3	V_2O_5
69.18	13.29	9.55	2.85	1.33	0.58	2.07	1.37	0.13	0.29

Table 2 Chemical compositions of the slags studied in the present work

No.	TiO_2	FeO	Ti_2O_3
1	80.8	9.20	10
2	79.00	9.00	12
3	77.21	8.79	14
4	75.41	8.59	16

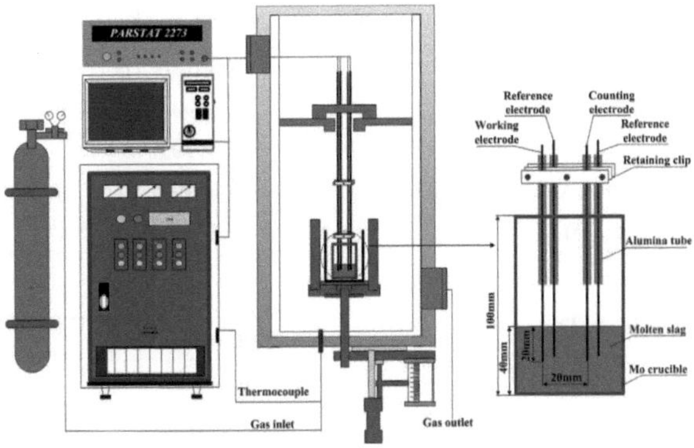

Fig. 1 Experimental setup for the electrical conductivity measurement and dimensions of the crucible and electrodes

the CaF_2 melt was used to calibrate cell constant, with its electrical conductivity value has been generally determined to be 6.40 s/cm at 1600 °C by many careful experiments [12]. The resistance of the slag in each temperature is computed by the average value of reality part in the Nyquist plot when its image part is zero. The electrical conductivity of the melts can be obtained by the following equation:

$$K = C/R_x \tag{1}$$

where C is the cell constant and R_x is the resistance of slag. There are many factors influence the measurement accuracy, such as the applied frequency, the height of melt, the immersion depth of electrode, and position of electrodes were also carefully considered in this paper.

Samples Preparation

The samples for electrical conductivity measurement were synthetized with analysis pure grade reagents that is shown in Table 3. the compositions except Ferrous Oxalate were firstly claimed at 1273 K in a muffle furnace to remove all the impurities such as hydroxide and moisture, while the Ferrous oxide was carefully obtained through decomposition of Ferrous Oxalate under strict condition of absence of air. Besides, the weighted chemicals were thoroughly mixed in desired proportion and pressed into cylindrical pellets and roasted at 1473 K for 2 h aiming at sintering and having a certain conductivity, which was favorable to pre-melting and homogenization of slag in a vacuum arc furnace before resistance measurement. After that about 220 g of sample was packed into Mo crucible to obtain a 40 mm-deep slag bath. The argon was poured into the whole furnace during both pre-melting and heating processes to achieve the atmosphere requirements and remove the potential traces of oxygen. The three dimensions of the molybdenum

Table 3 Oxides and gas used in the present work

Material	Purity (pct)	Supplier
Titania (TiO_2)	≧99.5	Chengdu Kelong
Titanium sesquioxide (Ti_2O_3)	≧99.5	Chengdu Kelong
Ferrous oxalate (FeC_2O_4)	≧99.5	Tianjin Zhiyuan
Argon (Ar)	99.999	Chongqing Zhongfa

Table 4 Dimensions of molybdenum crucible and electrode

Crucible (unit: mm)		Electrode (unit: mm)	
Inner diameter	44	Diameter	2
Wall thickness	2	Length	600
Inner depth	98	–	–
Base thickness	2	–	–
Height in total	100	–	–

crucible and electrodes are given in Table 4, the slag samples were heated to reach a temperature of above 1973 K and held at that point for 2 h for complete melting state (the duration time was demonstrated sufficient in preliminary studies of thermal equilibrium).

Procedure for Electrical Conductivity

The electrodes were initially placed at a level of 10 mm above the solid slag layer. The crucible and electrodes were vertically aligned to keep experimental errors from a slight deviation in immersion depth. Once the slag were fully molten, the position of the electrodes were adjusted to gradually approach the slag layer, the sharp change of resistance data indicated the touch between electrodes and slag and then this specific position was selected as a base for further immersion depth of electrodes inside the Mo crucible. The electrical conductivity measurements were conducted along with the cooling of temperature, duration time in each selected temperature was long enough to obtain a homogeneous slag melt and stable resistance value. After electrical conductivity measurement, the slag sample, along with the crucible was cooled with the furnace, and thereafter the slag samples were taken out from crucible to grind into fine powder for X-ray analysis. All of the electrical conductivity measurement works were carried out under a slightly positive pressure of high-purity argon atmosphere.

Results and Discussion

The variation of electrical conductivity as a function of temperature for ternary TiO_2–FeO–Ti_2O_3 slags and the compositional effect of Ti_2O_3 on electrical conductivity are showed in Fig. 2. The mass ratio of TiO_2 to FeO was fixed at 2.515. As can be drew from the Fig. 2 that the specific electrical conductivity value generally decrease with decreasing the temperature except where there is mainly a peak between 1740 and 1820 K. The slags have a high magnitude of order of 80–180 s/cm^{-1} in the whole experimental temperature range from 1523 to 2013 K and the Ti_2O_3 exhibits very strong effect on increasing the electrical conductivity of high-titania slag since the specific values averagely increase by 55 s/cm^{-1} with corresponding Ti_2O_3 content range from 10 to 16 wt%.

XRD Analysis of Slag

The main phases after naturally cooling down with furnace, as showed in Fig. 3, are consistently anosovite (Fe, Mg)$_x$Ti$_y$O$_5$ and rutile phases, the anosovite is

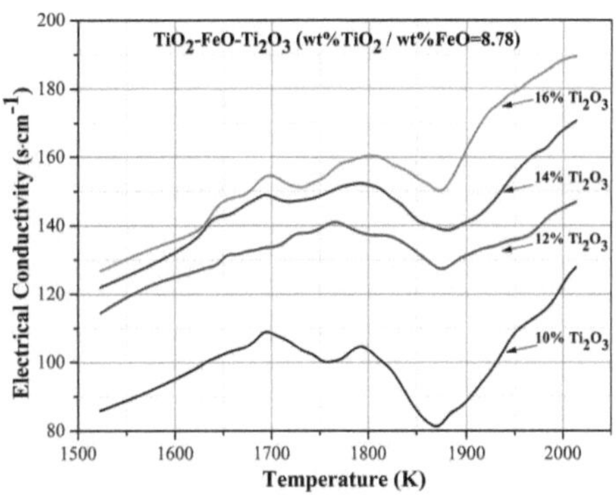

Fig. 2 The effect of Ti_2O_3 on the electrical conductivity in the TiO_2–FeO–Ti_2O_3 slag system

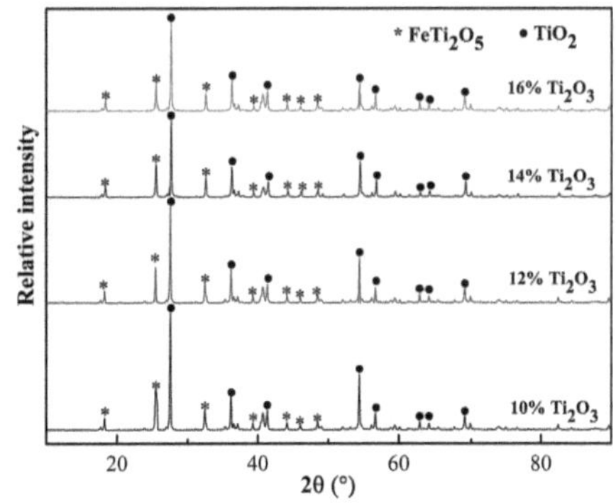

Fig. 3 The XRD pattern of TiO_2–FeO–Ti_2O_3 slag cooled with the furnace

pseudobrookite type, which is basically in accordance with the phase detecting result of Sorel slags studied by Pistorius [13].

Theoretical Phases Calculation by FactSage

The Fig. 4 describes the theoretical phases changes results calculated by *FactSage*. It can be seen that with the decrease of temperature, the mass proportion of slag phase plunges rapidly and disappears when it is just around 1600 K, whereas the rutile phase gradually increases. The pseudobrookite experiences a noticeable tendency of increasing along with the cooling of temperature, after peaking at 1600 K, its content gradually falls back to a very low level and disappears when the temperature reaches approximately under 1200 K. The $Ti_{20}O_{39}$ see a significant increase with the increase of Ti_2O_3 content just from 10 to 16%. Besides, some attention should also be drawn from Fig. 4 the appearance of a little amount of Fe, which may be related to of the reduction effect of that Ti_2O_3 on FeO.

The increase of electrical conductivity might also be dependent of enhancement of bimolecular reaction and electron hoping between Ti^{3+} and Ti^{4+} and the formation of Magneli phase.

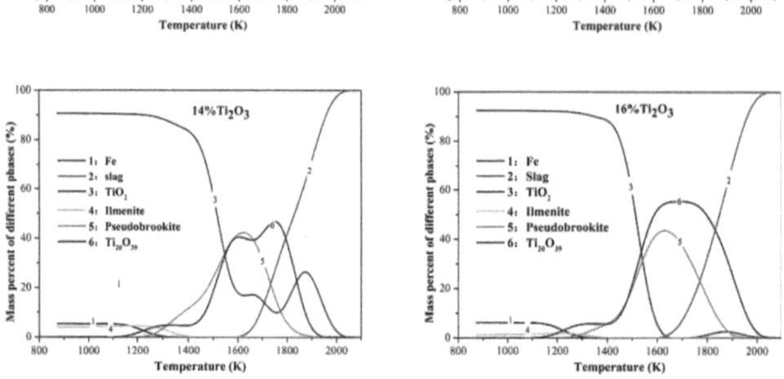

Fig. 4 Phases changes of ternary TiO_2–FeO–Ti_2O_3 slag

Conclusions

The electrical conductivity of ternary TiO_2–FeO–Ti_2O_3 slag was experimentally measured, and the following conclusions were obtained.

(1) The ternary TiO_2–FeO–Ti_2O_3 slag have a high magnitude of the order of 80–180 s/cm^{-1} in the whole experimental temperature range from 1523 to 2013 K.
(2) Ti_2O_3 exhibits very strong effect on increasing the electrical conductivity of high-titania slag.
(3) The main phases after naturally cooling down with furnace are consistently rutile and anosovite (Fe, Mg)$_x$Ti$_y$O$_5$ phases.

Acknowledgements The authors are especially grateful to the support of the Natural Science Foundation of China NSFC (Grant No. 51374262).

References

1. S.C. Britten, U.B. Pal, Solid-state amperometric sensor for the in-situ monitoring of slag composition and transport properties. Metall. Mater. Trans. B **31**(4), 733–753 (2000)
2. Y.M. Gao, X.M. Guo, K.C. Chou, Unpolluted short circuit electrochemical reduction analysis for molten oxide slags. Chin. J. Nonferrous Met. **16**(3), 530–535 (2006)
3. N.A. Fried, G.K. Rhoads, D.R. Sadoway, Electrochim. Acta. no. 46, 3351–3358 (2001)
4. Y.H. Shi, J.C. Wang, Electrical conductivity of TiO_2 containing slag. Iron Steel Vanadium Titanium 1 (1987)
5. J. Liu et al., Electrical conductivity and electronic/ionic properties of TiO x –CaO–SiO_2 slags at various oxygen potentials and temperatures. Metall. Mater. Trans. B **47**(1), 798–803 (2015)
6. G.Q. Gao, Y.C. Yang, D.X. Yang, Determination of viscosity and melting point of titanium slag. Iron Steel Vanadium Titanium 1, 009 (1987)
7. Z.J. Zhao, E.Q. Ma, J.L. Yu, Behavior of Al_2O_3 in titanium slag. Iron Steel Vanadium Titanium **23**(3), 36–38 (2002)
8. A.M. Amer, Alkaline pressure leaching of mechanically activated rosetta ilmenite concentrate. Hydrometallurgy **67**(1), 125–133 (2002)
9. Ivo Toromanoff, Fathi Habashi, The composition of a titanium slag from sorel. J. Less Common Met. **97**, 317–329 (1984)
10. X. Lu et al., Application of AC impedance for measuring resistance of slag. Chin. J. Nonferrous Met. **10**(3), 437–439 (2000)
11. S. Wang et al., The conductivity and the crystallization of perovskite (CaTiO3) from Ti-bearing blast furnace slag studied by A. C. impedance method. ISIJ Int. **39**(11), 1116–1119 (1999)
12. S. Hara, Electrical conductivity of molten slags for electro-slag remelting. Trans. Iron Steel Inst. Jpn. **23**(12), 1053–1058 (1983)
13. P.C. Pistorius, C. Coetzee, Physicochemical aspects of titanium slag production and solidification. Metall. Mater. Trans. B. **34**(5), 581–588 (2003)

Part VI
Materials Processing and Plasma Processing

DuraStell PTA Cladding for Wear Application

Qingjun Zheng and Robert Vasinko

Abstract Cladding by plasma transferred arc (PTA) welding is widely used for enhancing the surface properties of a metallic component, in particular wear, high temperature and corrosion resistances. This paper presents the application of Kennametal Stellite PTA cladding technology, especially the DuraStell PTA process technique, for tungsten carbide composite cladding for boiler tube application. DuraStell PTA process gives low heat input to the substrate and fast deposition rate. The cladded boiler tube samples showed more uniform carbide dispersion in the cladding, less base metal dilution, and insignificant effect on the microstructure and mechanical properties of the substrate base metal, in comparison of the traditional PTA and other arc welding process.

Keywords PTA · Wear · Carbide cladding · Fast deposition · Boiler tubes

Introduction

Extensive research and development have been conducted in thermal plasma processing for extraction of metals, synthesis of advanced material, destruction of hazardous wastes, and hardfacing/cladding [1–13]. Thermal plasma is essentially an ionized form of the gas with electrons and the charged particles in thermal equilibrium. Thermal plasma can be generated in a plasma generating device, called plasma torch kit, by converting electricity into thermal energy. In a plasma torch, an electric arc is established between two electrodes in the presence of a gas, which partially ionizes and becomes electrically conductive. While thermal plasma can be generated by alternating current (AC), radio-frequency (RC) and other discharges as well, direct current (DC) torch is the most researched and used in industry, owing to its better consistency, less noise, easier control, lower electrode and power consumption. Based on the anode structure, there are two types of DC torches,

Q. Zheng (✉) · R. Vasinko
Kennametal Inc., 1600 Technology Way, Latrobe, PA 16150, USA
e-mail: jack.zheng@kennametal.com

S. Wang et al. (eds.), *Applications of Process Engineering Principles in Materials Processing, Energy and Environmental Technologies*,
The Minerals, Metals & Materials Series, DOI 10.1007/978-3-319-51091-0_33

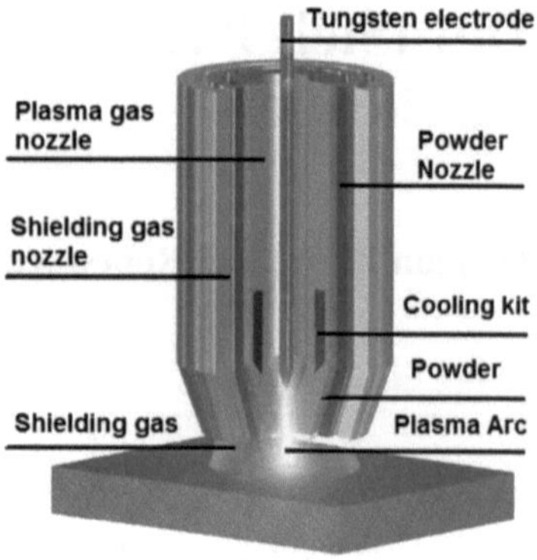

Fig. 1 Schematic of a transferred plasma DC torch for hardfacing

transferred arc and non-transferred arc. The difference between the two is that the anode is part of the torch structure in a non-transferred plasma DC torch while the conductive work piece, that is the material to be treated, is the anode in the transferred plasma mode. Figure 1 illustrates a transferred plasma DC torch, in which the plasma is generated by employing a non-consumable tungsten alloy electrode, a water-cooled constriction nozzle, plasma gas, and shield gas in the torch kit. Transferred plasma arc has higher energy efficiency but requires the work piece to be electrically conductive, therefore often selected for welding, hardfacing, spray coating, remelting, and refining of metallic materials. Non-transferred plasma DC torch is bulkier and less energy efficient but can be used for treating non-conductive materials, therefore good for processing of ceramic materials, hazardous wastes, ore processing etc., where the materials to be treated are not required to be electrically conductive.

Plasma transferred arc (PTA) cladding is essentially similar to transferred plasma arc welding (PAW). The difference between PAW and PTA is that PAW feeds metal wires whereas PTA deposits the cladding materials in the powder form. In addition to the plasma generating torch kit, a turn-key PTA cladding system includes the power console, control panel, chiller, welding chuck/work platform, powder feeder and software although the system design may vary dependent on the specific application and the suppliers. Figure 2 illustrates a turn-key PTA cladding system, Kennametal Stellite Starweld™ 400. The system can give output up to 450 A, with programmable logic and AC servo motors for accurate motion and process control.

PTA is a versatile process, enabling deposition of a variety of alloys and alloy matrix composites on the metal based substrates. The cladding deposited by PTA can have lower penetration, dilution and heat affected zone (HAZ) than conventional arc welding technologies like tungsten inert gas (TIG) welding and metal inert gas (MIG) welding. The thickness of the typical PTA cladding ranges from 0.5 to 3 mm in one pass. PTA process can easily be automated, providing high reproducibility. Owing to these advantages, PTA is popular for depositing tungsten carbide composite, Stellite alloy and nickel alloy claddings for wear, high

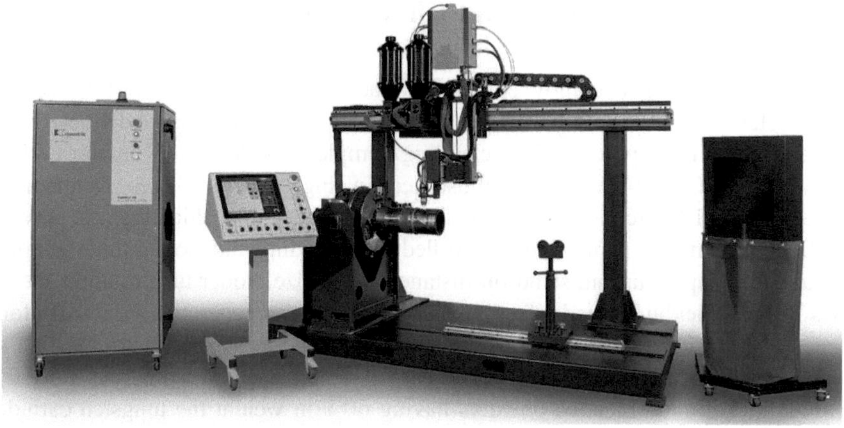

Fig. 2 Starweld™ 400 PTA system

temperature and/or corrosion applications. Examples of such applications include mud motor bearings and stabilizers in oil/gas drilling, drilling components in mining, refinery tubes, and valve components etc. PTA is also used for Inconel alloy cladding with a large variability of composition and uses around the world [9, 10]. Compared with the emerging laser cladding, the limitation for the traditional PTA process is that the energy input to the substrate is higher and less-concentrated, potentially causing more distortion in geometry and degradation in properties of the cladding and substrate base metals. Different from the traditional PTA process, Kennametal DuraStell PTA process is a finely tuned stringer bead PTA process with precise process control. It allows for faster deposition, shorter arc dwelling time, and lower heat input. The advantages of the DuraStell PTA process include less component distortion, lower dilution, better control of the coating thickness and higher deposition rate. DuraStell PTA process can be used for cladding heat sensitive cladding materials such as non-magnetic alloys and heat sensitive substrates. Since the low heat input, the post-cladding heat treatment often used for recovering the microstructure and properties of substrate alloys in traditional welding may become unnecessary. Owing to the fast deposition, DuraStell process also gives lower operation cost. These characteristics make DuraStell PTA process potentially a cost-effective alternative to the laser cladding. Compared with the traditional PTA process, DuraStell PTA process requires precise process control and the process development for different cladding-substrate material system and the part geoemetry. This paper reports the application of DuraStell PTA process for applying tungsten carbide composite cladding on the boiler tubes for hanger and super heater applications. Failure of the boiler tubes due to erosion and corrosion from the burning wastes such as fly ashes is one of the leading causes for the forced outages in coal-fired boilers.

Test Sample Preparation and Characterization

Tungsten carbide-nickel alloy composite cladding was applied on SA-210 A1 carbon steel and SA213-347H stainless steel tubes of 2 in. in outside diameter with the DuraStell PTA process. The cladding sample was deposited on the Stellite Starweld™ 400A PTA system illustrated in Fig. 2, which uses Kennametal Excalibur II PTA torch and argon gas for generating plasma, shielding and powder feeding. The major parameters controlled for cladding the boiler tube samples include the output current, stand-off distance, pitch size, boiler tube rotation speed, flow rate of shielding, powder feeding and center plasma-generating gas etc. Before depositing, the cladding sections of the boiler tubes were cleaned by grit blasting to remove the oxides, oil and grease. The cladding material was the tungsten carbide-nickel alloy blended powders having 60% in weight the tungsten carbide, supplied by Kennametal Stellite. After cladding, the boiler tubes were cooled on the welding chuck without additional heat treatment.

Post-cladding characterization is for understanding the cladding appearance, cladding quality, interfacial bonding, and effect of the DuraStell PTA process on the substrate base metal. The visual examination was to check if there were any apparent welding defects at the cladding surface like cracking, open voids, and sluggish weld puddles. Then samples were taken from different locations of the cladded tubes including the starting, cladding body, and stopping points. Mechanical testing includes Rockwell hardness according to ASTM standard E18, microhardness according to ASTM E92, bending test according to ASTM E190, and the abrasion test according to ASTM G65. The bending test samples were taken from the middle of cladded tubes. Some of the characterization and testing works were conducted by David N. French Metallurgists.

Results and Discussion

Tungsten Carbide-Ni Alloy Composite Cladding on SA-210 A1 Boiler Tubes

Figure 3 shows a tungsten carbide-nickel alloy composite cladded SA-210 A1 carbon steel tube sample. The cladding was deposited by the DuraStell PTA process with the surface speed about 5 m/min. This deposition rate is faster than the traditional stringer bead PTA process, in which the surface speed is less than 2 m/min. The cladding appears to be uniform in cladding thickness and free of cracks, voids, sluggish welding puddles and bumps at the cladding surface. No obvious distortion was observed.

Figure 4 depicts the metallography of the carbide composite cladding at the starting point, middle, and stopping end of the cladding. The samples were cut and polished along the tube length. The cladding is bonded metallurgically to the steel

Fig. 3 A tungsten carbide-nickel alloy composite cladded SA-210 A1 carbon steel tube by DuraStell PTA process with the sampling locations marked

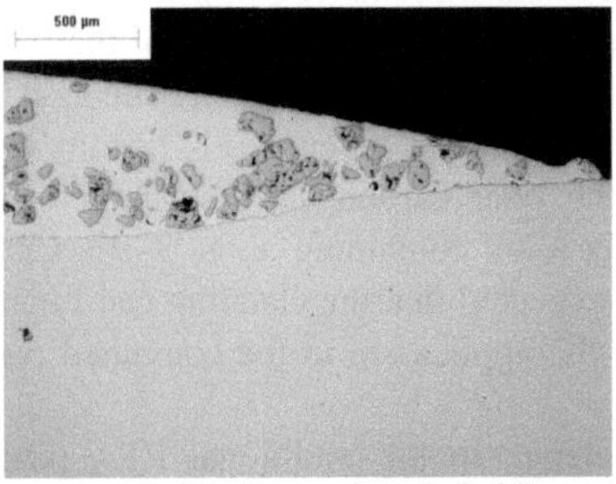

a) Starting point of cladding

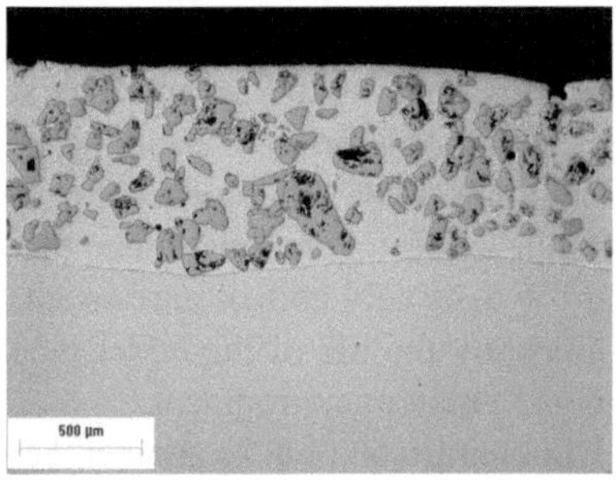

b) Middle of cladding

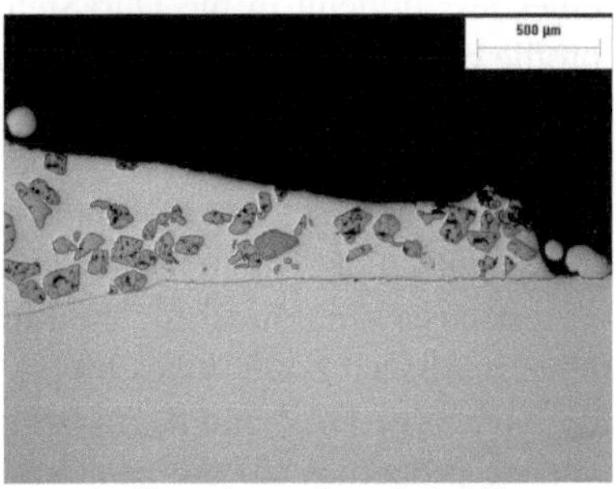

c) Stopping point of cladding

Fig. 4 Metallographic images of the tungsten carbide–Ni alloy composite cladding on a SA210-A1 boiler tube sample by DuraStell PTA process

substrate and is fully densified. Substantial disbonding was not discovered between the welding passes or at the substrate-cladding interface either in the welding toes or in the middle of the cladding. The tungsten carbide particles in the cladding are in angular shape with sharp edges, indicating little dissolution of the carbide particles into the matrix alloy. The tungsten carbide particles are uniformly dispersed through the thickness of cladding without obvious settling, as shown in Fig. 4. This is a significant improvement over the traditional PTA process, in which the settling of tungsten carbide is a key challenge [10]. Settling of the carbide particles produces a tungsten carbide deficient zone of lower wear resistance at the top of the cladding. The settling of tungsten carbide in the traditional PTA process is caused by the slow solidification of the matrix alloys and the difference in the densities between the tungsten carbide and nickel alloy matrix and [10]. Compared with the traditional PTA process, the DuraStell PTA process deposits the composite at a faster speed, shorter arc dwelling time and less heat input to the substrate. The solidification of matrix alloy is much faster in the DuraStell PTA process and thus the heavier carbide particles do not have enough time to settle. Hence, uniform distribution of carbide particles can be gained through the cladding thickness.

Microhardness was measured on the tungsten carbide particles and the matrix alloy in the cladding by Vickers indentation at 0.5 kg load (HV0.5). HV0.5 for the tungsten carbide was given as 1277 kg/mm^2 in average. The HV0.5 of the matrix alloy was around 420 kg/mm^2. The abrasion resistance of the composite cladding measured in volume loss according to ASTM G65 was given as 11–12 mm^3/6000 revolutions, about 40 times better than the bare carbon steel substrate, about 420 mm^3/6000 revolutions. Since the matrix alloy containing about 7–9% Cr and over 3% Si is very corrosion resistant, it is expected that the cladding can improve significantly the life of the boiler tubes for the application in the combined corrosion and erosion environment.

Dilution/dissolution is another key challenge in the traditional PTA process [10, 11]. Benefiting from the less heat input and fast solidification, controlling the dilution/dissolution becomes less difficult in the DuraStell PTA process. Figure 5 shows the composition profile of Fe in the cladding across the fusion line analyzed by EDS. This semi-quantitative analysis shows that the content of Fe reduces from interface to the cladding surface and is stabilized at about 5% in weight in the cladding at about 300 μm to the interface. Dissolution control is important for the super heater and hanger tube application, where the substrate mechanical properties and the surface corrosion resistance are critical for the life [14].

Since the high heat input, hardfaced boiler tubes by the traditional PTA and arc welding processes usually have large HAZ and degradation. Post heat treatment is thus often used to recover the properties for the heat treatable substrates, but the heat treatment increases the operation cost. For those non-heat treatable substrate

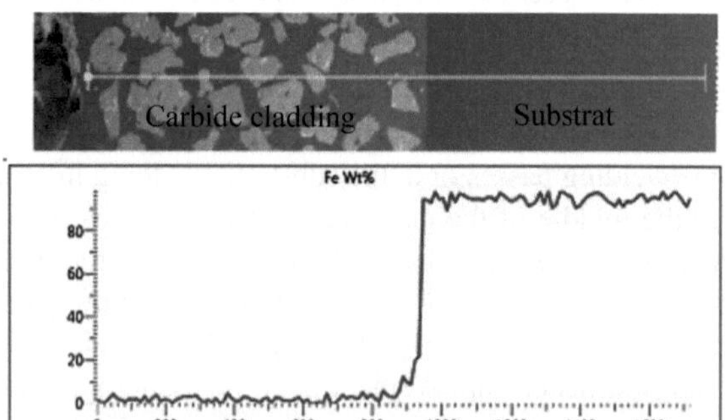

Fig. 5 Composition profile of Fe across the interface by EDS for the tungsten carbide composite cladding on SA-210 A1 carbon steel by DuraStell PTA process

materials, the intensive heat input can cause irreversible degradation in the microstructure and properties. In comparison, the impact of the DuraStell PTA process on the substrate tube is insignificant. The bending tests were conducted according to ASTM E190 to determine the cracking susceptibility into the substrate. The preselected bend angles are 30°, 60°, 120° and 180°. Figure 6 shows the macroscopic view of the metallographic image of the bend test samples. All the bending test samples exhibited cracking normal to the bending direction in the cladding, as expected. While the number of cracks appeared increasing with the increasing bend angles, no significant cracking was observed in the substrate even when bending to 180°. This suggests that the substrate is still relatively resistant to crack propagation post the cladding.

Figure 7 shows the microstructure of the cladded boiler tube near the interface by the DuraStell PTA process. The HAZ was measured at averaging 630 μm in the depth, which is insignificant. No noteworthy carburized layer, decarburized layer, or abnormal microstructure was observed in the HAZ. The microhardness in the

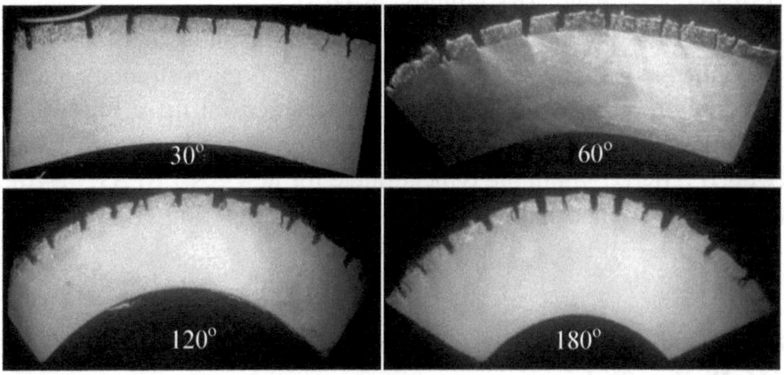

Fig. 6 Macroscopic view of the bend test samples bent at different degrees for the tungsten carbide-Ni alloy composite cladded SA-210 A1 boiler tube samples

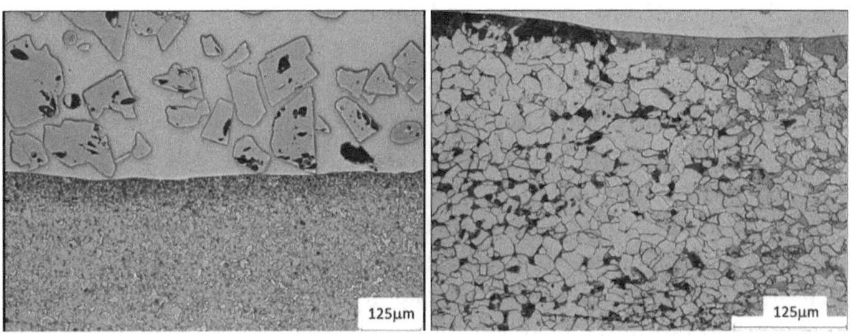

a) At lower magnification b) At higher magnification

Fig. 7 Microstructure near the interface for the SA-210 A1 boiler tube cladded by DuraStell PTA process at **a** lower and **b** at higher magnifications

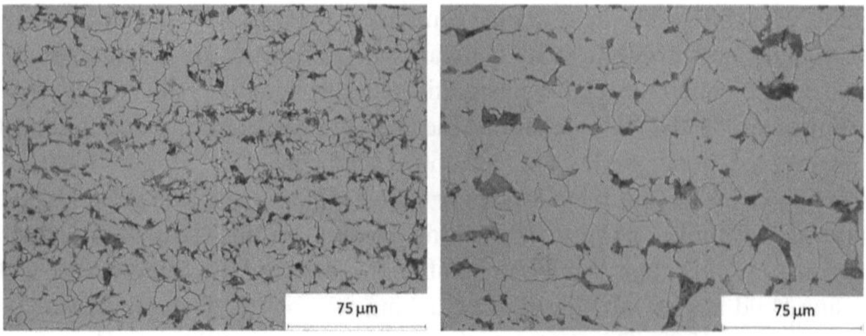

Fig. 8 Microstructure of the steel at the mid-wall for the cladded and bare SA-210 A1 boiler tubes, whereas the cladding was applied by the DuraStell PTA process

HAZ was measured and given as 203 kg/mm^2 in HV0.5 in average, slightly higher than that of the base metal, 163 kg/mm^2 in HV0.5.

Figure 8 compares the microstructure of steel at the mid-wall of the cladded tube and the bare uncladded tube. As shown, the mid-wall microstructure after cladding was still ferrite and pearlite. Hardness was measured on these mid-wall metallographic samples. The cladded sample showed the average hardness value about 74 in Rockwell B (HRB), within the typical range for new SA-210 A1, 72–78 in HRB while it is slightly higher than the bare tube, given by 70 in HRB.

Tungsten Carbide-Ni Alloy Composite Cladding on SA-213 TP347H Boiler Tubes

The tungsten carbide-Ni alloy composite cladding was also applied on SA-213 TP347 boiler tubes by using the DuraStell PTA process. Besides visual examination, metallography, SEM, bending and microhardness, the characterization also includes sensitization test for understanding the effect of DuraStell PTA on the corrosion properties of the base stainless steel.

Figure 9 depicts the metallographic images of the cladding at the welding toes and cladding body. Similar to the case for SA-210 A1 carbon steel, the tungsten carbide particles are uniformly dispersed in the cladding without obvious settling. No significant local disbonding was detected between the cladding and substrate base metal or between the welding passes. As shown by the composition profile of Fe given in Fig. 10, the dilution of substrate in the cladding is insignificant, limited within about 350 μm in the cladding to the interface.

Similar to the case for SA-210 A1 carbon steel tube, all bend test samples of the cladded SA-213 TP347 boiler tubes exhibited cracking in the cladding, but the cracks terminated at the interface. Figure 11 shows a metallographic image of the

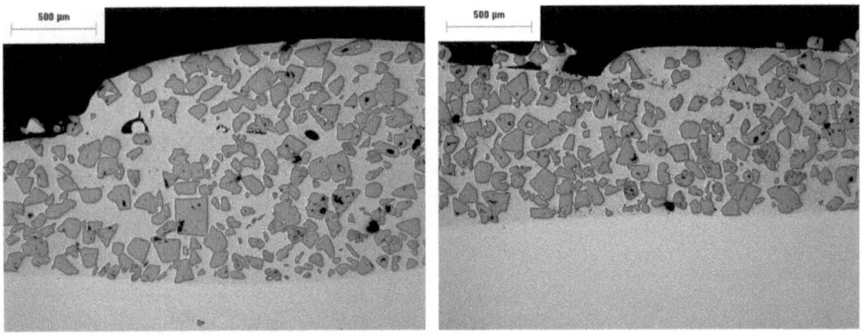

a) At the welding toe b) In the middle of cladding

Fig. 9 Metallographic images of the tungsten carbide-Ni alloy composite cladding on SA-213 TP347 boiler tubes by the DuraStell PTA process

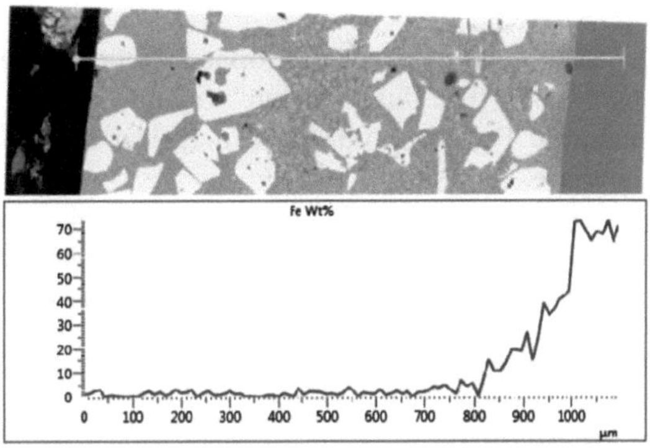

Fig. 10 Composition profile of Fe measured with EDS for the tungsten carbide-Ni alloy composite cladding on SA-213 TP347 boiler tubes by DuraStell PTA process

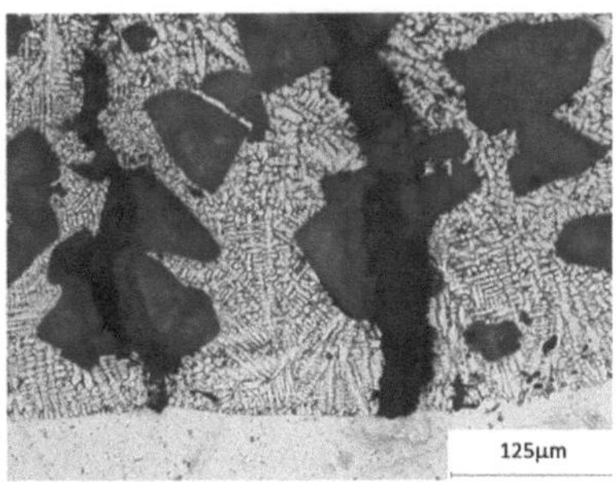

Fig. 11 Microstructure of the tungsten carbide-Ni alloy composite cladded SA-213 TP347H tubes by DuraStell PTA process near the interface after bending to 180°

bend samples near the interface at a higher magnification when bending to 180°. That the cracks did not penetrate into the substrate suggests that the cladded substrate is still crack resistant and the tendency of cracking is minimal from the cladding into the substrate. No evidence of cladding spalling or cracking along the interface implies that the interface bonding is strong. The hardness was uniform in

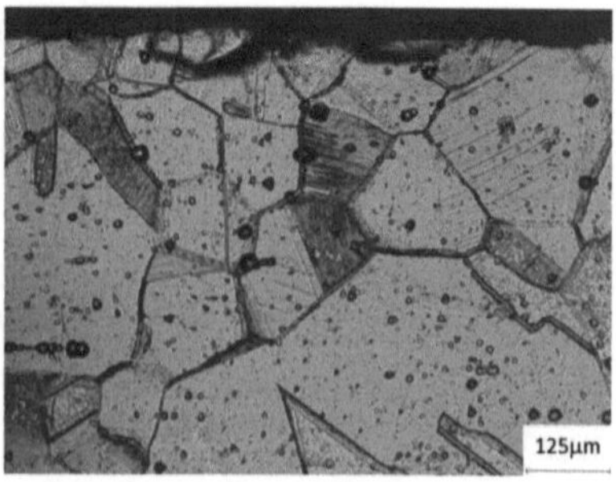

Fig. 12 Microstructure of the cladded SA-213 TP347H tube sample near the interface after sensitization test

the zone next to the bonding interface and at the mid-wall, which gave 218 and 214 kg/mm^2 in average in HV0.5 respectively. The average hardness on the mid-wall is 74 in HRB for the cladded tube, slightly higher than the bare tube, averaging 70 in HRB, but still within the typical range for new SA-213 TP347H, 73–87 HRB.

Sensitization test was performed on the cladded SA-213 TP347H tubes and the bare tubes. The dual structure was observed in the both cladded and bare tube samples. Figure 12 illustrated a typical image of the cladded sample near the interface.

As shown in Fig. 12, no complete ditches around the grains were observed, indicating no carbide precipitation at the grain boundary. The microstructure at the mid-wall showed the same condition as that near the interface.

Conclusions

The DuraStell PTA process may be a cost-effective alternate to laser cladding for applying high quality tungsten carbide cladding on the boiler tubes since the fast deposition rate, short arc dwelling time, and low heat input. Compared with traditional PTA and arc welding, DuraStell PTA process offers the advantages including.

(1) Fast consolidation of the matrix alloys which can prevent the settling of carbide particles in the cladding and give uniform carbide distribution through the cladding thickness.
(2) Low dilution/dissolution of substrate base metal into the cladding, important for retaining the wear and corrosion resistance of the cladding materials.
(3) Minor impact on the microstructure and mechanical properties of substrate base metals, critical for those substrates requiring high mechanical strength and/or susceptible to sensitization.

The boiler tube samples cladded by the DuraStell PTA process appear providing satisfactory results for application in the combined erosive and corrosive environment. The cladding gives high hardness and abrasion resistance. The cladded boiler tube samples have strong metallurgical bonding at the interface, insignificant dissolution and heat-affected zone. The bending tests indicate that the tendency of crack penetrating into the substrate is low. The sensitization tests performed on SA-213 TP347H stainless steel tubes suggest that the cladded tubes are not susceptible to intergranular corrosion or intergranular stress corrosion cracking. In summary, the tungsten carbide-Ni alloy composite cladding applied by the DuraStell PTA process showed the promise for use in the combined erosive and corrosive environment of super heater and hanger boiler tubes.

Acknowledgements The authors are pleased to acknowledge the support in testing and characterization of the cladded boiler tube samples by David. N. French Metallurgists.

References

1. P.R. Taylor, S.A. Pirzada, Thermal plasma processing of materials: a review. Adv. Perform. Mater. **1**, 35–50 (1994)
2. S. Niyomwas, B. Wu, R.G. Reddy, Modeling of materials synthesis in thermal plasma reactor. in *Materials Processing in the Computer Age III*, ed. by V.R. Voller et al. (TMS, Warrendale, PA, USA, 2000), pp. 199–210
3. R.G. Reddy, L.V.M. Antony, The thermal plasma processing of fine powders. J. Metals **55**(3), 19–22 (2003)
4. P.R. Taylor, W. Wang, Reverse-polarity direct current plasma-driven electro-reduction of refractory metals in molten oxide melts. Miner. Metall. Process. **21**(2), 103–109 (2004)
5. L. Tong, R.G. Reddy, Synthesis of titanium carbide nano-powders by thermal plasma. Scripta Mater. **52**(12), 1253–1258 (2005)
6. R.B. Heimann, *Plasma-Spray Coating: Principles and Applications* (Wiley, Hoboken, NJ, 2008)
7. R.P. Webber, Nozzle for plasma torch and method for introducing powder into the plasma plume of a plasma torch. US Patent 4, 866, 240 (1989)
8. R.L. Deuis, J.M. Yellup, C. Subramanian, Metal-matrix composite coatings by PTA surfacing. Compos. Sci. Technol. **58**(2), 299–309 (1998)
9. T. Swingley, K. Luer, R. Webber, J.B. Wu, Abrasion-resistant weld overlay. US Patent 20090032501 A1 (2009)
10. P.F. Mendeza, N. Barnesa, K. Bell et al., Welding processes for wear resistant overlays. J Manuf. Processes **16**(1), 4–25 (2014)
11. F. Fernandes, T. Polcar, A. Loureiro, A. Cavaleiro, Effect of the substrate dilution on the room and high temperature tribological behaviour of Ni-based coatings deposited by PTA on grey cast iron. Surf. Coat. Technol. **281**(15), 11–19 (2015)
12. M. Jones, U. Waag, The influence of carbide dissolution on the erosion–corrosion properties of cast tungsten carbide/Ni-based PTAW overlays. Wear **271**(9–10), 1314–1324 (2011)
13. C.S. Wu, L. Wang, W.J. Ren, X.Y. Zhang, Plasma arc welding: process, sensing, control and modeling. J. Manuf. Processes **16**(1), 74–85 (2014)
14. R. Dooley, E. Wiertel, A survey of erosion and corrosion resistant materials being used on boiler tubes in waste-energy boilers. in *Proceedings of the 17th Annual North American Waste-to-Energy Conference, NAWTEC17* (Chantilly, VA, USA, 2009), pp. 1–5

Production of SiMn-Alloys by Natural Gas and Carbon Black

Xiang Li and Merete Tangstad

Abstract Currently, the main carbonaceous material for Mn-alloys production is solid carbon, i.e. coke, charcoal and petroleum coke, which is produced with large CO_2 emissions. Due to the simple purification, high carbon activity and low CO_2 emissions, natural gas has some advantages compared with conventional reducing agents. The aim of the present project is to develop a new technology for production of Mn-alloys using natural gas as a source of carbon. Experimental results indicated that injection of CH_4 gas directly into melted slag reduced slightly the MnO and SiO_2. However, the metal yield was much less than the theoretical value due to a relative low residence time of CH_4 in the slag. The alternative approach was to use carbon black as reductant, produced by CH_4 decomposition. The preliminary experimental results showed a high reduction rate using carbon black.

Keywords Silicomanganese · Methane · Slag · Carbon black

Introduction

Silicomanganese (SiMn) alloys are produced by smelting of manganese ore in a submerged arc furnace (SAF) which requires high consumption of energy, and proceeds at high temperatures. In this process, raw materials containing manganese ore, quartzite, high carbon ferromanganese (HC-FeMn) slag, and coke are fed into the furnace. High manganese oxides in the ore are first reduced to MnO by CO gas in the upper level of furnace, which is called pre-reduction zone. Subsequently, the manganese ores and fluxes (CaO, MgO etc.) melt together forming a slag, and enter the coke bed zone of furnace, where MnO and SiO_2 are reduced to metal by solid carbon at 1500–1700 °C, as described by Reactions (1) and (2) [1].

X. Li · M. Tangstad (✉)
Department of Materials Science and Engineering, Norwegian University of Science and Technology, 7491 Trondheim, Norway
e-mail: merete.tangstad@ntnu.no

S. Wang et al. (eds.), *Applications of Process Engineering Principles in Materials Processing, Energy and Environmental Technologies*,
The Minerals, Metals & Materials Series, DOI 10.1007/978-3-319-51091-0_34

$$SiO_2(l) + 2C = Si(l) + 2CO(g) \tag{1}$$

$$MnO(l) + C = Mn(l) + CO(g) \tag{2}$$

Though this technology is world leading today, there is a continuous effort to develop more economical and environmental friendly processes. The solid carbon used for SiMn alloys production is typically coke which is produced under CO_2 emissions. CO_2 emissions for coke production varies from 1 ~ 4 t CO_2/t coke, depending on the processing temperature of coking [2]. Low temperature processes give low CO_2 emission. High temperature processes cause high CO_2 emission. High temperature metallurgical coke is typically used in ferromanganese industry.

The lower carbon activity of traditional solid carbon demands a high temperature for reduction. Secondly, impurities enter the SiMn alloys from the carbon materials [3, 4]. Due to the low impurities and high carbon activity, natural gas is advantageous compared with conventional reducing agents. In addition, natural gas could be considered the most environmentally friendly fossil fuel, because it has the lowest CO_2 emissions per unit of energy [5, 6].

Intensive studies have been conducted in the reduction of meal oxides by methane. Direct-reduced iron (DRI), is produced from direct reduction of iron ore by a reducing gas produced from natural gas. This process has been applied in industry for decades [7]. Using methane-containing gas, reduction of silicon, chromium and titanium oxides can proceed at much lower temperatures. Under standard conditions, methane is unstable at temperatures higher than 550 °C. At appropriate CH_4/H_2 ratio and temperature, carbon activity in the methane-containing gas can be well above unity (relative to graphite), which provides favourable thermodynamic conditions for reduction to occur at relatively low temperatures [8–11]. Anacleto et al. [12] studied the reduction of manganese ores and oxides by methane-hydrogen-argon gas mixtures. In a non-isothermal reduction, MnO_2 was reduced to Mn_3O_4 by H_2 and further to MnO by hydrogen at temperatures of 620 °C. The reduction of MnO to manganese carbide started at 760 °C and was completed at about 1200 °C. However, this gas-solid reduction process is difficult to apply industrially due to the carbide products being difficult to separate from the remaining oxides.

The aim of the present work is to study the feasibility of reduction of liquid MnO slag at higher temperatures by CH_4. In this case, CH_4 is injected into melted SiMn slag as reductant. The slags and produced metals compositions are analyzed for establishing the reactivity of CH_4. In addition, carbon black, which can be produced by CH_4 decomposition, is also used in the experiments as reductant.

Experimental

The SiMn slag used in this study was synthetic slag. High purity (chemical grade) fine powders of MnO, SiO_2, CaO, MgO, Al_2O_3 and Fe_2O_3 were mixed in a plastic jar with zirconia balls for 1 h. Table 1 presents the weight composition of mixed raw materials.

CH_4 gas with purity of 99.999% was supplied by AGA industrigasser AS (Oslo, Norway) in 50 L gas cylinders.

The experiments were run in a 75 kW induction furnace. A sketch of the experimental set up is shown in Fig. 1. The raw material mixture (1 or 1.5 kg) is fed into a graphite crucible with inner diameter of 11 cm and a height of 40 cm. The temperature is measured with a C-type thermocouple, which was protected by a graphite sheath. The furnace was heated to 1200 °C with a heating rate of approximately 40 °C/min and held at 1200 °C for 30 min. The furnace was then heated to the target temperature at a heating rate of approximately 20 °C/min and held at this temperature for a certain holding time. The reduction was stopped by lifting the crucible out from the furnace and quenching it in air. At 1200 °C, methane was injected via a gas lance. Two different gas lances made from graphite were used in the experiments, as shown in Fig. 1. Lance A has 2 gas outlet holes, and the distance between holes and the bottom of the crucible is 2.5 cm. In the experiment using lance A, the raw materials weight was 1 kg and the slag depth

Table 1 Weight composition of mixed raw materials

	MnO	SiO_2	CaO	MgO	Al_2O_3	Fe_2O_3	Total
wt%	37.3	31.2	8.7	3.5	12.2	7.3	100.0

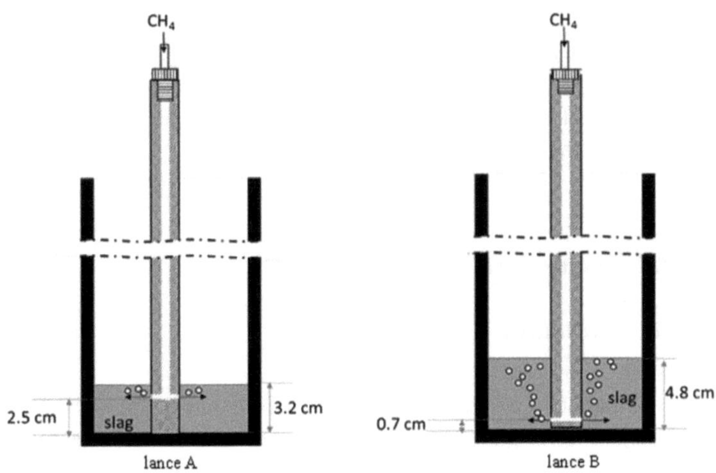

Fig. 1 Schematic of experimental set up

Table 2 Experimental conditions

Condition		Exp #	Temperature °C	Dwell time min	Slag weight g	CH_4 flow rate L/min	Lance
Experiment scheme A	with CH_4	1	1600	60	1000	3	A
	blank	2	1600	60	1000	–	–
	with CH_4	3	1700	60	1000	3	A
	blank	4	1700	60	1000	–	–
Experiment scheme B	with CH_4	5	1600	180	1500	1.5	B
	blank	6	1600	180	1500	–	–

was 3.2 cm. Lance B has 4 gas outlet holes, and the distance between the holes and the bottom of the crucible is 0.7 cm. In the experiments using lance B, the raw material weight was 1.5 kg and the slag depth was 4.8 cm. Several slag samples are taken during the experiment with a steel rod at different temperatures or times. After cooling, the crucible is cut. Slag and metal are separated, weighted and subsequently analyzed by SEM and EPMA.

Table 2 summarises experimental conditions. For each experiment scheme, a blank test without gas injection was operated for the purpose of comparison.

The samples were mounted in a mould by Epoxy resin, ground and polished carefully. Polished samples were coated with carbon film to enhance conductivity during SEM/EDS and EPMA observation. SEM images were recorded by field-emission scanning electron microscopy (FESEM, Zeiss Ultra 55 LE, Oberkochen, Germany) operated at 15 kV. The chemical composition of the slags was analyzed by an energy-dispersive X-ray spectrometer (EDS) mapping or point analysis. The chemical compositions of the slags were also analyzed by an electron probe X-Ray micro analyzer (EPMA), (JEOL JXA-8500F, Tokyo, Japan). In the microprobe the sample is bombarded with a finely focused electron beam. The emitted x-rays are analyzed quantitatively using Wavelength dispersive spectroscopy (WDS).

The carbon contents in the reduced metals were analyzed by combustion-IR (LECO CS-200, St. Joseph, USA). Other metal elementals contents in the reduced metals were analyzed by an X-ray fluorescence (XRF) spectrometer (SPECTRO XEPOS, Kleve, Germany).

Results and Discussions

Methane as Reductant

In Table 2, compared to experiment Scheme A, scheme B applied more slag to increase slag depth, lower gas flow rate to increase CH_4 residence time in the slag, a new design (Lance B) with more and lower gas outlet holes was used. All these measures were done to increase CH_4/slag contact rate.

The final slag compositions of various experiments were analysed by EPMA and summarized in Table 3. Although the MnO contents were slightly reduced, the MnO content are much higher than the equilibrium content of less than 5%MnO at 1600 and 1700 °C, indicating that the reduction rates were slow in these experiments.

The metal yield (shown in Table 4) and the slag composition with Lance A gave very little reduction with methane. Most of the reduction was attributed to the graphite crucible. One possible reason is that part of CH_4 was leaking from the surface of the graphite tube. Flames was observed on the surface of lance. The main reason is however that the CH_4 residence time in the slag was too short. Most of CH_4 escaped from the slag and burned subsequently on the surface of the slag.

Based on the first series of experiments measurements were carried out to increase CH_4 residence time in the slag, such as increasing slag depth by using more raw material (1.5 kg), lower gas flow rate (1.5 L/min) and modification of the gas lance. In the experiment using Scheme B, the introduction of CH_4 gave twice the metal yield (96.7 g) compared to the blank tests (45.3 g). The total extent of the reduction was still quite low.

Figure 2a, b present the MnO and SiO_2 contents during the reduction in the experiment using Scheme B, respectively. The MnO content increased with temperature, from 1200 to 1600 °C, due to the reduction of FeO at these low temperatures. At maximum temperature, the MnO content dropped faster in the test

Table 3 Measured slag composition after each experiment by EPMA

Exp #	Composition, wt%						
	MnO	SiO_2	CaO	MgO	Al_2O_3	FeO	Total
1 A	37.0	34.5	9.3	3.9	13.5	0.4	98.6
2 no CH_4	37.8	33.4	9.5	3.9	14.2	0.8	99.7
3 A	35.8	35.2	9.7	4.0	13.9	0.2	98.9
4 no CH_4	37.0	33.4	9.8	4.0	14.7	0.3	98.6
5 B	34.6	35.8	10.1	4.2	14.1	0.1	98.9
6 no CH_4	37.1	34.7	9.7	3.9	13.3	0.5	99.2

Table 4 Total measured metal composition for each experiment by XRF

Exp #	Metal yield g	Composition, wt%				
		Mn	Si	Fe	C	Total
1 A	30.0	15.2	0.2	78.4	5.2	99.0
2 no CH_4	27.1	6.7	0.2	86.2	5.7	98.8
3 A	51.9	24.8	0.8	67.6	6.2	99.3
4 no CH_4	44.7	21.3	0.7	71.4	5.8	99.2
5 B	96.7	31.9	1.0	60.2	6.0	99.1
6 no CH_4	45.3	12.7	0.5	80.4	5.6	99.2

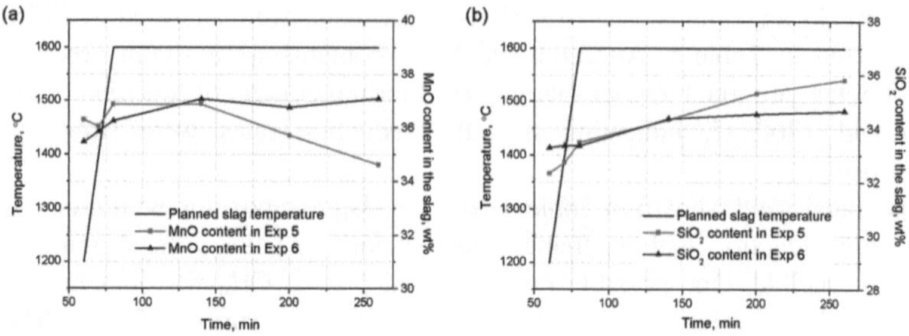

Fig. 2 MnO (**a**) and SiO_2 (**b**) contents changes during the reduction in experiment scheme B. Planned temperature profile is shown

with CH_4 injection than the blank test, indicating a faster reduction rate. The SiO_2 content shows a rising trend due to the reduction of MnO and FeO reduction. Due to the faster reduction of MnO, the SiO_2 will increase faster using CH_4.

The measured metal yield and metal composition by XRF was summarized in Table 4. It is noted that the introduction of methane into the slag did improved the Mn and Si content in the produced alloys. However, Mn and Si content is not even close to the standard of commercial SiMn alloys or the thermodynamic equilibrium composition. Both Mn and Si contents were much lower than the equilibrium values.

Carbon Black as Reductant

Although CH_4 reaction with slag increased in the experiment using Scheme B, about 88.8% of CH_4 was still wasted, based on the calculation of theoretical metal yield, which should be 460 g. Short residence of CH_4 gas in the slag is believed to be responsible for such low utilization of CH_4.

An alternative approach suggested in this study is based on the use of natural gas as a source of carbon black and hydrogen for production of SiMn alloys. The carbon black will be produced by decomposition of methane:

$$CH_4(g) = C + 2H_2(g) \quad (3)$$

This process provides a high purity carbon black and improves the economy of hydrogen production from natural gas. Hydrogen produced in this procedure will be used in the pre-reduction of high manganese oxides at temperature below 700 °C, which is much lower than in the CO reduction atmosphere [12].

One preliminary experiment was run to study the reduction rate by carbon black. 1000 g of synthetic slag mixed with 225 g of carbon black (particle size approximately 1 μm), heated to 1600 °C for 60 min in the induction furnace. A comparison test was also run using 225 g of coke lumps (size range 1.5 ~ 3 cm).

Table 5 Weight balance for experiment 7 and 8

Exp #	Carbon source	Slag in g	Slag out g	Metal out g	Unreacted powder g
7	Carbon black	1000	289.1	222.1	137.0
8	Coke lumps	1000	881.0	25.0	–

Table 6 Total measured slag and metal compositions for experiment 7 and 8

Exp #	Slag composition, wt%		Metal composition, wt%				
	MnO	SiO_2	Mn	Si	Fe	C	Total
7	2.5	35.4	66.5	16.5	13.6	2.2	98.8
8	34.8	35.7	28.4	0.5	65.6	5.1	99.6

The mass balance of the two experiments are presented in Table 5. The metal yield from the reduction experiment with carbon black was significantly higher than the one with coke, indicated a much higher reduction activity of carbon black. At the end of reduction, 137 g of unreacted powder was collected from top layer of crucible. It was consisted with molten slag droplets, unmelted slag, as well as unreacted carbon black.

The slag and metal compositions for the two experiments are summarized in Table 6. The MnO content in the final slag was 2.8 wt%, which is close to, and within the uncertainty of the experiments, to the equilibrium value (4.3 wt%). The produced metal was a typical commercial SiMn alloy.

Conclusions

It was confirmed that injection of CH_4 gas into melted slag improved the reduction of MnO and SiO_2 in SiMn slag. However, the metal yield was far from the theoretical value due to a relative low residence time of CH_4 in the slag and thus a low reaction rate. Most of CH_4 passed the slag and burned subsequently on the surface of slag, before it reacted with the slag or converted to carbon black. Alternative approach was to use of carbon black as reductant, which was produced by CH_4 decomposition. The preliminary experiment results showed a high reduction rate by using carbon black.

Acknowledgements This research is supported under the Norwegian Research Council (project No. 224950).

References

1. S.E. Olsen, M. Tangstad, T. Lindstad, *Production of Manganese Ferroalloys* (Tapir Academic Press, Trondheim, Norway, 2007)
2. C. Xu, D.Q. Cang, A brief overview of low CO_2 emission technologies for iron and steel making. J. Iron. Steel Res. Int. **17**, 1–7 (2010)
3. T. Rosenqvist, *Principles of Extractive Metallurgy* (Tapir Academic Press, Trondheim, Norway, 2004)
4. F.J. Vasko, D.D. Newhart, A.D. Strauss, Coal blending models for optimum coke making and blast furnace operation. J. Oper. Res. Soc. **56**, 235–243 (2005)
5. O. Ostrovski, G. Zhang, Reduction and carburization of metal oxides by methane-containing gas. AIChE J. **52**, 300–310 (2006)
6. X. Li et al., Carbothermal reduction of quartz in methane-hydrogen-argon gas mixture. Metall. Mater. Trans. B **46**, 2384–2393 (2015)
7. J. Kopfle, R. Hunter, Direct reduction's role in the world steel industry. Ironmaking Steelmaking **35**, 254–259 (2008)
8. G. Zhang, O. Ostrovski, Reduction of titania by methane-hydrogen-argon gas mixture. Metall. Mater. Trans. B **31**, 129–139 (2000)
9. J. Zhang, O. Ostrovski, Cementite formation in CH_4–H_2–Ar gas mixture and cementite stability. ISIJ Int. **41**, 333–339 (2001)
10. R.J. Lngbottom, O. Ostrovski, E. Park, Formation of cementite from titanomagnetite ore. ISIJ Int. **46**, 641–646 (2006)
11. R.J. Longbottom et al., Stability of cementite formed from hematite and titanomagnetite ore. Metall. Mater. Trans. B **38**, 175–184 (2007)
12. N. Anacleto, O. Ostrovski, S. Ganguly, Reduction of manganese ores by methane-containing gas. ISIJ Int. **44**, 1615–1622 (2004)

Effect of Flux Ratio on the Products of Self Propagating High Temperature Synthesis-Casting in the WO_3–Si–Al System

S. Niyomwas and T. Chanadee

Abstract The influence of flux CaO_2–Al ratio on the Self propagating high temperature synthesis (SHS)-casting reaction products in WO_3–Si–Al system has been investigated in this study. Incorporation of CaO_2–Al amounting to 1–5 wt% of total charge has been found to be adequate to form WSi_2 intermetallic with a maximum hard-W_5Si_3 secondary phase. The micro-hardness of tungsten silicides intermetallic product increased significantly with CaO_2–Al ratio. The as-SHS-casting products were characterized by X-ray diffraction (XRD) and scanning electron microscope (SEM) with energy dispersive x-ray (EDX) technique.

Keywords Self propagating high temperature synthesis-casting • Phase separation • Tungsten silicide • Flux

Introduction

Tungsten silicides are well known as a good high temperature structural materials due to their high melting temperature (2700 ± 4 °C), excellent creep resistance, good strength, excellent oxidation resistance at elevated temperatures and it also adherent SiO_2 scale at high temperatures [1, 2]. However, tungsten silicides also have a potential as coating materials for low-resistance contacts, gate electrodes and interconnections in microelectronic and semiconductor devices because of the high conductivity and thermal stability of its tetragonal phase [3–5].

Conventionally, tungsten silicides have been processed by various methods, including mechanical alloying (MA), direct heating of mixtures of stoichiometric

S. Niyomwas (✉)
Faculty of Engineering, Department of Mechanical Engineering,
Prince of Songkla University, Hat Yai, Songkhla 90112, Thailand
e-mail: sutham.n@psu.ac.th

T. Chanadee
Faculty of Science, Department of Materials Science and Technology,
Prince of Songkla University, Hat Yai, Songkhla 90112, Thailand

S. Wang et al. (eds.), *Applications of Process Engineering Principles in Materials Processing, Energy and Environmental Technologies*,
The Minerals, Metals & Materials Series, DOI 10.1007/978-3-319-51091-0_35

amounts of pure W and Si elements in vacuum or inert atmosphere to above 1700 °C. However, these methods are costly and have limitations on purity of the product [6–8].

Elaboration of new cost effective methods or simplification of existing ones seems to be of current importance in so-called "Self propagating high temperature synthesis (SHS)". SHS is potentially an energy-efficient process to produce in-situ refractory ceramics, ceramic composites, and intermetallic compounds. Recently, a new approach to the "SHS-casting" of bulk ceramic materials has been reported. In this approach, exothermic reactions are utilized to achieve high temperatures well above the melting points of reaction products [9, 10].

In this communication, we report on the effects of flux CaO_2–Al content on the SHS reaction products of WO_3–Si–Al system, including phase compositions, morphology and mechanical properties.

Experimental

Powder of WO_3 (Fluka Analytical, purity mass 99.9%) and Si lump (Sigma-Aldrich, purity mass 98.5%) have been used as starting materials for oxides. Powder of CaO_2 (Sigma-Aldrich, purity mass 75%) has been used as cum slag fluidizer as well as supply of additional reaction heat to the system. Fine powder of Al (HiMedia laboratories, purity mass 93%) has been used as a reducing agent. These powders were weighted to produce a composition to cause the chemical reaction (1).

$$\begin{aligned} &x_1(\mathrm{WO}_{3(s)} + \mathrm{Si}_{(s)} + \mathrm{Al}_{(s)}) + x_2(\mathrm{CaO}_2 + \mathrm{Al})_{(s)} \\ &\quad \rightarrow x_3(\mathrm{WSi}_2 + \mathrm{W}_5\mathrm{Si}_3)_{(s)} + x_4(\mathrm{Al}_2\mathrm{O}_3) + x_5(\mathrm{CaO} + \mathrm{Al}_2\mathrm{O}_3)_{(s)} \end{aligned} \tag{1}$$

The precursor powders have been homogeneously mixed and dry-blended thoroughly in the desired ratio for 60 min in a ball-mixing mill with ZrO_2 spheres at room temperature. Then, the powder mixture was uniaxially pressed without binder, using Herzog, TP20P with green density in the range of 50–60% of the theoretical value into cylindrical pellets compact with the diameter of 25.4 mm. The compacts were inserted in a 26 mm inner diameter silica sand tube (97% SiO_2) and placed on a 3-layer graphite plate with 15 mm diameter hole at the center. Then, the compact and graphite plate was transferred into a SHS reactor. The reaction chamber was evacuated with a vacuum pressure of 70 mmHg for 5 min and filled with argon gas up to 0.3 MPa. This operation was repeated at least twice in order to ensure an inert environment during the reaction revolution. The compact was ignited at one end by tungsten filament. The details of experimental setup and the reaction phenomena were given in Fig. 3. After the reaction, a distinctly separated intermetallic bottom is then recovered by mechanical breaking the top slag layer.

The metallographic sample of tungsten silicides intermetallic was prepared using standard mechanical polishing procedures. The microstructure was investigated using a scanning electron microscope, SEM (JSM-5800LV, JEOL, Japan) with an energy dispersive x-ray spectrometer, EDX (ISIS 300, Oxford, England). Phase compositions were indentified by X-ray diffraction, XRD (PHILIPS, X' Pert MPD, Cu Kα, Netherlands). The density was measured according to the Archimedes' principle. The micro-hardness values were measured under a 4.9 N load by a Vickers hardness tester (High Wood, HWDM-3, Japan).

Results and Discussion

Thermodynamic Investigation

The equilibrium compositions of the WO_3–Si–Al system with varied CaO_2–Al ratios at different temperatures were calculated using "HSC chemistry®" computer software based on Gibbs free energy minimization method [10] and the results are shown in Fig. 1. The formation of WSi_2 and W_5Si_3 intermetallic was initiated at above 1500 and 2000 °C, respectively with a decrease in the amount of reactants.

A prerequisite for SHS casting materials is that the attained combustion temperature (T_c) must be higher than the melting point (T_m) of combustion products: $T_c > T_m$. In the presence of unavoidable heat loss, the real combustion temperature is not enough to complete melt formation, and it has a short melted life time. Thermodynamic analysis of many systems has shown that, in order to elevate T_c, an some amount of CaO_2–Al mixture is added to react to provide the supply of additional reaction heat to the system. As follows from Fig. 2, the addition of CaO_2–Al mixture gives a marked increase in adiabatic temperature (T_{ad}) [11]. The calculated adiabatic temperatures were much higher than melting points of the products and so the produced the intermetallics and oxides formed a superheated liquid.

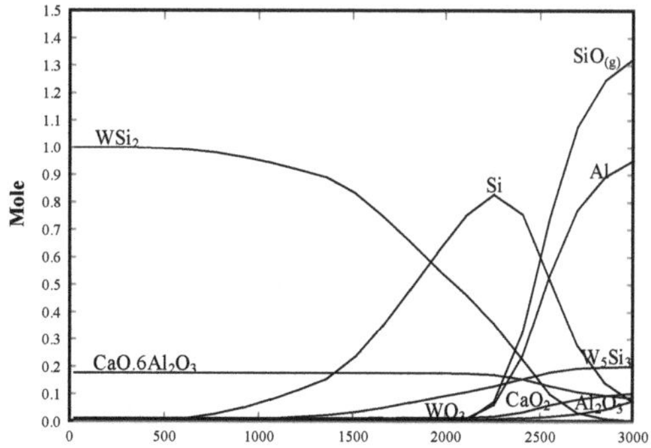

Fig. 1 Equilibrium composition of WO_3–Si–Al reactant system in Ar gas atmosphere

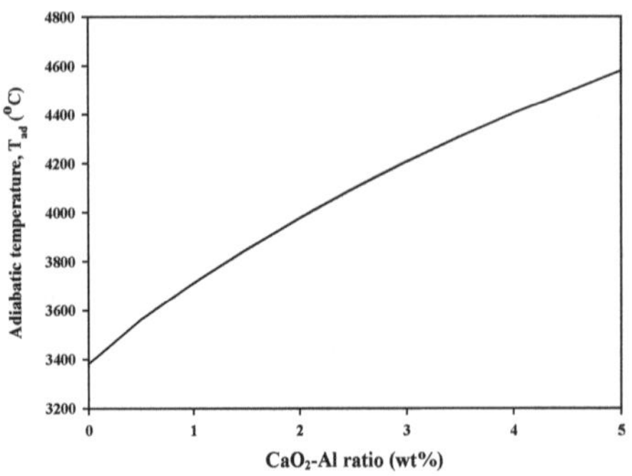

Fig. 2 Calculated values of adiabatic temperature versus relative amount of (CaO_2–Al) mixture added to WO_3–Si–Al mixtures

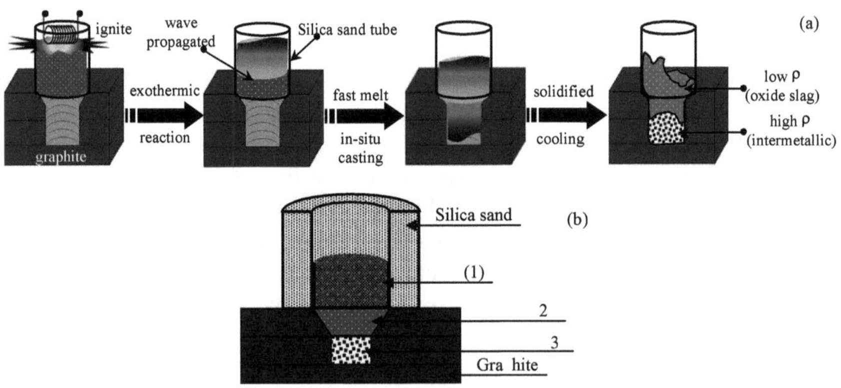

Fig. 3 **a** The short-term phenomena in-situ SHS-casting reaction of WO_3–Si–Al reactant compact through product and **b** product phase layers after cooling period

SHS-Product Characteristics

Because the reaction temperature was extremely high, the oxide reactant compact was melted, cast and subsequently solidified and cooled at the silica sand tube wall or inside a graphite substrate cavity where the whole synthesis process lasts only seconds. The products were separated to two immiscible layers: heavy intermetallic compound at the bottom and light oxide ceramic slag on the top which was controlled by buoyancy force and the different density of the products [12], as shown cross-section schematic mechanism in Fig. 3a.

It is found that the use of the silica sand tube helped the phase separation of the oxide and intermetallic product. Figure 3b shows the presence of three approximate layers after cooling down process: a Al_2O_3 deposited on silica sand surface (layer 1), calcium aluminates in the middle zone (layer 2), and an intermetallic at the bottom

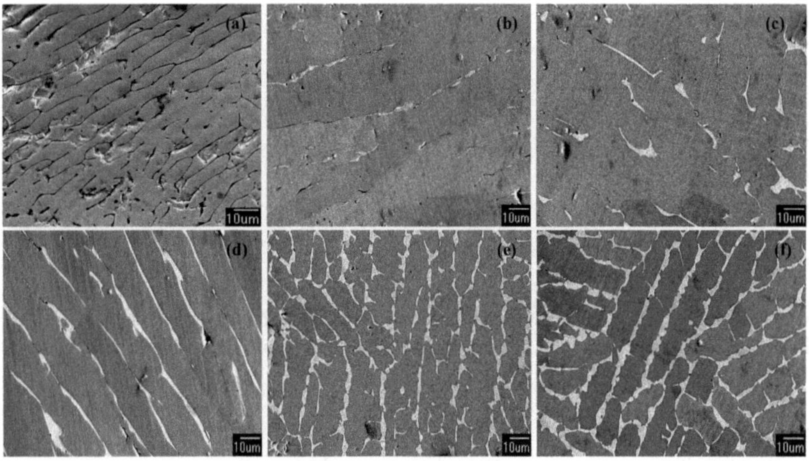

Fig. 4 Microstructure and local composition of SHS-tungsten silicides intermetallic products at various CaO_2–Al ratio: **a** without CaO_2–Al **b** 1 wt%, **c** 2 wt%, **d** 3 wt%, **e** 4 wt% and **f** 5 wt%

(layer 3). The nucleation on layer (1) is due to the formation of solid solution between silica sand and Al_2O_3 (SiO_2–Al_2O_3) which was shown that this phenomenon is controlled by surface tension and wet ability as well as enthalpy of formation ($\Delta H_f^\circ = -108.74$ kJ/mol). While, liquid calcium aluminate on top of the intermetallic is solidified in the cooling process at the same time. Since the thermal expansion coefficients of CaO–Al_2O_3 compounds (20×10^{-7} °C^{-1}) [13] are much smaller than that of tungsten silicides intermetallic (9.3×10^{-6} °C^{-1}) [14], thermal stresses at the interface between calcium aluminates and tungsten silicides intermetallic were so large that it did not bind together.

Polished sections (Fig. 4a–f) showed the following microstructural constituents: dark-gray matrix, precipitates of a white-gray phase, and pores. With no CaO_2–Al, the microstructure consisted of overlapping narrow-layer grains with many pores (Fig. 4a). A significant change in microstructure is observed with increase in the amount of CaO_2–Al in the precursor mixture. Figure 4b–d shows a white-gray cellular precipitates embed into the dark-gray matrix while, cross-link grain microstructure of white-gray precipitates with higher CaO_2–Al content is dominantly formed as shown in Fig. 4e–f.

In order to determine the chemical composition of SHS-cast product, locally resolved EDX microanalysis was applied during scanning SEM investigations. The EDX spectrum in Fig. 5a shows this SHS-cast product contains both tungsten (W) and silicon (Si) as alloying elements. In addition, we see that the tungsten and silicon are in the same area of the two-phase region present in Fig. 5b represented by dark-gray matrix (α) and the white-gray precipitates (β). The EDX quantitative analyses for α and β in Figure 5b show that the approximate alloy contents at each region contains 79.88% W-20.12% Si corresponds to WSi_2 and 85.82% W-14.18% Si coresponds to W_5Si_3, respectively.

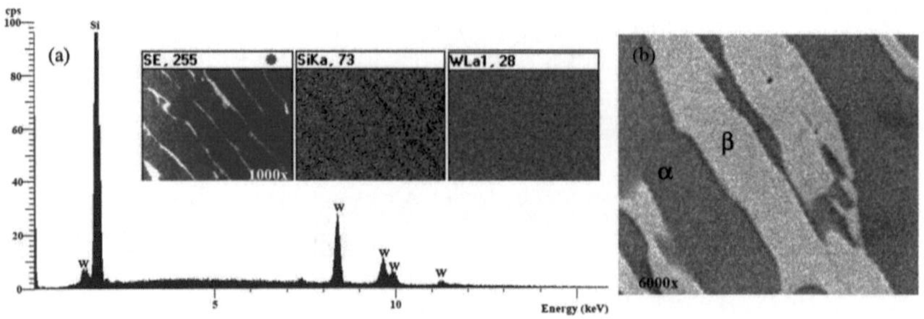

Fig. 5 SEM micrographs and EDX data for the SHS-tungsten silicides intermetallic ingot surface. **a** EDX spectrum and elemental mapping from the entire area of Fig. 4b and d SEM micrograph magnifying from the entire area of Fig. 4f

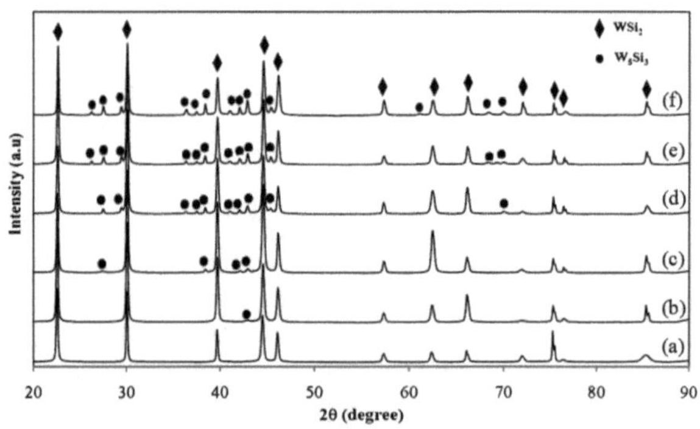

Fig. 6 XRD pattern of SHS-tungsten silicides intermetallic products at various CaO_2–Al ratio: **a** without CaO_2–Al **b** 1 wt%, **c** 2 wt%, **d** 3 wt%, **e** 4 wt% and **f** 5 wt%

Table 1 Semi-quantitative analysis of SHS-tungsten silicides intermetallic products related hardness treated with different CaO_2–Al ratio

CaO_2–Al ratio (wt%)	Phase composition (%)		Micro-hardness (HV)
	WSi_2	W_5Si_3	
0	100	–	795.1
1	99.72	0.28	900.4
2	98.4	1.6	906.3
3	92	8	924.3
4	87.86	12.14	955.2
5	82.54	17.46	989.9

Figure 6 and Table 1 show the evolution of XRD patterns of tungsten silicides intermetallic product as a function of CaO_2–Al ratio. According to the XRD results, without the addition of CaO_2–Al, products only appear as WSi_2 intermetallic phase. When CaO_2–Al ratio increase, the major phase is WSi_2 and W_5Si_3 is the secondary phase. It is believed that the introduction of CaO_2–Al mainly contributed to the formation of WSi_2 + W_5Si_3 intermetallic. Because of a reaction temperature much above the W-Si eutectic temperature (~2284 °C) by which, on cooling, a liquid transforms into WSi_2 + W_5Si_3 solid phases at the same time.

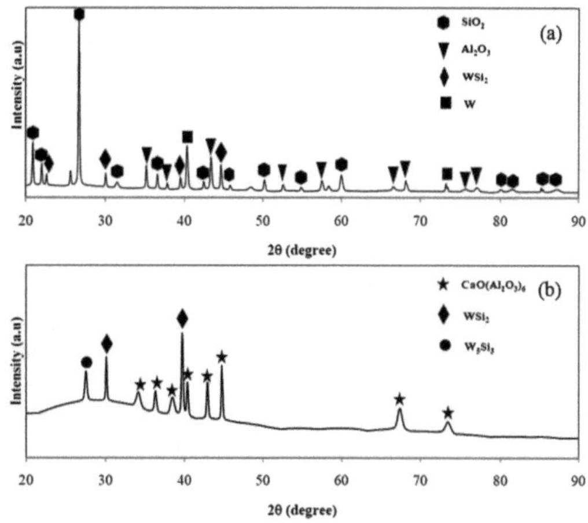

Fig. 7 XRD pattern of ceramic slag at different region **a** silica sand tube surface and **b** top of intermetallic ingot

Table 1 includes the hardness variation for the tungsten silicides intermetallic produced at different ratios of CaO_2–Al. The Vickers micro-hardness of the prepared tungsten silicides intermetallic specimens is tested with 4.9 N load. The hardness of the tungsten silicides intermetallic prepared without CaO_2–Al addition is about 795.1 HV. However, it increases with the increasing of CaO_2–Al ratio from 1–5 wt%. These results indicated that the micro-hardness increases as the W_5Si_3 content increases. Therefore, it can be said that microstructures are affected by the use of the CaO_2–Al additive. The density of SHS-tungsten silicides intermetallic products are 5.7 g/cm^3 with no CaO_2–Al and about 8.5 g/cm^3 with 1 wt% and up. From the aforementioned results, there must be a significant effect of the addition of CaO_2–Al on the microstructure.

Typical XRD patterns of the ceramic slag are shown in Fig. 7. Major intensity peaks of SiO_2 with Al_2O_3 and amorphous $CaO{\cdot}6Al_2O_3$ appears in layer (1) and layer (2), respectively.

Conclusions

A study on the influence of CaO_2–Al content on phase separation phenomena, microstructural development, phase composition and hardness of tungsten silicides intermetallic reveals the following. (1) The extent of phase separation depended on adiabatic temperature as well as reaction temperature. (2) The incorporation of CaO_2–Al additive was helpful to increase the melted life time and complete phase separation. (3) The phase compositions of the WO_3–Si–Al precursor system with 1–5% wt of CaO_2–Al included a major WSi_2 intermetallic with hard W_5Si_3 secondary phase. (4) Application of CaO_2–Al resulted in a SHS-cast tungsten silicides intermetallic product having improved micro-hardness.

Acknowledgements This work has been carried out with the support from Prince of Songkla University Research Fund (Funding No. ENG-550115S) and the Ceramic and Composite Materials Engineering Research Group (CMERG), Materials Engineering Research Center (MERC), Faculty of Engineering, Prince of Songkla University (PSU), Thailand.

References

1. M. Alam, S. Saha, B. Sarma, D. Das, Formation of WSi_2 coating on tungsten and its short-term cyclic oxidation performance in air. Int. J. Refract. Met. Hard. Mater. **29**, 54–63 (2011)
2. D. Oh, H. Kim, J. Yoon, I. Ko, I. Shon, Synthesis of dense WSi_2 and WSi_2-*x*vol.%SiC composites by high-frequency induction heated combustion and its mechanical properties. Met. Mater. Int. **12**(4), 307–316 (2006)
3. V. Glebovsky, S. Ermolov, V. Motuzenko, E. Stinov, Thin silicide films deposited from cast silicide targets. Mater. Lett. **37**, 44–50 (1998)
4. O. Pluchery, G. Lucazeau, V. Fortuna, F. Meyer, R. Madar, Raman study of tungsten disilicide formation in thin films. Microelectron. Eng. **37**(38), 543–550 (1997)
5. J. Callanan, R. Weir, E. Westrum, Transition metal silicides and tellurides: crystal structure, heat capacities, and derived thermodynamic properties from absolute zero to 2200 K. Pure Appl. Chem. **69**(11), 2289–2294 (1997)
6. P. Feng, A. Farid, X. Wang, I. Humail, X. Qu, Mechanically activated reactive synthesis of refractory molybdenum and tungsten silicides. Int. J. Refract. Met. Hard. Mater. **26**, 173–178 (2008)
7. M. Chandrasekharaiah, J. Margrave, The disilicides of tungsten, molybdenum, tantalum, titanium, cobalt, and nickel, and platinum monosilicide: a survey of their thermodynamic properties. J. Phy. Chem. Ref. Data. **22**(6), 1459–1468 (1993)
8. G. Fei, L. Liu, X. Ding, L. Zhang, Q. Zheng, Preparation of nanocrystalline intermetallic compounds $WSi_2$and $MoSi_2$ by mechanical alloying. J. Alloy. Compd. **229**, 280–282 (1995)
9. G. Liu, J. Li, Y. Chen, Phase separation in melt-casting of ceramic materials by high-gravity combustion synthesis. Mater. Chem. Phys. **133**, 661–667 (2012)
10. S. Niyomwas, Preparation of Al reinforced with TiB_2–Al_2O_3–Fe_xAl_y composites derived from natural ilmenite. Int. J. Self-Propag. High-Temp. Synth. **19**(2), 150–156 (2010)
11. V. Sanin, D. Andreev, D. Ikornikov, V. Yukhvid, Cast intermetallic alloys by SHS under high gravity. Acta Phys. Pol. A **120**, 331–335 (2011)
12. T. Chanadee, S. Niyomwas, J. Wannsin, Effect of Ar gas pressure on phase separation of tungsten silicides intermetallic compound in-situ self propagating high temperature synthesis-casting process. Adv. Mater. Res. **626**, 143–146 (2013)
13. D. Beall, S. Lakhwani, Fabrication of low thermal expansion calcium aluminate articles. U.S. Patent No. 6,809,051, 26 Oct 2004
14. N. Matyushenko, Thermal expansion of tungsten and molybdenum disilicides. Powder. Metall. Met. Ceram. **3**, 15–17 (1964)

Synthesis of Chrysin Based Cationic Lipids: Plasmid Delivery and Transgene Expression

Bhavani Kedika, Venkatagiri Noole, Krishna Thotla and Krishna Reddy Chepyala

Abstract Cationic lipids are attractive candidates for transgene delivery and expression and are safer alternatives to viral delivery vehicles. Here, we report the design, synthesis, characterization and transgene expression efficacies of a series of chrysin derived cationic lipids. A series of ten lipids of chrysin was synthesized by treating the 7-hydroxy group with dibromoalkane and further substitution of the bromo derivative with alkyl amines. The synthesized lipids were characterized spectroscopically. Parallel screening of the synthesized lipids for in vitro plasmid delivery expressing luciferase led to identification of several leads that demonstrated higher transgene expression when compared to the standard. Transgene expression efficacy was dependent on the chain length of the lipid. The influence of the various physicochemical properties on transgene expression efficacies was investigated. This study indicates that lipids derived from chrysin results in high levels of transgene expression and may be attractive for gene therapy applications.

Keywords Chrysin · Plasmid · Transgene expression · Lipids · Gene therapy

Introduction

Gene therapy mediated by the delivery of genetic materials into target cells to treat various human diseases like cancer, infectious diseases and immunodeficiency is a promising clinical modality [1–6]. Many challenges associated with the delivery of

B. Kedika (✉) · V. Noole · K. Thotla · K.R. Chepyala
Department of Chemistry, Osmania University, Hyderabad 500007, Telangana, India
e-mail: kedikabhavani@gmail.com

V. Noole
e-mail: giri9.chemistry@gmail.com

K. Thotla
e-mail: tkrishnachem1@gmail.com

K.R. Chepyala
e-mail: ckreddy58@gmail.com

S. Wang et al. (eds.), *Applications of Process Engineering Principles in Materials Processing, Energy and Environmental Technologies*,
The Minerals, Metals & Materials Series, DOI 10.1007/978-3-319-51091-0_36

genes include enzymatic instability, low cellular uptake and inability to escape from the endosomes. Hence, development of efficient vehicles to deliver genes into target cells is very important. Major types of gene delivery vectors include viral and non-viral vectors. Among the non-viral vectors many properties associated with cationic lipids such as their handy synthesis, low immune response and good safety made them deeply investigated. Cationic lipids interact with negatively charged DNA through electrostatic interaction and condense it into lipid/DNA aggregates which are called lipoplexes. The positive charges on the lipoplexes interact with negatively-charged cell membrane residues. There are a variety of cationic lipids developed [7–12], but there is still space for the development of efficient and safe delivery vectors.

Chrysin (5,7-dihydroxyflavone or 5,7-dihydroxy-2-phenyl-4Hchromen-4-one) belongs to the flavone class of the ubiquitous 15-carbon skeleton natural polyphenolic compounds collectively called flavonoids. Chrysin found in several plants, mushroom and honeycomb is extensively studied flavone. It is known to have broad spectrum of biological activities including anticancer, anti-inflammatory, antioxidant, antibacterial, anti-diabetic and anti-HIV [13]. But low solubility and relatively poor absorption in the intestine are the main drawbacks that limit its potential applications. To improve the bioavailability and efficacy of chrysin many efforts had been made by researchers on designing its analogs and conjugates. Several articles and reviews on biological activity of synthetic derivatives of chrysin have been reported earlier [14–16]. To our knowledge chrysin based cationic lipids had never been reported, hence in this study we designed and synthesized a series of ten chrysin based cationic lipids which differ in the chain length and studied their in vitro transfection studies in multiple cultured animal cell lines. It is demonstrated that some of these chrysin based lipids exhibit higher transfection efficiency and low cytotoxicities (Fig. 1).

Experimental

Synthesis

Synthetic routes for preparing cationic lipids **1–10** are shown schematically in Scheme 1. Details of the synthetic procedures, purifications and spectral

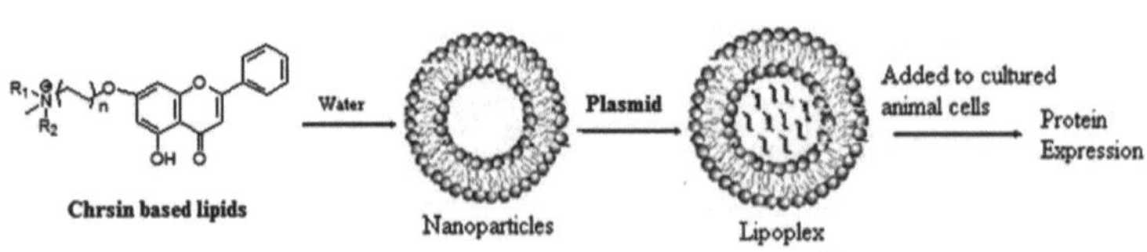

Fig. 1 Schematic Representation of in vitro transgene expression

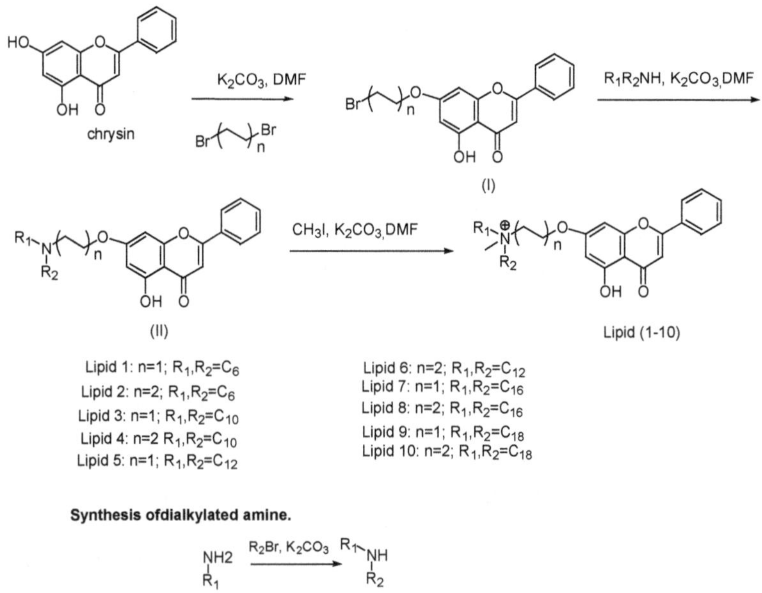

Scheme 1 Schematic representation of synthesis of lipids **1–10**

characterizations of the lipid **2** as well as its synthetic intermediates shown in Scheme 1 is described below. For the remaining lipids the similar procedure is followed.

Synthesis of Lipid 2: (Scheme 1)

(a) Synthesis of 7-(4-bromobutoxy)-5-hydroxy-2-phenyl-4H-chromen-4-one (I): The synthesis of 7-(4-bromobutoxy)-5-hydroxy-2-phenyl-4H-chromen-4-one is carried out according to the reported procedure [17]. Briefly in a flask equipped with N_2 atmosphere and charged with 100 mL of acetone, added 29 mmol of Chrysin and 1 eq. of 1,4-dibromobutane in the presence of 1.05 eq. of potassiumcarbonate. The reaction mixture was refluxed for 24 h until no starting material left as monitored by Thin Layer Chromatography. After the reaction completion, the reaction mixture was concentrated, cooled at room temperature and diluted with ethyl acetate (100 mL) and washed with water (2 × 75 mL). The organic phase was separated and treated with Magnesium Sulphate and then concentrated under vacuum to furnish a yellowish white colored compound Yield: 83%.

Synthesis of N, N-di-n-dihexadecyl amine:

In a 100 ml round bottomed flask equipped with N_2 atmosphere and charged with 25 mL of acetone, added 5 mmol of hexadecylamine and 1 eq. of hexadecylbromide in the presence of 1.05 eq. of potassiumcarbonate. The reaction mixture was stirred at room temperature for 24 h until no starting material left as monitored by TLC. After the reaction completion, the reaction mixture was concentrated, cooled at room temperature and diluted with ethyl acetate (100 mL) and washed with water (2 × 75 mL). The organic phase was separated and treated with Magnesium Sulphate and then concentrated under vacuum to furnish a white colored solid.

(b) Synthesis of 5-hydroxy-2-phenyl-7-(4-(N,N-di-n-hexadecylamino)butoxy)-4H-chromen-4-ones:

In a flask charged with 50 mL of Acetonitrile was added 2.5 mmol of compound (I) and appropriate N, N-di-n-dihexadecyl amine (1.5 eq.) and the reaction mixture was refluxed for 10–38 h until the complete consumption of starting material as detected by TLC. After the completion of the reaction, the reaction mixture was treated with ice and the resulting solid was filtered and washed with water (2 × 25 mL). The residue was purified with a silica gel column chromatography and was eluted with dichloromethane: methanol (40:1) to afford corresponding product.

(c) Synthesis of 5-hydroxy-2-phenyl-7-(4-(N,N-di-n-hexadecylammonium chloride) butoxy)-4H-chromen-4-ones:

In a 25 mL round bottomed flask, excess methyl iodide is added to intermediate tertiary amine 5 mmol prepared in step (b). The reaction mixture was stirred at room temperature for 12 h. The solvent from the reaction mixture was evaporated and dried. The residue upon column chromatographic purification using 60–120 mesh size silica gel, and 2% methanol in chloroform as eluent followed by chloride ion exchange (using Amberlyst A-26 with chloroform as eluent) afforded the pure title lipid **2** as a yellowish white solid.

Hydrodynamic Diameter and Zeta Potential of Lipids

The sizes and the surface charges (zeta potentials) of liposomes were measured by photon correlation spectroscopy and electrophoretic mobility on a Zeta sizer $3000HS_A$ (Malvern, U.K.). The sizes were measured in Dulbecco's modified Eagle's medium (DMEM) with a sample refractive index of 1.59 and a viscosity of 0.89 cP. The zeta potential was measured using the following parameters: viscosity, 0.89 cP; dielectric constant, 79; temperature, 25 °C; F(Ka), 1.50 (Smoluchowski); maximum voltage of the current, V. Measurements were done 10 times with the zero-field correction. The potentials were calculated by using the Smoluchowski approximation.

Transfection Biology

Cells were seeded at a density of 15,000 cells for CHO (Chinese hamster ovary) and A-549 (Human lung carcinoma cells) per well in a 96-well plate 18–24 h before the transfection. 0.3 μg (0.91 nmol) of plasmid DNA was complexed with varying amounts of lipids in plain DMEM medium (total volume made upto 100 μL) for 30 min. The charge ratios were varied from 1:1 to 8:1 over these ranges of the lipids. Just prior to transfection, cells plated in the 96-well plate were washed with PBS (2 × 100 μL) followed by the addition of lipoplexes. After 4 h of incubation, 100 μL of DMEM with 20% FBS was added to the cells. The medium was changed to 10% complete medium after 24 h and the reporter gene activity was estimated after 48 h. The cells were lysed and the β-galactosidase activity per well was estimated by adding 50 μL of 2X-substrate solution [1.33 mg/mL of ONPG, 0.2 M sodium phosphate (pH 7.3) and 2 mM magnesium chloride] to the lysate in a 96-well plate. Absorbance of the product ortho-nitrophenol at 405 nm was converted to β-galactosidase units by using a calibration curve constructed using pure commercial β-galactosidase enzyme.

Toxicity Assay

Cytotoxicity of lipids **1–10** were assessed by the 3-(4,5-dimethylthiazol-2-yl)-2,5-diphenyltetrazolium bromide (MTT) reduction assay as described earlier. The cytotoxicity assay was performed in 96-well plates by maintaining the same ratio of number of cells to amount of cationic lipid, as used in the transfection experiments. MTT was added 24 h after the addition of cationic lipids to the cells followed by 3 h of incubation. Results were expressed as percentage viability = [A_{540} (treated cells) − background/A_{540} (untreated cells) − background] × 100.

Discussion

Chemistry

The details of the synthetic procedures for generally for all the lipids shown in Scheme 1are described for **lipid 2** in the experimental section. Cationic lipids were synthesized by the bromination of the 7-hydroxy group of chrysin with dibromo compounds available commercially which upon further reaction with N, N-di-n-alkyl amines yielded the intermediate tertiary amines. The resulting tertiary amine intermediates upon quaternization with excess methyl iodide followed by chloride ion exchange over Amberlyst-26 yielded lipids **1–10**. Structures of all the synthetic intermediates and final lipids shown in Scheme 1 are confirmed by ^{1}H NMR and molecular ion peaks in their ESI mass spectra. The purity of the final lipids was confirmed by elemental (C, H, N and S) analysis.

Nanosizes and Global Surface Charges of the Lipoplexes

The nanosizes and the global surface charges of the liposomes were measured in order to characterize the present lipids physicochemically. The experiment is carried out on liposomes of lipids **1–10** and cholesterol as colipid using a dynamic laser light scattering instrument equipped with ξ-sizing capacity. These measurements were made in the presence of Dulbecco's modified Eagle's medium (DMEM). The results reveal that in general the sizes of liposomes were in the range from 100–300 nm (Table 1). The global surface potentials of liposomes made from representative cationic lipids **1–10** using cholesterol as co-lipid were found to be positive in presence of DMEM (Table 1).

In Vitro Transfection Studies

The relative in vitro gene delivery efficacies of lipids **1–10** in CHO and A-549 cells across the lipid:DNA charge ratios of 8:1 to 1:1 using cholesterol as co-lipid are summarized in Figs. 2 and 3. *p*CMV-SPORT-*β*-gal plasmid DNA was used as the reporter gene. The transfection efficiencies of the lipids **1–10** were compared with that of the commercial formulation, Lipofectamine-2000. The transfection results

Table 1 Hydrodynamic diameter (nm) and Zeta Potentials (mV) of lipids 1–10. Sizes (average hydrodynamic diameter) were measured by laser light scattering technique using Zetasizer nanoseries 500 (Malvan Instruments, UK)

Lipid	Hydrodynamic diameter (nm)	Zeta potential (mV)
1	129.2 ± 2.2	33.7 ± 1.0
2	149.6 ± 21.1	12.8 ± 2.1
3	205.1 ± 10.6	10.1 ± 2.5
4	119.4.3 ± 09	19.3 ± 10.3
5	279.0 ± 11.6	6.9 ± 3.2
6	135.6 ± 21.0	11.5 ± 3.5
7	309.0 ± 2.0	5.4 ± 2.8
8	324.1 ± 7.1	8.5 ± 1.8
9	265.8 ± 15.0	6.7 ± 3.3
10	125 ± 4.9	12.8 ± 5.5

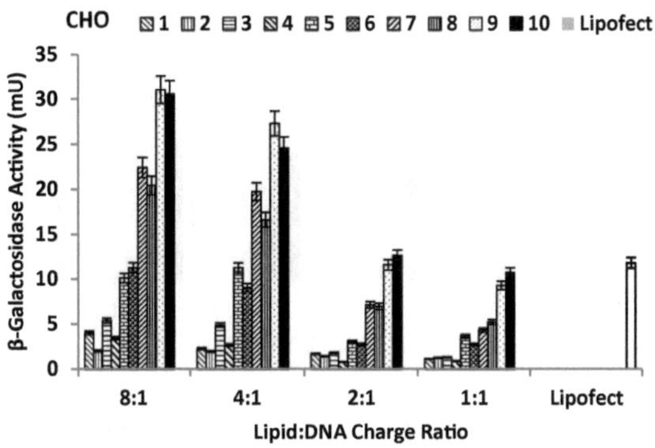

Fig. 2 In vitro gene delivery efficiencies of lipids **1–10** in CHO cells. Units of β-galactosidase activity were plotted against the varying lipid:DNA charge ratios (8:1-1:1). The transfection efficiencies of the lipids were compared to that of commercial formulation. All the lipids were tested on the same day, and the data presented are the average of three experiments performed on three different days

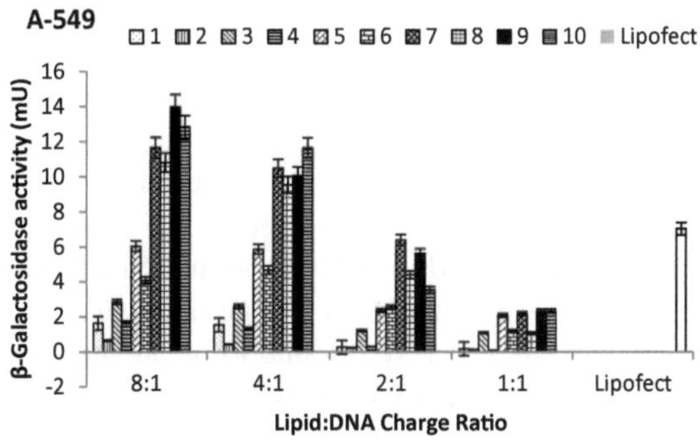

Fig. 3 In vitro gene delivery efficiencies of lipids **1–10** in A549 cells. Units of β-galactosidase activity were plotted against the varying lipid:DNA charge ratios (8:1-1:1). The transfection efficiencies of the lipids were compared to that of commercial formulation. All the lipids were tested on the same day, and the data presented are the average of three experiments performed on three different days

reveals that in the case of presently studied lipids **1–10**, lipids with smaller hydrophobic group are in general proved to be more efficient when compared to lipids with longer hydrophobic group. This difference in transfection profiles of lipoplexes prepared from varying chain lengths has been reported earlier [18, 19]. The highest transfection efficiencies of all the transfection efficient lipids were found to be at lipid:DNA charge ratios of 8:1 and 4:1. The spacer length between 7-O position of chrysin and the quaternary ammonium group was also found to influence the transfection results. Taken together, the transfection profiles for the presently described chrysin based cationic amphiphiles summarized in Figs. 2 and 3

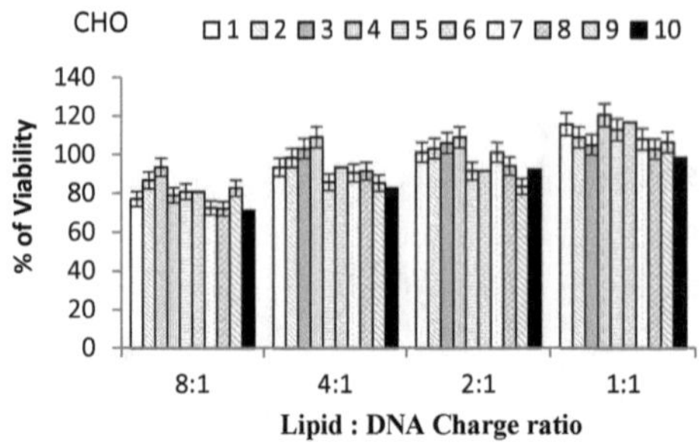

Fig. 4 Representative percent cell viabilities of lipids **1–10** in CHO types of cells using MTT assay. The absorption obtained with reduced formazan with cells in the absence of lipids was taken to be 100. The data presented are the average values of three independent experiments

convincingly demonstrate that these new lipids may lead to developments of efficient delivery agents in the field of non-viral gene delivery.

Toxicity Studies

MTT-based cell viabilities of lipids **1–10** were evaluated in CHO cell lines across the entire range of lipid:DNA charge ratios used in the actual transfection experiments. Cell viabilities of all the lipids **1–10** were found to be remarkably high (more than 85%) particularly up to the lipid:DNA charge ratios of 4:1 (Fig. 4). Thus, the lipids **1–10** which are chrysin based were found to be nontoxic and the contrasting in vitro gene transfer efficacies of lipids (Figs. 2 and 3) are unlikely to originate from varying cell cytotoxicities of the lipids.

Conclusions

The findings described provide the first experimental evidence of chrysin head group based cationic lipids in liposomal gene delivery. Cationic amphiphiles containing the same head group functionality but differ only in the chain length are designed and synthesized. In vitro findings delineated that lipids with longer chain lengths showed better transfection than lipids with shorter chain lengths in multiple cultured mammalian cells. Structural investigation studies of these lipids clearly demonstrate the differences in transfection profiles.

Acknowledgements Financial support for this work from Department of Science and Technology (DST-SERB, Fast-Track Young. Scientist), Government of India, New Delhi is gratefully acknowledged.

References

1. R.J. Zhan et al., The effect of side-chain functionality and hydrophobicity on the gene delivery capabilities of cationic helical polypeptides. Biomaterials **35**, 3443–3454 (2014)
2. M. Candolfi et al., Gene therapy-mediated delivery of targeted cytotoxins for glioma therapeutics. Proc. Natl. Acad. Sci. U. S. A. **107**, 20021–20026 (2010)
3. J.L. Wang et al., A gene nanocomplex conjugated with monoclonal antibodies for targeted therapy of hepatocellular carcinoma. Biomaterials **33**, 4597–4607 (2012)
4. S.K. Cho et al., Acid-degradable core-shell nanoparticles for reversed tamoxifen-resistance in breast cancer by silencing manganese superoxide dismutase (MnSOD). Biomaterials **34**, 10228–10237 (2013)
5. L. Yin et al., Supramolecular self-assembled nanoparticles mediate oral delivery of therapeutic TNF-α siRNA against systemic inflammation. Angew. Chem. Int. Ed. Engl. **52** (22), 5757–5761 (2013)
6. C.E. Thomas, A. Ehrhardt, M.A. Kay, Progress and problems with the use of viral vectors for gene therapy. Nat. Rev. Genet. **4**, 346–358 (2003)
7. W.H. Qiao, M. Zhou, Hydroxyl- modified cationic lipids with a carbamate linkage as gene delivery vehicles. Eur. J. Lipid Sci. Technol. **115**, 483–489 (2013)
8. A. Bajaj, P. Kondiah, S. Bhattacharya, Design, synthesis, and in vitro gene delivery efficacies of novel cholesterol-based gemini cationic lipids and their serum compatibility: a structure-activity investigation. J. Med. Chem. **50**, 2432–2442 (2007)
9. S.B. Zhang et al., Cationic lipids and polymers mediated vectors for delivery of siRNA. J. Control. Release **123**, 1–10 (2007)
10. K. Ewert et al., Efficient synthesis and cell-transfection properties of a new multivalent cationic lipid for nonviral gene delivery. J. Med. Chem. **45**, 5023–5029 (2002)
11. L. Wang, R.C. MacDonald, New strategy for transfection: mixtures of medium-chain and long-chain cationic lipids synergistically enhance transfection. Gene Ther. **11**, 1358–1362 (2004)
12. Y. Kaneda, Y. Tabata, Non-viral vectors for cancer therapy. Cancer Sci. **97**, 348–354 (2006)
13. X. Song et al., Synthesis of novel amino acid derivatives containing chrysin as anti-tumor agents against human gastric carcinoma MGC-803 cells. Med. Chem. Res. **24**, 1789–1798 (2015)
14. Y. Liu et al., Synthetic derivatives of chrysin and their biological activities. Med. Chem. Res. **23**(2), 555–563 (2014)
15. S.F. Nabavi et al., Neuroprotective effects of chrysin: from chemistry to medicine. Neurochem. Int. **90**, 224–231 (2015)
16. W. Apilak et al., Probing the origins of anticancer activity of chrysin derivatives. Med. Chem. Res. **24**(5), 1884–1892 (2015)
17. R.V. Patel et al., Synthesis of coumarin based 1, 3, 4-oxadiazol-2ylthio-N-phenyl/benzothiazolyl acetamides as antimicrobial and antituberculosis agents. Med. Chem. Res. **22**, 195–210 (2013)
18. B. Majeti et al., In vitro gene transfer efficacies of N, N-dialkylpyrrolidinium chlorides: a structure-activity investigation. J. Med. Chem. **48**, 3784–3795 (2005)
19. Y.V. Mahidhar, M. Rajesh, A. Chaudhuri, Spacer-arm modulated gene delivery efficacy of novel cationic glycolipids: design, synthesis, and in vitro transfection biology. J. Med. Chem. **47**, 3938–3948 (2004)

Part VII
Energy Storage and Engineering Issues

Corrosion Mechanism of Haynes 230 Alloy with Ni Crucible in $MgCl_2$–KCl

Yuxiang Peng and Ramana G. Reddy

Abstract The long-term corrosion behavior of Haynes 230 (H230) alloy in $MgCl_2$–KCl eutectic salt was evaluated at 1273 K for 500 h in Ni crucible in argon gas. The results showed H230 alloy has good corrosion resistance. In addition, corrosion was noted to occur by depletion of Cr, which was deposited on the Ni crucible. Based on the calculation of Gibbs energy of Ni-Cr alloy and exchange current density of Cr in $MgCl_2$–KCl, the Tafel model was constructed to investigate the corrosion potential and rate. The galvanic corrosion can be explained as the predicted results were in agreement with the experimental data.

Keywords Haynes 230 alloy · $Mgcl_2$–KCl · Corrosion · Ni–Cr alloy · Tafel

Introduction

As solar energy is emphasized as one of the promising energy sources to satisfy the large requirement in both industry and daily life [1], molten salts have been proposed as a material for solar thermal energy storage to increase the energy conversion efficiency [2]. At low temperature, nitrate salt presents very good thermal properties: good heat capacity, low melting point and light weight property [3–9]. However, this salt can only be served at low temperature [10]. And for the salts serving at high temperature, carbonate salts have a high vapor pressure [11] and fluoride salts are very corrosive to the structural alloys [12]. Chloride salt seems to be a possible candidate for a heat transferring fluid not only because it has comparable thermal properties to other salts, but also it is less corrosive [12]. In addition, the cost of chloride salt is much cheaper especially for the $MgCl_2$–KCl salt [12]. Thus, using $MgCl_2$–KCl salt will be more economical in the solar power application.

Y. Peng · R.G. Reddy (✉)
Department of Metallurgical and Materials Engineering, The University of Alabama, Tuscaloosa, AL 35487, USA
e-mail: rreddy@eng.ua.edu

S. Wang et al. (eds.), *Applications of Process Engineering Principles in Materials Processing, Energy and Environmental Technologies*, The Minerals, Metals & Materials Series, DOI 10.1007/978-3-319-51091-0_37

In fact, $MgCl_2$–KCl served as the first heat transferring fluid in the nuclear industry [13]. However, the widespread applicability of $MgCl_2$–KCl is still limited mainly because of its corrosive property. Ni proved to be stable in molten chloride salt while Cr and Mn are dissolute into the salt in a large amount [14–16]. However, a material designed for containing the chloride molten salt was alloyed with at least 6 atom % of Cr to minimize the corrosion damage in the air at high temperature [17]. Thus, many studies focus on the corrosion rate of different Ni–Cr alloys to rank their corrosion resistance in molten chloride salt. It is reasonable to rank the corrosion rate of alloys in the same condition, but by ignoring the effect of the crucible, the real corrosion rate is not obtained. Containers used in the static corrosion experiments are made by quartz, nickel, or alumina as they are very stable in the molten chloride salt [12, 18]. However, Cr will be dissolved into the fluoride salts and then reacted with the crucible. For the corrosion of Incoloy-800H alloy in LiF–NaF–KF (FLiNaK), Olson [19] found that Cr-rich film was formed on the graphite crucible and that film was identified as a Cr_7C_3 compound. Meanwhile, Qiu [20] found Cr-rich film was not only obtained in the graphite crucible, but also in the nickel crucible. In fact, containers may accelerate the corrosion rate of alloys in the chloride salt [21]. Comparing with stainless steel 316 and Al_2O_3 crucibles, a quartz crucible resulted in the largest weight loss of alloys in $MgCl_2$–KCl molten salt [12]. Ozeryanaya [21] indicated the corrosion of alloys within crucible might be an electrochemical process. However, more information is needed to understand the corrosion process. The driving force and effect on the potential and corrosion rate are still under debate. Thus, the corrosion mechanism between alloys and crucible is necessary to fully understand the corrosion behavior in the molten chloride salt.

As H230 presents an excellent structural property [22], the present experiment was performed to evaluate the corrosion behavior of H230 in $MgCl_2$–KCl molten salt with a Ni crucible at 1273 K. The purpose is to investigate the corrosion rate and corrosion mechanism of H230 in this chloride molten salt. Thus, a galvanic corrosion model was constructed based on the calculation of Gibbs energy of Ni–Cr alloy and exchange current density of Cr in $MgCl_2$–KCl molten salt. Moreover, the result shows the theoretical calculation is comparable to the experimental data.

Experimental Procedure

$MgCl_2$–KCl was chosen in this research not only because it presents excellent thermal properties, but also it is less corrosive to the alloy [12]. $MgCl_2$–KCl salt used in this experiment is of extremely high purity and it was kept in the glove box to isolate water and oxygen in the air. The effect of a trace amount of impurities in $MgCl_2$–KCl is ignored as $MgCl_2$–KCl is more corrosive to the H230 alloy. In addition, galvanic corrosion between crucible and alloy may also be very significant. Table 1 lists the composition of the H230 alloy tested in this research. H230 alloy was selected for this research based on its structural properties and its history in serving at high temperatures [19]. Corrosion experiments were performed in a Ni

Table 1 Composition (mol%) of H230 alloy tested in this research

Haynes 230	Al	Mo	C	Ni	Co	Si	Cr	W	Fe	Mn
mol%	0.9	0.8	0.5	63.8	0.2	0.7	26.6	4.8	1.1	0.6

crucible because the purpose is to investigate the effect of a Ni crucible on the corrosion behavior, even Ni was proved to be very stable in chloride salt [15].

H230 samples were ground with SiC grinding paper and finally polished to the mirror-finished surface with W0.5 Diamond Polishing Lapping Grind Paste to ensure the exact weight change and corrosion rate in the chloride salt. The dimensions and weight of samples were carefully measured after polishing. According to the method used by Tao [23], samples were tied with Ni wire and positioned at the bottom of the crucible. The other end of the Ni wire was spot-welded to the inner surface of Ni-crucible to facilitate the loading of salt powder and ensure all surfaces contacting with $MgCl_2$–KCl salt. The process of loading salts was performed inside a glove box with pure Argon gas to isolate the effect of air. Ni crucible was inserted into the Al_2O_3 tube and the tube was transferred to the furnace after well sealing. The furnace was pumped to ultra-vacuum and purged with Argon gas twice. The temperature was heated to 1273 K and a small positive pressure was carefully maintained inside the chamber to prevent air from going inside the tube. After 500 h of exposure, the furnace temperature was lowered to room temperature. H230 samples were taken out of the crucible by crumbling the salt to the piece. The retrieved H230 samples were cleaned with DI water three times. Moreover, dimensions and weight were measured again to evaluate the corrosion rate. In addition, samples were characterized by scanning electron microscopy (SEM) with energy-dispersive X-ray spectroscopy (EDS).

Results and Discussion

Weight Loss and EDS Analysis

Two H230 alloy coupons were selected for this long-term corrosion test. Moreover, the results of weight change of the H230 alloys after corrosion in molten $MgCl_2$–KCl at 1273 K for 500 h are shown in Table 2 [24].

As shown in Table 2, the average corrosion rate of H230 was 0.939 ± 0.0289 mg/cm^2/day in $MgCl_2$–KCl at 1273 K. According to the EDS result reported by Tao [24], the surface composition of H230 alloy after corrosion was changed as shown in Table 3. As shown in Table 3, Cr content was decreased significantly and as a result the concentrations of other elements were increased.

It was a safe assumption that only Cr was corroded in the molten salt after ignoring the effect of a trace amount of other elements. In addition, $CrCl_3$ shows more negative Gibbs energy (−824.4 kJ/mol) than $CrCl_2$ (−620 kJ/mol), Cr in

Table 2 Weight loss of H230 after 500 h corrosion in $MgCl_2$–KCl at 1273 K [24]

Alloy	L (mm)	W (mm)	T (mm)	Surface (cm^2)	W (g)	WL (g)	Rate ($mg/cm^2/day$)
H230	11.85	9.90	1.30	2.912	1.3155	0.0587	0.96765
H230	11.80	9.90	1.40	2.944	1.4766	0.0558	0.90978

Table 3 EDS analysis of surface of H230 alloy corrosion in $MgCl_2$–KCl for 500 h [24]

Element	Al	Cr	Ni	W
Weight%	1.68	3.34	82.86	12.12
Atomic%	3.88	4.01	88	4.11

chloride salt might be $CrCl_3$ instead of $CrCl_2$. Therefore, the corrosion rate in weight loss was converted to current density, which was $10^{-4.217\pm0.0134}$ A/cm^2.

The Cr alloy in the surface was assumed to be a ternary alloy and atomic% of Al was attributed to Ni and W to calculate the activity of the Cr. Thus, the composition on the surface was assumed as 0.04 Cr-0.90 Ni-0.06 W. The activity of Cr was obtained from thermodynamic modeling, which was 0.0102 in this assumed ternary alloy. In addition, after the long-term corrosion test, the activity of Cr was equal on the surface of H230 alloy, in the molten salt, and on the surface of Ni crucible. The relationship is shown in Eq. (1).

$$a_{Cr}\,on\,surface\,of\,H230 = a_{Cr}\,in\,chloride\,salt = a_{Cr}\,on\,surface\,of\,\,Ni\,crucible \quad (1)$$

Therefore, the activity of Cr on the surface of the Ni crucible was 0.0102. As Qiu [20] indicated, there was only Cr and Ni on the Ni crucible after corrosion. Thus the composition formed on the surface of Ni crucible might be $Ni_{0.943}Cr_{0.057}$ with the known activity of Cr.

Thermodynamic Calculation of Ni–Cr Alloy

The driving force of the depletion of Cr was the Gibbs energy of Cr alloying with Ni. Moreover, the prediction of the Gibbs energy of Ni-Cr solution is shown in Eqs. (2)–(5) [25].

$$G^s_m = x_{Ni}\,G^s_{Ni} + x_{Cr}\,G^s_{Cr} + RT\,ln(x_{Ni}\,lnx_{Ni} + x_{Cr}\,lnx_{Cr}) + \Delta^{ex}G^s_m \quad (2)$$

$$\Delta^{ex}G^s_m = x_{Cr}\,x_{Ni}\,\varepsilon_{fcc} \quad (3)$$

$${}^0\varepsilon_{fcc} = 8030 - 12.8801\,T\,(J) \quad (4)$$

$$^{1}\varepsilon_{fcc} = 33080 - 16.0362\,T\,(J) \tag{5}$$

where x is the mole-fraction of the element (Cr or Ni), R is the universal gas constant, and T is the temperature in K. The quantity ΔG_i^s is the molar Gibbs energies of solution or element "i" with the *s* structure in the nonmagnetic state. Because Ni-Cr alloy is an fcc structure with little amount of Cr at 1273 K, $\Delta^{ex}G_m^s$ is the excess Gibbs energy and ε_{fcc} is the binary interaction parameter for the fcc structure. The interaction parameter calculation is shown in Eq. (6) [26]:

$$\varepsilon_{fcc} = \sum_{i=0} (x_{Ni} - x_{Cr})^{i*i}\varepsilon_{fcc} \tag{6}$$

x_{Ni} and x_{Cr} is the composition in the sublattice. However, this 2 sublattice model can be used directly with x_{Ni} and x_{Cr} in the Ni-Cr alloy [26]. As the polynomial of $^{0}\varepsilon$ and $^{1}\varepsilon$ are defined, ε_{fcc} can be written as:

$$\varepsilon_{fcc} = 8030 - 12.8801\,T + (x_{Ni\,in\,fcc} - x_{Cr\,in\,fcc}) * (33080 - 16.0362\,T) \tag{7}$$

Based on the known Gibbs energy of Ni (−64538 J/mol) and Cr (−53124 J/mol) at 1273 K, the Gibbs energy of $Ni_{0.943}Cr_{0.057}$ was calculated, which was −66048.34 J/mol. The Gibbs energy of the alloying reaction for 1 mol Cr is shown in Eq. (8).

$$\mathrm{Cr} + \tfrac{0.943}{0.057}\mathrm{Ni} == \tfrac{1}{0.057}\, Ni_{0.943}\, Cr_{0.057} \quad \Delta G^{\circ} = -37.91\,\mathrm{kJ/mol} \tag{8}$$

The negative Gibbs energy of the reaction indicates Cr and Ni were spontaneously formed as $Ni_{0.943}Cr_{0.057}$ alloy at 1273 K.

Galvanic Corrosion with $CrCl_3$ Diffusion

As H230 alloy and Ni crucible were connected by Ni wire in the molten $MgCl_2$–KCl electrolyte in the test, the galvanic corrosion model was constructed to predict the corrosion mechanism. The theoretical potential of the selective dissolution of the element (Cr) from the alloy was calculated from Nernst equation given by Eq. (9) as:

$$\mathrm{E} = E^{o} - \left(\frac{RT}{nF}\right) ln\, J \tag{9}$$

where E^o is the standard potential calculated from the standard Gibbs energy data ($\Delta G^o = -nFE^o$), *n* is the number of transferred electrons, *F* is Faraday constant, *R* is the universal gas constant, *T* is the temperature in K, *J* is the ratio of activity of oxidized state to the activity of reduced state. *J* is calculated considering $a(Cl^{-}) = 1$ and $a(M_{alloy,s})$ is the activity of Cr in the alloy before corrosion test. Assuming

H230 is a ternary alloy, elements with a little amount are accounted to tungsten (0.266 Cr-0.638 Ni-W). Thus, the activity of Cr was 0.244 in the initial H230 alloy from the thermodynamic calculation. As $CrCl_3$ may be soluble in the chloride salt [27], the activity of $CrCl_3$ was unknown in this calculation. However, the activity of $CrCl_3$ was assumed to be 1 first in the calculation. Even the activity of $CrCl_3$ will affect the potential of cathode and anode, the corrosion potential between these two electrodes remained as the same. In addition, there is no influence on the corrosion rate, either. The corrosion potential of H230 from the potentiodynamic polarization curve was 2.24 V versus LiCl/Li in $MgCl_2$–KCl at 1123 K [18]. From theoretical calculation, equilibrium potential of $CrCl_2$/Cr was 1.993 V versus LiCl/Li. For $CrCl_3$/Cr, the potential was 2.238 V versus LiCl/Li. Even though there were many species contributing to the corrosion potential during potentiodynamic polarization experiment, the major corroded element was Cr. Thus, assuming it was $CrCl_3$ diffused from H230 alloy to Ni crucible seems more reasonable. Therefore, the galvanic equations at 1273 K are shown in Eqs. (10)–(12):

$$\text{Anode: Cr(bcc in H230)} + 3Cl^- = \text{Cr}Cl_3 + 3e^- \quad \text{E} = 8.9896\text{V} \tag{10}$$

$$\text{Cathode: Cr}Cl_3 + \frac{0.0943}{0.057}Ni(fcc) + 3e^- = \frac{1}{0.057}Ni_{0.943}Cr_{0.057}(fcc) + 3Cl^- \quad \text{E} = -8.9102\text{V} \tag{11}$$

$$\text{Total: Cr(bcc in H230)} + \frac{0.943}{0.057}\text{Ni(fcc)} = \frac{1}{0.057}Ni_{0.943}Cr_{0.057}(fcc) \quad \text{E} = 0.0794\text{V} \tag{12}$$

Calculation of Exchange Current Density of $CrCl_3$/Cr in $MgCl_2$–KCl

Exchange current density (i_0) of $CrCl_2$/Cr in LiCl–KCl salt at 773 K with 3.366×10^{-2} mol/L of $CrCl_2$ is 4.75×10^{-4} A/cm^2 [28]. Based on the method of minimization of Gibbs energy, there was about 10^{-6} mol/L Cr dissolved to the molten chloride salt if only the intrinsic corrosion of H230 is considered. Then, i_0 of 10^{-6} mol/L $CrCl_2$ was calculated based on the relationship between concentration and i_0 [28], which was $10^{-6.04}$ A/cm^2. In addition, i_0 of $CrCl_3$/Cr and $CrCl_2$/$CrCl_3$ in $MgCl_2$–KCl at 1123 K were obtained from Tafel experiment, which were 1.6×10^{-5} A/cm^2 and 1×10^{-4} A/cm^2 respectively [18]. Reduction of $CrCl_3$ to Cr can be separated as two steps [28], which are shown in Eqs. (13 and 14).

$$CrCl_3 + 3e^- = Cr \quad \text{i}_{01} = 1.6 \times 10^{-5}\text{A/cm}^2 \tag{13}$$

$$CrCl_3 + e^- = CrCl_2 \quad \text{i}_{02} = 1 \times 10^{-4}\text{A/cm}^2 \tag{14}$$

$$CrCl_2 + 2e^- = Cr + 2Cl^- \quad \text{i}_{03} \tag{15}$$

The transferring time of the total reaction was equal to both of the two separated steps at equilibrium potential. Thus the relationship between these three exchange current densities is shown in Eq. 16.

$$\frac{1}{i_{01}} = \frac{1}{i_{02}} + \frac{1}{i_{03}} \quad (16)$$

Based on the calculation of Eq. (16), i_0 of $CrCl_2/Cr$ at 1123 K was obtained, which was 1.9×10^{-5} A/cm^2. As the salts used were similar in these two experiment performed at 773 and 1123 K [18 and 28], i_0 of $CrCl_2/Cr$ was assumed to be the same in these two systems. Therefore, the relationship between temperature and exchange current density was obtained by using these two data points. And the relationship is shown in Eq. (17), known as the Arrhenius equation.

$$\log(i_0) = -3271.42 * \frac{1}{T} - 1.8079 \quad (17)$$

Based on the equation, i_0 of $CrCl_2/Cr$ at 1273 K is about $10^{-4.38}$ A/cm^2 with 10^{-6} mol/L $CrCl_2$. Assuming the difference between the log (i_0) of $CrCl_3/Cr$ and $CrCl_2/Cr$ at 1123 K is the same as at 1273 K, i_0 of $CrCl_3/Cr$ was obtained at 1273 K, which was $10^{-4.45}$ A/cm^2.

Calculation of Corrosion Potential and Corrosion Rate

Corrosion potential and corrosion rate are investigated with the Tafel curve model. Tafel slope is calculated by the Eq. (19), where R is the universal gas constant, T is the temperature in K, n is the transferred electron number, F is the Faraday constant, and the parameter alpha is assumed to be 0.5. Thus, the calculated Tafel slope is ±0.168.

$$\beta_{cathode} = -2.303\frac{RT}{anF} \quad and \quad \beta_{anode} = 2.303\frac{RT}{(1-a)nF} \quad (19)$$

Therefore, a Tafel curve was constructed based on the known i_0 and Tafel slope, which was shown in Fig. 1.

As shown in Fig. 1, the corrosion potential (E_{corr}) was about 2.323 V versus LiCl/Li, and corrosion rate (i_{corr}) was about $10^{-4.22}$ A/cm^2. Comparing with the experimental data ($10^{-4.217\pm0.0134}$ A/cm^2), the predicted data seemed reasonable. The comparison of corrosion rate of H230 at different temperature with different crucibles is listed in Table 4.

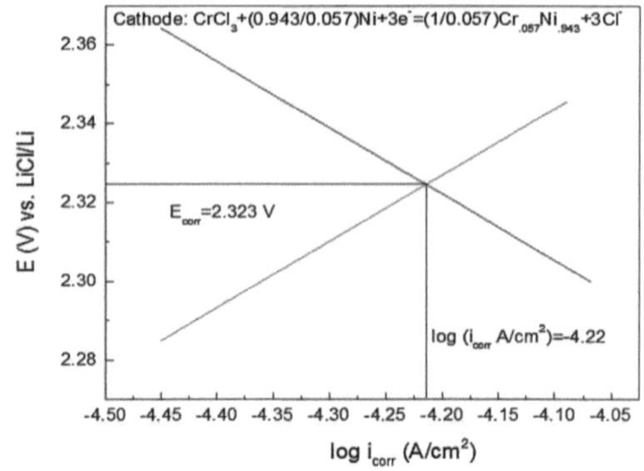

Fig. 1 Tafel curve of H230 corrosion in $MgCl_2$–KCl at 1273 K

Table 4 Comparison of corrosion rate of H230 in different temperature with Ni and quartz crucible

Alloy	i_{corr} at 1273 K in Ni crucible (A/cm^2) present work	i_{corr} at 1123 K in Ni (A/cm^2). Source [29]	i_{corr} at 1123 K in quartz (A/cm^2). Source [12]
H230	$10^{-4.217\pm0.0134}$	$10^{-4.26\pm0.21}$	$10^{-3.99\pm0.15}$

From Table 4, corrosion rate of H230 in present research is reasonable as it is larger than which at 1123 K. Increased temperature results in the increased corrosion rate. Sridharan [12] indicated that corrosion rate of alloy in quartz crucible is much larger than which in Al_2O_3 and SS-316 crucibles. According to the present study, the corrosion rate of H230 alloy in quartz will also be larger than which in Ni crucible. This may be due to the different corrosion mechanism and different driving force of depletion of Cr in different crucibles. Therefore, this tafel modeling is only available for the alloy corrosion in the Ni crucible. In our modeling, only galvanic corrosion is considered as the corrosion mechanism. However, in the experiment, the weight loss of alloy is contributed by both galvanic corrosion and intrinsic corrosion, which could provide a reason why there is a little difference between modeling and experiment.

Conclusions

In conclusion, corrosion of H230 in $MgCl_2$–KCl at 1273 K was studied in this paper. The experimental corrosion rate was 0.939 ± 0.0289 mg/cm^2/day, and converted current density was $10^{-4.217\pm0.0134}$ A/cm^2. The Gibbs energy of formation of 1/0.057 mol $Ni_{0.943}Cr_{0.057}$ was −37.91 kJ at 1273 K, which indicated the reaction of Cr alloying with Ni was spontaneous. In addition, deposition of Cr on Ni crucible can be explained as galvanic corrosion according to the negative Gibbs energy of formation. Moreover, exchange current density of $CrCl_3$/Cr in $MgCl_2$–KCl at 1273 K was predicted based on the Arrhenius equation with two reasonable assumptions. With the calculated current density of $CrCl_3/Cr$, the Tafel model was constructed at 1273 K. And the predicted corrosion rate was $10^{-4.22}$ A/cm^2, which was comparable to the experimental data. However, the tafel prediction in this study is only available for the corrosion in $MgCl_2$–KCl molten salt with Ni crucible. The other crucible, like quartz may change the driving force of depletion of Cr from alloys. Thus, the corrosion rate with quartz crucible may be faster than which with Ni crucible.

References

1. P. Sabharwall, M. Ebner, M. Sohal, P. Sharpe, M. Anderson, K. Sridharan, J. Ambrosek, L. Olson, P. Brooks, Molten salts for high temperature reactors: University of Wisconsin molten salt corrosion and flow loop experiments–Issues identified and path forward. Idaho National Laboratory Report INL/EXT-10-18090, US, Idaho Falls, ID (2010)
2. A. Steinfeld, R. Palumbo, Solar thermochemical process technology. Encycl. Phys. Sci. Technol. **15**(1), 237–256 (2001)
3. D. Mantha, T. Wang, R.G. Reddy, Thermodynamic modeling of eutectic point in the $LiNO_3$–$NaNO_3$–KNO_3–$NaNO_2$ quaternary system. Solar Energy Mater. Solar Cells (SOLMAT) **118**, 18–21 (2013)
4. T. Wang, D. Mantha, R.G. Reddy, Novel low melting point quaternary eutectic system for solar thermal energy storage. J. Appl. Energy **102**, 1422–1429 (2013)
5. T. Wang, D. Mantha, R.G. Reddy, Thermodynamic properties of $LiNO_3$–$NaNO_3$–KNO_3–$2KNO_3$–$Mg(NO_3)_2$ system. Thrmochemica Acta **551**, 92–98 (2013)
6. R.G. Reddy, T. Wang, D. Mantha, Thermodynamic properties of potassium nitrate—magnesium nitrate compound [$2KNO_3.Mg(NO_3)_2$]. Thermochemica Acta **531**, 6–11 (2012)
7. T. Wang, D. Mantha, R.G. Reddy, Thermal stability of the eutectic composition in $LiNO_3$–$NaNO_3$–KNO_3 ternary system used for thermal energy storage. Solar Energy Mater. Solar Cells (SOLMAT) **100**, 162–168 (2012)
8. D. Mantha, T. Wang, R.G. Reddy, Thermodynamic modeling of eutectic point in the $LiNO_3$–$NaNO_3$–KNO_3 ternary system. J. Phase Diagrams Diffus. (JPED) **33**, 110–114 (2012)
9. T. Wang, S. Viswanathan, D. Mantha, R.G. Reddy, Thermal conductivity of the ternary eutectic $LiNO_3$–$NaNO_3$–KNO_3 salt mixture in the solid state using a simple inverse method. Solar Energy Mater. Solar Cells (SOLMAT) **102**, 201–207 (2012)
10. H.E. Reilly, G.J. Kolb, An evaluation of molten-salt power towers including results of the solar two project. Sandia National Labs Report SAND2001-3674, Sandia National Labs, US, Albuquerque, NM; US, Livermore, CA (2001)
11. S. Passerini, Optical and chemical properties of molten salt mixtures for use in high temperature power systems. Ph.D. Thesis, Massachusetts Institute of Technology, Massachusetts, US, 2010
12. K. Sridharan, M. Anderson, T. Allen, M. Corradini, Liquid salts as media for process heat transfer from VHTR's: forced convective channel flow thermal hydraulics, materials, and coating. DOE Report No. DOE/ID14826, University of Wisconsin, Madison, US (2012)
13. D.F. Williams, Assessment of candidate molten salt coolants for the NGNP/NHI heat-transfer loop. Oak Ridge National Laboratory Report ORNL/TM-2006/69, Oak Ridge, Tennessee, US (2006)
14. M.S. Sohal, M.A. Ebner, P. Sabharwall, P. Sharpe, Engineering database of liquid salt thermophysical and thermochemical properties. Idaho National Laboratory Report INL/EXT-10-18297, Idaho Falls. ID, US (2010)

15. H. Susskind, F.B. Hill, L. Green, S. Kalish, L.E. Kukacka, W.E. McNulty, Jr. E. Wirsing, Corrosion studies for a fused salt-liquid metal extraction process for the liquid metal fuel reactor. Brookhaven National Lab Report BNL-585, Upton, NY, US (1960)
16. S.N. Flengas, T.R. Ingraham, Electromotive force series of metals in fused salts and activities of metal chlorides in 1: 1 molar KCl-NaCl solutions. J. Electrochem. Soc. **106**(8), 714–721 (1959)
17. M. Anderson, K. Sridhara, T. Allen, P. Peterson, Liquid salt heat exchanger technology for vhtr based applications. Report NEUP-Project-09-777, University of Wisconsin, Madison, WI, US; University of California at Berkeley, CA, US; Battelle Energy Alliance, LLC, US (2012)
18. L.C. Olson, D.B. Garcia, R. Fuentes, R.M. Marinez, J. Gray, H.S. Cho, Z.J. Van, R. Reddy, Fundamental corrosion studies in high-temperature molten salt systems for next generation concentrated solar power systems. Savannah River National Laboratory Report SRNL-STI-2013-00507, Savannah, GA, US (2013)
19. L.C. Olson, J.W. Ambrosek, K. Sridharan, M.H. Anderson, T.R. Allen, Materials corrosion in molten LiF–NaF–KF salt. J. Fluor. Chem. **130**(1), 67–73 (2009)
20. J. Qiu, Y. Zou, G. Yu, H. Liu, Y. Jia, Z. Li, P. Huai, X. Zhou, H. Xu, Compatibility of container materials with Cr in molten FLiNaK salt. J. Fluor. Chem. **168**, 69–74 (2014)
21. I.N. Ozeryanaya, Corrosion of metals by molten salts in heat-treatment processes. Met. Sci. Heat Treat. **27**(3), 184–188 (1985)
22. K. Sadananda, P. Shahinian, Creep crack growth behavior of several structural alloys. Metall. Trans. A **14**(7), 1467–1480 (1983)
23. T. Wang, D. Mantha, R.G. Reddy, Novel high thermal stability LiF–Na_2CO_3–K_2CO_3 eutectic ternary system for thermal energy storage applications. Sol. Energy Mater. Sol. Cells **140**, 366–375 (2015)
24. T. Wang, R.G. Reddy, Corrosion of nickel based alloys in ultra high temperature heat transfer fluid. High Temperature Materials and Processes (2016)
25. W. Huang, Y.A. Chang, Thermodynamic properties of the Ni–Al–Cr system. Intermetallics **7**(8), 863–874 (1999)
26. B. Sundman, Modification of the two-sublattice model for liquids. Calphad. **15**(2), 109–119 (1991)
27. A. Cotarta, J. Bouteillon, J.C. Poignet, Electrochemistry of molten LiCl–KCl–$CrCl_3$ and LiCl–KCl–$CrCl_2$ mixtures. J. Appl. Electrochem. **27**(6, 199), 651–658
28. D. Inman, J.C. Legey, R. Spencer, The electrochemistry of chromium in molten $LiCl^+$ KCl. J. Electroanal. Chem. Interfacial Electrochem. **61**(3), 289–301 (1975)
29. L.C. Olson, D.B. Garcia, R. Fuentes, R.M. Marinez, J. Gray, H.S. Cho, Z.J. Van, R. Reddy, Fundamental corrosion studies in high-temperature molten salt systems for next generation concentrated solar power systems. Savannah River National Laboratory Final Report; Savannah, GA, US (2016)

Conceptualization of Doped Black P Thin Films for Potential Use in Photovoltaics with Validation from First Principle Calculations

Sayan Sarkar, Prashant K. Sarswat, Weizhi Zeng and Michael L. Free

Abstract Black Phosphorus was recently discovered as a 2D material and it exemplified huge prospects for application in electronics, optoelectronics, photonics and catalysis. It is a direct band semiconductor with tunable band gap varying from 0.35 eV (bulk) to 2 eV (single layer). But the layer dependent band gap of black phosphorus restricts it to make use of the full spectra of sunlight for absorption. First principle calculations were performed using 'Quantum Wise Atomistic Toolkit' package with Perdew-Burke-Ernzerhof (PBE) exchange-correlation for black phosphorus doped with non-metallic dopants of B, S and Se and metallic dopants of Ti, Mg and In. It has been found that doping with non-metallic impurities result in a small diminution of the band edges (0.96–0.57 eV), whereas doping with metallic impurities decrease the band edge drastically (0.85–0.025 eV). Moreover, it was also observed that dopants with an even number of valence electrons incorporated a metallic nature of black phosphorus, whereas dopants with odd number of valence electrons further attributed to its semiconducting features. Thus in a holistic discernment, black phosphorus with the incorporation of some dopants in appropriate amounts can absorb more visible range of the optical spectra, making it ideal for use in photovoltaics.

Keywords Black-Phosphorus · DFT · Band structure · Dielectric function

Introduction

In recent years, phosphorene or few-layer black phosphorus has established itself as a monumental member of the 2D materials family. Being a combination of highly tunable band gap and substantial carrier mobility [1], it has surpassed graphene and other TMDCs (Transitional Metal Dichalcogenides) for potential application in

S. Sarkar (✉) · P.K. Sarswat · W. Zeng · M.L. Free
Department of Metallurgical Engineering, University of Utah, Salt Lake City, UT 84112, USA
e-mail: sayan.ju92@gmail.com

S. Wang et al. (eds.), *Applications of Process Engineering Principles in Materials Processing, Energy and Environmental Technologies*,
The Minerals, Metals & Materials Series, DOI 10.1007/978-3-319-51091-0_38

optoectronics, electronics. Phosphorene exhibits a strong covalent bonding within its 2d plane but its interlayer interactions are meagre. Thus it was first successfully fabricated through mechanical exfoliation by exploitation of its weak Van der Waals interactions [2]. However, unlike the other members of the 2D materials family, the fascinating properties of black phosphorus lie in its anisotropy. The dihedral angle and the bond length are distinctively different from those in the zigzag and in the armchair direction. In addition, the local structural configuration is bilayer in the zigzag direction while the structure is striated in the armchair direction. This structural anisotropy results in the genesis of unique properties in phosphorene like optical anisotropy, Peierls-distortion [3], anisotropic thermoelectric properties and mechanical properties including Poisson's ratio and Young's modulus [4, 5].

However, in our present investigation we have endeavored to focus on a relatively unexplored attribute of phosphorene which is its anisotropic optical spectrum. In fact, Phosphorene follows two distinct dispersion relations along the armchair (Fermi-Dirac distribution) and zigzag direction (Schrodinger equation). Here we systematically propose from first principle that by the incorporation of different metallic and non-metallic impurities the band edges of pristine black phosphorus can be redshifted so that it can make use of the entire optical spectra of the sun. Further, by accounting for the anisotropy of the optical spectrum we can obtain different absorption edges along the zigzag and armchair direction. The implementation of this principle will serve as a new platform for the application of black phosphorus in photovoltaics.

Experimentation

The simulations were carried out based on density functional theory (DFT) using 'QuantumWise Atomistix Toolkit' package. For obtaining the optical spectrum of monolayer of black phosphorus along with the monolayers of phosphorene doped with impurities, the generalized gradient approximation of Perdew–Burke–Ernzerhof (PBE) [6] was used for obtaining the optical spectrum of monolayer of black phosphorus along with the monolayers of phosphorene doped with impurities. Moreover, the interactions of valence electrons with ionic cores were analyzed by projector augmented wave (PAW) pseudopotentials [7]. The density cut-off was maintained at 180 Rydberg. $8 \times 8 \times 1$ k-point sampling was used for the Brillouin zone integration in the calculation of the optical spectrum and Van der Waals correction factor of Grimme-DFT 2 was used, which subsequently increases the convergence of our calculation.

Results and Discussion

The DFT optimized band structures are shown in the Fig. 1. Figure 1a shows the intrinsic anisotropy in phosphorene. Figures 1b–d show the monolayers of phosphorene doped with 10% S, B and Se respectively while Figs. 1e, f doped with 5% Ga and In respectively. We are now interested in the observation of the optical spectrum and the influence of doping on the absorption states. Only the imaginary part of the dielectric function is plotted since the real part can be derived from Kramers−Kronig relation. The first peak position in the pristine phosphorene monolayer layer was observed at ~2.31 eV for the electric field polarization along the zigzag direction, compared to only ~0.62 eV for the armchair electric field polarization (shown in Fig. 2a), which conclusively proves that phosphorene is more optically transparent along the zigzag direction for the electromagnetic spectrum as compared to the armchair direction. We observe an interesting phenomenon that due to the incorporation of doping of metallic and non-metallic impurities there is a reduction in the band edge and the band gap is red shifted. Thus the doped phosphorene can now utilize more portions of the optical spectra for energy conversion. The red-shift was however greatest due to the doping of B among non-metals and due to the doping of In among the metals. In case of all the

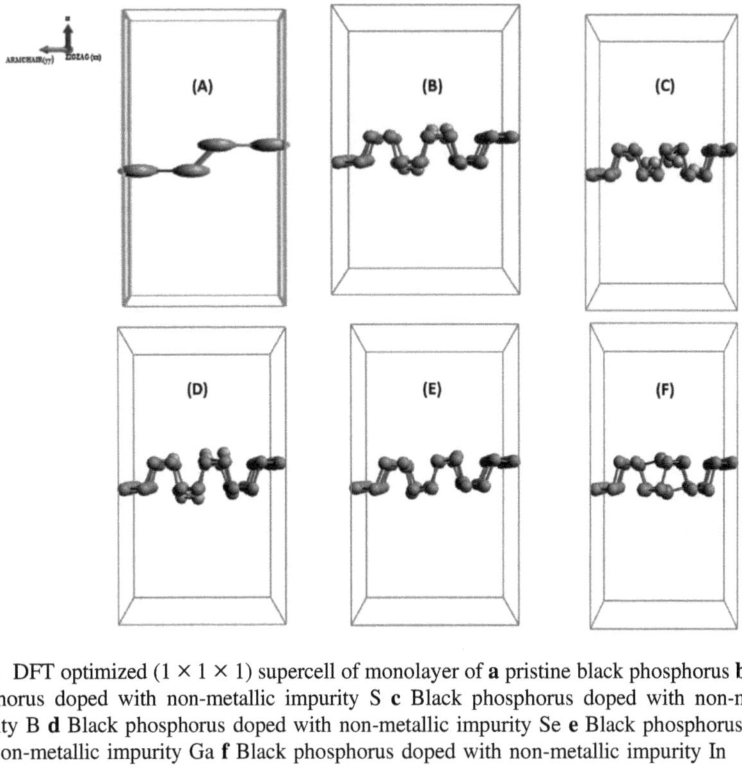

Fig. 1 DFT optimized (1 × 1 × 1) supercell of monolayer of **a** pristine black phosphorus **b** Black phosphorus doped with non-metallic impurity S **c** Black phosphorus doped with non-metallic impurity B **d** Black phosphorus doped with non-metallic impurity Se **e** Black phosphorus doped with non-metallic impurity Ga **f** Black phosphorus doped with non-metallic impurity In

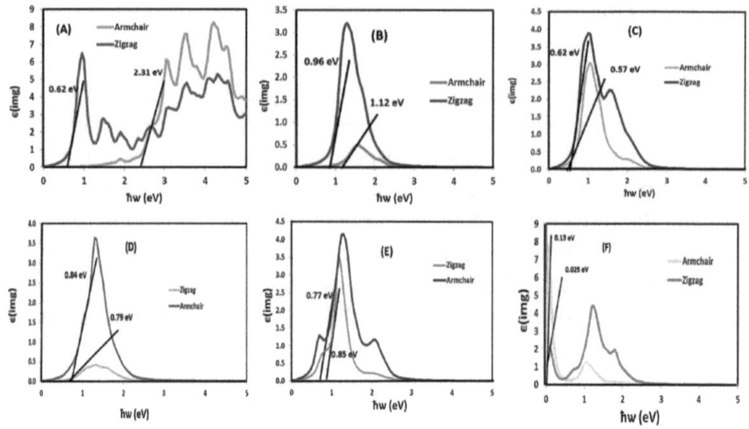

Fig. 2 Variation of frequency-dependent dielectric function in **a** phosphorene monolayer along with the monolayers of phosphorene doped with impurities of **b** S **c** B **d** Se **e** Ga and **f** In for the incident light distributed along the zigzag and armchair direction. The tangent lines represent first absorption peak manifesting the positions of band edges for the corresponding cases. The extrapolated values are shown in the bordered boxes indicate the values of the band edges which are listed in the table below

Table 1 Variation of the band absorption edge for monolayer black phosphorus and black phosphorus doped with impurities along zigzag and armchair direction

Doping elements		Band edge in zigzag direction (eV)	Band edge in armchair direction (eV)
Pure phosphorus		2.31	0.62
Non-metals	Sulfur	0.96	1.12
	Boron	0.62	0.57
	Selenium	0.79	0.84
Metal	Gallium	0.85	0.77
	Indium	0.025	0.13

monolayers of black phosphorus doped with impurities, the red shift in the first peak position for the armchair direction was even more predominant than the zigzag polarization. This fact conclusively proves that optical anisotropy of black P Is even conserved after doping.

Owing the variation of the band edges due to doping over a considerably wide range (shown in Table 1), we can conclude that black phosphorus when doped with various metallic and non-metallic elements can absorb light over a wide range. Since the absorption coefficients are directly related to the imaginary part of the dielectric function, they also follow the same trend. In summary we can infer that the synergistic effects of doping of various metallic and non-metallic impurities in black phosphorus results in better tenability of the band gap and the absorption of light in the visible range, which should act as motivation for related research in future.

Conclusions

In summary, it can be concluded from first principle calculations that incorporation of various metallic and non-metallic dopants in black phosphorus facilitates better tenability of the band gap. There is a wide variation of band absorption edges due to doping and it should allow better absorption of light in the visible range. This conceptualization of doping of various metallic and non-metallic dopants in black phosphorus may lay the platform for its optimum use in photovoltaics and optoelectronics.

References

1. L. Li, Y. Yu, G.J. Ye, Q. Ge, X. Ou, H. Wu, D. Feng, X.H. Chen, Y. Zhang, Black phosphorus field-effect transistors. Nat. Nanotechnol. **9**, 372–377 (2014)
2. H. Liu, A.T. Neal, Z. Zhu, Z. Luo, X. Xu, D. Tománek, P.D. Ye, Phosphorene: an unexplored 2D semiconductor with a high hole mobility. ACS Nano **8**, 4033–4041 (2014)
3. A. Maity, A. Singh, P. Sen, Peierls transition and edge reconstruction in phosphorene nanoribbons. arXiv preprint arXiv:14042469 (2014)
4. L. Wang, A. Kutana, X. Zou, B.I. Yakobson, Electro-mechanical anisotropy of phosphorene. Nanoscale **7**, 9051–9746 (2015)
5. J.W. Jiang, H.S. Park, Mechanical properties of single-layer black phosphorus. J. Phys. D Appl. Phys. **47**, 38 (2014)
6. J.P. Perdew, K. Burke, M. Ernzerhof, Generalized gradient approximation made simple. Phys. Rev. Lett. **77**, 3865 (1996)
7. G. Kresse, D. Joubert, From ultrasoft pseudopotentials to the projector augmented-wave method. Phys. Rev. B **59**, 1758 (1999)

Energy Efficiency and Sustainability in Steel Production

L.E.K. Holappa

Abstract Iron and steel making plays a significant role in global energy consumption and carbon dioxide emissions. The target of limiting the global warming by 2050 is extremely challenging for this energy-intensive branch of industry. In order to be responsible for its own share in cutting CO_2 emissions, great advancements must be done. A brief history, present situation and different scenarios are discussed. Possibilities to decrease CO_2 emissions in current processes via improved energy efficiency, alternative process routes, energy sources and recycling are examined. On-going and planned efforts and programs of steel producers and institutions, as well as trends of energy generation in the long run are reviewed and evaluated.

Keywords Steel industry · Energy consumption · Energy saving · Sustainability · Carbon dioxide emissions

Introduction

Sustainability is a common worldwide aim today involving all fields of human activity. Applied to industry it means economically and socially sound course of actions, which are environmentally protective and sustainable in the long term. From the technological viewpoint, energy consumption has a central role. The global energy usage was estimated as 13.5 Btoe (billion tons oil equivalent) in 2013 [1]. Respectively, the anthropogenic CO_2 emissions were 36 Bt in 2014 of which the steel production was responsible for about 6.6% [2, 3]. These figures reveal the central role of energy in metallurgical processes.

L.E.K. Holappa (✉)
Department of Materials Science and Engineering,
Aalto University School of Chemical Technology, Vuorimiehentie 2,
PO Box 16200, 00076 Espoo, Aalto, Finland
e-mail: lauri.holappa@aalto.fi

S. Wang et al. (eds.), *Applications of Process Engineering Principles in Materials Processing, Energy and Environmental Technologies*,
The Minerals, Metals & Materials Series, DOI 10.1007/978-3-319-51091-0_39

The overall progress of steel production during the last 150 years is shown in Fig. 1 [3]. In the early 19th century, the world annual steel production was only a few million tons. After the breakthrough of new technologies, converter and open hearth processes, it exceeded 30 Mt in 1900. In 1927, the steel production was 100 Mt and 200 Mt in 1951. Then a "new industrial revolution" took place with innovative novel processes, and extensive investments in steel industry were performed in Japan, Soviet Union, United States and South Korea in the vanguard and the steel production attained the level of 700 Mt/year in the 1970s (the record 749 Mt in 1979). Then the growth stagnated due to economic crises and political changes until the turn of the millennium, when it attained 850 Mt in 2000. That was the first signal of the last "boom" with China as the major motor. The current production record 1670 Mt was attained in the year 2014 [3].

China's production has now reached a "saturation level", whereas other developing countries (India, Brazil and Russia in the forefront) have increased steel production remarkably. In the near future, the growth in consumption will happen in developing countries. Still ten years earlier many scenarios forecast continuous growth up to 3000 Mt/year in 2050. Today, after the recession period and stabilization in China, the scenarios are more conservative for example, the estimate by World Steel Association is 2500 Mt in 2050 [3]. In this contribution that figure is used.

At present, over 73% of crude steel is made via converter processes based on blast furnace hot metal. About 26% is produced in electric furnaces utilizing recycled steel scrap as the iron source, with small share of direct reduced iron (DRI + HBI in Fig. 1). The open-hearth process, which was important more than half a century has almost disappeared to 0.5%.

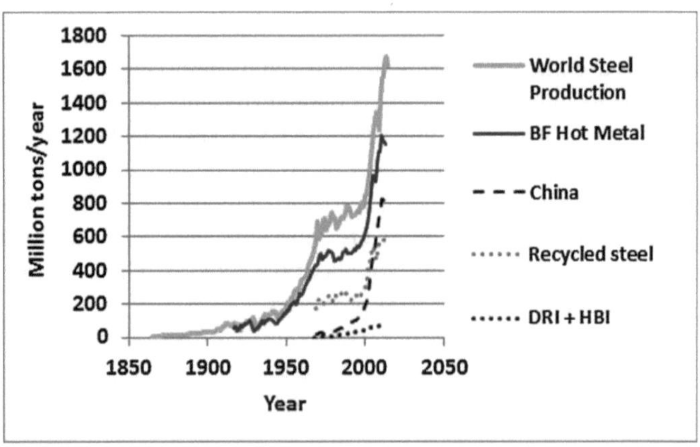

Fig. 1 World production of steel, blast furnace hot metal and direct reduced iron from 1860 to 2015. Even the steel production in China and estimated fraction of recycled steel are shown [3, 4]

Carbon Dioxide Emissions

Anthropogenic greenhouse gas (GHG) emissions have drastically increased especially during the past half-century due to the rapid growth of population, expanded industrialisation and rise of the standard of living. The observed growth can be seen in Fig. 2. The scale in the figure is in Gt C/year (gigaton carbon). The conversion to CO_2 takes place by multiplying with 3.667, thus the current emission per year is about 35 Gt CO_2. Carbon dioxide is the most important GHG and its content in the atmosphere has increased from the level 300 ppm/1950 to current 400 ppm. The relation of GHG to global climate warming is indisputable. The red line in Fig. 2 represents the growth of CO_2 emissions until the year 2050, if no radical changes in the climate policy were done [5, 6]. This would also mean rapid climate warming which could carry catastrophic consequences. The United Nations' Intergovernmental Panel on Climate Change (IPCC) has stated that CO_2 concentration must be stabilised at 450 ppm to have a fair chance of avoiding global warming above 2 °C, which was set as a limit at the COP 21/CMP 11 Conference in Paris Dec. 2015 [7].

To achieve the target of "stop the climate change" will require a 40–70% emissions cut by 2050, compared to 2010 level, and zero emissions by the end of the century. The green line in Fig. 2 presents a 50% reduction by 2050.

Transfer of the overall global demands for elimination of CO_2 emissions sets great challenges to the steel industry. The share 6.6% of the all emissions corresponds to 2.4 Gt CO_2/year. These are, however, only the direct emissions from iron and steel making. If also indirect emissions incl. energy generation e.g. electricity is included the number is around 3 Gt/year [7]. The specific emission is around 1.8 t CO_2/t steel, respectively. If we choose a scenario with 2.5 Gt steel and a target for CO_2 emissions 1.25 Gt CO_2/year, we get a specific CO_2 emission of 0.5 t CO_2/t steel. What should we do in order to attain this level until 2050? The author

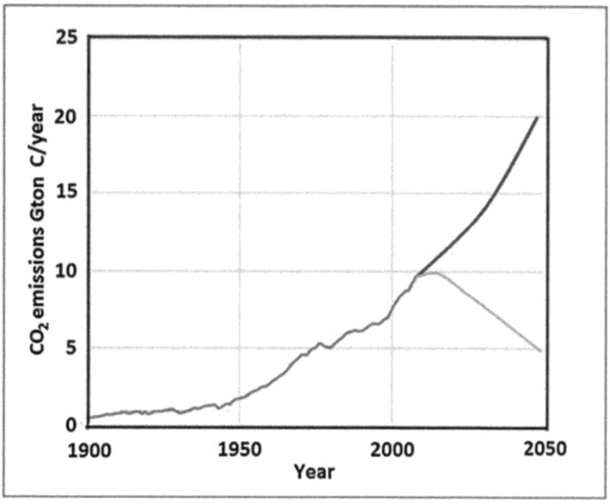

Fig. 2 Total CO_2 emissions from fossil fuel, cement production and gas flaring from 1900 to 2012 and high "business as usual" *red* Greenhouse Gas Emission line (RCP8.5) as well as *low green* GHG line (RCP3-PD). Redrawn based on the data from references [5, 6]

considered the problem in 2011 by analyzing the steel production practices in different countries, by comparing with BAT technologies (Best Available Practice), by estimating emissions from different energy sources including electricity as well as the role of energy saving, low-carbon and carbon-free new innovative technologies [8, 9]. This contribution is based on these previous studies but is an up-dated, generalized version. Five key factors are discussed, which enable to solve the problem of drastic reduction of CO_2 from steel production.

The *first key factor* is improved energy efficiency. A comprehensive study by IEA/OECD analyzed steel industries in different countries and published improvement potential for each country, respectively [10]. The study was based on data of 2006, and showed the worldwide average of 4.1 GJ/t steel corresponding to 20% energy saving. The saving potential varied from 1.4 to 8.7 GJ/t in different countries the greatest potentials being in China, Ukraine, Russia, India and Brazil. In spite of great advancements done in ten years, the World Steel Association has still the target of energy saving with 15–20% [3].

The *second key factor* is recycling. Steel recycling was extremely valued in the Iron Age before industrialization. Even reuse by remanufacturing other new products from "scrap" was common. When steel became mass product, its price fell, reuse almost disappeared and interest to return scrap to steel producers weakened. However, collection of scrap and delivery to steel plants has been duely organized in industrialized countries, and recycling rate is moderate. Most of purchased scrap is used in electric arc furnaces, and smaller share in converters in which it is used as a coolant, typically 15–25% of the iron charge. Until the turn of the century the share of EAF increased being about 34% in 2001 but then the rapid growth in China, which based on BF + BOF route, changed the ratio and now the EAF's share is only 26% whereas BOF has almost 74%, respectively [3]. The EAF share varies in different countries (100% in some small countries, 60% in United States, 40% in EU, 6% in China). The recent amounts of scrap in steel making are shown in Fig. 3. Purchased scrap means external "obsolete" scrap. Another group of scrap is internal "own" scrap which can be counted as the Remainder of Scrap use minus Purchased scrap.

Recently, the principle of "Circular economy" has become popular. Intensified use of scrap is a self-evident goal. All scenarios assume significant growth in availability and use of steel scrap. A simple explanation is that the abrupt growth of steel production in the early 2000 raised also the use of steel to a new level. In different applications, the steel components have different lifetime. Anyway, the rise in production should reflect in scrap availability few decades later. Consequently amount of scrap will strongly increase after the 2020s [11, 12]. The scrap ratio is estimated at 50% level in 2050. This means a substantial growth in EAF steel making whereas BOF production would stay approximately on its present level on the global scale [3, 13]. However, regionally e.g. in China availability of scrap will increase and partial transfer from BF + BOF route to EAF will happen. A general increasing demand will incite recycling and raise collection rates.

Comparison of energy efficiency of different processes (e.g. ore-based and scrap-based steel production) can be done when the specific energy consumption

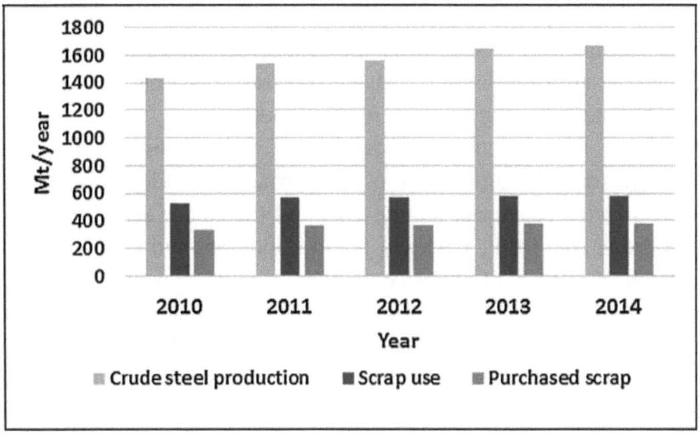

Fig. 3 Scrap use for steel production and amount of purchased scrap as well as total world steel production 2010–2014 [11]

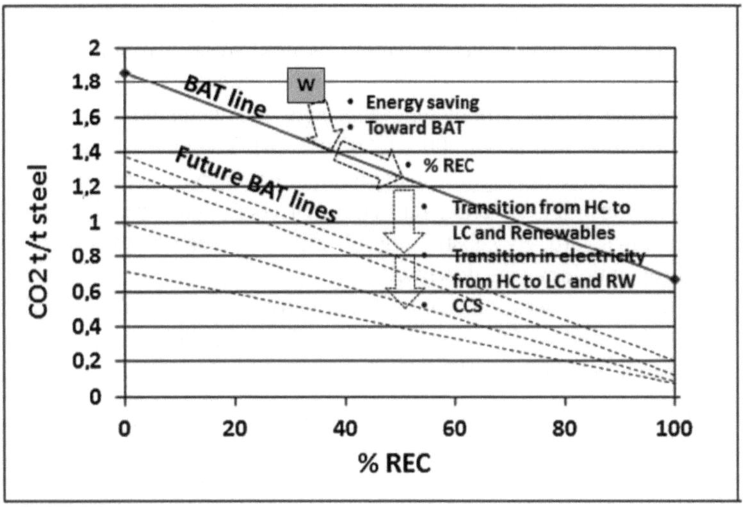

Fig. 4 Prospective scenario showing how the specific CO_2 emissions from iron and steel production can be decreased from the current level (W) to the target value (1.8 → 0.5 t CO_2/t steel). *% REC* Recycled steel ratio, *W* Current global situation, solid BAT line = present BAT function. *Dotted lines* show possible future relations. CCS = CO_2 Capture and Storage

(GJ/t steel) is examined against the recycling ratio (percentage of scrap from the total Fe input) [8, 9]. Respectively, the specific CO_2 emissions in ton/ton steel can be presented, which was done in Fig. 4. The BAT line was drawn based on the data by Worrell et al. [14]. The current position of "world steel" is CO_2 1.8 t/t steel versus 34% REC. The BAT line is attained by 15% reduction in emissions which can be achieved by improving energy efficiency (1st arrow in Figure). The influence of increasing scrap ratio is shown by the 2nd arrow along the BAT line. If only these two actions will be done by 2050, the specific emissions would be on the level 1.2–1.3 t CO_2/t steel. Further actions are thus needed.

The *third key factor* concerns reduction of CO_2 emissions by utilizing less emitting energy sources. Direct emissions in iron and steel production come from fossil energy sources (coal, oil, natural gas) and their derivatives (coke, coke oven gas, off-gas from BF, BOF, EAF etc.). Emissions are reduced by energy saving actions. Transfer from high-C fuels coal/coke to low-C fuels like natural gas have a positive effect and the BAT line settles on a lower level. Injection of hydrogen rich LC fuels in blast furnaces is one possible way to reduce CO_2. A short survey of new technologies follows later. Except for direct emissions even indirect emissions should be considered. The most important is electricity generation, which is discussed as the *fourth key factor.* As it is well-known electricity is not a "clean" emission-free commodity. Depending on the type of power station i.e. primary energy source, the specific CO_2 emissions can be over 1000 g CO_2/kWh in a coal power station whereas in a gas power station the corresponding figure is around 500 g CO_2/kWh. Concerning non-fossil technologies, solar, wind, nuclear and hydro power stations the emissions are roughly 70, 30–50, 20 and <10 g CO_2/kWh, respectively [15, 16]. When electric energy is used in steel production, the electricity generation method plays a significant role in emissions and it will have even bigger role in the future. About ten years ago, the weighted world average of electricity generation was about 500 g CO_2/kWh [8]. Today, it is appr. 400 g CO_2/kWh, but e.g. in China, the biggest steel producing country, it is much higher due to coal power stations. By lowering this emission coefficient, the BAT line would come further downward. By substituting low emission electricity for fossil fuel, the BAT line turns anticlockwise downward.

The *fifth key factor* was allocated for innovative new technologies, which will facilitate in achieving the low emission target. There are several national or international research programs worldwide acting in this field.

- ULCOS program (Ultra Low Carbon Dioxide Steelmaking) by a European consortium with several projects. One is TGR-BF (Top Gas Recycling Blast Furnace) with CO_2 removal. It can save coke and reduce CO_2 emissions by 50% when CCS is applied. CCS (CO_2 capture and Storage) is initially under development for coal power stations to decrease emissions by removing CO_2 from off-gases and then by depositing it into geological formations [17]. HIsarna is a new smelting reduction technology with reduced CO_2 emissions by 80% compared to average blast furnace with CCS, 20% reduction without CCS. ULCORED is a direct reduction process utilizing reducing gas produced from natural gas. Even iron production by electrolysis of iron ore was planned. Of these projects Top Gas Recycling and HIsarna have proceeded to scaling up stage at the moment [18–20].

- COURSE50 project (CO_2 Ultimate Reduction in Steelmaking process by innovative technology for cool Earth 50) in Japan [21–23]. The goal is to establish technologies, which contribute to a mitigation of approximately 30% in CO_2 emissions at integrated steel plants by 2050. Two technologies are developed: (1) Intensified hydrogen reduction of iron ore using coke oven gas to curb carbon input in blast furnaces, and (2) Sequestration of CO_2 in the BF gas through the chemical absorption method and physical adsorption method by the effective utilization of unused waste heat in the integrated steel plants.
- POSCO project: In Korea, POSCO is investigating adaptation of CCS to the Finex smelting reduction processes. Also CO_2 capture from a blast furnace is under development parallel to the ULCOS program [24, 25].
- In North America, the American Iron and Steel Institute (AISI) and the US Department of Energy (DOE) are managing a CO_2 breakthrough program to develop a flash ironmaking process utilizing iron ore fines and low-carbon gas for reduction. The process could bypass ore agglomeration and coke making which belong to the traditional iron making. Removal of CO_2 from off-gas or primary reducing gas are options, which could decrease emissions to a low level [26, 27].
- Renewable energy is a "hot topic" in today's discussion but it is not at all a new item. Until 18th century iron was produced in charcoal blast furnaces or bloomeries. Then coal/coke overtook its position. There was a great necessity for this innovation namely disappearance of forests in Great Britain. Today we have a new necessity, "stop the global warming", and biomass is one possible lot in the total solution. It is unrealistic to load too big anticipation on bio energy (charcoal, biogas) in metallurgy. The role will be additional at least in big scale industry. The topic has been actively investigated worldwide. In Brazil charcoal is used in ferroalloys industry and in pilot iron BFs. In Australia comprehensive studies have been done to evaluate biomass resources, production technologies for charcoal, bio-oil and biogas and their usage in iron and steel making [28, 29]. A life cycle assessment of charcoal substitution for coal/coke showed 25% reduction in GWP (Global Warming Potential) compared to conventional integrated BF-BOF route [29]. Current experiences and extensive studies show that utilization of biomass in iron and steel making is technically possible. However, the costs are relatively high and wide-ranging applications wait for higher CO_2 emission allowance prices.
- The steel industry in China is mostly quite new and thus technically modern. Today the biggest problem is to resolve the problem of excessive capacity [30]. On the other hand serious environmental problems force to improve energy systems, coal power stations are renovated to eliminate particle emissions and to improve efficiency. Corresponding requirements are set to steel industry as well. Positive progresses have been attained e.g. in energy consumption on BF-BOF route. Ongoing and future actions aim at new types of fuel, energy conservation and use of waste heat from slag [30].

- Recently, it was released that in Nordic countries a consortium of three companies representing steel producers, mining and energy (SSAB, LKAB and Vattenfall). Recently launched a project to develop a CO_2 emission free iron production [31]. The project was named HYBRIT (Hydrogen Breakthrough Ironmaking Technology). Hydro power, wind energy and bio mass were mentioned as potential energy sources.

Concluding Remarks

The present global target to stop the climate warming by cutting CO_2 emissions has set big challenges to all sectors of the society. Steel industry as an energy intensive branch is ahead of really demanding tasks to readjust to the future goals. The rapid growth of production during the last 20 years has increased the total emissions, although great progresses have been done to cut specific energy consumption per ton steel and specific CO_2 emissions, respectively.

By 2050, the world steel production is forecast to grow by about 50% to the level 2.5 billion tons/year. At the same time, the CO_2 emissions should decrease by 60% to the level 1.25 billion tons CO_2/year. This means a target of specific emissions 0.5 t CO_2/t steel, when the current world average is 1.8 t CO_2/t steel. The prime means to attain this target are the following.

The first task is to raise the energy efficiency to the current BAT level, which means 15–20% decrease in specific energy consumption in average. In many countries and steel plants the deviation is much larger and the problems and necessary improving operations are readily identifiable.

The foreseeable prominent increase of scrap availability is a central positive factor to reduce specific energy consumption and CO_2 emissions. According to scenario evaluations, the mean scrap ratio will increase from 34 to 50% on the increasing steel production level. This will mean remarkable growth in electric steel making. Certain transfer from BF—BOF into EAF will happen at least in China.

The third and fourth key actions concern energy. By transferring from high carbon HC-fuels and reductants to LC- and renewable energy sources in the processes and in electricity generation, respectively, great reductions in CO_2 emissions can be attained.

Finally, new innovative processes, which aim at low carbon footprint by combining afore-mentioned means and utilizing renewable energy sources, are under development and some of them will attain industrial stage in the near few decades and thus influence for the common goal.

References

1. IEA, Energy and climate change, p. 200. https://www.iea.org/publications/freepublications/publication/WEO2015SpecialReporton.pdf
2. CO2-Earth 2016, https://www.co2.earth/daily-co2
3. World Steel Association 2016, http://www.worldsteel.org/statistics/statistics-archive.html
4. R. Haslehner, B. Stelter, N. Osio, Steel as a model for a sustainable metal industry in 2050. 07 Oct 2015. Categories: Industrial Products & Processes, Sustainability. https://www.bcgperspectives.com/content/articles/metals-mining-sustainability-steel-as-model-for-sustainable-metal-industry-2050/
5. M. Meinshausen, S.J. Smith, K. Calvin, J.S. Daniel, M.L.T. Kainuma, J.-F. Lamarque, K. Matsumoto, S.A. Montzka, S.C.B. Raper, K. Riahi, A. Thomson, G.J.M. Velders, D.P.P. van Vuuren, The RCP greenhouse gas concentrations and their extensions from 1765 to 2300. **109,** 213 (2011). doi:10.1007/s10584-011-0156-z
6. The global carbon budget 2012, p. 41. http://www.globalcarbonproject.org/carbonbudget/archive/2012/CarbonBudget_2012.pdf
7. United Nations Climate Change Conference, https://en.wikipedia.org/wiki/2015_United_Nations_Climate_Change_Conference
8. L. Holappa, Toward low carbon metallurgy in iron and steel making. in *Proceedings of the Guthrie Honorary Symposium*, Montreal, 6–9 June 2011, pp. 248–254
9. L. Holappa, Toward sustainability in ferroalloys and steel production. in *Fray International Symposium on Metals and Materials Processing in a Clean Environment*, Cancun Mexico, Nov 27–Dec 1 2011. vol. 5: Environmental, Health, Policy, Legal, Management and Social Issues, pp. 203–220
10. IEA/OECD 2009, Energy technology transitions for industry, strategies for the next industrial revolution. http://www.oecd-ilibrary.org/energy/energy-technology-transitions-for-industry_9789264068612-en
11. BIR Bureau of International Recycling, World steel recycling in figures 2010—2014, p. 40. http://bdsv.org/downloads/weltstatistik_2010_2014.pdf
12. J. Morfeldt, W. Nijs, S. Silveira, The impact of climate targets on future steel production -an analysis based on a global energy system model. J. Clean. Prod. **103**, 469–482 (2015)
13. R.L. Milford, S. Pauliuk, J.M. Allwood, D.B. Müller, The roles of energy and material efficiency in meeting steel industry CO_2 targets. Environ. Sci. Technol. **47**, 3455–3462 (2013)
14. E. Worrell, L. Price, M. Neelis, C. Galitsky, N. Zhou, World best practice energy intensity values for selected industrial sectors. Ernest Orlando Lawrence Berkeley National Laboratory, LBNL-62806 REV. 2, pp. 51 (2008)
15. C. Hussy, E. Klaassen, J. Koornneef, F. Wigand, International comparison of fossil power efficiency and CO_2 intensity—Update 2014. EcoFys, p. 95http://www.ecofys.com/files/files/ecofys-2014-international-comparison-fossil-power-efficiency.pdf
16. W. Graus, E. Worrell, Methods for calculating CO_2 intensity of power generation and consumption: A global perspective. Energy Policy **39**, 613–627 (2011)
17. Global CCS Institute, Advantages and disadvantages of major CO_2 capture technologies. https://hub.globalccsinstitute.com/publications/technology-options-co2-capture/advantages-and-disadvantages-major-co2-capture
18. J.-P. Birat, Steel and CO_2—the ULCOS program, CCS and mineral carbonation using steelmaking slag. http://www.ulcos.org/en/docs/Ref01__Birat_slag_finaal.pdf
19. Y. Yang, K. Raipala, L. Holappa, Ironmaking, in *Treatise on process metallurgy* ed. by S. Seetharaman, vol. 3: Industrial Processes, Part 1 Ferrous Process Metallurgy ed. by L. Holappa, 2–88
20. M. Abdul Quader, A. Shamsuddin, S.Z. Dawal, Y. Nukman, Present needs, recent progress and future trends of energy-efficient ultra-low carbon dioxide (CO_2) steelmaking (ULCOS) program. Renew. Sustain. Energy Rev. **55**, 537–549 (2016)

21. S. Tonomura, N. Kikuchi, N. Ishiwata, S. Tomisaki, Y. Tomita, Concept and current state of CO_2 ultimate reduction in the steelmaking process (COURSE50) aimed at sustainability in the japanese steel industry. J. Sustain. Metall. **2**, 191–199 (2016)
22. K. Nishioka, Y. Ujisawa, S. Tonomura, N. Ishiwata, P. Sikstrom, Sustainable aspects of CO2 ultimate reduction in the steelmaking process (COURSE50 Project), part 1: hydrogen reduction in the blast furnace. J. Sustain. Metall. **2**, 200–208 (2016)
23. M. Onoda, Y. Matsuzaki, F.A. Chowdhury, H. Yamada, K. Goto, S. Tonomura, Sustainable aspects of ultimate reduction of CO_2 in the steelmaking process (COURSE50 Project), part 2: CO_2 capture. J. Sustain. Metall. **2**, 209–215 (2016)
24. POSCO; Carbon Report 2011, Toward a sustainable society, p. 53. https://www.iea.org/media/weowebsite/ebc/POSCOCarbonReport2011.pdf
25. World Steel Association, Taking carbon capture and storage a step further. https://www.worldsteel.org/media-centre/Steel-news/Taking-carbon-capture-and-storage-a-step-further-.html
26. The American Iron and Steel Institute, Technology roadmap research program for the steel industry. Final Report 31 Dec 2010, p. 36. https://www.steel.org/~/media/Files/AISI/Making%20Steel/TechReportResearchProgramFINAL.pdf
27. H.Y. Sohn, Y. Mohassab, Development of a novel flash ironmaking technology with greatly reduced energy consumption and CO_2 emissions. J. Sustain. Metall. **2**, 216–227 (2016)
28. T. Norgate, N. Haque, M. Somerville, S. Jahanshahi, Biomass as a source of renewable carbon for iron and Steelmaking. ISIJ Int. **52**(8), 1472–1481 (2012)
29. S. Jahanshahi, J.G. Mathieson, M.A. Somerville, N. Haque, T.E. Norgate, A. Deev, Y. Pan, D. Xie, P. Ridgeway, P. Zulli, Development of low-emission integrated steelmaking process. J. Sustain. Metall. **1**, 94–114 (2015)
30. J. Zhang, Z. Liu, K. Li, G. Wang, K. Jiao, T. Yang, Current status and prospects of chinese steel industry. in *Scanmet V Conference*, 12–15 June 2016, Luleå, Sweden. p. 15
31. SSAB, LKAB, Vattenfall; CO_2-emission free ironmaking. Press conference, 17 pp. 4 April 2016. http://materialsbusinesscenter.se/wp-content/uploads/2016/09/160404-SSAB-CO2-emission-free-ironmaking-Short-version.pdf

Application of Surface Effect on Metallurgical Processes

K.C. Chou

Abstract The surface effect is one of fundamental topics in physical chemistry that has been studied over decades. However, its effect has been neglected in metallurgy researches and industry applications in the past years. With the progress of science and technology in metallurgy, the consideration of surface effect to the metallurgical processes has been paid more and more attention. In this report, it is shown that, how this factor affects the physicochemical properties of metallurgical processes and how to derive some new and more correct theoretical formulae for its application. Besides, some new technique related to the application of this factor in the experiment has also been developed for both fundamental research and industry applications. It is hoped that these progresses will promote some new developments both in fundamental and industrial applications for metallurgy processes.

Keywords Surface tension · Melting point · Heating and cooling curve

Introduction

The object of metallurgical process is to separate and extract required elements and/or pure compounds from mineral resources or waste disposal materials through a series of physicochemical processes including phase transformation and/or chemical reactions. For these processes, it is normal that one has to deal with a series of liquid phases including pure liquids or liquid solutions. In the other words, one has to consider solid-liquid reactions. In general, the less the particle sites, the faster the chemical reaction. It is well known that one has to consider the size effect when the particle is very small. Unfortunately this effect has been neglected in the past years due to less theoretical researches and insufficient practical works in the

K.C. Chou
University of Science and Technology Beijing, Beijing, China

K.C. Chou (✉)
Shanghai University, Shanghai, China
e-mail: kcc126@126.com

S. Wang et al. (eds.), *Applications of Process Engineering Principles in Materials Processing, Energy and Environmental Technologies*,
The Minerals, Metals & Materials Series, DOI 10.1007/978-3-319-51091-0_40

metallurgical processes. At present we have entered into the nano-world and many new technologies have become stronger and stronger, it is the time to consider the size effect in our metallurgical process in both theoretical and experimental applications.

Though the surface effect has been studied almost over one century, the theoretical progress of which is still very limited, on the other hand, the practical applications are not developed, especially in the metallurgical process. In this report we will like to give two examples to show how to improve a theoretical research concerning the melting point depression for nano size materials used in metallurgical process. We will also develop a practical experimental method for measuring the melting process of glass material such as slag based on a theoretical consideration. Both of them are developed in our laboratory.

A New Theoretical Formula for Melting Point Depression

The relation between the melting point dropping and the particle size can be tracked back to the end of 19 century by Gibbs-Thomson, from which the equilibrium conditions between a solid particle and a liquid one has been studied and its relation formula has been derived as,

$$\theta = 1 - \frac{2M\sigma_{sg}}{\rho_s r_s \Delta H_m}\left(\sigma_{sg} - \sigma_{lg}\left(\frac{\rho_s}{\rho_l}\right)^{2/3}\right) \tag{1}$$

where θ represents the ratio of melting point T of small particle to that of big size material,

$$\theta = \frac{T}{T_\infty} \tag{2}$$

ΔH_m melting enthalpy, M molecular weight, ρ_s, ρ_l densities for solid and liquid phase respectively, r_s particle radius, σ_{sg}, σ_{lg} surface tension between solid-gas and liquid-gas interface respectively. [1–9].

The actual situation in most melting processes is that, the solid particle melting happens within a liquid thin layer.

$$\theta = 1 - \frac{2M}{\rho_s \Delta H_m}\left(\frac{\sigma_{sg}}{r_l - \delta} - \frac{\sigma_{lg}}{r_l}\left(1 - \frac{\rho_s}{\rho_l}\right)\right) \tag{3}$$

where δ represents the thickness of the liquid layer outside the solid corn. The other symbols are the same as above. Of course the Eq. (3) is more accurate than the Eq. (1), since it is closer to the practical situation [2, 5–7, 10–12].

Nevertheless, the above treatments are still having some problems, for instance, (i) this treatment has ignored the influence of outside pressure that sometimes cannot be neglected, especially when the density of liquid state is different from solid one; (ii) more important is that, this formula is not easy to be examined from experiments since the liquid layer outside from solid core is very difficult to measure; (iii) in this treatment it has ignored the "δ" thickness in some middle step. In this report let us to do some research to overcome all of these defects [8, 12–15]. Recently in our group, we have developed some new formulae from which all of above defects can be overcome. Our calculation formula for the particle size impression is [16]

$$\theta = 1 - \frac{M}{\Delta H_m}\left(\left(\frac{1}{\rho_s} - \frac{1}{\rho_l}\right)\Delta P + \frac{2}{R_s^0}\left(\frac{\sigma_{sl}}{\rho_s(1-\varphi)^{\frac{1}{3}}} + \frac{\sigma_{lg}\left(\frac{1}{\rho_s} - \frac{1}{\rho_l}\right)}{\left(1-\left(1-\frac{\rho_s}{\rho_l}\right)\varphi\right)^{\frac{1}{3}}}\right)\right) \quad (4)$$

where

$$\varphi = 1 - \left(\frac{r_s}{R_s^0}\right)^3 \quad (5)$$

R_s^0 means the original radius of particle at beginning of melting, φ the ratio of reacted portion to total portion that should start from zero to unit.

From Eq. (4), it can be seen that, there are totally three portions within the right-hand side parenthesis. In the most cases, the effect of the second portion should be bigger than that of the third one. Therefore one can conclude that, (i) if the $\rho_s > \rho_l$ (that should be the most normal situations in the nature world), larger ΔP should cause the melting point drop; when $\rho_s = \rho_l$ the pressure effect could be negligible; (ii) when the react portion φ is going bigger and bigger the melting point should become lower and lower. The advantages and more applications of Eq. (4) will be discussed.

A New Technique for Measuring the Melting Rang of Glass Materials

A large number of metallurgical reactions are occurring in the liquid state, for which one needs to know the melting points, reaction heat, heat capacity,…and so forth, and in which the melting point is a very important parameter when one wants to control the reaction processes. It is also well known that, there are two kinds of materials in metallurgical processes, one has typical melting point such as pure compounds, pure alloys, while another of them are only with a temperature change range instead of a fixed temperature when the temperature rising. The later one is also with a smaller heat effect. Most glass materials belong to the later one. Some slags, mixed raw materials, scraps might also show the glass characteristics that will prevent us from obtaining these melting point data. Therefore, it will be a

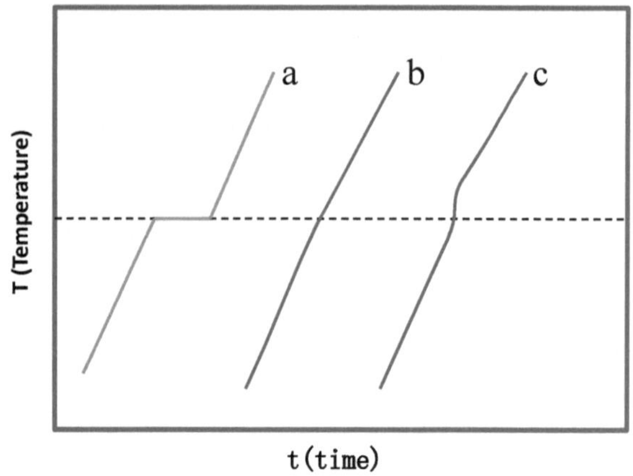

Fig. 1 A typical heating curve with different size sample. **a** Normal size crystal materials. **b** Normal size glass materials. **c** Fine powder glass materials

meaningful subject for us to design new experimental methods or develop calculation theory for obtaining these data.

Figure 1a shows a typical heating curve (or cooling curve) that can be used as an experimental method to decide the melting point of a pure compound, in which, the curve is with a "horizontal segment" that means a process of melting is set in motion, the length of this segment represents the quantity of heat absorbed or released, the longer the segment the more the heat absorbed. One can even determine a melting point only relying on a "turning point" that should be a case with less heat effect. Nevertheless, for some glass materials one cannot find a horizontal segment, instead, only a piece of curve with a variable tangent. If this change is so "evident" that one can still judge that there should be a phase change point. Unfortunately some glass materials such as certain slag, molten salt, the heat released are so small that will be very difficult to find this "turning point", instead an approximate straight line appear in the plot (Fig. 1 curve b). The reader might ask what we can do under this case.

In our laboratory we have established a theory to solve this problem. The basic idea comes from the following analysis. For a pure material, the heating curve or cooling curve has a horizontal segment due to the heat of phase transformation. According to Gibbs phase rule, the freedom degree, $f = c - \varphi + 2$, when a phase transformation happens, $f = 0$, that means a horizon line should appear (Fig. 1a). However, when glasses are melted without two phase region, the degree of freedom, $f \neq 0$, no horizontal line appears. If there is still a small heating effect; a "turning point" should be expected. Nevertheless, if this heat effect is so small that one cannot find this "turning point", under this situation a smooth curve with a small changing of tangent should be expected (Fig. 1b). Under this situation one will be difficult to find the "glass softening point".

In order to solve this problem, we have derived a formula of tangent for heating curve in the melting process of glass, that is [17]

$$\frac{dT}{dt} = \frac{\frac{dQ_{p1}}{dt} + \frac{3\sigma}{r\rho} w_0 \xi}{w_0(C_p^l - (C_p^l - C_p^s)\xi)} \tag{6}$$

where T is temperature of the system, t time, $\frac{dQ_{p1}}{dt}$ the intensity of heat supplied from surrounding, C_p^s , C_p^l heat capacity of solid and liquid sample per gram, r the radius of particle, σ surface tension, ρ and w_0 density and weight of glass, ξ a parameter indicating how many percentage of solid particle has been melted, that is related to the outside heat supplied (i.e.$\frac{dQ_{p1}}{dt}$). From Eq. (6) it is easy to see that, for a big pieces of glass, "r" goes to infinitive, the tangent of heating curve should be simplified to

$$\frac{dT}{dt} = \frac{dQ_{p1}}{w_0(C_p^l - (C_p^l - C_p^s)\xi)dt} \tag{7}$$

and one couldn't find tangent varying as the system heated from low temperature to high temperature. However, if one uses a bunch of small balls instead of a big piece of glass, the situation will be changed at the glass softening point. Under this situation one has to use Eq. (6) instead of Eq. (7). As a result, $\frac{dT}{dt}$ becomes larger and larger as "r" getting smaller and smaller due to $\frac{3\sigma}{r\rho} w_0 \xi$ can't be neglect at this time. It is also easy to find from Eq. (6) that, the factors density and surface tension do have their influence either. The more the surface tension "σ" and the less the density weight "ρ", the larger the tangent $\frac{dT}{dt}$ value.

The analysis describing above will give us an opportunity to measure the "glass softening point". If some glass materials are difficulty to find the softening point due to smaller heat effect that could be solved with the following treatments. Before glass melting, let us grind the solid molten slag or glass materials into a small powder as tiny as possible". Thus a miracle appears, when this powder is heated, a turning point appears around the softening point.

Discussion and Conclusion

The surface effect is a very important factor in physicochemical processes. Though it has been studied over a century, there are still not matured enough to be studied and applied to practical systems especially in the field of metallurgical processes. With the progresses of science and technology, especially with the coming era of Nanotechnology, the theoretical and application studies in this topic are become more and more important.

In this report, we have discussed some new progresses in our laboratory relating the applications of surface effects on both theoretical and experimental applications for metallurgical process: (i) though the melting point depression is an old topic in physicochemical process, there are still some defects both in theoretical analysis and practical measurement. Our new progresses in this topic might be useful for their further progresses. (ii) In order to control the metallurgical process, it is significant to know the slag or molten salt data such as melting point. For a glass material, the softening point is not easy to be measured. In this report a new technical method has been introduced to measure this data based on a theory of surface effect. We sincerely hope this idea would be able to stimulate more new techniques in our metallurgical process and other scientific and technical fields.

Acknowledgements The authors like to express his acknowledgement to National Science Foundation of Science for their kind support under the Contract No: 51474141.

References

1. P. Pawlow, The dependency of the melting point on the surface energy of a solid body. (Supplement.). Zeitschrift Fur Physikalische Chemie–Stochiometrie Und Verwandtschaftslehre **65**, 545–548 (1909); published online EpubFeb
2. H. Reiss, I.B. Wilson, The effect of surface on melting point. J. Colloid Sci. **3**, 551–561 (1948); published online Epub1948 (10.1016/s0095-8522(48)90048-8)
3. D. Turnbull, Formation of crystal nuclei in liquid metals. J. Appl. Phys. **21**, 1022–1028 (1950); published online Epub1950 (10.1063/1.1699435)
4. K.J. Hanszen, Theoretische untersuchungen uber den schmelzpunkt kleiner kugelchen—ein beitrag zur thermodynamik der grenzflachen. Z. Phys. **157**, 523–553 (1960); published online Epub1960 (10.1007/bf01340711)
5. C.R.M. Wronski, Size dependence of melting point of small particles of tin. Br. J. Appl. Phys. **18**, 1731–1737 (1967); published online Epub1967 (10.1088/0508-3443/18/12/308)
6. C.J. Coombes, Melting of small particles of lead and indium. J. phys. F Met. phys. **2**, 441–449 (1972); published online Epub1972 (10.1088/0305-4608/2/3/013)
7. P. Buffat, J.P. Borel, Size effect on melting temperature of gold particles. Phys. Rev. A **13**, 2287–2298 (1976); published online Epub1976 (10.1103/PhysRevA.13.2287)
8. H. Saka, Y. Nishikawa, T. Imura, Melting temperature of in particles embedded in an al matrix. Philos. Mag. A **57**, 895–906 (1988); published online EpubJun
9. K.K. Nanda, S.N. Sahu, S.N. Behera, Liquid-drop model for the size-dependent melting of low-dimensional systems. Phys. Rev. A **66** (2002); published online EpubJul (10.1103/PhysRevA.66.013208)
10. S.L. Lai, J.Y. Guo, V. Petrova, G. Ramanath, L.H. Allen, Size-dependent melting properties of small tin particles: Nanocalorimetric measurements. Physi. Rev. Lett. **77**, 99–102 (1996); published online EpubJul 1 (10.1103/PhysRevLett.77.99)
11. S.L. Lai, J.R.A. Carlsson, L.H. Allen, Melting point depression of Al clusters generated during the early stages of film growth: Nanocalorimetry measurements. Appl. Phys. Lett. **72**, 1098–1100 (1998); published online EpubMar 2 (10.1063/1.120946)
12. K.C. Chou, Under preparing
13. T. Bachels, H.J. Guntherodt, R. Schafer, Melting of isolated tin nanoparticles. Phys. Rev. Lett. **85**, 1250–1253 (2000); published online EpubAug 7 (10.1103/PhysRevLett.85.1250)

14. K. Dick, T. Dhanasekaran, Z.Y. Zhang, D. Meisel, Size-dependent melting of silica-encapsulated gold nanoparticles. J. Am. Chem. Soc. **124**, 2312–2317 (2002); published online EpubMar 13 (10.1021/ja017281a)
15. P.R. Couchman, W.A. Jesser, Thermodynamic theory of size dependence of melting temperature in metals. Nature **269**, (1977)
16. K.C. Chou, et al., To be published
17. K.C. Chou, To be published

Part VIII
Modeling and Simulation

Metal Silicides for High-Temperature Thermoelectric Application

M.R. Bogala and Ramana G. Reddy

Abstract Transition metal silicides have good potential for converting waste heat into electrical energy at higher temperatures. Transition metal silicides are used as thermoelectric alloys at higher temperatures because of their low toxicity, high natural abundance, good transport properties, and better thermal stabilities. In this article, thermodynamic, and thermoelectric properties of promising metal silicides are reviewed, calculated, and compared for waste heat recovery application. Higher manganese silicide (Mn_4Si_7) was produced from easy one-step self-propagating high-temperature synthesis (SHS) procedure of arc-melting technique. Characterization of the alloy phases were done using X-ray diffraction (XRD). In this study, lattice constants, Vickers microindentation hardness, and thermoelectric (TE) properties (Seebeck coefficient, electrical conductivity, thermal conductivity, and figure of merit) of Mn_4Si_7 alloy were experimentally determined, and compared with the reported literature values.

Keywords Metal silicides · HMS (Mn_4Si_7) · SHS synthesis · Arc-melting · Gibbs energy · Thermoelectric properties

Introduction

Thermoelectric (TE) alloys are used for inter-conversion of energy i.e., thermal energy to electrical energy (Seebeck effect) [1] or electrical energy to thermal energy (Peltier effect) [2]. Thus, the TE devices operate not only in the power generation mode, but also in the refrigeration mode. Figure 1a depicts the schematic of a typical TE device, and respective components. The temperature gradient

M.R. Bogala · R.G. Reddy (✉)
Metallurgical and Materials Engineering, University of Alabama,
360 H.M. Comer Hall, 245 7th Avenue, Tuscaloosa, AL 35487, USA
e-mail: rreddy@eng.ua.edu

M.R. Bogala
e-mail: mrbogala@crimson.ua.edu

S. Wang et al. (eds.), *Applications of Process Engineering Principles in Materials Processing, Energy and Environmental Technologies*,
The Minerals, Metals & Materials Series, DOI 10.1007/978-3-319-51091-0_41

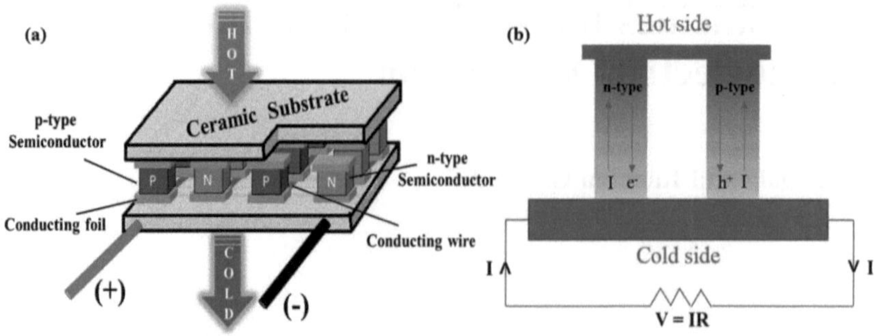

Fig. 1 A schematic of **a** TE device with components, and **b** its working principle

is introduced between the hot, and cold sides of a TE device. The two substrate plates are made-up of ceramic material such as AlN. An array consisting of *p*-type, and *n*-type semiconducting segments are sandwiched between the ceramic substrates of TE device. These segments are inter-connected electrically in series, and thermally in parallel by conducting foil/strips that are made-up of metals like Cu or Ni. Each connected pair of the *p*-type, and *n*-type semiconducting segments serves as a 'thermocouple'. Thus, both the Seebeck effect, and the Peltier effect can be realized with such configuration. The temperature difference between the hot and cold sides of a TE device establishes the Seebeck voltage. The *p*-type and *n*-type semiconducting segments have positive, and negative Seebeck coefficients respectively. As a result, the type, the concentration, and the mobility of charge carriers also differ for each element. Holes (h^+) are the main carriers in *p*-type semiconductors, while electrons ($e-$) are the predominant charge carriers in *n*-type semiconductors. As shown in Fig. 1b, *p*-type and *n*-type segments are joined together to form a TE couple. Both holes and electrons flow from the hot side to the cold side of a TE device, thus establishing electric voltage in the circuit [3].

The performance of a TE device is given by dimensionless figure-of-merit (ZT), which is obtained by using Eq. 1 as follows:

$$ZT = \frac{S^2\sigma}{\kappa}T \tag{1}$$

where S, σ and κ are Seebeck coefficient (μVK^{-1}), electrical conductivity ($Ohm^{-1}m^{-1}$) and total thermal conductivity ($Wm^{-1}K^{-1}$) respectively, and T is the absolute temperature (K) [3–6]. Although Pb- and Te-based TE materials have good ZT (1–1.5), they are toxic, costly, low abundance, and have lower melting point. In order to reduce the manufacturing cost of the TE devices, a good direction for TE research is to start with the low-cost, and high-abundant raw materials [3, 6]. The TE materials constituting the rock forming elements (Si, Mg, Al, Na, K, Ca, Ti, Mn, Fe, C, O, S and P) are preferred over the rare metals such as Te and Sb. Therefore, a paradigm-shift in bulk TE material research include: (a) consideration

Table 1 A list of metal silicides studied as thermoelectric alloys

TE alloy	Type	Constituent metal (M)
Metal silicides	MSi	Na, Ca, Sr, Co and Ba
	M_2Si	Mg, Ca and Sr
	MSi_2	Ca, Ti, Cr, Mn, Fe, Co, Sr, Zr, Mo, Ru, Ba, W, Re and Os
	Other	Ca_3Si_4, Ca_5Si_3, $Mg_{2-x}Ca_xSi$, $MnSi_{2-x}$, $RuSi_{2-x}$, Ru_2Si_3, Os_2Si_3, $ReSi_{1.75}$, Rh_3Si_4, Rh_4Si_5, $IrSi_3$, Ir_3Si_5, $Mg_{127}Si_{64}$, $Mg_{127}Al_1Si_{64}$, $Mg_{129}Si_{63}$, $Mg_{128}Al_1Si_{63}$, $Mg_{127}Si_{64}$, $Co_{1-x}Pt_xSi$, $Co_{1-x}Ni_xSi$, $Co_{1-x}Pd_xSi$, $Fe_{1-x}Os_xSi_2$ and $Ru_{2-x}Os_xSi_3$

Table 2 Properties of selected thermoelectric metal silicides at 300 K

Metal Silicides	$M.P^a$ (K)	ρ^b (g/cm^3)	$-\Delta_f H^{oc}$ (kJ/mol)	Type	κ_l^d (W/mK)	μ_i^e (cm^2/Vs)	E_g^f (eV)	ZT_{max}
Mg_2Si	1358	1.98	26	n	~8	65	0.7	0.9
$CrSi_2$	1763	4.98	108	p	6.8	0.15–15	0.7	0.25
$MnSi_{1.7}$	1430	5.18	33	p	2.9	40	0.66	0.9
β-$FeSi_2$	1490	4.93	74	n, p	4.0	2–4	0.87	^{g}0.4, ^{h}0.2
CoSi	1700	6.56	100	n	~1.5	~43	0.01	0.2
Ru_2Si_3	1970	6.96	134	n, p	4.0	10–29	1.1	^{g}0.4, ^{h}0.2
$ReSi_{1.75}$	2213	10.44	70	p	~5.5	~105	0.15	0.8

^{a}*M.P* melting point, $^b\rho$ density, $^c\Delta_f H^o$ heat of formation, $^d\kappa_l$ thermal conductivity, $^e\mu_i$ carrier mobility, ^fE_g band gap, ^{g}*(n-type)*, ^{h}*(p-type)*

of low-cost materials like metal silicides, and (b) engineering the TE properties (increasing S/σ, and decreasing κ) for waste-heat recovery applications.

Several semiconducting metal silicides have been examined as high temperature TE alloys, and some of them are listed in the Table 1 [6–9]. In 1958, Nikitin conducted first thermoelectric property study on silicides by measuring the electrical conductivity, and thermal power at different temperatures [10]. The author revealed that $CrSi_2$, MnSi, $MnSi_2$, and CoSi are promising materials for the TE application at mid/high temperatures. Then, several other researchers have shown interest in developing metal silicides as TE materials. Metal silicides are attractive due to (a) low-cost, (b) high abundance in earth's crust, (c) eco-friendly, (d) high mechanical strength, and (e) good chemical stability [6, 7, 9].

Extensive research has been conducted on alkaline earth metal silicide (Mg_2Si), and transition metal silicides ($MnSi_{1.73}$, $ReSi_{1.75}$, β-$FeSi_2$, Ru_2Si_3, $CrSi_2$, and CoSi) because of good *ZT* at mid/high temperatures [6, 7, 11, 12]. Table 2 lists important properties of selected metal silicides. Mg_2Si exhibits relatively low melting point, low density, low heat of formation, and high *ZT* in comparison to the transition metal silicides. $MnSi_{1.7}$ and $ReSi_{1.75}$ are other high performing TE materials with ZT_{max} comparable to that of Mg_2Si. Figure 2 shows the crystal structure of unit cell of (a) Mg_2Si, (b) $MnSi_{1.7}$ and (c) β-$FeSi_2$, and multiple unit cell of

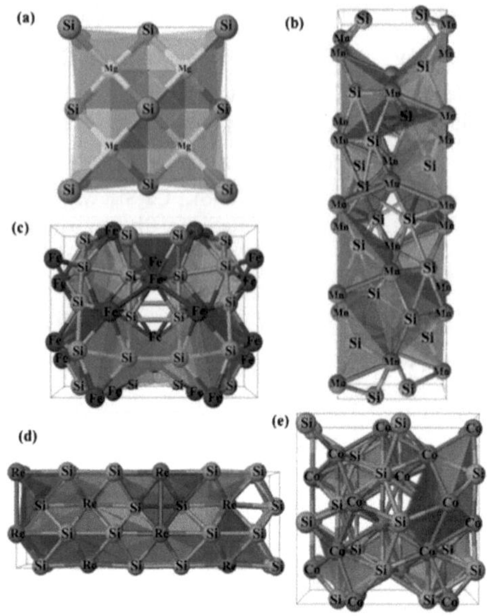

Fig. 2 Crystal structures of unit cell: **a** Mg_2Si, **b** $MnSi_{1.7}$, **c** β-$FeSi_2$, multiple unit cell: **d** $ReSi_{1.75}$, and **e** CoSi

$ReSi_{1.75}$ and (e) CoSi [13]. From the polyhedron structures, the proportion of metallic component (green-shaded region) in Mg_2Si structure is relatively high, while compared to the corresponding region (other than sand-colored portion) in the structures of transition metal silicides. Therefore, the area ratio of metallic shaded region to silicon region is proportional to the molar ratio of metal to silicon in the metal silicides.

The structural framework also influences the thermoelectric properties of the metal silicides. As shown in the Fig. 3a, the Seebeck coefficients of six selected metal silicides (Mg_2Si, $CrSi_2$, $MnSi_{1.73}$, β-$FeSi_2$, CoSi, and Ru_2Si_3) are plotted from 300 to 1030 K. Mg_2Si, β-$FeSi_2$, Ru_2Si_3, and CoSi have negative *S*-values (*n*-type), while $MnSi_{1.7}$, $CrSi_2$, and Ru_2Si_3 have positive *S*-values (*p*-type) [6, 7, 11, 12]. Ru_2Si_3 contain high *S*-values and therefore, it can be used as both *n*-type ($T < \sim 550$ K), and *p*-type ($T > \sim 550$ K) TE material. Other alloys with better *S*-values are Mg_2Si, and $MnSi_{1.7}$. Figure 3b displays the effect of temperature on the σ-values of the metal silicides [6, 7, 11, 12]. CoSi and $CrSi_2$ have very high σ-values, and Ru_2Si_3 have the least σ-value. With the increase in temperature, an increase in σ-value for Ru_2Si_3 and β-$FeSi_2$, while decrease in the σ-value of CoSi and $MnSi_{1.7}$ was observed. The σ-value of Mg_2Si and $CrSi_2$ decreases up to 700 K, and then, a further increase in the temperature results in an increase in their electrical conductivities. Metal silicides are characterized by low thermal conductivities. As shown in Fig. 3c, the effect of temperature on the κ-values of Mg_2Si, $CrSi_2$, $MnSi_{1.7}$, β-$FeSi_2$, CoSi, and Ru_2Si_3. $MnSi_{1.7}$ exhibited the lowest κ-values, followed by Ru_2Si_3, $FeSi_2$, and Mg_2Si at all temperatures [6, 7, 11, 12]. CoSi and $CrSi_2$ have relatively higher thermal conductivities. The selected metal silicides

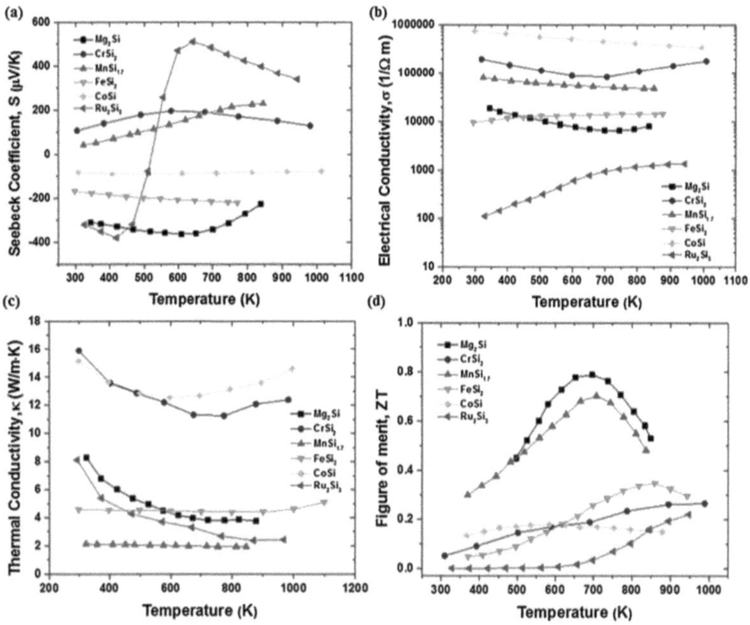

Fig. 3 Effect of temperature on **a** Seebeck coefficient, **b** electrical conductivity, **c** thermal conductivity and **d** figure of merit of Mg_2Si, $CrSi_2$, $MnSi_{1.73}$, β-$FeSi_2$, CoSi, and Ru_2Si_3

have good *ZT* from 500 to 800 K. As shown in the Fig. 3d, the *ZT* vs. *temperature* plot of Mg_2Si, $CrSi_2$, $MnSi_{1.7}$, β-$FeSi_2$, CoSi, and Ru_2Si_3 [6, 7, 11, 12]. Mg_2Si and $MnSi_{1.7}$ have high *ZT* values, while other metal silicides exhibited relatively low *ZT* values. Although Ru_2Si_3, $FeSi_2$, and $CrSi_2$ have low *ZT* at $T < 600$ K, the TE performance of these silicides increases with further increase in the temperature. No appreciable change in the *ZT*-value of CoSi was observed with the increase in temperature.

As shown in the Fig. 4, both the calculated band structure (a and b), and density of states (c and d) of Mg_2Si and Mn_4Si_7 respectively, reveal the existence of narrow indirect energy band gaps for both the alloys. The energy band gap for Mg_2Si is 0.218 eV. Also, the formation energy of Mg_2Si is −0.160 eV per atom [14]. Similarly, the energy band gap of Mn_4Si_7 is 0.779 eV with the Fermi energy level located next to the peak ($\Delta E = E - E_f = \sim 0$). Unlike the band structure of Mg_2Si, the band structure of Mn_4Si_7 is much more complicated with larger number of energy levels in both valance and conduction bands. The evaluation also predicts that the energy required for the formation of Mn_4Si_7 is −0.433 eV per atom [14].

Higher manganese silicide (HMS or $MnSi_{2-x}$) assume Nowotny chimney ladder structure, in which Mn atoms form the chimney, and Si atoms spiral around them as the ladder [15]. As shown in the Fig. 2b earlier, the crystal structure of Mn_4Si_7, and the structures of other higher order HMS, namely $Mn_{11}Si_{19}$, $Mn_{15}Si_{26}$, and $Mn_{27}Si_{47}$ are obtained by varying the length of Mn_4Si_7 along the *c*-axis.

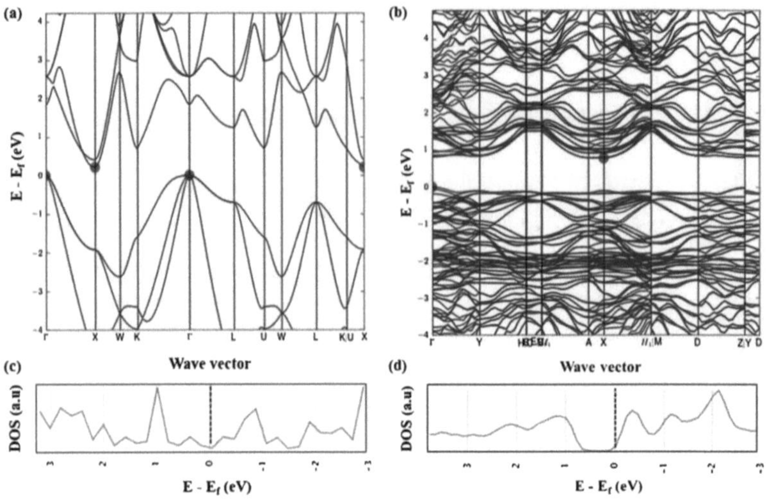

Fig. 4 Calculated energy band structures (**a**, **c**), and density of states (**b**, **d**) of Mg_2Si and Mn_4Si_7 respectively

Besides having complicated crystal structure, HMS shows distinct anisotropy of TE properties. HMS alloys have good potential for TE heat recovery from 573–873 K. Zaitsev et al. have shown that the doping of HMS alloys with the metals (Al, Mo, and Ge) can further improve the TE performance of the HMS alloys [16].

In the current study, HMS phase (Mn_4Si_7) was chosen for further experimental studies based on Gibbs energy calculations, and available literature data on thermoelectric properties of selected binary metal silicides. Synthesis, characterization, testing, and thermoelectric properties of Mn_4Si_7 were measured using arc-melting, X-ray diffraction, Vickers microhardness, and *ZT*-Scanner instrumental techniques respectively.

Materials and Methods

All the raw materials and instruments needed for the preparation, and characterization of the TE alloys were purchased from various vendors. Elemental powders of manganese (−325 mesh, 99.3%), and silicon (−325 mesh, 99.5%) were purchased from Alfa Aesar, MA, U.S.A. Weighing and mixing of stoichiometric amounts of as-received elemental powders of Mn, and Si were carried out in a Labconco glove box (Kansas City, MO, U.S.A) under ultra-high pure argon gas atmosphere. Appropriate amounts of the raw materials were weighted using the microbalance, and the powders were mixed uniformly in Pyrex glass vials using Fischer Scientific Vortex tool. After the mixing step, about 1–3 g of the powder mixture (4Mn + 7Si) was placed in a cylindrical stainless steel dye of 1.3 cm inner diameter and then, the mixture was subjected to a uni-axial force of 5.8 ksi using Carver press at room temperature for 2 min.

Arc-Melting and SHS Method

Manganese silicide (Mn_4Si_7) samples were synthesized by placing the cold-pressed pellets (4Mn + 7Si) in small bell jar arc-melt furnace (ABJ338, Materials Research Furnaces Inc.), and heating up of the pellets under ultra-high pure argon atmosphere. Figure 5 shows the images of (a) the electric arc furnace, and (b) its open bell jar chamber that was used for the synthesis of Mn_4Si_7 samples. Figure 5c shows the schematic of electric arc furnace with important parts labeled. The arc-furnace consists of Cu stringer or handle with its inside end attached to a pointed thorium-doped tungsten rod electrode. The Mn_4Si_7 sample was placed onto custom-made cavities of water-cooled copper hearth plate (8.5 cm diameter). The 304L stainless steel bell jar chamber was closed with the metal clamps and then, it was water-cooled to maintain the temperature of the chamber below 50 °C throughout the experiment. An evacuation pump (56 LPM) was used to create the vacuum inside the chamber. Prior to the start of the experiments, the furnace pressure was reduced to 30 mm Hg to remove any ambient air present inside the chamber. The compartment was refilled with inert argon gas for at least three times. The thorium-doped tungsten rod tip of the water-cooled stringer was placed over

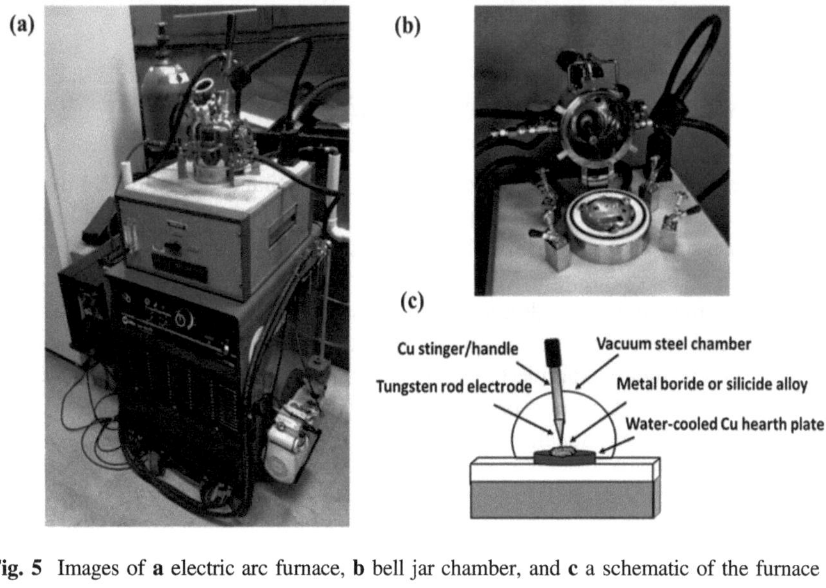

Fig. 5 Images of **a** electric arc furnace, **b** bell jar chamber, and **c** a schematic of the furnace

the oxygen getter (titanium sponge), and an electric-arc was struck to melt the titanium metal. This process removes any residual oxygen from the chamber. Later, under a standard positive argon pressure of 2 psig and by passing a current of 200 A, an electric-plasma arc was struck on one-edge of the pellet (4Mn + 7Si). When the temperature of the pellet reached the formation temperature of Mn_4Si_7, self-propagating high-temperature synthesis (SHS) of Mn_4Si_7 took place, and the pellet was completely converted to a button-shaped ingot (product alloy). After the initial melting of the pellet, the ingot button was flipped up-side down, and re-melted for at least 2–3 times in order to obtain homogeneous composition of Mn_4Si_7 alloy.

X-Ray Diffraction and Lattice Constants

The arc-melted samples were analyzed for the presence of different alloy phases using XRD (model Philips X'Pert MPD) instrument. The instrument uses monochromatic radiation from Cu-Kα source with a wavelength of 1.540 Å. The MPD instrument was operated at a voltage of 45 kV and a current of 40 mA. The samples were spread uniformly on a glass mounting slide holder, and the XRD data was acquired at different diffraction angles ($2\theta°= 20°–120°$) using the scan rate of 0.01° per sec. High-Score plus software was used to adjust the baseline threshold of the XRD spectra. The spectral analysis was made manually by comparing with the standard PCPDF#04-010-6000 card of Mn_4Si_7. Using the XRD data of two major reflections, the crystal lattice constants ($a = b$ and c) of Mn_4Si_7 was calculated from Bragg's law of diffraction. Furthermore, the experimental lattice constants were compared with the corresponding values available in the literature.

Hardness and Thermoelectric Properties

The synthesized Mn_4Si_7 alloys were subjected to mechanical testing. Buhler high quality hardness tester was used to perform the Vickers microindentation hardness test on three samples of Mn_4Si_7 alloys. A load of 1000 gf was applied on the samples for the test duration of 15 s. The Vickers microhardness (H_v) was measured by averaging the indentation dimension of the sample micrograph in both *x*- and *y*-directions, and from the unit conversion (length to force) table of the instrument manual. The arc-melted sample (3–6 mm height and 10–30 mm^2 area) of Mn_4Si_7 was examined for the transport properties (S, σ, and κ), and ZT measurements from 300 to 700 K using ZT-Scanner (TMTE Inc.).

Gibbs Energy Minimization

In a multicomponent system, the phase equilibrium is calculated by minimizing the total Gibbs energy 'G' of all individual components that participates in the equilibrium. The total Gibbs energy is given by Eq. 2 as:

$$G = \sum\nolimits_{i=1}^{p} n_i G_i^o = \text{minimum} \quad (2)$$

where n_i and G_i^o are the number of moles, and Gibbs energy of individual phase 'i' [17]. Thermodynamic tools were used to model the Gibbs energy of binary transition metal silicides. Change in Gibbs energy of twelve binary silicides ($NbSi_2$, Nb_5Si_3, $CoSi_2$, CoSi, Co_2Si, Ti_5Si_3, TiSi, $TiSi_2$, MnSi, Mn_5Si_3, Mn_3Si, and $MnSi_{1.73}$) were evaluated using 'Reaction Equations' module of HSC 7.1 software tool [18]. The change in Gibbs energy of the selected silicides were calculated from 25 to 1500 °C at 1 bar pressure. Using 'Compound' module of Factsage 6.3 tool [19], the temperature-dependent Gibbs energy function, *G(T)*, is given by Eq. 3. The parameters (A, B, C, D, and E) of the temperature polynomial function, *G(T)*, was estimated for nine selected metal silicides (CoSi, Ti_5Si_3, TiSi, $TiSi_2$, MnSi, Mn_5Si_3, Mn_3Si, $NbSi_2$, and Nb_5Si_3).

$$G(T), \frac{\text{kJ}}{\text{mol}} = A + B \times T + C \times T^2 + D \times T^{-1} + E \times T \ln T \quad (3)$$

Results and Discussion

Change in Gibbs Energy of Metal Silicides

Thermodynamic modeling tools [18, 19] were used to model the thermodynamic properties of the metal silicides. In particular, Gibbs energy minimization was performed on twelve binary silicides ($NbSi_2$, Nb_5Si_3, $CoSi_2$, CoSi, Co_2Si, Ti_5Si_3, TiSi, $TiSi_2$, MnSi, Mn_5Si_3, Mn_3Si, and $MnSi_{1.73}$) using HSC tool over a wide temperature range of 0–1500 °C [18]. Thermodynamic calculations were conducted to estimate the change in Gibbs energy of selected metal silicides and also to determine their relative thermal stabilities.

Figure 6 shows the change in Gibbs energy of selected silicides from 0 to 1500 °C. The negative value of change in Gibbs energy suggests that the metal silicides are stable at higher temperatures. Among all the metal silicides studied, Ti_5Si_3 and Nb_5Si_3 shows more negative change in Gibbs energy value and thus, they are relatively more thermally stable transition metal silicides. Furthermore, FactSage tool was used to determine the Gibbs energy as a temperature polynomial function for nine silicides (CoSi, Ti_5Si_3, TiSi, $TiSi_2$, MnSi, Mn_5Si_3, Mn_3Si, $NbSi_2$, and Nb_5Si_3). The Gibbs energy function is obtained from Eq. 3, and the parameters

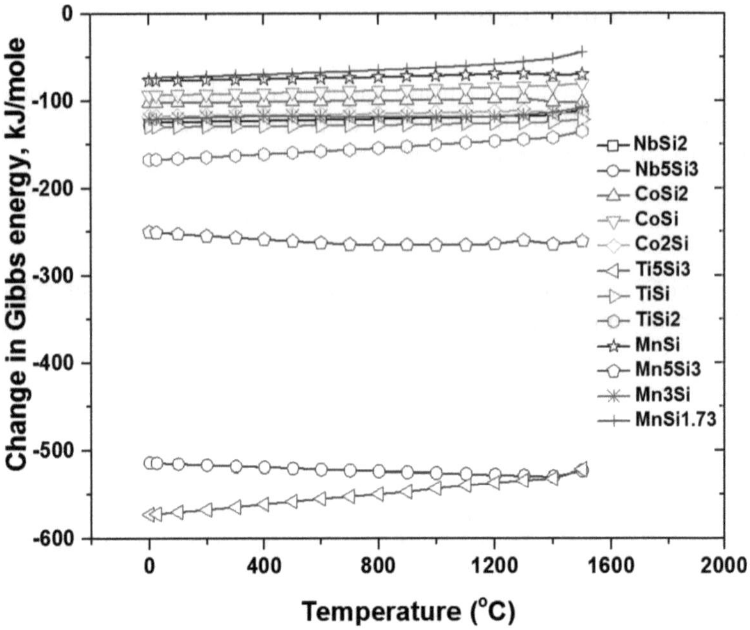

Fig. 6 Change in Gibbs energy of selected metal silicides at different temperatures

(A, B, C, D, and E) of the polynomial were estimated for each silicide in Table 3. Based on thermal stability and energy band gap calculations (Figs. 6 and 4c), and available literature on TE properties (Fig. 3) of metal silicides, the HMS phase (Mn_4Si_7) was selected for the experimental studies.

Synthesis of Mn_4Si_7

Pellet (4Mn + 7Si) was arc-melted, at an applied current of 200 A under partial Ar gas pressure of 2 psi, to synthesize Mn_4Si_7 alloy using facile single-step self-propagating high-temperature synthesis (SHS) method. The reaction of Eq. 4 is exothermic, and the formation of the Mn_4Si_7 product is very favorable with the initiation of an electric arc or heat at the edge of the pellet (4Mn + 7Si), enough to start the self-propagating reaction.

$$4\mathrm{Mn} + 7\mathrm{Si} = \mathrm{Mn_4Si_7} \quad \Delta H_{298.15\mathrm{K}} = -227.57\,\mathrm{kJ} \tag{4}$$

The synthesized Mn_4Si_7 alloys were characterized using XRD technique. Figure 7 shows the XRD spectrum, and an inset image of Mn_4Si_7 alloy synthesized using the arc-melting technique. No other binary Mn–Si alloy phases were identified from the XRD spectra. Therefore, formation of pure single-phase of

Table 3 Gibbs energy function parameters of selected silicides

Silicides	A (J/mol)	B (J/mol.K)	C (J/mol.K^2)	D (J K/mol)	E (J/mol.K)
CoSi	−90,238	311.080	−0.00414	200,333	−53.520
Ti_5Si_3	−646,360	1122.32	−0.02238	1,004,160	−196.43
TiSi	−146,382	279.770	−0.00571	271,960	−48.110
$TiSi_2$	−193,045	406.080	−0.00815	−1,477,660	−69.330
MnSi	−92,278	276.720	−0.00926	246,494	−46.290
Mn_5Si_3	−269,849	1136.90	−0.02708	980,102	−201.35
Mn_3Si	−144,109	534.880	−0.02396	386,148	−93.200
$NbSi_2$	−158,529	359.420	−0.00768	139,745	−63.170
Nb_5Si_3	−514,695	1033.52	−0.01539	753,956	−189.15

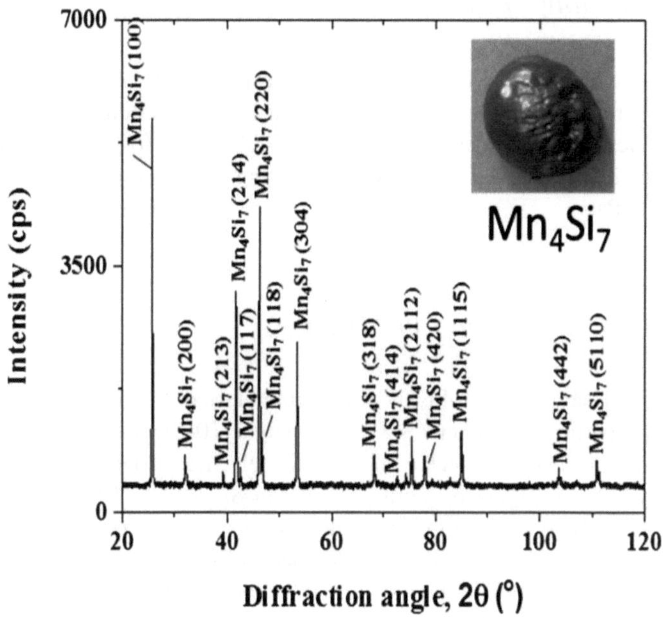

Fig. 7 XRD spectrum and an image of Mn_4Si_7 obtained from arc-melting method

simple/primitive tetragonal Mn_4Si_7, with the space group of *P-4c2* (116), was confirmed by comparing with the standard data (PCPDF#04-010-6000).

Lattice Constants of Mn_4Si_7

The lattice constants, *a* (=*b*) and *c*, for the tetragonal crystal lattice of Mn_4Si_7 were obtained from Eq. 5 as:

$$\frac{1}{d^2} = \frac{h^2 + k^2}{a^2} + \frac{l^2}{c^2} \tag{5}$$

where '*d*' is the spacing between the adjacent miller indices (*hkl*) planes. The lattice constants (*a*, *b*, and *c*) for Mn_4Si_7 was calculated from the Eq. 5, by using the two strong XRD reflections corresponding to (100) and (220) reflections of three arc-melted samples. The values of *a* (=*b*) and *c* for Mn_4Si_7 lattice obtained were 0.5525 ± 0.0003 nm, and 1.7489 ± 0.0055 nm respectively. These values are in very good agreement with the corresponding reported lattice constants of PCPDF#04-010-6000 for Mn_4Si_7, $a = b = 0.5526$ nm, and $c = 1.7516$ nm.

Hardness Testing of Mn_4Si_7

The arc-melted sample of Mn_4Si_7 was further tested for the mechanical properties. The Vickers microindentation hardness of Mn_4Si_7 was tested using Buhler high quality hardness tester with an applied load of 1000 gf and for time interval of 15 s. The average microhardness (H_v) measured using three samples of Mn_4Si_7 was 10.8 ± 0.7 GPa.

Thermoelectric Properties of Mn_4Si_7

The arc-melted samples of Mn_4Si_7 was further tested for measurement of thermoelectric properties. Specimen ($\sim$3–6 mm height, 10–30 mm^2 cross-section area) were chosen to measure Seebeck coefficient (*S*), electrical conductivity (σ), thermal conductivity (κ), and figure of merit (*ZT*) of Mn_4Si_7 using *ZT*-Scanner instrument [20] at 300–700 K. As shown in Fig. 8, the thermoelectric properties of arc-melted Mn_4Si_7 sample was plotted at different temperatures, and compared with the corresponding values reported in the literature [6, 7, 11, 12]. Higher manganese silicide (Mn_4Si_7) alloy phase exhibited good positive Seebeck coefficient (Fig. 8a), which suggests that the arc-melted Mn_4Si_7 alloy is a *p*-type semiconductor. From Fig. 8c, the low thermal conductivity values of Mn_4Si_7 also indicates the possibility of more phonon-phonon scattering taking place within the sample. However, the arc-melted Mn_4Si_7 sample displayed poor electrical conductivity (Fig. 8b), while compared to the corresponding reported σ-values [6, 7, 11, 12] for the alloy. Therefore, as shown in Fig. 8d, the reduced electrical conductivities may have resulted in lowering of figure of merit values (maximum $ZT = 0.05$) for the *p*-type Mn_4Si_7 alloy, in comparison to the literature *ZT* values. Another reason for the low TE performance of the arc-melted Mn_4Si_7 may be due to the high porosity or low density of the samples. Further improvements in the synthesis, and processing methods may enhance the *ZT* of *p*-type Mn_4Si_7 alloy to much higher values.

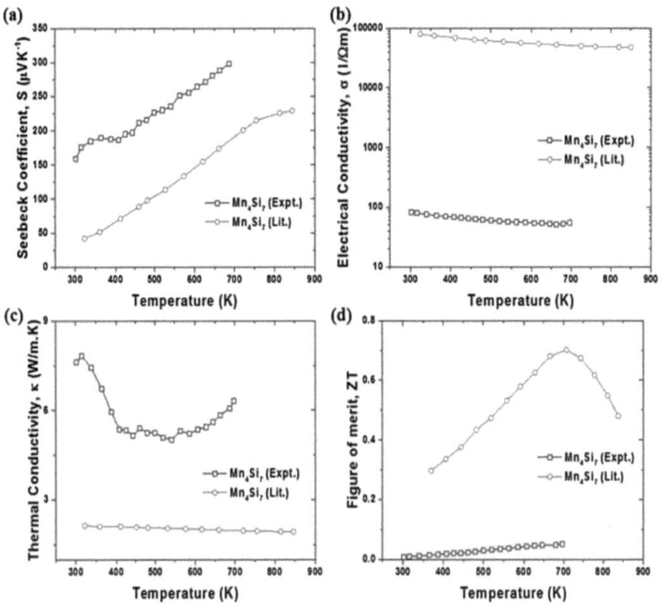

Fig. 8 Experimental and reported thermoelectric properties **a** Seebeck coefficient, **b** electrical conductivity, **c** thermal conductivity, and **d** figure of merit of Mn_4Si_7

Conclusions

In the current study, efforts were made to investigate binary metal silicides as potential thermoelectric alloys through initial screening of the alloys via thermodynamic modelling studies, and available thermoelectric property data from the literature. Gibbs energy minimization method was used to study the thermal stabilities of selected metal silicides from 0 to 1500 °C. Higher manganese silicide (Mn_4Si_7) was chosen for the experimental studies. A simple and one-step self-propagating high-temperature synthesis (SHS) method was used to produce ingots of Mn_4Si_7 from the electric arc-melter furnace. Characterization of the alloy phases was done using the XRD spectrum, and the mechanical property was measured from Vickers microindentation hardness tests. Furthermore, thermoelectric properties (S, σ, κ and ZT) of Mn_4Si_7 samples were determined using the ZT-Scanner instrument. Mn_4Si_7 exhibited low thermal conductivity values. Good positive Seebeck coefficients were obtained for Mn_4Si_7 sample. A maximum ZT of 0.05 was achieved for the p-type Mn_4Si_7 alloy synthesized from the arc-melting technique.

Acknowledgements The authors are thankful to the financial support from the National Science Foundation (NSF) agency, Grant No. DMR-1310072 and American Cast Iron Pipe Company (ACIPCO) during the course of the project.

References

1. T.J. Seebeck, Ueber die magnetische Polarisation der Metalle und Erze durch Temperaturdifferenz. Ann. Phys. **82**, 253–286 (1826)
2. J.C.A. Peltier, Nouvelles expériences sur la caloricité des courants électrique. Ann. Chim. Phys. **56**, 371–386 (1834)
3. F.D. Rosi, Thermoelectricity and thermoelectric power generation. Solid-State Electron. **11**, 833–868 (1968)
4. G.J. Snyder, E.S. Toberer, Complex thermoelectric materials. Nat. Mater. **7**, 105–114 (2008)
5. T.M. Tritt, H. Böttner, L. Chen, Thermoelectrics: direct solar thermal energy conversion. MRS Bull. **33**, 366–368 (2008)
6. D.M. Rowe, *Thermoelectrics Handbook: Macro to Nano* (CRC press, 2005)
7. M.I. Fedorov, V.K. Zaitsev, The features of silicide thermoelectrics development. in *6th European International Conference on Thermoelectrics*, I–11, 1–6 (2008)
8. H. Lange, Electronic properties of semiconducting silicides. Physica Status Solidi (b) **201**, 3–65 (1997)
9. M.R. Bogala, R.G. Reddy, Phase stability of thermoelectric alkaline earth metal borides and silicides. TMS2015 Suppl. Proc. **77**, 623–633 (2015)
10. E.N. Nikitin, Study of temperature dependencies of electrical conductivity and thermal power of silicides. Zhurnal Tekhnicheskoj Fiziki **28**, 23 (1958)
11. M.I. Fedorov, Thermoelectric silicides: past, present and future. J. Thermoelectr. **2**, 51–60 (2009)
12. M.I. Fedorov, G.N. Isachenko, Silicides: materials for thermoelectric energy conversion. Jpn. J. Appl. Phys. **54**, 7S2 (2015): 07JA05
13. P. Villars, (Landolt-Börnstein Database), Physical & chemical data in materials science. Springer Materials, http://www.springermaterials.com/docs/index.html. Accessed on 05 Jan 2016
14. A. Jain et al., Commentary: the materials project: a materials genome approach to accelerating materials innovation. APL Mater. **1**, 011002 (2013). https://materialsproject.org/materials/
15. J.M. Higgins et al., Higher manganese silicide nanowires of nowotny chimney ladder phase. J. Am. Chem. Soc. (JACS) **130**(47), 16086–16094 (2008)
16. V.K. Zaitsev, K.A. Rakhimov, A.E. Engalychev, Higher manganese silicide: material for thermoelectric converters. Geliotekhnika **3**, 15–17 (1983)
17. U.R. Kattner, The thermodynamic modeling of multicomponent phase equilibria. JOM **49** (12), 14–19 (1997)
18. Outotec HSC Chemistry, Pori, Finland. Available at: http://www.hsc-chemistry.com
19. The FactSage system. Available at: http://www.factsage.com
20. ZT-Scanner, Thermoelectric Materials and Technology (TEMTE Inc., Canada), http://www.temte.ca/ZT-Scan.html. Accessed on 15 May 2016

CFD Modeling of Slag-Metal Reactions and Sulfur Refining Evolution in an Argon Gas-Stirred Ladle Furnace

Q. Cao, L. Nastac and A. Pitts

Abstract Accurate prediction of the desulfurization behavior of steel is of great importance for process control during ladle metallurgical furnace (LMF) steel refining. A CFD model capable of simulating multiphase flow with bulk and free surface turbulence phenomena and desulfurization behavior in the gas-stirred ladle has been developed. The fluid flow behavior shows that slag and metal try to entrap each other, forming a two-phase mixing zone. The resulting substantial contact area between the two phases will provide very favorable kinetic conditions for chemical reactions. For reaction kinetics model, the reactions at the metal/slag bath interface as well as the entrapped droplet surface are included. Predicted results show that the sulfur content changing with time in the ladle agrees well with the industrial-scale experimental measurements.

Keywords 3D CFD modeling · Ladle metallurgical furnace · Fluid flow characteristics · Desulfurization kinetics

Introduction

The high levels of desulfurization of molten steel can effectively decrease the amount of surface defects [1, 2], hydrogen induced cracking [3] and sulfides [4, 5] of the ultimate steel products. With the higher requirements of steel quality, there is an increasing demand for low-sulfur steel [6]. Therefore, the improvement of desulfurization efficiency becomes one of the most important objectives in ladle metallurgical furnace (LMF) refining operations.

Q. Cao · L. Nastac (✉)
Department of Metallurgical and Materials Engineering, The University of Alabama, Box 870202, Tuscaloosa, AL 35487, USA
e-mail: lnastac@eng.ua.edu

A. Pitts
Nucor Tuscaloosa, Tuscaloosa, AL 35401, USA

S. Wang et al. (eds.), *Applications of Process Engineering Principles in Materials Processing, Energy and Environmental Technologies*,
The Minerals, Metals & Materials Series, DOI 10.1007/978-3-319-51091-0_42

The argon gas-stirring is of enormous importance during LMF operations in steelmaking [7–9]. The bubbly plume flow is produced by flowing inert gas through the ladle bottom plug, which is used to promote flotation of inclusions, and to enhance slag-metal reactions [10–13]. There are many factors that influence the efficiency of sulfur removal such as the compositions of the liquid steel and slag, slag layer thickness, gas flow rate, the arrangement of bottom porous plugs, etc. [14]. Therefore, in addition to sampling, it is necessary to be able to predict the status of the refining process to adjust the operations in practical LMF process [15].

A considerable amount of mathematical investigations was carried out to study the fluid flow in argon gas-stirred LMF using different CFD models [16–19]. The desulfurization behavior can be simulated by incorporating the thermodynamic reactions into CFD model in ANSYS Fluent [13, 20]. The first model describing desulfurization behavior was developed by El-Kaddah et al. [21] in 1981. The most significant limitation of their model is that the effect of the slag phase on desulfurization was neglected. Later, Jonsson et al. [14] developed a three-phase model of desulphurization in the LMF by combining 2-D fluid-flow with the thermodynamics of desulfurization. Their simulation results showed that desulfurization rate mainly depends on the transfer rate of sulfur from the metal to the slag/steel interface. Recently, Lou et al. [13] proposed a three-dimensional CFD-simultaneous reaction coupled model to describe the desulfurization behavior in a gas-stirred ladle. The simultaneous slag/metal reaction rates at the interface were calculated using the kinetics parameter based on film theory and the interfacial reaction kinetic equilibrium. However, their model only included the steel and gas phases without considering the effects of slag layer, which plays a significant role in steel refining in ladle metallurgy [22].

The objective of the present work is to develop and refine a fully transient three-dimensional three-phase mathematical model in the argon gas-stirred ladle using Ansys's Fluent commercial CFD package. This model can predict the three phase flow characteristics and desulfurization kinetics considering the effect of liquid flow on the desulfurization.

Model Description

CFD Model

The CFD model consists of a multiphase VOF model in conjunction with momentum, energy, and species transfer models as well as the k-epsilon realizable turbulence model in ANSYS Fluent [23, 24].

Reaction Model

The user-defined function (UDF) is used to specify reaction rates for slag-metal phase heterogeneous reactions in every fluid cell that contains slag and steel interface. The reactions at the metal/slag bath interface as well as the entrapped droplet surface are included. The slag system ($CaO–Al_2O_3–SiO_2–MgO$) is used in refining process. Based on the slag-metal reaction, the desulfurization process in the ladle can be expressed by the following reaction:

$$(O^{2-}) + [S] = [O] + (S^{2-}) \tag{1}$$

where ( ) indicates the species in the slag, and [ ] represents species in the liquid steel bath [25].

The overall desulfurization reaction rate V_S (the change in the concentration of sulfur in steel) can be written as:

$$V_S = \rho_m k_{eff,S} \frac{A}{V} \left\{ [wt\%S] - \frac{(wt\%S)}{L_S} \right\} \tag{2}$$

where the (wt%S) indicates the mass fraction of sulfur in slag phase, [wt%S] is the local mass fraction of S in liquid steel, ρ_m is the density of liquid steel. $k_{eff,S}$ characterizes the overall mass transfer coefficient of element S. L_S is the distribution ratio of element S between slag and metal phases at equilibrium. A and V are the interface area between slag and metal and the steel volume in the fluid cell at the slag/steel interface, respectively.

The sulfur distribution ratio L_S can be calculated by considering the chemical composition of liquid steel and slag, as follows [20]:

$$lgL_S = lg\frac{(wt\%S)^*}{[wt\%S]^*} = -\frac{935}{T} + 1.375 + lg(C_{S^{2-}}) + lgf_S - lga_O^* \tag{3}$$

where the superscript symbol * represents a value of reaction equilibrium at the slag–metal interface. T is the temperature at slag–metal interface. a_O^* and f_S are the activity of oxygen and activity coefficient of sulfur in liquid metal, respectively. $C_{S^{2-}}$ is the sulfide capacity, which describes the potential ability of an arbitrary homogeneous molten slag to remove sulfur [20].

The oxygen activity in the steel can be solved through the equilibrium reactions between slag and metal [26]. To simplify the model we assume that the equilibrium between dissolved aluminium and oxygen in the liquid steel and alumina in the top slag controls the oxygen activity, since the oxygen affinity of aluminium is very high compared to other alloying elements [14, 27]. The reaction is

$$2\underline{Al} + 3\underline{O} = Al_2O_3(s) \tag{4}$$

$$\Delta G^0 = -1205115 + 386.714\ \mathrm{T} \tag{5}$$

The equilibrium constant K for Eq. (4) is

$$K = exp\left(\frac{-\Delta G^0}{RT}\right) = \frac{a_{Al_2O_3}}{a_{Al}^2 a_O^3} \tag{6}$$

$$loga_{Al_2O_3} = \frac{\{-0.275(\%CaO) + 0.167(\%MgO)\}}{(\%SiO_2)} + 0.033(\%Al_2O_3) - 1.560 \tag{7}$$

$$a_i = f_i[wt\%Y_i] \tag{8}$$

The aluminum activity is the product of f_{Al} and the aluminum content in the liquid steel. The activity coefficient f_i of a dissolved element i in steel can be calculated by Wagner's equation written as [21]

$$logf_i = \sum_j e_i^j [wt\%Y_j] \tag{9}$$

where i, j are S, C, Si, Al, Mn and O; e_i^j is the interaction parameters between these elements in liquid steel. The aluminum activity a_{Al} is the product of f_{Al} and the aluminum content in the liquid steel.

The sulfide capacity $C_{S^{2-}}$ is determined by using Young's model [13, 28]. The overall mass transfer coefficient of element i is calculated as

$$k_{eff,i} = \frac{k_{m,i}k_{s,i}L_i\rho_s}{k_{s,i}L_i\rho_s + \rho_m k_{m,i}} \tag{10}$$

where $k_{m,i}$, $k_{s,i}$ are the mass transfer coefficient of species i in liquid steel and slag, respectively. The mass transfer coefficient could be calculated through the Kolmogorov theory of isotropic turbulence as follows:

$$k_{m,i} = cD_{m,i}^{0.5}\left(\frac{\varepsilon_l}{\nu}\right)^{0.25} \tag{11}$$

$$k_{s,i} = cD_{s,i}^{0.5}\left(\frac{\varepsilon_l}{\nu}\right)^{0.25} \tag{12}$$

where c is a constant and is 0.4 for this work [13]. $D_{m,i}$ and $D_{s,i}$ are diffusion coefficients of species i in liquid steel and slag, respectively. The diffusivity of species in slag are generally two orders of magnitude lower than that of steel species. Diffusion coefficient of species in steel is assigned to 7.0×10^{-9} m^2/s [13].

Numerical Procedure

In this work, the fluid flow–desulfurization coupled model was solved using the commercial CFD software ANSYS Fluent combined with UDF. The geometry of the LMF is shown in Fig. 1. It has two off-centered plugs (diameter 0.092 m). The height of the ladle is 3.5 m and the initial thickness of the slag layer is 0.150 m. The argon flow rate was 0.051 m^3/s. All the computations were done at 1885 K. To validate and correct the model, typical industrial tests were carried out in LMF at Nucor Steel Tuscaloosa, United States. The chemical composition of the initial steel and slag are shown in Table 1.

Results and Discussion

Figure 2 illustrates the velocity vector for the steel and slag phases. The flow buoyancy from the injected gas causes high upward-directed steel velocities in the ladle. Then at the steel surface, downward convection forces the steel to return from the open eye back to the steel bath. When the steel flow reaches the ladle wall, it is directed downward along the ladle wall toward the bottom, causing circulation loops in the ladle. It is obvious that steel movement can result in radial velocities and fluctuation in the slag. And the slag and metal try to entrap each other, which forms a two-phase mixing zone.

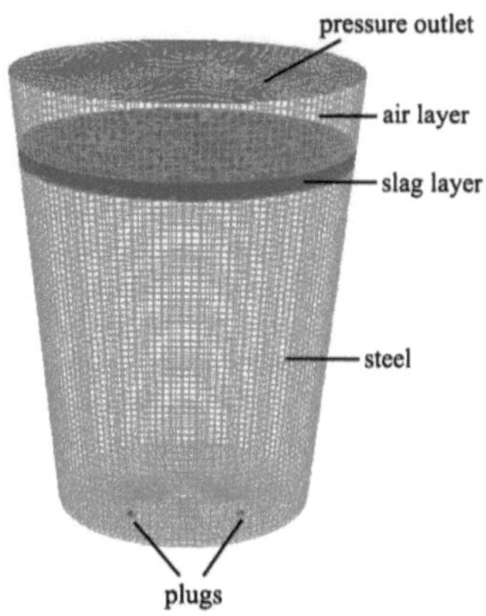

Fig. 1 Meshed geometry of the LMF system

Table 1 Chemical composition of slag and liquid steel at the initial time

Steel [wt%]					Slag (wt%)				
(S)	(Al)	(Si)	(Mn)	(C)	(Al_2O_3)	(SiO_2)	(CaO)	(MgO)	(FeO)
0.0246	0.0604	0.004	0.661	0.085	36	3.7	53.3	5.5	1.5

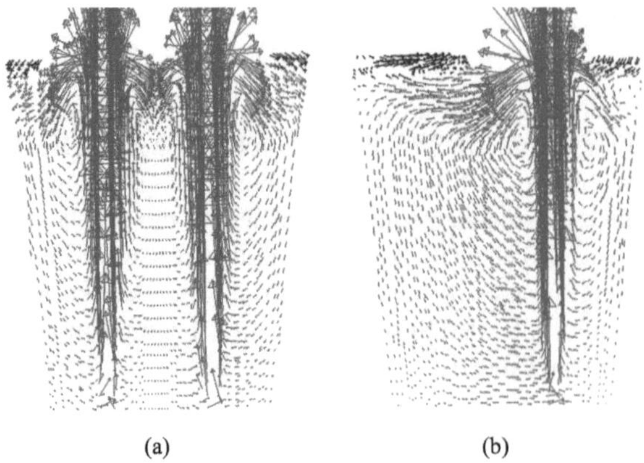

Fig. 2 The velocity vector of steel/slag phases **a** of the two-plug cross section, **b** of the one-plug cross section

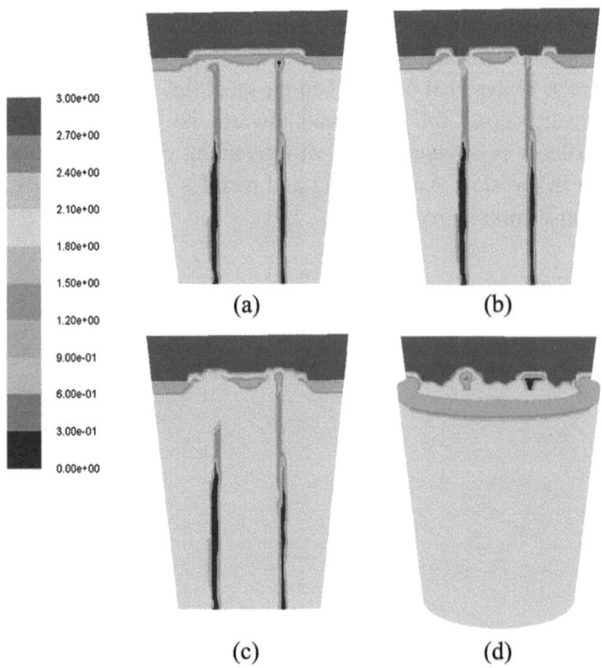

Fig. 3 **a–c** The contour of argon/steel/slag/air phases during the slag eye formation of the two-plug cross section: 2.3, 2.4, 2.5 s. **d** The phase contour of the steel/slag phase in the ladle with the two-plug cross section

Figure 3a–c show the slag eye formation process under argon gas flow. As the argon gas flow impinges the slag intermittently, the upwelling argon gas forces the steel upward and overcomes the slag tension, forming open slag eyes. The fluctuation pattern of liquid surface at the slag/air free surface is displayed in Fig. 3d. Figure 4a shows the predicted open slag eyes. The predicted diameter of the slag eyes varies in the range of 0.5–1 m. Their size is in reasonable agreement with experimental observations at Nucor (see Fig. 4b).

Figure 5 shows the desulfurization reaction rate in the ladle. It can be observed that reaction occurs in the slag/steel interaction region. Figure 6 compares the calculated average sulfur content in the steel phase with the measured data at Nucor Steel as a function of time. It can be seen that the predicted sulfur contents agree well with the measured data, which indicates that the current model can accurately predict the desulfurization kinetics in gas stirred ladle.

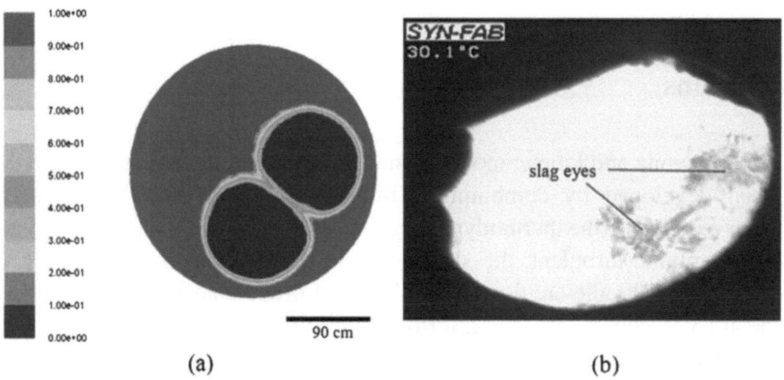

Fig. 4 **a** Volume fraction profile for the slag phase, **b** open slag eye observation in the real ladle

Fig. 5 Predicted desulfurization reaction rate [unit: kmol/(m^3s)] at 50 s in the ladle

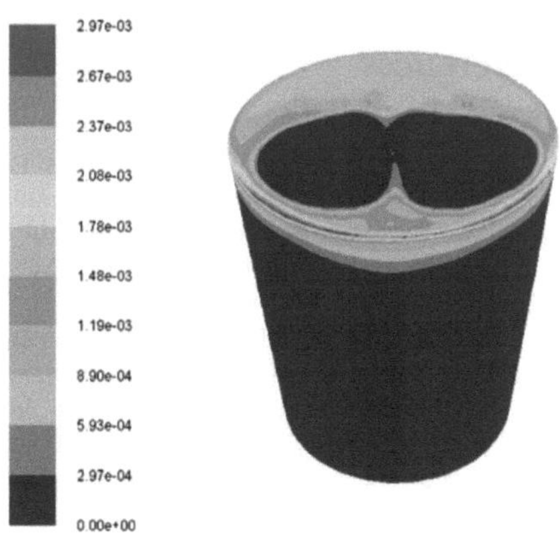

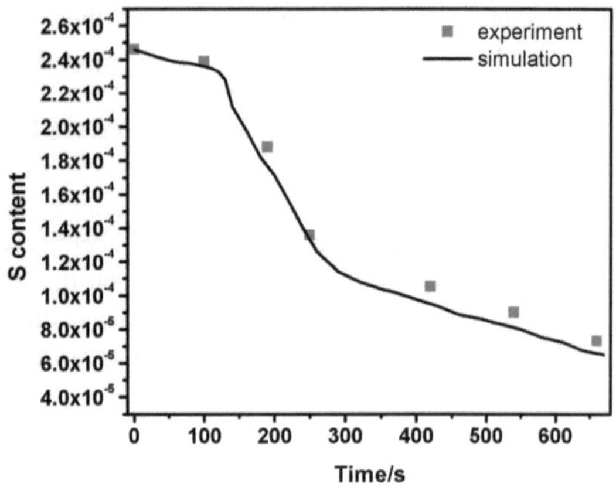

Fig. 6 Predicted and measured mass fraction of sulfur in the steel with time

Conclusions

A thermodynamic and kinetic model of sulfur refining in the argon gas-stirred ladle has been developed by combining a three-dimensional fluid-flow model with equations expressing the thermodynamics of desulfurization. Using the CFD model, the bubbly plume turbulent, the slag eyes and the circulation loops of the liquid flow in gas-stirred ladle are described. The fluid flow behavior shows that the slag and metal try to entrap each other, forming a two-phase mixing zone. The resulting substantial interface area between the two phases provides very favorable kinetic conditions for the chemical reactions. For the desulfurization kinetics model, the reactions at the metal/slag bath interface as well as the entrapped droplet surface are included. The predicted sulfur content changing with time in the ladle agrees reasonably well with the measured data.

References

1. J. Diao, B. Xie, S.S. Wang, Research on slag modifying agents for CaO–Mg based hot metal desulphurisation. Ironmaking Steelmaking **36**(7), 543–547 (2009)
2. P.K. Iwamasa, R.J. Fruehan, Effect of FeO in the slag and silicon in the metal on the desulfurization of hot metal. Metall. Mater. Trans. B **28B**, 47–57 (1997)
3. D. Takahashi, M. Kamo, Y. Kurose, H. Nomura, Deep steel desulphurisation technology in ladle furnace at KSC. Ironmaking Steelmaking **30**(2), 116–119 (2003)
4. D.S. Vinoo, D. Mazumdar, S.S. Gupta, Optimisation and prediction model of hot metal desulphurisation reagent consumption. Ironmaking Steelmaking **34**(6), 471–476 (2007)
5. W. Lv, Z. Xie, Z. Mao, P. Yuan, M. Jia, Hybrid modelling for real-time prediction of the sulphur content during ladle furnace steel refining with embedding prior knowledge. Neural Comput Appl. **25**, 1125–1136 (2014)
6. J. Lehmann, M. Nadif, Interactions between metal and slag melts: steel desulfurization. Rev. Mineral. Geochem. **73**(1), 493–511 (2011)

7. D. Mazumdar, R.I.L. Guthrie, The physical and mathematical modelling of gas stirred ladle systems. ISIJ Int. **35**(1), 1–20 (1995)
8. K. Krishnapisharody, G.A. Irons, A model for slag eyes in steel refining ladles covered with thick slag. Metall. Mater. Trans. B **46**(1), 191–198 (2015)
9. X.M. Yang, C.B. Shi, M. Zhang, G.M. Chai, F. Wang, A thermodynamic model of sulfur distribution ratio between CaO–SiO_2–MgO–FeO–MnO–Al_2O_3 slags and molten steel during LF refining process based on the ion and molecule coexistence theory. Metall. Mater. Trans. B **42**(6), 1150–1180 (2011)
10. O.J. Ilegbusi, J. Szekely, The modeling of gas-bubble driven circulations systems. ISIJ Int. **30** (9), 731–739 (1990)
11. Y.H. Liu, Z. He, L.P. Pan, Numerical investigations on the slag eye in steel ladles. Adv. Mech. Eng. **6**, 834103-1–834103-6 (2014)
12. W.T. Lou, M.Y. Zhu, Numerical simulation of gas and liquid two-phase flow in gas-stirred systems based on Euler-Euler approach. Metall. Mater. Trans. B **44**(5), 1251–1263 (2013)
13. W.T. Lou, M.Y. Zhu, Numerical simulation of slag-metal reactions and desulfurization efficiency in gas-stirred ladles with different thermodynamics and kinetics. ISIJ Int. **55**, 961–969 (2015)
14. M.A.T. Andersson, L.T.I. Jonsson, Pär G. Jönsson, A thermodynamic and kinetic model of reoxidation and desulphurisation in the ladle furnace. ISIJ Int. **40**, 1080–1088 (2000)
15. P.G. Jönsson, L.T.I. Jonsson, The use of fundamental process models in studying ladle refining operations. ISIJ Int. **41**(11), 1289–1302 (2001)
16. D. Guo, G.A. Irons, Modeling of gas-liquid reactions in ladle metallurgy: part II. Numerical simulation. Metallur. Mater. Trans. B **31B**(12), 1458–1464 (2000)
17. P. Sulasalmi, A. Krn, T. Fabritius, J. Savolainen, CFD model for emulsification of slag into the steel. ISIJ Int. **49**(11), 1661–1667 (2009)
18. S.T. Johansen, F. Boysan, Fluid dynamics in bubble stirred ladles: part II. Mathematical modeling. Metallur. Mater. Trans. B **19B**(10), 755–764 (1988)
19. E. Ramström, Mass Transfer and slag-metal reaction in ladle refining-a CFD approach, pp. 1–31. Licentiate Thesis, Royal Institute of Technology, Stockholm, Sweden, 2009
20. M.A.T. Andersson, P.G. Jönsson, M. Hallberg, Optimisation of ladle slag composition by application of sulphide capacity model. Ironmaking Steelmaking **27**(4), 286–293 (2000)
21. N. El-Kaddah, J. Szekely, Mathematical model for desulphurization kinetics in ar-stirred ladles. Ironmaking Steelmaking **2**, 269–278 (1981)
22. E. Pretorius, *Fundamentals of EAF and Ladle Slags and Ladle Refining Principles*, (Bakers Refractories, 2015)
23. Fluent 6.3, User's Guide Manual Fluent Inc. and Ansys's Fluent (2006), http://ansys.com/
24. M.V. Annaland, N.G. Deen, J.A.M. Kuipers, Numerical simulation of gas bubbles behaviour using a three-dimensional volume of fluid method. Chem. Eng. Sci. **60**(11), 2999–3011 (2005)
25. D. Roy, P.C. Pistorius, R.J. Fruehan, Effect of silicon on the desulfurization of al-killed steels: part I. Mathematical model. Metallur. and Mater. Trans. B **44B**(10), 1086–1094 (2013)
26. P.G. Jonsson, L.T.I. Jonsson, The use of fundamental process models in studying ladle refining operations. ISIJ Int. **41**, 1289–1302 (2001)
27. M. Andersson, M. Hallberg, L. Jonsson, P. Jönsson, Slag–metal reactions during ladle treatment with focus on desulphurisation. Ironmaking Steelmaking **29**, 224–232 (2002)
28. R.W. Young, J.A. Duffy, G.J. Hassall, Z. Xu, Use of Optical basicity concept for determining phosphorus and sulfur slag metal partitions. Ironmaking Steelmaking **19**, 201–219 (1992)

Numerical Study of the Fluid Flow and Temperature Distribution in a Non-transferred DC ARC Thermal Plasma Reactor

Yudong Li and Ramana G. Reddy

Abstract Numerical modeling of the thermal plasma process was carried out based on the thermal plasma reactor in our lab and confirmed using experimental data. The inlet boundary conditions of a non-transferred DC arc thermal plasma reactor were used in modeling the temperature and fluid flow distribution in the reactor. Different mesh grid sizes were used to confirm the model is independent of grid size. Temperature profile and gas flow distribution in the thermal plasma reactor were developed by the computational fluid dynamics (CFD) with ANSYS Fluent. The predicted temperatures are in good agreement with the experimentally measured temperatures in the reactor. The influence of plasma torch input power as well as the plasma gas flux on the temperature distribution was investigated using this model. The influence of power input and gas flow rate on temperature and velocity distributions are not independent. Generally, higher power input and lower gas flow rate will give rise to the temperature increase in the reactor.

Keywords Thermal plasma · Material plasma synthesis · Computational fluid dynamics · Arc jet flow · Plasma model

Introduction

Thermal plasma processing (TPP) techniques [1] are widely used for spraying, coating, synthesis, sintering, extractive metallurgy and waste treatment. Due to its high energy content and high energy density, TPP brings large opportunities for materials processing. It has a lot of advantages in synthesizing materials, such as, enhanced kinetics because of plasma and gaseous state reactions, lower activation

Y. Li · R.G. Reddy (✉)
Department of Metallurgical and Materials Engineering,
The University of Alabama, Tuscaloosa, AL 35487, USA
e-mail: rreddy@eng.ua.edu

S. Wang et al. (eds.), *Applications of Process Engineering Principles in Materials Processing, Energy and Environmental Technologies*,
The Minerals, Metals & Materials Series, DOI 10.1007/978-3-319-51091-0_43

energy in the plasma state and high purity synthesizing because of its clean atmosphere. Moreover, the thermal plasma reactor (TPR) in our lab integrated a quench tube for shock quenching which will quench the product in a short amount of time forming nanoscale products. There are numerous reports on the production of high temperature ceramics [2, 3] and composites by TPP such as TiC nano-powders [4], TiC/TiN-Al(Ti) ultrafine composite powders [5], and SiC nano-powders [6]. Due to the high temperature and the set up of TPR in our lab, it is not possible to measure the gas flow and temperature distribution. However, the computational fluid dynamics (CFD) can be used as a powerful tool to investigate the flow pattern and heat transfer inside the TPR and to gain a better understanding of the physical phenomena of TPP.

Plasma jet flow is highly complex due to high temperature and high velocity. There are numerous investigations that have been done in modeling of the plasma spray process using CFD. Mashayak et al. has done CFD modeling of the thermal plasma process for waste treatment [7]. They used steady state incompressible Navier-Stokes equations with a standard k-ε turbulence model to account for the fluid flow. Lorcet et al. has done the kinetics modeling with CFD of biomass gasification process using TPR [8]. Matveev et al. has done TPP modeling using CFD for coal gasification process using a hybrid plasma torch [9]. Fan et al. has investigated the effect of operation parameters [10], including current and flow rates of primary and secondary gas, on coating quality in plasma spray process. Mankelevich et al. did modeling of the dc arc jet process in CVD reactors [11], which is basically very similar to TPP, using a two-dimensional model. Williamson et al. has done modeling of the high-velocity and high-temperature plasma jet [12]. Agon et al. has developed a three-dimensional CFD model to investigate the thermophysical properties of gas mixture effect on fluid flow in a hybrid H_2O/Ar plasma jet flow process [13].

In this work, we developed a three-dimensional CFD model based on the DC arc TPR in our lab which described the fluid flow and gas-solid-liquid coupled heat transfer phenomenon in the TPP. The plasma inlet boundary conditions are approximated using the known DC arc torch input power and the torch gas flow rate. Experimental data of the temperature measured at different thermocouple places were used to justify the model. Using the velocity profile as well as the temperature distribution inside TPR, we are able to have better understanding of the material synthesis process.

Experimental Method

The model used in this study was developed to study the thermal plasma synthesis system in our lab, which contains a water cooling system, plasma generating system (with a non-transferred PT-50C plasma torch in DC arc mode), particle feeding system, reactor chamber and temperature measurement system. Figure 1 shows the photo and schematic diagram of the TPR. The TPR mainly consists of three zones.

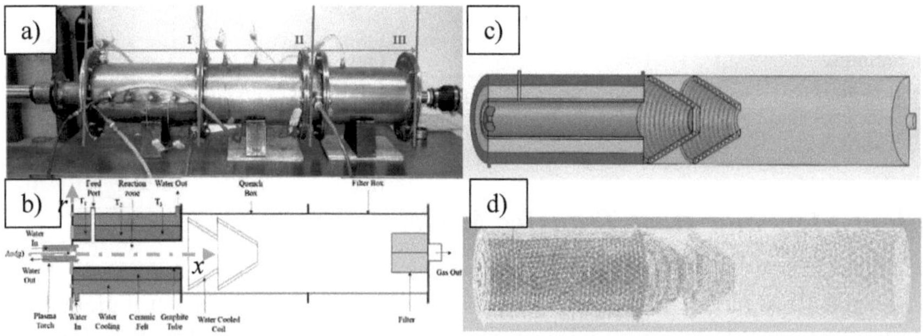

Fig. 1 Photograph (**a**) and schematic diagram (**b**) of the TPR (**c**) and (**d**) are Geometry and one computational mesh used in this study, respectively

Fig. 2 Temperature and velocity distribution at nozzle exit

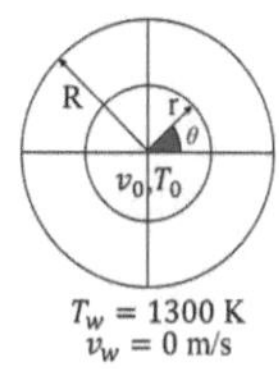

They are the reaction zone (x = 0–450 mm), Zone **I**; the quenching zone (x = 450–900 mm), Zone **II**; the filter zone (x = 900–1350 mm), Zone **III**. Outer shell of the whole reactor is made of 316L stainless steel. Zone **I** consists of five parts: a plasma gun port for connecting a plasma gun, a powder feeding port, a water-cooling jack ($r = \pm 87$–115 mm) with cooling water inlet port and outlet port, an insulation layer ($r = \pm 52$–87 mm) and a graphite tube ($r = \pm 45$–52 mm). Zone **II** and Zone **III** are basically two stainless steel chambers with two cone-shape water-cooled copper quench coils and a cloth filter, respectively.

Geometry and Computational Mesh

The three-dimensional geometry and a representative computational mesh used in this work are shown in Fig. 2, which used same dimension parameters as the TPR to ensure geometrical similarity for modeling [14]. All features, including three layers as we mentioned above, at Zone **I** are included in the model. The cooling coil at the downstream of Zone **I** have large effect on fluid flow as well as heat transfer. So we considered the design of coils including a cap of the first coil which is in contrast to the previous study [15]. Connections of these two cone-shape cooling coils will not have a large influence and thus be reasonably neglected.

Table 1 A summary of governing equations

Continuity	$\nabla \cdot (\rho \vec{v}) = 0$	(1)
Momentum	$\nabla \cdot (\rho \vec{v}\vec{v}) = -\nabla p + \nabla \cdot \left(\mu[(\nabla \vec{v} + \nabla \vec{v}^T) - \frac{2}{3}\nabla \cdot \vec{v}I]\right) + \rho \vec{g}$	(2)
Energy	$\nabla \cdot (\vec{v}(\rho E + p)) = \nabla \cdot (k\nabla T)$	(3)

Fluid Flow and Heat Transfer Model

The fluid flow model in this work is assumed to be steady state turbulent flow. A standard *k*-ε turbulence model was used to account for the turbulent flow [16, 17]. This model is capable of reproducing the overall flow field characteristics, including temperature, velocity and so on, with fair accuracy [17, 18]. Even though this model may not be suitable for all plasma jet flow conditions, it offers a practical compromise between simplicity and much more complicated models. Thus, considering our study is mainly focused on reproducing temperature and velocity profile, the standard *k*-ε turbulence model is feasible for this study. Governing equations are summarized in Table 1 [19].

Boundary Conditions

In this study, the flow of plasma from nozzle exit was defined by means of temperature and velocity. The profiles of temperature and velocity can be expressed by the following functions [12, 13, 20, 21]:

$$T = (T_0 - T_w)\left[1 - \left(\frac{r}{R}\right)^{n_T}\right] + T_w \tag{4}$$

$$v = v_0\left[1 - \left(\frac{r}{R}\right)^{n_v}\right] \tag{5}$$

where T_0 and v_0 are temperature and velocity at the center of nozzle exit respectively. T_w is the temperature at the nozzle outlet wall. Considering the nozzle is made of copper, 1300 K is a reasonable approximation of T_w [13]. Variables in Eqs. (4) and (5) are illustrated in Fig. 3.

Since we can only measure the plasma gun input power and the gas flow rate, we need to calculate for four variables, i.e. T_0, n_T, v_0 and n_v,based on Eqs. (4) and (5) respectively. We chose n_v value of 1.688 and n_T of 2.5 to solve for the value of v_0 and T_0 at different conditions in order to have a similar velocity distribution based on both experimental and numerical velocity distributions [13, 21, 22]. Calculated velocity distributions and temperature distribution are used as boundary conditions for the nozzle inlet by the User Defined Function (UDF). The UDF in this work also contains the customization of gas density considering materials properties used in this work are the same as those described in our previous work [5, 15, 23].

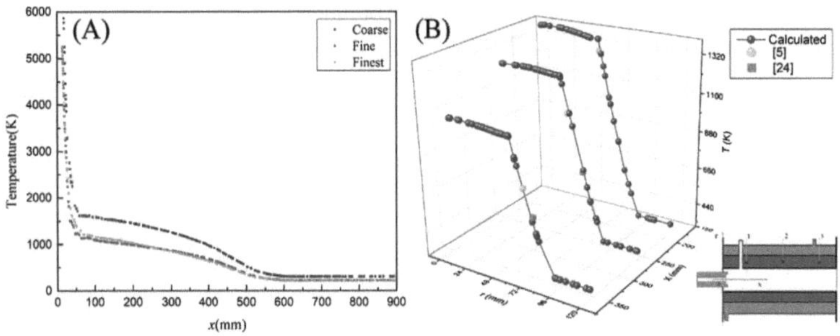

Fig. 3 **A** is temperature profile on *x*-axis using different mesh and **B** is the validation of the model

Grid Independence Study

In the grid independence study, we constructed three different computational meshes. They consist of 977,084 (coarse), 1,417,707 (fine) and 2,156,051 (finest) hexahedral cells respectively. In the grid independence study, the plasma power input was set to be 25 kW. Plasma gas flow rate was 6SCFM. Feeding gas flow rate was set to be 7.5 LPM. Temperature profiles along the axis of Zone I and Zone II downstream the nozzle exit of three different grid sizes are compared (at r = 0 mm, x = 0–900 mm, denoted as axis, also see Fig. 1). Results are shown in Fig. 3A. We can see that, using coarse grid will produce huge difference in the temperature distribution along x-axis. But, the fine grid agreed with the finest grid reasonably well. Due to the computational cost consideration, in this study we will use the fine mesh grid to achieve reasonable results reliability.

Model Verification

Simulations were done using a plasma torch input power described previously [5, 24], i.e. 25.2 kW, for verification of the model. The radical temperature distributions at different thermocouple positions were calculated. The thermocouple positions are listed in Table 1.

The calculated temperature was compared with experimental data as shown in Fig. 3B and Table 2. There's not much difference between the calculated and experimentally measured temperature in previous works at different thermocouple points. So we can conclude that the model we developed in this work has a good agreement with experiment data and thus can be used for prediction of the temperature and velocity profile.

Table 2 Thermocouple positions and comparison of calculated temperature with experimental measurements in previous work [5, 24]

Thermocouple position	Radius (*r*/mm)	Length (*x*/mm)	T_{Exp}	T_{Calc}	References
1	52.4	165.1	1273	1319.70	[5]
2	57.4	254.9	1023	1078.03	
3	62.4	342.9	773	771.22	
2′	70.0	255.0	698	725.05	[24]
3′	70.0	343.0	625	586.78	

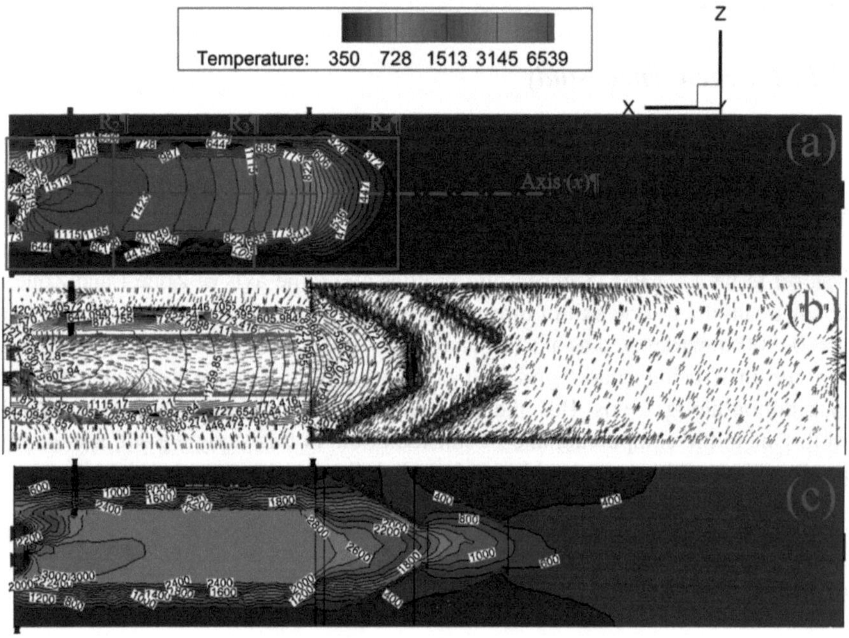

Fig. 4 Temperature profile **a** and velocity vector distribution **b** of 25.2 kW and 6SCFM; **c** is the temperature distribution in the model without cap at the first coil [15]

Results and Discussion

Temperature and Velocity Distribution

Figure 4a, b shows the temperature distribution and velocity distribution of a power input of 25.2 kW and plasma gas flow rate of 6 SCFM along a center slice of the TPR. We can see that from the first quenching coil to the outlet at Zone III, the fluid flow and temperature distribution is almost uniform and do not vary much from those before the first cooling coil.

In comparing with previous work [15], this model includes a cap at the first coil which was omitted by previous work, shown in Fig. 4c. Comparing Fig. 4a, c, we can see that with a cap the temperature is lowered at Zone I. And the temperature calculated in this work is closer to what is measured experimentally than previous tests.

Plasma Input Power Effect

Plasma input power (P) has a great effect on the synthesis process in TPR. With all other boundary conditions fixed as stated in the boundary condition section, at plasma gas flow rate of 6SCFM, the input power of 18.2, 22.4 and 25.2 kW were used in the calculation which is the same as the experimental operation parameters in previous work [5, 25]. The input power effect on temperature distribution was investigated as shown in Fig. 5A, B. As shown in Fig. 5, plasma power input has a large effect on the temperature distribution inside Zone I. The temperature will generally increase with the increasing power input. Zone I can be divided into three different sections, R2 (T > 1500 K), R3 (1500 K > T > 1000 K) and R4 (T < 1000 K). As seen that with the power input increase, section R2 and R3 enlarged, which is good for high temperature favorable chemical reactions in material synthesizing.

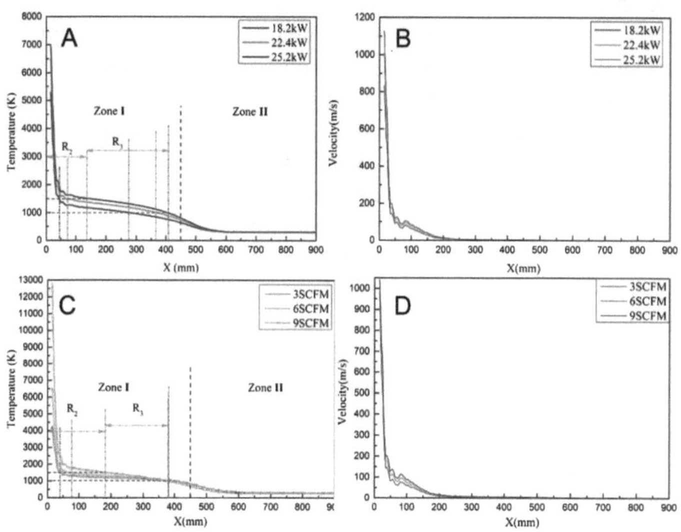

Fig. 5 Temperature and velocity distribution on axis at different plasma power (**A**, **B**) and different plasma gas flow rates (**C**, **D**)

Plasma Gas Flow Rate Effect

Plasma gas flow rate (or the plasma gun input gas flow rate, denoted as Q) turned out to be a very important factor influencing the material synthesis process in this study. However, to our knowledge, people have rarely studied the effect of plasma gas (argon gas in this study) flow rate. No work has been done considering the plasma gas flow rate effect both experimentally and numerically.

In this study, three different cases were calculated using different Q, 3SCFM, 6SCFM and 9SCFM. The power input in this study is fixed at 22.4 kW. After calculations were done of these three cases, temperature distributions on x-axis at different input power are plotted for comparison as shown in Fig. 5C, D.

Considering both axial temperature distributions, we can see that as the plasma gas flow rate decreased from 9SCFM to 3SCFM, section R2 is increasing while section R4 remains the same. Lower velocity may increase the residence time of particle inside the reactor which is good for vaporization of particles as well as reaction kinetics. Thus, we can conclude that, lower the plasma gas flow rate will increase the residence time and enhance the reaction kinetics due to the increasing of temperature.

Conclusions

In this study, a numerical model was developed for thermal plasma reactor (TPR). The geometry used in this model was simplified based on the TPR from our lab. Different mesh grid sizes were used to confirm the model is independent of grid size. Data predicted by the model were compared with experimental data. The influence of plasma torch input power and plasma gas flow rate on the TPR was investigated using this model. The numerical results suggested that higher input power will result in higher temperature which is an advantage in producing materials which are high temperature favorable. The model also implied that plasma gas flow rate is a crucial factor that can influence the material synthesis process. Lower plasma gas flow rate is preferred considering both its temperature profile and velocity distribution. Though the modeling, we can see that the influence of power input and gas flow rate on temperature and velocity distribution is not independent but rather coupled. So in modeling the material synthesis in TPR, we should simultaneously consider the effects of both plasma power input as well as gas flow rate on temperature and velocity distributions.

Acknowledgments The authors are thankful for the financial support from National Science Foundation (NSF) agency (Grant No. DMR-1310072), American Cast Iron Pipe Company (ACIPCO) and The University of Alabama during the course of the current research project.

References

1. M.I. Boulos, Thermal plasma processing. IEEE Trans. Plasma Sci. **19**(6), 1078–1089 (1991)
2. H.-C. Lee, S. Dhage, M.S. Akhtar, D.H. Kwak, W.J. Lee, C.-Y. Kim, O.-B. Yang, A simulation study on the direct carbothermal reduction of SiO_2 for Si metal. Curr. Appl. Phys. **10**(2), S218–S221 (2010)
3. N. Thakkar, R.G. Reddy, Thermal plasma production of B4C nanopowders. J. Manuf. Sci. Prod. **7**(2), 87–100 (2006)
4. L. Tong, R.G. Reddy, Synthesis of titanium carbide nano-powders by thermal plasma. Scr. Mater. **52**(12), 1253–1258 (2005)
5. S. Niyomwas, B. Wu, R.G. Reddy, *Synthesis of Fe-TiN Composites by Thermal Plasma Processing*. Ultrafine Grained Materials (as held at the 2000 TMS Annual Meeting, 2000), pp. 89–98
6. L. Tong, R.G. Reddy, Thermal plasma synthesis of SiC nano-powders/nano-fibers. Mater. Res. Bull. **41**(12), 2303–2310 (2006)
7. S.Y. Mashayak, *CFD Modeling of Plasma Thermal Reactor for Waste Treatment* (Purdue University West Lafayette, 2009)
8. H. Lorcet, D. Guenadou, C. Latge, M. Brothier, G. Mariaux, A. Vardelle, kinetics modeling of biomass gasification under thermal plasma conditions. Application to a refractory species: the methane. International symposium on plasma chemistry, Germany (Bochum, 2009)
9. I.B. Matveev, S. Serbin, Modeling of the coal gasification processes in a hybrid plasma torch. IEEE Trans. Plasma Sci. **35**(6), 1639–1647 (2007)
10. F. Qunbo, W. Lu, W. Fuchi, Modeling influence of basic operation parameters on plasma jet. J. Mater. Process. Technol **198**(1), 207–212
11. Y.A. Mankelevich, M. Ashfold, A. Orr-Ewing, Measurement and modeling of $Ar/H_2/CH_4$ arc jet discharge chemical vapor deposition reactors II: Modeling of the spatial dependence of expanded plasma parameters and species number densities. J. Appl. Phys. **102**(6), 063310 (2007)
12. R. Williamson, J. Fincke, D. Crawford, S. Snyder, W. Swank, D. Haggard, Entrainment in high-velocity, high-temperature plasma jets.: Part II: computational results and comparison to experiment. Int. J. Heat Mass Transf. **46**(22), 4215–4228 (2003)
13. N. Agon, J. Vierendeels, M. Hrabovský, A. Murphy, G. Van Oost, Interaction of a H_2O/AR plasma jet with nitrogen atmosphere: effect of the method for calculating thermophysical properties of the gas mixture on the flow field. Plasma Chem. Plasma Process. **35**(2), 365–386 (2015)
14. L. Yan, L. Yudong, Z. Ting'an, F. Naixiang, Research on the penetration depth in aluminum reduction cell with new type of anode and cathode structures. JOM **66**(7), 1202–1209 (2014)
15. Y. Li, R. Reddy, Experimental and numerical investigation of thermal plasma synthesis of silicon. J. Manuf. Sci. Prod. **15**(4), 345–354 (2015)
16. J. Ramshaw, C. Chang, Computational fluid dynamics modeling of multicomponent thermal plasmas. Plasma Chem. Plasma Process. **12**(3), 299–325 (1992)
17. C. Chang, J. Ramshaw, Numerical simulations of argon plasma jets flowing into cold air. Plasma Chem. Plasma Process. **13**(2), 189–209 (1993)
18. J. Fincke, C. Chang, W. Swank, D. Haggard, Entrainment and demixing in subsonic thermal plasma jets: comparison of measurements and predictions. Int. J. Heat Mass Transf. **37**(11), 1673–1682 (1994)
19. A. Fluent, Release 14.0, User Guide, Ansys, Inc., Lebanon, US (2011)
20. H. Lorcet, M. Brothier, D. Guenadou, C. Latge, A. Vardelle, Modeling bio-oil gasification by a plasma process. High Temp. Mater. Proc.: An Int. Q. High-Technol. Plasma Process. **14**(1–2) (2010)
21. M. Hrabovsky, V. Kopecký, V. Sember, T. Kavka, O. Chumak, M. Konrad, Properties of hybrid water/gas DC arc plasma torch. IEEE Trans. Plasma Sci. **34**(4), 1566–1575 (2006)

22. J. Fincke, D. Crawford, S. Snyder, W. Swank, D. Haggard, R. Williamson, Entrainment in high-velocity, high-temperature plasma jets. Part I: experimental results. Int. J. Heat Mass Transf. **46**(22), 4201–4213 (2003)
23. M. Ramachandran, R.G. Reddy, Thermal plasma synthesis of SiC. Advances in Manufacturing **1**(1), 50–61 (2013)
24. R.G. Reddy, L.V. Antony, Processing of SiC nano powders using thermal plasma technique. In *Proceedings of the International Conference on Nanotechnology: Scientific Challenges and Commercial Opportunities*, (Rhode Island, pp. 17–18, 2003)
25. L. Tong, R.G. Reddy, In situ synthesis of TiC-Al (Ti) nanocomposite powders by thermal plasma technology. Metall. Mater. Trans. B **37**(4), 531–539 (2006)

Part IX
Thermodynamics and Kinetics

Thermodynamic Studies on the Mg-B System Using Solid State Electrochemical Cells

Muhammad A. Imam and Ramana G. Reddy

Abstract This study provides a reasonable evaluation of the Mg-B binary system. Magnesium and Boron (Mg-B) alloys were investigated to determine their thermodynamic properties employing solid state electrochemical cells based on CaF_2 electrolyte represented as $(-)\mathrm{Pt, Ar}/\{\mathrm{Mg} + \mathrm{CaMgF_4}\}\|\mathrm{CaF_2}\|\{[\mathrm{Mg} - \mathrm{B}]_{\mathrm{alloy}} + \mathrm{CaMgF_4}\}/\mathrm{Ar, Pt}(+)$ Investigations were performed over the temperature range of 773–873 K to measure the electromotive force (EMF), which was used to derive the partial Gibbs Free energies of the alloys in the composition range of $X_B = 0.07$ to 0.95. The activities of Mg were also calculated from partial Gibbs Free energies. The activities of Mg in MgB_2, MgB_4, and MgB_7 were expressed as a function of temperature respectively, $\ln a_{Mg} = 4.27–9.32 \times 10^3/T$, $\ln a_{Mg} = -7.54 + 6.67\times10^3/T$, and $\ln a_{Mg} = 4.93–19.57 \times 10^3/T$, where T is the temperature in K. From this expression, the activity of Mg for these intermediate phases can be extrapolated at a higher temperature to get the accurate phase boundaries for the Mg-B system.

Keywords EMF · CaF_2 electrolyte · MgB_2 · MgB_4 · MgB_7 · Activity of Mg

Introduction

The binary Mg-B system drew attention due to the MgB_2 superconducting phase at 39 K [1]. This binary system contains other intermediate compounds such as MgB_4 and MgB_7. Some other intermediate compounds were reported in the literature such as Mg_3B_2 [2], MgB_6 [3], and MgB_{12} [3]. However, they have not been confirmed

M.A. Imam · R.G. Reddy (✉)
Department of Metallurgical and Materials Engineering,
The University of Alabama, Tuscaloosa, AL 35487, USA
e-mail: rreddy@eng.ua.edu

S. Wang et al. (eds.), *Applications of Process Engineering Principles in Materials Processing, Energy and Environmental Technologies*,
The Minerals, Metals & Materials Series, DOI 10.1007/978-3-319-51091-0_44

yet in any further study. In addition, it is reported that these phases are the result of using low purity materials in the early experiments or the mixture of stable borides and/or boron [4]. The phase diagram of the Mg-B system was reported by Massalski [5], which was extensively based on Spear's [6] assessment, confirming the three intermediate phases MgB_2, MgB_4, and MgB_7. Recently, Liu et al. [7] and Kim et al. [8] re-assessed the Mg-B binary phase diagram using CALPHAD and ab initio calculations. Their study reproduced the intermediate phases reported by Massalski. Moreover, their study did not include the Mg_3B_2, MgB_6, and MgB_{12} phases. The recent thermodynamic CALPHAD based calculation (Thermo-Calc and PANDAT) also corroborate the literature [9, 10]. All boride phases are reported as a line compound with no significant deviation from the stoichiometry [11]. The phase diagram of Mg-B system has been obtained from Thermo-Calc is shown in Fig. 1 [9]. Different phases, crystallographic structures, and decomposition temperature are summarized in Tables 1 and 2.

Thermodynamic data has been limited due to very few experimental phase equilibria studies performed on the Mg-B binary system. This is due to the experimental difficulties in working with Mg, which is highly reactive and possesses a high vapor pressure at high temperatures [12]. Cook et al. have carried out vapor pressure measurements using Knudsen effusion vacuum thermogravimetry over the temperature range of 873–1123 K and enthalpy measurements using a calorimetry method [13]. Brutti et al. have performed vapor pressure measurements

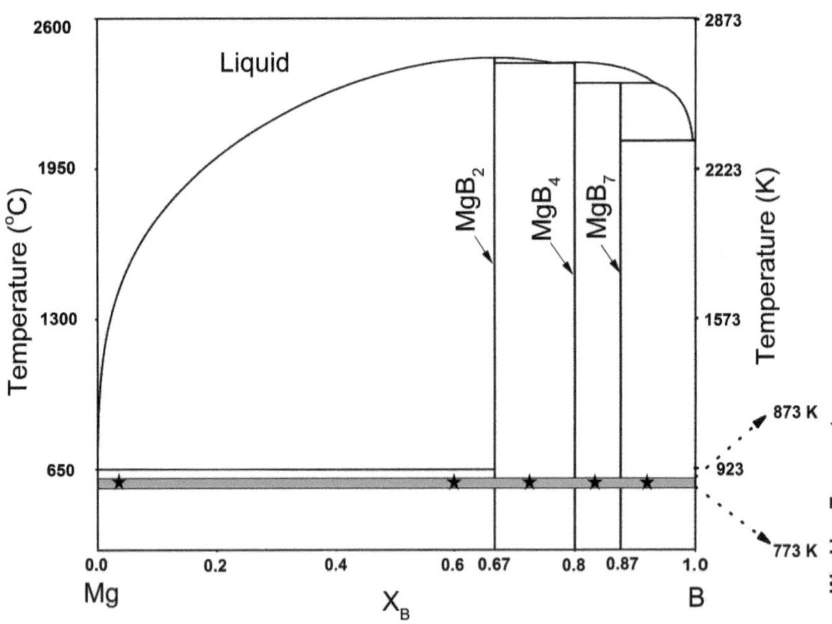

Fig. 1 Binary Mg-B phase diagram [9]

Table 1 Different phases and their crystallographic structures in the Mg-B system [15]

Phase	Approximate composition (X_B)	Pearson symbol	Prototype/crystal structure	Hermann mauguin
(Mg)	0–0.66	*hP*2	Mg/HCP	$P6_3/mmc$
MgB_2	0.67	*h*P3	AlB_2	P6/mmm
MgB_4	0.8	*o*P20	MgB_4	Pnma
MgB_7	0.87	*o*I64	MgB_7	Imma
(B)	>0.88	*h*R12	ZrCl	R-3m

Table 2 Comparison of phase decomposition temperature from different literature sources for Mg-B system

Phase	Massalski et al. [5] (Calc.) °C	Liu et al. [7] (Calc.) °C	Kim et al. [8] (Calc.) °C	Cook et al. [13] (Exp.) °C
MgB_2	1550	1545	1174	1268
MgB_4	1830	1735	1273	
MgB_7	2150	2150	2509	

using a Knudsen spectrometry technique over the range of 883 to 1154 K [14]. It is interesting to mention that the essential difference in phase decomposition values reported in Table 2 might be due to the use of a different thermodynamic database employed in a CALPHAD method and limited availability of experimental data. Using the high-temperature range data reported by Cook et al. and Brutti et al. and avoiding the high vapor pressure of Mg; we resolve to execute our thermodynamic measurement in the temperature range 773–873 K (Fig. 1) in a pure solid state electrochemical cell using CaF_2 as the solid electrolyte.

After obtaining an equilibrium state in a galvanic cell, the thermodynamic data would be more reliable and accurate. This equilibrium thermodynamic data obtained from the solid state electrochemical cell can be used in the optimization of the phase boundary calculations for the Mg-B system.

Experimental

Materials Preparation

Elemental magnesium powder and boron powder (both 99.99% pure, metal basis, purchased from Alfa Aesar, USA) were mixed thoroughly using different molar ratios of boron (X_B = 0.07, 0.4, 0.73, 0.84, and 0.95) to produce five different alloys. The mixing process was performed in an inert atmosphere glove box and then transferred into a sealed jar. After that, jar milling was conducted for 24 h to obtain a homogenous mixture of boron and magnesium. These mixtures were made into 13 mm diameter pellets under a uniaxial load of 5000 psi for 5 min using a

Carver cold pressing unit. The pellets were wrapped in Ta foil and transferred into a vacuum-sealed quartz tube with Cu getter to avoid oxidation. The pellets were then placed in a Fischer Scientific box furnace for sintering at 600 °C for 168 h. After that, a electrochemical cell electrode was made with a mixture of an equimolar ratio of CaF_2 and MgF_2 (both 99.99% pure, metal basis, purchased from Alfa Aesar, USA) with elemental Mg and previously prepared alloys for reference and working electrode respectively.

These mixtures were also made into a 13 mm diameter pellet under a uniaxial load of 5000 psi for 5 min using a Carver cold pressing unit. The pellets were wrapped in Ta foil and transferred into a vacuum-sealed quartz tube with Cu getter to avoid oxidation. The pellets were then placed in a Fischer Scientific box furnace for sintering at 600 °C for 336 h. It needs to be mentioned that equimolar ratio of CaF_2 and MgF_2 was used to prepare $CaMgF_4$.

Solid State Electrochemical Cell

A schematic diagram of the solid-state galvanic cell used in this study is shown in Fig. 2. Reddy et al. have described the detailed experimental procedure for the determination of phase stability of Ti-Al and Nb-Al binary system [16–19]. As shown in Fig. 2, the solid-state galvanic cell was in a single compartment alumina tube which was purged with ultra-high purity argon. Anhydrous calcium sulfate (dierite) was used in the argon purged line to remove the trace amount of moisture. A Cu gettering furnace temperature was maintained at 958 K before the galvanic cell compartment to remove the residual oxygen from the argon gas. Cu getters were also used in the bottom of the alumina tube to ensure no further oxidation. In addition, the chamber was evacuated using a vacuum pump. The assembly of the electrodes and the electrolyte (inset of Fig. 2) was placed in the isothermal zone of the vertical resistance furnace to maintain a constant temperature. A type K thermocouple was used to measure the cell temperature accurately. The reference electrode, single crystal CaF_2 electrolyte, and working electrodes were sandwiched between the Pt discs. This setup was pressed by alumina discs with the spring tightened alumina pressing tube. One end of the Pt/Rh wire was connected to the Pt discs while the other end was attached to the Keithly 2700 to obtain the EMF from the cell. When the cell voltage variation was within ± 0.1 mv over the period of 7–9 h, it was characterized as a steady voltage. That steady/stable voltage indicated that the system was in an equilibrium state.

Results and Discussion

The solid state electrochemical cells based on single crystal CaF_2 electrolyte is represented as

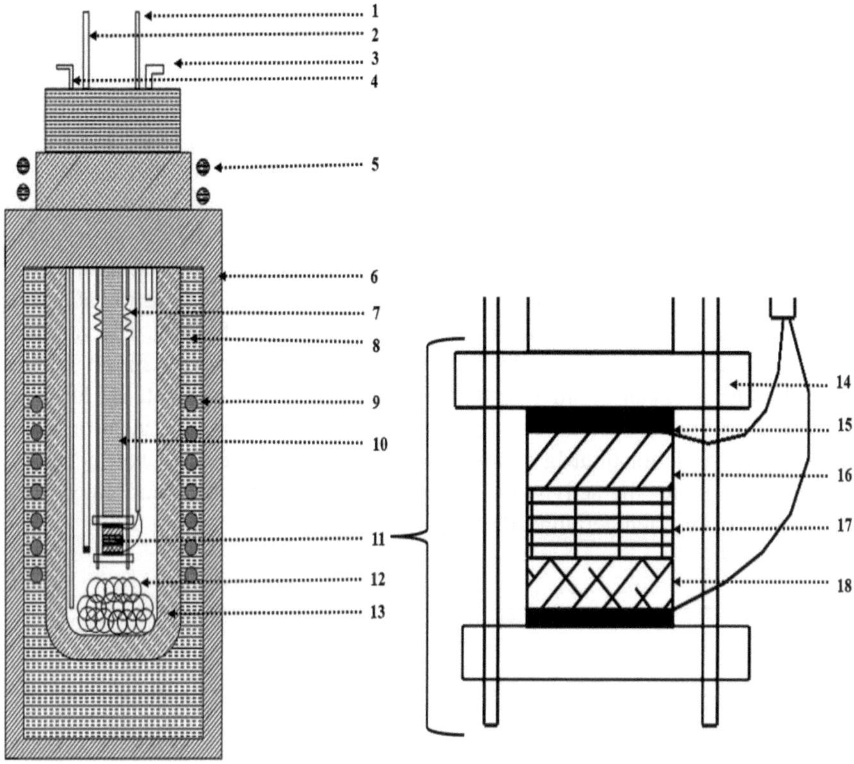

Fig. 2 Schematic diagram of solid state electrochemical galvanic cell: *1* Pt/Rh wires with alumina sleeve, *2* sample thermocouple (type k), *3* gas outlet, *4* gas inlet, *5* water cooled brass flange, *6* vertical tubular furnace, *7* spring, *8* fire brick, *9* heating coil, *10* alumina pressing tube, *11* galvanic cell assembly, *12* Cu-gutter, *13* alumina tube, *14* alumina support disc, *15* Pt disc, *16* working electrode, *17* CaF_2 electrolyte, and *18* reference electrode

$$(-)\text{Pt, Ar}/\{\text{Mg} + \text{CaMgF}_4\}\|\text{CaF}_2\|\{[\text{Mg} - \text{B}]_{\text{alloy}} + \text{CaMgF}_4\}/\text{Ar, Pt}(+) \quad (1)$$

The half-cell reactions for the galvanic cell are explicitly,

$$(-)\textit{electrode}: \text{Mg(Solid)} + \text{CaF}_2 + 2\text{F}^- = \text{CaMgF}_4 + 2e^- \quad (2)$$

$$(+)\textit{electrode}: \text{CaMgF}_4 + 2e^- = [\text{Mg} - \text{B}]_{alloy} + \text{CaF}_2 + 2\text{F}^- \quad (3)$$

The Eqs. (2) and (3) gives the net cell reaction simply,

$$\text{Mg(Solid)} = [\text{Mg} - \text{B}]_{alloy} \quad (4)$$

The EMF of the proceeding galvanic cell reaction is written as

$$\Delta\overline{G}_{Mg} = RT \ln a_{Mg} = -nFE \tag{5}$$

where n is the number of electrons participating in the half cell reactions, F is the Faraday's constant (F = 96,500 C/mol), and E is the EMF of the cell in volts (V). The natural logarithm of activity of magnesium (ln a_{Mg}) is calculated for the selected alloy compositions over the temperature range of 773–873 K. As we know that activity remains constant in two-phase regions. Therefore, the terminal activity values of the two-phase regions can be determined/estimated. The terminal activity of Mg reflects the single line compound of MgB_2, MgB_4, and MgB_7. The variations of ln a_{Mg} with inverse temperature for different phases are shown in Fig. 3. As seen from the Fig. 3, it can be inferred that the higher the boron content, the lower the activity of Mg in the system. This figure can be used to obtain the equation (linear fit, R-Square >94) for ln aMg for different phases, which are tabulated in Table 3. These equations for the activity can be used for the entire temperature range up to the decomposition temperature of any particular phase tabulated in Table 2.

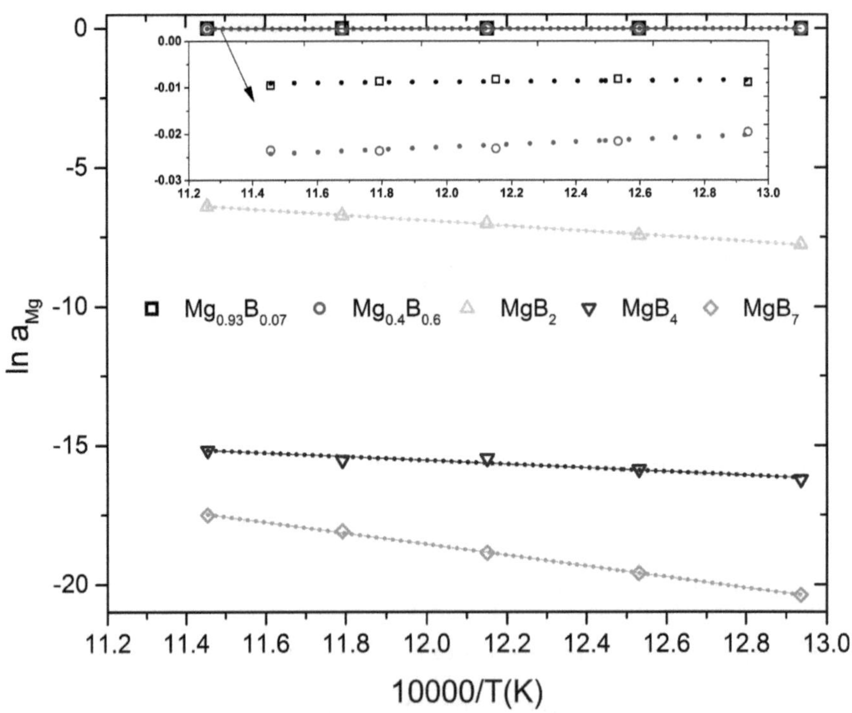

Fig. 3 Variation of ln a_{Mg} versus $10^4/T$ measured by solid state electrochemical galvanic cell

Table 3 Summarized ln a_{Mg} expressions as a function of temperature for different phases/alloys

Alloy/intermetallic phases	ln a_{Mg} (773–873 K)
$Mg_{0.93}B_{0.07}$	−0.0143 + 4.63/T
$Mg_{0.4}B_{0.6}$	−0.056 + 27.415/T
MgB_2	4.27–9.32 × 10^3/T
MgB_4	−7.54 + 6.67 × 10^3/T
MgB_7	4.93–19.57 × 10^3/T

Conclusions

In summary, the Mg-B binary system was evaluated using a solid-state galvanic cell over the temperature range of 773–873 K. The activity of Mg for different alloys was derived from the measured EMF data. The activity values for Mg in intermetallic phases, such as MgB_2, MgB_4, and MgB_7, were determined as a function of temperature. From this expression, we can extrapolate the activity of Mg for these intermediate phases at a higher temperature to get the accurate phase boundary for Mg-B system, which will be useful in constructing a more reliable Mg-B binary diagram.

Acknowledgements Authors gratefully acknowledge the financial support, Grant No. DMR-1310072, of the National Science Foundation (NSF).

References

1. J. Nagamatsu, N. Nakagawa, T. Muranaka, Y. Zenitani, J. Akimitsu, Superconductivity at 39 K in magnesium diboride. Nature **410**(6824), 63–64 (2001)
2. W.C. Johnson, Hydrides of boron and silicon (stock, alfred). J. Chem. Educ. **11**(4), 256 (1934)
3. Y. Markovskii, Y.D. Kondrashev, G.V. Kaputovskaya, O sostave i khimicheskikh svoistvakh boridov magniya. Zh. Obshch. Khim **25**, 433–444 (1955)
4. R. Naslain, A. Guette, P. Hagenmuller, Crystal chemistry of some boron-rich phases. J. Less Common Met. **47**, 1–16 (1976)
5. T.B. Massalski, H. Okamoto, P. Subramanian, L. Kacprzak, W.W. Scott, *Binary Alloy Phase Diagrams* (American Society for Metals Metals Park, Ohio, OH, USA, 1986)
6. K. Spear, *Rare Earth-Boron Phase Equilibria. In Boron and Refractory Borides* (Springer, Berlin, Germany, 1977), pp. 439–456
7. Z.K. Liu, Y. Zhong, D. Schlom, X. Xi, Q. Li, Computational thermodynamic modeling of the Mg-B system. Calphad **25**(2), 299–303 (2001)
8. S. Kim, D. Stone, J.I. Cho, C.Y. Jeong, C.S. Kang, J.C. Bae, Phase stability determination of the Mg–B binary system using the CALPHAD method and ab initio calculations. J. Alloy. Compd. **470**(1), 85–89 (2009)
9. J.O. Andersson, T. Helander, L. Höglund, P. Shi, B. Sundman, Thermo-Calc & DICTRA, computational tools for materials science. Calphad **26**(2), 273–312 (2002)
10. S.L. Chen, S. Daniel, F. Zhang, Y. Chang, X.Y. Yan, F.Y. Xie, R. Schmid-Fetzer, W. Oates, The PANDAT software package and its applications. Calphad **26**(2), 175–188 (2002)

11. S. Bohnenstiehl, M. Susner, S. Dregia, M. Sumption, J. Donovan, E. Collings, Experimental determination of the peritectic transition temperature of MgB_2 in the Mg–B phase diagram. Thermochim. Acta **576**, 27–35 (2014)
12. P.W. Gilles, Vapor pressure of the chemical elements. J. Am. Chem. Soc. **86**(24), 5702–5703 (1964)
13. L. Cook, R. Klein, W. Wong-Ng, Q. Huang, R. Ribeiro, P. Canfield, Thermodynamics of MgB_2 by calorimetry and Knudsen thermogravimetry. IEEE Trans. Appl. Supercond. **15**(2), 3227–3229 (2005)
14. S. Brutti, A. Ciccioli, G. Balducci, G. Gigli, P. Manfrinetti, A. Palenzona, Vaporization thermodynamics of MgB_2 and MgB_4. Appl. Phys. Lett. **80**(16), 2892–2894 (2002)
15. P. Villars, K. Cenzual, R. Gladyshevskii, *Handbook* (Walter de Gruyter GmbH & Co KG, Berlin, Germany, 2015)
16. S. Parida, R. Reddy, Thermodynamic properties of $CaTiF_5$ (s). J. Chem. Thermodyn. **39**(6), 888–892 (2007)
17. R. Reddy, A. Yahya, L. Brewer, Thermodynamic properties of Ti–Al intermetallics. J. Alloy. Compd. **321**(2), 223–227 (2001)
18. P. George, S. Parida, R. Reddy, Thermodynamic studies on the system Nb-Al. Metall. Mater. Trans. B **38**(1), 85–91 (2007)
19. S. Kumar, R. Reddy, L. Brewer, Phase equilibria in Ti_3Al-Nb alloys at 1000 °C. J. Phase Equilib. **15**(3), 279–284 (1994)

An Investigation on the Kinetics and Mechanism of Alkali Reduction of Mine Waste Containing Titaniferous Minerals for the Recovery of Metals

S. Parirenyatwa, L. Escudero-Castejon, S. Sanchez-Segado, Y. Hara and A. Jha

Abstract In a world where declining ore grades are increasingly common, it has become necessary to process low-grade feedstock. Carbothermic reduction in the presence of alkali (Na_2CO_3) has been adapted to beneficiate waste that contains titaniferous minerals (TiO_2 ca. 12 wt%), in order to recover valuable constituents such as TiO_2, Fe and V_2O_5. The waste from vanadium metal processing has environmental legacy as it leaves nearly 1 wt% V_2O_5 process waste, which is environmentally problematic due to V^{5+} ions in contact with water and soil. This investigation focuses on the kinetics and mechanism for alkali reduction of mineral waste bearing 10–12 wt% TiO_2, which we studied in the 1073–1323 K range. The thermogravimetric analysis (TGA) technique was used to record weight loss data. Two distinct regimes demonstrated mixed-control kinetics: (1) at low temperatures the activation energy was found to be 199 kJ mol^{-1}, which corresponds to the outward diffusion of O^{2-} ions; and (2) at high temperatures the calculated value was 130 kJ mol^{-1}, which is consistent with the activation energy for the outward diffusion of Fe^{2+} ions. The metallic iron, sodium titanate and sodium aluminosilicate phases that formed were characterised using X-ray powder diffraction (XPRD) and scanning electron microscopy (SEM) techniques, and their significance for metal recovery is explained.

Keywords Mine waste · Titaniferous minerals · Alkali reduction · Activation energy · Sodium titanate

S. Parirenyatwa (✉) · L. Escudero-Castejon · S. Sanchez-Segado · Y. Hara · A. Jha
Faculty of Engineering, School of Chemical and Process Engineering,
Institute for Materials Research, University of Leeds, Leeds LS2 9JT, UK
e-mail: pm12sp@leeds.ac.uk

S. Wang et al. (eds.), *Applications of Process Engineering Principles in Materials Processing, Energy and Environmental Technologies*,
The Minerals, Metals & Materials Series, DOI 10.1007/978-3-319-51091-0_45

Introduction

Titanium is an abundant metal in the Earth's crust and the minerals of industrial significance are ilmenite ($FeTiO_3$), leucoxene ($Fe_2O_3.TiO_2$) and rutile (TiO_2) [1]. Most titanium production, approximately 90%, is directed towards making TiO_2 pigment and the rest is focused on Ti metal [2]. The chloride and sulphate processes are the two methods utilised for the production of pigment-grade TiO_2. The chloride route is increasingly favoured due to its economic advantages and the fact that it produces less waste [3]. However, in order to minimise waste this process requires high-grade feedstock such as natural rutile, with more than 95 wt% TiO_2. The limited amounts of this mineral has led to the development of processes to beneficiate titaniferous minerals, via the Becher route [4] and smelting in electric arc furnace [5], both of which are now used on a commercial scale.

The Becher process proceeds according to the following steps: (i) oxidation in air/O_2 (Eq. 1), (ii) reduction in the rotary kiln at 1473 K (Eq. 2), (iii) catalytic aeration for iron precipitation (Eq. 3) and (iv) acid leaching to remove any residual iron, to obtain synthetic rutile with 92% wt. of TiO_2. However, this method requires feedstocks with more than 50 wt% TiO_2 content and iron is treated as a waste product [6].

$$4FeTiO_3 + O_2(g) \rightarrow 2Fe_2O_3 \cdot TiO_2 + 2TiO_2 \tag{1}$$

$$Fe_2O_3 \cdot TiO_2 + 3CO(g) \rightarrow 2Fe + TiO_2 + 3CO_2(g) \tag{2}$$

$$4Fe + 3O_2(g) \rightarrow 2Fe_2O_3 \tag{3}$$

Smelting of ilmenite in the presence of carbon and a fluxing agent (Eq. 4) at 1923 K minimises the amount of process waste in which iron oxide is reduced to pig iron and the Ti oxide reports to a complex molten slag [2]. However, this method is energy-intensive and the high operating temperature means that furnace linings must be replaced regularly. It is only possible to run this process where the energy cost is quite favourable The Ti-slag must be further processed in order to obtain a marketable high-titania product that can be used as a feedstock in the chloride process. In order to make the smelting process economically viable, the ilmenite starting material must contain at least 35 wt% TiO_2 and a sufficiently high iron content [7].

$$FeTiO_3 + C \rightarrow Fe + TiO_2 + CO(g) \tag{4}$$

Studies have shown that the alkali roast-leach process can be used to upgrade titaniferous minerals [8]. The ilmenite mineral is mixed with alkali salts and roasted in an oxidizing atmosphere, as shown in Eq. 5. The roast calcine is water leached to selectively solubilise iron ferrite. The water leach residue containing sodium titanate is leached in acid media to produce high-grade TiO_2 (97 wt%). Although lower temperatures are required during the alkaline roasting, iron is still considered as a processing waste.

$$FeTiO_3 + 1.5Na_2CO_3 + 0.25O_2(g) \rightarrow Na_2TiO_3 + 0.5Na_2Fe_2O_4 + 1.5CO_2(g) \quad (5)$$

Alkali Reduction of Ilmenite

The Na-Fe-Ti-O-C system at 1323 K, shown in Fig. 1, points out that metallic iron can co-exist with sodium titanate under reducing conditions. However, the presence of molten sodium titanates would inhibit the diffusion of CO gas, decreasing the reduction rate [9]. The major benefit of the process is that the reaction (Eq. 6) is thermodynamically feasible from temperatures as low as 1073 K [10]. Magnetic separation would enable the separation of magnetic iron from the non-magnetic sodium titanate phases. The alkaline reduction of ilmenite (42 wt% TiO_2 and 53.6 wt% Fe-O) was studied by El-Tawil et al. [11] but they only manged to achieve 85% metallisation of Fe at 1473 K. This low metallisation is a result of experiments carried out under oxidizing atmosphere and not the required reducing atmosphere, as depicted in Fig. 1. The un-reduced Fe was lost in the non-magnetic fraction as Na-Fe-Ti-O ternary compounds and hence the overall magnetic separation was poor.

$$FeTiO_3 + Na_2CO_3 + 2C \rightarrow Na_2TiO_3 + Fe + 3CO(g) \quad (6)$$

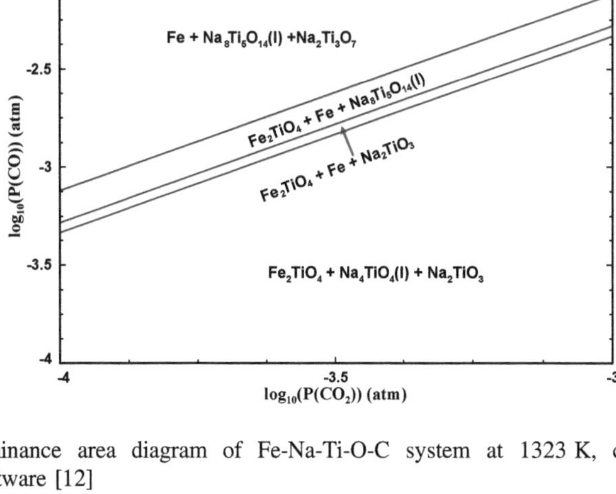

Fig. 1 Predominance area diagram of Fe-Na-Ti-O-C system at 1323 K, computed using FACTSage software [12]

Table 1 Chemical composition of the mine waste used for reduction experiments, as analyzed by X-ray fluorescence

	Fe_2O_3	TiO_2	SO_3	Na_2O	Al_2O_3	SiO_2	MnO	CaO	MgO	V_2O_5
wt%	66.5	11.7	2.1	5.6	5.0	5.8	0.3	1.6	0.4	0.3

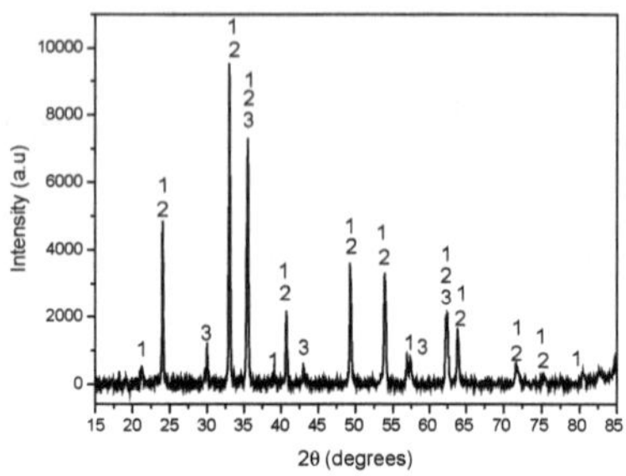

Fig. 2 XRPD pattern of as-received mine waste. *1* $FeTiO_3$ (04-012-1150), *2* Fe_2O_3 (04-002-7501), *3* Fe_3O_4 (04-013-9807)

This study focuses on the thermogravimetric analysis of carbothermic reduction of mine waste in the presence of alkali. The phases formed are characterised and the kinetics and mechanism of the reaction are investigated.

Experimental

A mine waste containing titaniferous minerals with the composition given in Table 1, has been used for this investigation. The mine waste was ground and thoroughly mixed with Na_2CO_3 and activated charcoal, before being placed in an alumina crucible that was suspended on the thermogravimetric balance using a stainless steel wire. The reactions were carried out isothermally at temperatures between 1073 and 1323 K. The TGA reaction tube was purged with argon gas at 0.5 l/min in order to maintain a reducing atmosphere. The reduced samples were analyzed using Scanning electron microscopy—energy dispersive X-ray spectroscopy (SEM-EDX) and X-ray powder diffraction techniques (XRPD). XRPD analysis (Fig. 2) of the as-received mine waste demonstrates that the dominant phases are ilmenite, hematite (Fe_2O_3), and magnetite (Fe_3O_4).

Results and Discussion

Phase Analysis

The XRPD patterns of the mine waste reduced in the presence of alkali in the temperature range of 1073–1323 K are presented in Fig. 3a–f. The main phase

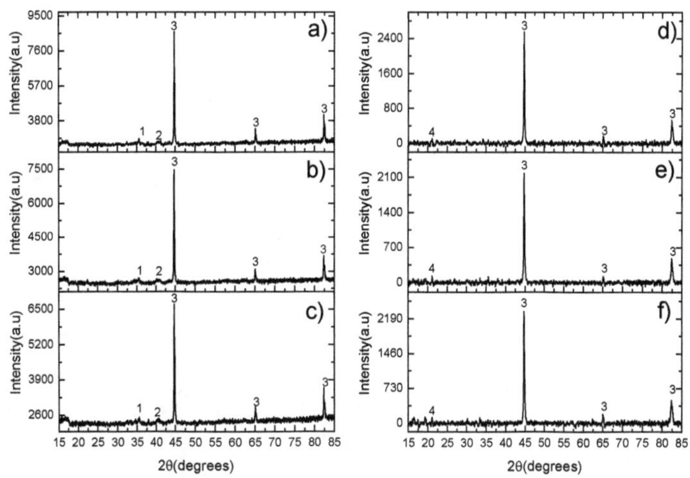

Fig. 3 XRPD patterns of mine waste reduced with alkali in the TGA at **a** 1073 K, **b** 1123 K, **c** 1173 K, **d** 1223 K, **e** 1273 K, **f** 1323 K. *1* Fe_2TiO_4 (04-006-0224), *2* Na_2TiO_3 (00-047-0130), *3* α-Fe (04-007-9753), *4* $NaAlSiO_4$ (00-011-0221)

present at all temperatures is α-Fe. Below 1173 K, there is a minor peak for ulvöspinel (Fe_2TiO_4) and above this temperature there are peaks representing sodium aluminosilicate ($NaAlSiO_4$). There are minor peaks representing sodium titanate in Fig. 3a–c. This shows that the reaction allows for the significant reduction of iron oxide present at low temperatures. These XRPD patterns broadly agree with the predominance diagram (Fig. 1), which predicted the co-existence of metallic Fe and sodium titanate. Figure 1 also demonstrated that the ulvöspinel phase is present when the reduction is incomplete.

Microstructure Analysis

Figure 4 is the backscattered SEM image from the alkali reduction of the mine waste at 1323 K. The atomic number contrast shows three different phases formed during the reaction. Elemental mapping indicates segregation between metallic Fe and sodium titanate phases. The large iron particles observed at the periphery of the particle, illustrate that the reaction proceeds via the shrinking core model [13]. However, the mapping also demonstrates the formation of a complex Na-Ti-Al-Si-O phase, formed as a consequence of the reaction between sodium aluminosilicates with Ti-rich phases [9]. The V and Na elemental maps overlap, implying that the majority of vanadium pentoxide (V_2O_5) reacts with alkali to form sodium vanadate. The distinct separation of Fe from Ti and V oxides means that magnetic separation of the reduced sample would allow for selective recovery of these metals. The Fe would report to the magnetic fraction, whereas Ti and V oxides would be present in the non-magnetic fraction.

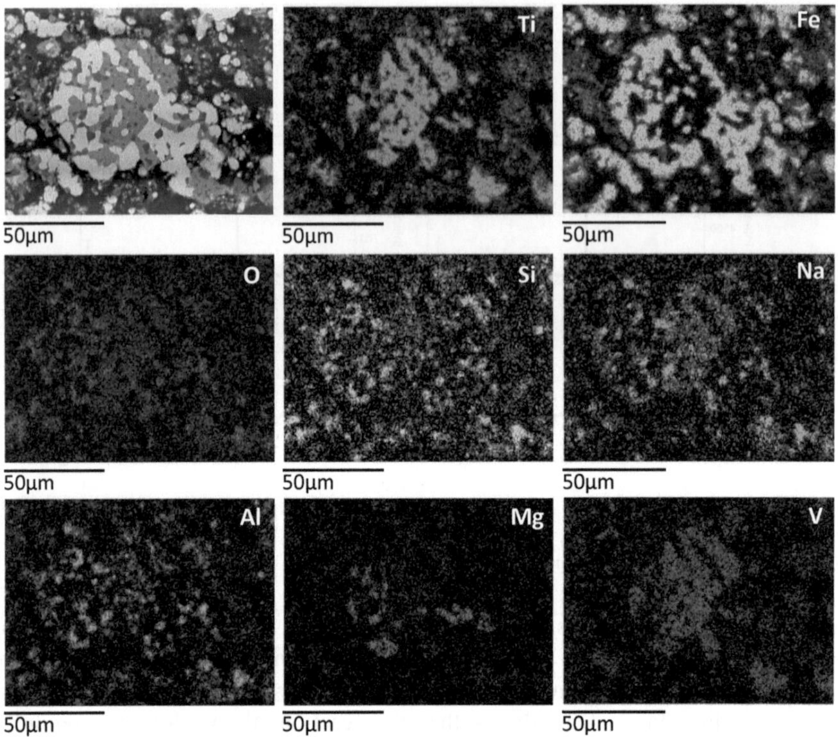

Fig. 4 Backscattered SEM image and elemental mapping of mine waste reduced with Na_2CO_3 at 1323 K in the TGA

Thermogravimetric Analysis

Solid state reactions may be controlled by the reaction at the interface or diffusion of the product and reactant species through the product layer. When the diffusion of species is faster than the chemical reaction, the process is controlled by the chemical reaction at the interface [14] and can be described by Eq. 7. Ginstling and Brounstein (GB) derived a model for reactions under diffusion control given by Eq. 8 [15], where X is the fraction reacted, t is the time in hours and k is the rate constant.

$$1 - (1 - X)^{\frac{1}{3}} = kt \tag{7}$$

$$1 - \frac{2}{3}X - (1 - X)^{\frac{2}{3}} = kt \tag{8}$$

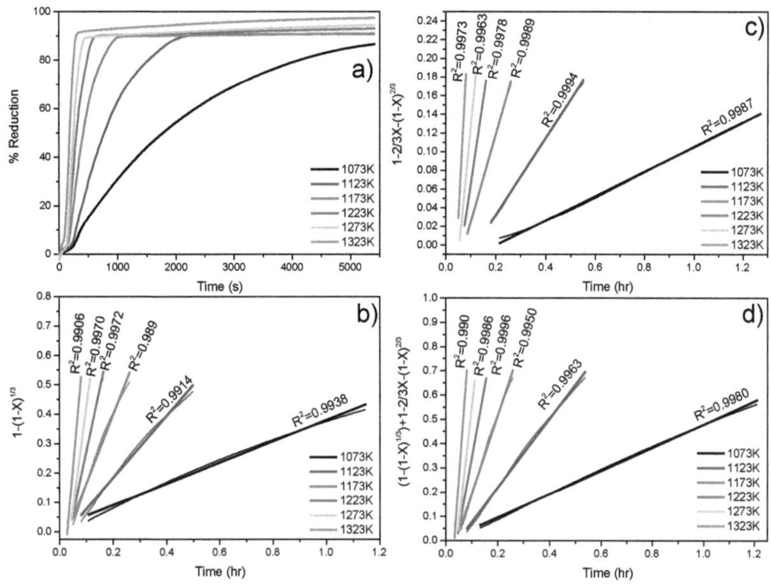

Fig. 5 **a** Isothermal carbothermic reduction with Na_2CO_3 of titaniferous mine waste from 1073 to 1323 K. Curve fitting using the **b** Interface, **c** G.B and the **d** combined Interface + G.B models

The plot of percent reduction (%R) against time for isothermal reduction in the temperature range of 1073–1323 K is presented in Fig. 5a. The %R was calculated as a percentage of the experimental weight loss over theoretical weight loss, as given in Eq. 9.

$$\%\text{Reduction} = \frac{\text{Experimental weight loss}}{\text{Theoretical weight loss}} \times 100 \tag{9}$$

The effect of temperature can be seen on the reaction; the %R achieved increases as the temperature rises and a maximum value of 97% reduction is attained at 1323 K. The rate of reaction is slowest at 1073 K, where the rate of weight change increases gradually and the %R achieved is below 90%. As the temperature increases the curves change from growing exponentially to increasing linearly. For isotherms 1173–1323 K, it can be noted that initially the reactions proceed rapidly before reaching a plateau where the reaction virtually stops.

The interface and G.B models were fitted to the isothermal weigh loss data obtained from the TGA experiments. Figure 5b, c are plots for the interface and G. B models, respectively. It can be observed that the experimental data fits well for both models, meaning that the reaction is under mixed-control. The fitting for the combined interface + G.B model is shown in Fig. 5d. During the carbothermic reduction of the mine waste in the presence of Na_2CO_3, Na^+ ions from Na_2CO_3

Table 2 Linear regression analysis of rate of constants at different temperatures using Interface, Ginstling and Brounstein (GB) and combined Interface + G.B models

Temp (K)	Interface model		G.B model		Combined Interface + G.B	
	k	Standard error	k	Standard error	k	Standard error
1073	0.3616	0.0015	0.1316	2.4694E−4	0.4784	0.0011
1123	1.0573	0.0080	0.4127	8.8008E−4	1.4170	0.0068
1173	2.3322	0.0282	0.9300	0.0039	3.2003	0.0267
1223	4.2246	0.0344	1.7919	0.015	5.8790	0.0175
1273	6.8793	0.0753	2.7610	0.0353	9.5606	0.0737
1323	10.0867	0.2249	5.0427	0.0786	14.5188	0.3542

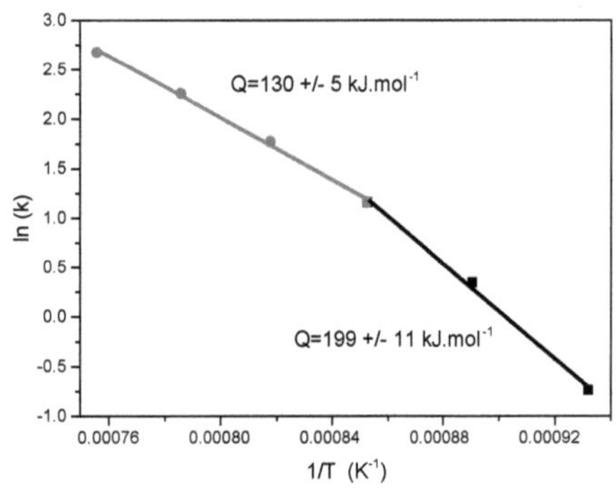

Fig. 6 Arrhenius plot of ln k versus 1/T for calculation of activation energy. The k values used were calculated using the mixed (Interface + G.B) model

enter the ilmenite lattice structure. Since the ionic sizes of Na^+ and Fe^{2+} are 0.97 and 0.74 Å, respectively, ionic substitution of Fe^{2+} by Na^+ is possible to form sodium titanate [16]. The derived rate constants are detailed in Table 2.

Arrhenius Plot and Activation Energy

The plot of the natural logarithm versus reciprocal temperature for the alkali reduction is presented in Fig. 6. The slope of ln k versus 1/T line changes at 1173 K, suggesting that there are two reaction regimes which are dependent on temperature during the reaction. At lower temperatures the activation energy was seen to be 199 ± 11 kJ mol^{-1}, which is consistent with the outward diffusion of O^{2-} ions, and this value is comparable with the reported activation energy of 190 kJ mol^{-1} [17]. The activation energy at higher temperatures is 130 ± 5 kJ mol^{-1}, as corresponds to the outward diffusion of Fe^{2+} ions, which has an activation energy within the range of 117–140 kJ mol^{-1} [18]. This is confirmed by the presence of Fe at the periphery of the particles as it is shown in Fig. 4.

Conclusions

1. The XRPD patterns and the microstructure of the reduced samples indicate that elemental iron and sodium titanate co-exist at the process conditions, as predicted in Fig. 1.
2. The microstructure shows that the reaction proceeded via the shrinking-core model with distinct segragation of Fe from Ti and V, allowing for their separation in subsequent process steps.
3. The kinetic analysis demonstrated that a maximum of 97% of reduction was achieved at 1323 K. The reaction models fitting revealed that the overral reaction was under mixed-control.
4. The ln k versus 1/T plot displayed that the overall reaction is governed by two regimes during the alkali reduction. Up to 1173 K, the activation energy was found to be 199 $\pm$ 11 kJ mol^{-1}, which correlates to the outward diffusion of O^{2-} ions. From 1173 to 1323 K, the calculated value was 130 kJ mol^{-1} which compares with the activation energy for the outward diffusion of Fe^{2+} ions.

Acknowledgements The authors acknowledge the financial support from the EPSRC standard grants (GR/T08074/01 and GR/L95977/01) and PhD studentships for research which was initiated in 1997 at the University of Leeds. AJ also acknowledges the support from the European Union's Marie Curie Fellowship grant number 331385 for Dr. Sanchez-Segado and from the NERC's Catalyst Grant reference NE/L002280/1. SP acknowledges the IoM^3 for their support in the form of the Stanley Elmore Fellowship.

References

1. G. Auer et al., *Pigments, Inorganic, 2. White Pigments*, in *Ullmann's Encyclopedia of Industrial Chemistry* (Wiley & KGaA, New York & Darmstadt, 2000)
2. J.E. Kogel, *Industrial Minerals & Rocks: Commodities, Markets, and Uses* (SME, New Delhi, 2006)
3. J.B. Rosenbaum, Titanium technology trends. JOM **34**(6), 76–80 (1982)
4. R. Becher et al. A new process for upgrading ilmenitic mineral sands. Aus. Inst. Mining. Met. Proc. (1965)
5. P. Pistorius, Ilmenite smelting: the basics. J. South Afr. Inst. Min. Metall. **108**(1), 35–43 (2008)
6. F. Cardarelli, *Materials Handbook: A Concise Desktop Reference* (Springer, London, 2008)
7. C. Murty, R. Upadhyay, S. Asokan, Electro smelting of ilmenite for production of TiO2 slag–potential of India as a global player. in *Proc. INFACON XI* (Deli, India, 2007) p. 18–21
8. S. Sanchez-Segado et al., Reclamation of reactive metal oxides from complex minerals using alkali roasting and leaching—an improved approach to process engineering. Green Chem. **17**(4), 2059–2080 (2015)
9. V. Tathavadkar, A. Jha. The effect of molten sodium titanate and carbonate salt mixture on the alkali roasting of ilmenite and rutile minerals. in *VII international conference on molten slags fluxes and salts* (2004)

10. A. Roine, H. Outokumpu, *Chemistry for Windows: Chemical Reaction and Equilibrium Software with Extensive Thermodynamical Database, Version 5.1, User's Guide* (Outokumpu Research Oy, Finland, 2002)
11. S. El-Tawil et al., Alkali reductive roasting of ilmenite ore. Can. Metall. Q. **35**(1), 31–37 (1996)
12. C. Bale et al., FactSage thermochemical software and databases. Calphad **26**(2), 189–228 (2002)
13. O. Levenspiel, *Chemical Reaction Engineering* (Wiley, New York, 1999)
14. W.D. Spencer, B. Topley, CCCLIV.—Chemical kinetics of the system $Ag_2CO_3 \rightleftharpoons Ag_2O + CO_2$. J. Chem. Soc. (Resumed), 2633–2650 (1929)
15. A.M. Ginstling, B.I. Brounshtein, The diffusion kinetics of reaction in spherical particles. J. Appl. Chem. USSR **23**, 1327–1338 (1950)
16. A. Lahiri, A. Jha, Kinetics and reaction mechanism of soda ash roasting of ilmenite ore for the extraction of titanium dioxide. Metall. Mater. Trans. B **38**(6), 939–948 (2007)
17. W.D. Kingery, H.K. Bowen, D.R. Uhlmann, *Introduction to Cermaics*. Wiley series on the science and technology of materials (Wiley, New York, 1976), p. 1032
18. A. Jha, S.J. Yoon, Formation of titanium carbonitride phases via the reduction of TiO2 with carbon in the presence of nitrogen. J. Mater. Sci. **34**(2), 307–322 (1999)

Empirical Activation Energies of MnO and SiO_2 Reduction in SiMn Slags Between 1500 and 1650 °C

P. Kim, T. Larssen, M. Tangstad and R. Kawamoto

Abstract The reduction rates of MnO and SiO_2 were investigated for various charge compositions. The empirical activation energies of MnO and SiO_2 reduction from SiMn slags were calculated between 1500 and 1650 °C under CO atmospheric pressure. The amounts of metal produced at different temperatures were compared, and the Arrhenius plots were described for each slag A, slag B and synthetic slag, which shows that the raw material have significant influence on the reduction rate. The rate of Mn-produced in slag A was faster than slag B despite of the relatively lower driving force. The lower rate constant in slag B implies that other kinetic factors, such as viscosity, affects the reduction rate. The estimated activation energies of MnO were 920, 304 and 975 kJ/mol MnO for slag A, slag B and synthetic slag, respectively. The rates of Si-produced were similar and slow for all slag types. However, the different activation energies of SiO_2 between industrial ores and synthetic materials implied that the trace elements can have catalytic effects toward SiO_2 reduction in SiMn slags. The estimated activation energies of SiO_2 reduction were around 800 kJ/mol for slag A and slag B.

Keywords SiMn · MnO · SiO_2 · Activation-energy · Kinetics

P. Kim (✉) · T. Larssen · M. Tangstad
Department of Materials Science and Engineering, Norwegian University of Science and Technology (NTNU), N-7491 Trondheim, Norway
e-mail: pyung.h.kim@ntnu.no

T. Larssen
e-mail: trineala@stud.ntnu.no

M. Tangstad
e-mail: merete.tangstad@ntnu.no

R. Kawamoto
Department of Materials Engineering, The University of Tokyo, Hongo, Bunkyo-ku, Tokyo 113-8656, Japan
e-mail: ryosukek@stud.ntnu.no

S. Wang et al. (eds.), *Applications of Process Engineering Principles in Materials Processing, Energy and Environmental Technologies*,
The Minerals, Metals & Materials Series, DOI 10.1007/978-3-319-51091-0_46

Introduction

Silicomanganese (SiMn) represents as an essential ingredient for steel producing industries due to its many contributions. Numerous studies have shown that Mn as an alloying unit enhances the strength, toughness and hardness of steel products, and both Mn and Si serves as an excellent deoxidizer to prevent porous structures [1–8].

While manganese thermodynamics have been investigated intensively [2], kinetic information in the SiMn process is however rather scarce. The lack of empirical kinetic information increases the ambiguousness of reduction mechanisms, and it is not clear how the different raw materials affect the reduction of MnO and SiO_2 in the SiMn process. The metal producing reactions in the SiMn process are described by the following reactions:

$$MnO_{(l)} + C = Mn_{(l)} + CO_{(g)} \tag{1}$$

$$SiO_{2(l)} + 2C = Si_{(l)} + 2CO_{(g)} \tag{2}$$

A recent study have shown that both MnO and SiO_2 reduction becomes significant above 1500 °C [9]. The mass loss observed from a TGA-type furnace, which indicates MnO and SiO_2 reduction, was insignificant until 1500 °C but increased drastically at higher temperatures. It was also tentatively perceived that the difference in Mn-source influences the reduction behaviors of SiMn charges. Therefore, the present study focuses on estimating the empirical activation energies of MnO and SiO_2 reduction in SiMn slags (MnO-SiO_2-CaO-MgO-Al_2O_3) between 1500 and 1650 °C using different Mn-sources.

Theoretical Considerations

Previous studies have shown that the reduction rate of MnO can be described by Eq. (3) [2, 10]. Assuming that the SiO_2 reduction in the SiMn slag system is also controlled by chemical reaction, a similar kinetic model for SiO_2 reduction can be considered by Eq. (4), which was presumed and used for estimating the activation energies in this work:

$$\begin{aligned} r_{MnO} &= k_{Mn} \cdot A \cdot \left(a_{MnO} - \frac{a_{Mn} \cdot p_{CO}}{K_T} \right) \\ &= k_{o,Mn} \cdot A \cdot e^{-E_{MnO}/RT} \cdot \left(a_{MnO} - \frac{a_{Mn} \cdot p_{CO}}{K_{T,Mn}} \right) \end{aligned} \tag{3}$$

$$r_{SiO_2} = k_{Si} \cdot A \cdot \left(a_{SiO_2} - \frac{a_{Si} \cdot p_{CO}^2}{K_T}\right) = k_{o,Si} \cdot A \cdot e^{-E_{SiO_2}/RT} \cdot \left(a_{SiO_2} - \frac{a_{Si} \cdot p_{CO}^2}{K_{T,Si}}\right) \tag{4}$$

Where r is the reduction rate (g/min), k is the rate constant (g/min cm^2), k_o is the frequency factor, A is the interfacial area (cm^2), E is the activation energy (kJ/mol), R is the gas constant, T is the temperature, a_{MnO}, a_{SiO_2} are the activity of MnO and SiO_2 in the slag phase, a_{Mn}, a_{Si} are the activity of Mn and Si, p_{CO} is the partial pressure of CO(g) and K_T is the equilibrium constant at temperature T.

The presumed models for MnO and SiO_2 reduction also imply that the driving force for reduction, which is the difference between actual and equilibrium activities, contribute to the reduction rates as seen in Eq. (4). Simplified models for activities of slag (MnO, SiO_2) and metal (Mn, Si) components in their respective melts have been recently studied and were expressed as Eqs. (5–8) [11]. These activities and thermodynamic equilibrium data from HSC Chemistry 7 [12] were used to estimate the driving forces of MnO and SiO_2 reduction at different temperatures of each slag composition.

$$\begin{aligned} a_{MnO} = C_{MnO} \cdot \exp(&0.0007576T - 123.7C_{MnO} + 30.14C_{SiO_2} + 47.84C_{MgO} + 49.54C_{CaO} - 47.96C_{Al_2O_3} \\ &+ 122.8C_{MnO}^2 - 67.78C_{SiO_2}^2 - 46.32C_{MgO}^2 - 47.68C_{CaO}^2 + 22.51C_{Al_2O_3}^2 + 78.35C_{MnO}C_{CaO} \\ &+ 77.56C_{MnO}C_{MgO} + 176.6C_{MnO}C_{Al_2O_3} + 101.2C_{MnO}C_{SiO_2} - 71.52C_{SiO_2}C_{CaO} - 70.58C_{SiO_2}C_{MgO} \\ &+ 27.35C_{SiO_2}C_{Al_2O_3} + 46C_{SiO_2}^3 - 92.97C_{CaO}C_{MgO} + 2.44C_{CaO}^3) \end{aligned} \tag{5}$$

$$\begin{aligned} a_{SiO_2} = C_{SiO_2} \cdot \exp(&-0.0003408T + 113.8C_{MnO} - 22.79C_{SiO_2} - 51.63C_{MgO} - 52.44C_{CaO} + 36.3C_{Al_2O_3} \\ &- 119.3C_{MnO}^2 + 42.56C_{SiO_2}^2 + 32.25C_{MgO}^2 + 30.12C_{CaO}^2 - 26.26C_{Al_2O_3}^2 - 82.725C_{MnO}C_{CaO} \\ &- 82.9C_{MnO}C_{MgO} - 155.2C_{MnO}C_{Al_2O_3} - 86.98C_{MnO}C_{SiO_2} + 86.21C_{SiO_2}C_{CaO} + 86.19C_{SiO_2}C_{MgO} \\ &- 23.06C_{SiO_2}C_{Al_2O_3} - 31.26C_{SiO_2}^3 + 69.45C_{CaO}C_{MgO} + 11.29C_{CaO}^3) \end{aligned} \tag{6}$$

$$\begin{aligned} a_{Mn} = C_{Mn} \cdot \exp(&0.0005382T - 37.41C_{Mn} - 2.966C_{Si} - 0.6835C_{Fe} + 39.52C_{Mn}^2 - 1.453C_{Si}^2 \\ &- 0.5561C_{Fe}^2 + 27.48C_{Mn}C_{Si} + 38.69C_{Mn}C_{Fe} + 0.214C_{Si}C_{Fe}) \end{aligned} \tag{7}$$

$$\begin{aligned} a_{Si} = C_{Si} \cdot \exp(&0.002464T + 10.3C_{Mn} - 1.081C_{Si} + 27.52C_{Fe} - 15.49C_{Mn}^2 - 3.713C_{Si}^2 - 34.66C_{Fe}^2 \\ &+ 1.324C_{Mn}C_{Si} - 47.01C_{Mn}C_{Fe} - 9.127C_{Si}C_{Fe}) \end{aligned} \tag{8}$$

where $C_{MnO}, C_{SiO_2}, C_{CaO}, C_{MgO}, C_{Al_2O_3}$ are the mass fraction of MnO, SiO_2, CaO, MgO, Al_2O_3 in the slag phase, and $C_{Mn}, C_{Si}, C_{Fe}, C_C$ are the mass fraction of Mn, Si, Fe, C in the metal phase, respectively.

Experimental Procedures

The focus of this work was to study the reaction rates and estimate the empirical activation energies of MnO and SiO_2 reduction in the SiMn slags. The characterization and preparation of SiMn charges, the TGA-type furnace and the experimental conditions are described in this section.

Three different SiMn charges based on different Mn-sources, such as ore A, ore B and fine synthetic powders, were used in this work and their charge profiles are described in Table 1. The synthetic charge was aimed to have the similar composition of the charge based on ore A. The sizes of raw materials were 0.6–1.6 mm. Each charge was thermodynamically considered to aim at approximately 40 mass pct SiO_2 in slag and 18 mass percent Si in metal phase [2].

The experiments were conducted by using a TGA-type furnace, which is schematically shown in Figure 1. The furnace can endure temperatures up to 1700 °C and the maximum heating rate is up to 25 °C/min. A mass balance is installed at the top and a Molybdenum (Mo)-wire was used to suspend the graphite crucible inside the furnace. Graphite crucibles (36 mm outer diameter, 30 mm inner diameter, 70 mm height and 61 mm deep) were used to contain each charge sample in this experiment.

Initially, the furnace was heated up to 1200 °C (+25 °C/min) and held for 30 minutes to secure complete pre-reduction [2]. Then, further heating (+4.5 °C/min) was done and stopped at targeted temperatures between 1500 and 1650 °C. The continuous temperature increase was used to simulate an industrial furnace operation. All experiments were conducted in CO atmosphere.

The weight loss (% mass loss) of each charge sample was recorded and data were logged every 5 s during the experiment. Lastly, each charge sample was prepared by mounting it in epoxy to be further analyzed. EDS analyses were done by using Zeiss-Supra 55VP to analyze slag compositions. The average slag composition from more than 3 points were used to calculate the metal compositions.

Results and Discussions

The chemical compositions of slag and metal with their respective activities (slag: a_{MnO} and a_{SiO_2}, metal: a_{Mn} and a_{Si}) at different temperatures are shown in Table 2, and the amount of produced metals, Mn and Si, are described in Fig. 2. Also, the driving forces of MnO and SiO_2 reduction are shown in Fig. 3. The amount of Mn and Si increased, as expected, with temperature.

It is shown in Fig. 2 that Mn was reduced faster from slag A, compared to slag B and the synthetic slag. The reduction rates are determined by two factors according to Eqs. (3) and (4), which are the driving force given by the activities in the slag and the rate constant (k) expressing all other kinetic parameters. At 1500 °C where hardly any reduction of Mn has occurred, the activity of MnO is higher for slag B,

Table 1 Chemical composition of raw materials and charge profiles

Material		MnO	MnO_2	SiO_2	Fe_2O_3	CaO	MgO	Al_2O_3	C	CO_2	H_2O	Total (wt %)
Ore A		32.69	33.22	5.77	15.06	6.26	1.1	0.26	0.27	3.52	1.55	99.7
Ore B		3.03	72.4	4.6	6.73	0.07	0.05	5.6		0.1	5	97.58
Q		0.14		93.85		0.09	0.05	1.19				95.32
HCS		35.23		25.45		18.45	7.53	12.3	0.4		2.2	101.56
C		0.04		5.6	0.86	0.42	0.22	2.79	87.68		15.5	113.11
L				1		54	1	0.27		47.5	1	104.77
Syn.	MnO	99	0.1								0.17	99.27
	SiO_2			99.5								99.5
	CaO					96–100.5					3.5	99.5–104
	MgO						98–100.5					98–100.5
	Al_2O_3							98			0.5	98.5

Charge	Ore A	Ore B	Q	HCS	C	L	Syn.						Total (g)
							MnO	SiO_2	CaO	MgO	Al_2O_3	Fe	
A	4		1.69	4	2.5								12.19
B		5	1.75		1.9	1.62							10.27
Syn.					2.5		3.79	2.9	0.98	0.34	0.52	0.42	11.45

Q Quartz, *HCS* High-carbon FeMn slag, *C* Coke, *L* Limestone, *Syn.* Synthetic

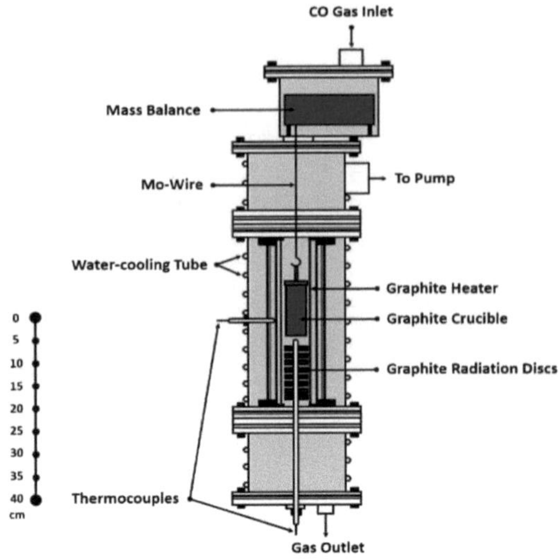

Fig. 1 Experimental setup: Schematic of TGA-type furnace

Table 2 Slag and metal compositions and their respective activities at different temperatures between 1500 and 1650 °C

Charge	Slag (wt%)							Metal (wt%)			
	MnO	SiO_2	CaO	MgO	Al_2O_3	a_{MnO}	a_{SiO2}	Mn	Si	a_{Mn}/K_T	a_{Si}/K_T
A.1500	42.5	35.0	13.0	3.7	5.8	0.21	–	34.6	2.9	0.002	–
A.1510	43.0	33.9	12.7	3.9	6.5	0.24	–	34.1	11.0	0.002	–
A.1520	42.1	35.2	12.6	3.9	6.3	0.21	–	39.4	3.2	0.003	–
A.1530	41.5	35.4	13.4	3.2	6.4	0.21	0.12	42.7	4.1	0.002	0.049
A.1540	46.3	34.1	10.9	4.7	4.0	0.18	0.14	55.5	4.0	0.003	0.043
A.1550	36.0	39.2	14.0	3.7	7.3	0.15	0.19	62.7	0.1	0.003	0.001
A.1560	33.1	40.9	15.0	3.8	7.1	0.12	0.22	67.3	0.2	0.003	0.001
A.1570	26.0	43.0	17.9	4.5	8.6	0.09	0.24	71.3	6.6	0.003	0.047
A.1580	21.0	43.9	19.5	5.5	10.2	0.07	0.23	71.9	9.7	0.002	0.075
A.1590	15.8	47.4	21.2	5.0	10.6	0.04	0.30	74.1	7.9	0.002	0.040
A.1600	11.8	45.7	23.5	6.9	12.2	0.03	0.22	72.4	12.4	0.002	0.082
B.1500	48.0	30.8	14.9	0.4	5.9	0.33	0.06	49.1	20.7	0.002	–
B.1550	47.3	31.5	14.2	0.8	6.2	0.32	0.07	51.8	18.5	0.001	0.905
B.1600	44.7	31.1	17.6	0.4	6.1	0.34	0.06	59.2	20.2	0.001	0.369
B.1650	40.4	34.9	17.7	0.4	6.6	0.24	0.10	66.5	13.9	0.001	0.046
Syn.1585	38.4	33.8	14.1	9.1	4.5	0.24	0.08	55.7	17.9	0.001	0.358
Syn.1615	25.6	40.6	18.0	10.4	5.5	0.10	0.16	69.1	12.0	0.002	0.060
Syn.1630	11.6	43.9	24.0	12.9	7.7	0.04	0.17	71.0	14.1	0.001	0.067
Syn.1650	5.3	42.0	28.2	15.8	8.7	0.02	0.10	70.0	16.8	0.001	0.073

Charge type. Temperature [°C]: ex, A. 1540

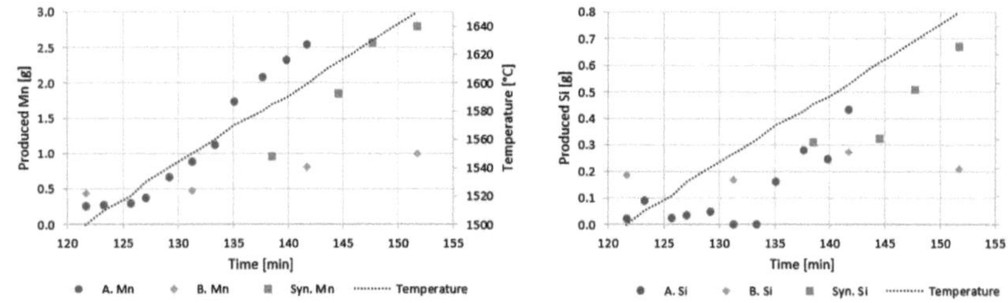

Fig. 2 Amount of Mn (*left*) and Si (*right*) of three different slags with process time and temperature

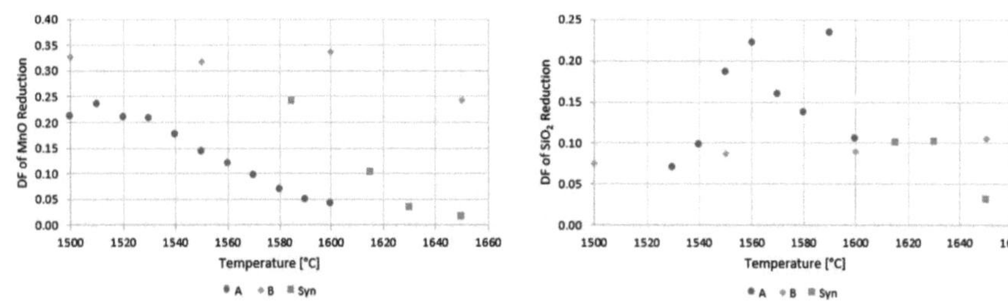

Fig. 3 Driving forces (DF) of MnO (*left*) and SiO_2 (*right*) of three different slags between 1500 and 1650 °C

and one would expect a higher reduction rate. Slag B should have been more favorable towards MnO reduction due to the higher driving force than slag A and synthetic slag. It was however, shown the opposite in Fig. 2. This implies that slag A has a higher rate constant than slag B. The rate constant is the factor which depends on the other kinetic factors, such as diffusion, viscosity and influence of trace elements and is affecting the reduction rate. The higher basicity of slag A would indicate a lower viscosity. The ratios of non-reducing oxides (CaO + MgO/Al_2O_3) are 2.6 and 2.0 for slag A (synthetic slag) and slag B, respectively. The difference of those ratios is not significant but further examination regarding viscosity change seems necessary.

The higher iron content in slag A charge should give a higher reduction rate due to the lower a_{Mn}/K_T value. This number is, however, so small that it will not affect the total driving force significantly. It has been previously observed that the formation of initial iron can lead to reduction of MnO [13]. Ore A had more iron than ore B. However, the activity of Mn (a_{Mn}/K_T) was higher in slag A than slag B. The reason for this seems to be from the very low reduction rate and hence small amount of reduced Mn in slag B. Also, the activities of Mn (a_{Mn}/K_T) were relatively low compared to the activities of MnO regardless of slag type. This implies that other kinetic factors should be considered.

No significant difference can be seen in the reduction rate of SiO_2 from the different slags. The temperature where Si starts to increase in Fig. 2 corresponds to

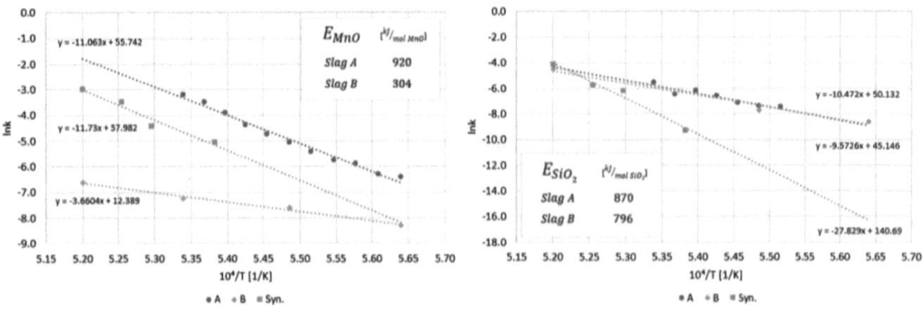

Fig. 4 Arrhenius plots of different slags and the estimated activation energies of MnO and SiO_2 reduction

the higher driving force in Fig. 3. The rapid increase of driving force seems to be from the change of slag composition due to the reduction of MnO.

The corresponding Arrhenius plots of different charges including the estimated activation energies of MnO and SiO_2 reduction are described in Fig. 4. While the rate constant is about the same for the SiO_2 reduction, it is seen that the rate constant for Mn is varying more. Slag A have a higher value, where synthetic slag was lower and slag B showed the lowest. This has been observed previously where the slags from industrial raw materials have higher reduction rate compared to synthetic materials, until the catalytic sulphur was added into the slag [14]. It was believed that a slow reduction rate may be beneficial in SiMn production, as the temperature then will be higher in the furnace, helping the silicon content.

The estimated activation energies of MnO for slag A and synthetic slag were approximately 920 and 975 kJ/mol, respectively. This was relatively higher than the activation energies previously reported in FeMn slags, which were between 350 and 370 kJ/ mol [2]. Both estimations showed that the MnO reduction is highly dependent on temperature in SiMn slags. The activation energy for slag B was however, lower than the first two slags. This gives implication that the transfer of Mn in slag B may be hindered by diffusion.

The difference between slag A and synthetic slag implies the influence of trace elements can have some impact on the MnO reduction rate. For example, sulphur can enhance the reduction rate despite of the unknown mechanism [14]. Using controlled amount of trace elements into synthetic slag will be studied further.

Estimated activation energies of SiO_2 for slags A and B were approximately 870 and 796 kJ/mol, respectively, which are quite in the same range.

Conclusions

The empirical activation energies of MnO and SiO_2 in SiMn slags between 1500 and 1650 °C were estimated in this study. Charge type based on different Mn-source have shown different results. The slag A have showed higher temperature dependency while slag B showed more diffusion controlled for MnO reduction. This can give implications to the different reduction rates of MnO in SiMn slags from using different Mn-source. For SiO_2 reduction, the difference between slags A and B compared to synthetic slag implies the influence of trace elements.

Acknowledgements The authors acknowledge the project funds provided by the Norwegian Research Council through the SFI: Metal Production projects.

References

1. International Manganese Institute (2014) www.manganese.org
2. S.E. Olsen, M. Tangstad, T. Lindstad, *Production of Manganese Ferroalloys* (Tapir Academic Press, Trondheim, 2007), pp.11–18
3. Y. Tomota, M. Strum, J. Morris Jr., The relationship between toughness and microstructure in Fe-high Mn binary alloys. Metall. Trans., *A* **18**(6) (1987)
4. D.K. Subramanyam, A.E. Swansiger, H.S. Avery, Austenitic manganese steels in properties and selection: irons, steels and high-performance alloys. ASM Int., 822–840 (1990)
5. G.F. Deev, V.V. Popovich, V.N. Palash, Role of iron sulfide in the formation of cracks in weld joints. Mater. Sci. **18**(3), 109–112 (1982)
6. J.R. Cain, Influence of sulphur, oxygen, copper and manganese on the red-shortness of iron. Technological papers of the bureau of standards no. 261. Department of Commerce, USA, 30 July 1924, pp. 327–335
7. S.I. Gubenko, A.M. Galkin, Nature of the red-shortness of steel. Metal Sci. Heat Treat. **26**(10), 732–737 (1984)
8. O. Grong, T.A. Siewert, G.P. Martins, D.L. Olson, A model for the silicomanganese deoxidation of steel weld metals. Metall. Trans. **17A**, 1797–1807 (1986)
9. P. Kim, J. Holtan, M. Tangstad. Reduction behavior of assmang and comilog ore in the SiMn process. Advances in molten slag, fluxes and salts: proceeding of the 10th international conference on molten slags, fluxes and salts (MOLTEN16), May 2016, pp. 1285–1292
10. O. Ostrovski, S.E. Olsen, M. Tangstad, M. Yastreboff, Kinetic modelling of MnO reduction from manganese ore. Can. Metall. Q. **41**(3), 309–318 (2002)
11. H. Olsen, A theoretical study on the reaction rates in the SiMn production process. Master's thesis, Department of Materials Science and Engineering (DMSE), Norwegian University of Science & Technology (NTNU), 2016
12. Outotec: HSC Chemistry 7. hsc-chemistry.com
13. J. Safarian, M. Tangstad, Slag-carbon reactivity. 12th international ferroalloy congress (INFACON), 2010, pp. 327–338
14. T.A. Skjervheim, S.E. Olsen, The rate and mechanism for reduction of manganese oxide from silicate slags. INFACON 7, Trondheim, Norway, June 1995, pp. 631–640

Experimental Evaluation of Thermodynamic Interactions Between Tellurium and Various Elements in Molten Iron

S. Ueda, Y. Matsuki and K. Morita

Abstract Tellurium has mainly been used as an alloying element for resulfurized free-machining steel. It globularizes MnS inclusions and improves both machinability and mechanical properties of such steel. This effect is quite sensitive to tellurium concentration. Therefore, its incorporation should be controlled accurately during the secondary refining process, using the thermodynamic properties of tellurium in molten iron. In the present study, Wagner's interaction parameters of various elements on tellurium were experimentally evaluated by vapor-liquid equilibration, and the vapor pressure of tellurium was controlled using the transpiration method. The values obtained were compared with those of oxygen and sulfur, and they were found to be correlated with the standard Gibbs energy of formation of the corresponding oxides, sulfides, and tellurides. The results indicate that the relations described here could provide estimates of interaction parameters that have not been experimentally determined thus far by using available thermodynamic quantities.

Keywords Steelmaking · Thermodynamics · Molten iron · Tellurium · Interaction parameters

Introduction

In the steelmaking field, tellurium has mainly been used as an alloying element for resulfurized free-machining steel since it can globularize MnS inclusions and improve both machinability and mechanical properties of such steel [1, 2]. As the effect of tellurium addition is reported to be sensitive to its concentration in molten steel [2], it is important to control this concentration precisely during the steelmaking processes, taking into consideration the loss of tellurium either by

S. Ueda (✉) · Y. Matsuki · K. Morita
Department of Materials Engineering, The University of Tokyo, 7-3-1 Hongo, Bunkyo-ku, Tokyo 113-8656, Japan
e-mail: ueda@wood3.t.u-tokyo.ac.jp

S. Wang et al. (eds.), *Applications of Process Engineering Principles in Materials Processing, Energy and Environmental Technologies*,
The Minerals, Metals & Materials Series, DOI 10.1007/978-3-319-51091-0_47

evaporation to the atmosphere or by elution into slags. In order to analyze or estimate such behavior, information about the thermodynamic stability of tellurium in molten steel is necessary. However, there has been little investigation into this field, possibly due to the difficulty in handling of tellurium at high temperatures. Furthermore, resulfurized free-machining steel has a variety of alloying elements and their concentrations vary depending on the application. Therefore, effects such additional elements have on the stability of tellurium in molten iron, often referred to as "interactions," must also be taken into consideration. In this study, we used a vapor-liquid equilibration technique in which the chemical potential, or partial pressure, of tellurium in the ambient atmosphere was fixed using the transpiration method in order to investigate the dependence of the solubility of tellurium on the concentrations of other alloying elements. These effects were quantitatively evaluated as Wagner's interaction parameters.

Experimental

Figure 1 is a schematic diagram of the experimental apparatus. In the upper hot-zone, granular tellurium (99.99%, 30 μm pass, 4 g) was heated to a constant temperature (683 K) within a graphite container (20 mm OD, 11 mm ID, and 50 mm H). A carrier gas (Ar + 3% H_2) was passed over it at a constant rate (200 mL/min) so that the tellurium vapor is generated at a constant partial pressure ($\sim 10^{-4}$ atm). This procedure is called the transpiration method, and is widely used as an experimental method for the measurement of vapor pressures of condensed matter [3]. The above carrier gas containing the tellurium vapor was then carried through a thin alumina tube (6 mm OD, 4 mm ID) to molten iron surfaces

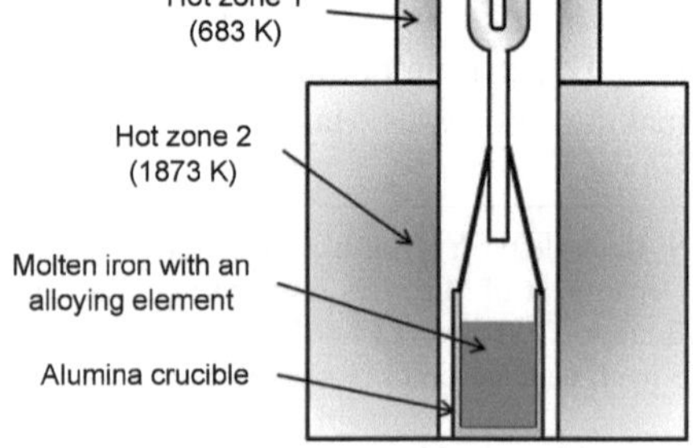

Fig. 1 Experimental apparatus for the vapor-liquid equilibration experiments using transpiration method to control the partial pressure of tellurium within ambient atmosphere of the instrument

(99.9%, 2.0 g) with known contents (up to 6 mass%) of alloying elements, heated within an alumina crucible (99.7%, 15 mm OD, 12 mm ID and 30 mm H) kept at a constant temperature (1873 K). After 1 h of equilibration, the whole gas lance was pulled up to the highest position inside the reaction tube so that the sample could be solidified quickly without any leakage of tellurium vapor to the outside.

The samples obtained from these experiments were dissolved in acid solutions and their chemical composition was analyzed using inductively coupled plasma atomic emission spectroscopy (ICP-AES) for tellurium, aluminum, chromium, copper, and molybdenum. For the analysis of tellurium, iron ions were removed from the aqueous solutions by a palladium co-precipitation method [4] beforehand, in order to avoid serious spectral interference. Sulfur contents in the samples were determined using the infrared absorption method after combustion.

Principle

In the limit of dilute solution, the relation between the partial pressure of tellurium and the solubility of tellurium in molten iron may be expressed by the following equation:

$$a_{\mathrm{Te}} = \gamma_{\mathrm{Te}} \cdot X_{\mathrm{Te}} = K \cdot P_{\mathrm{Te}} \tag{1}$$

Here, a_{Te} and γ_{Te} are the activity and Henrian activity coefficient of tellurium in molten iron, based on $X_{\mathrm{Te}} = 1$ standard, respectively. Term K is the equilibrium constant for the following reaction:

$$\mathrm{Te(g)} = \underline{\mathrm{Te}}(X,\ \mathrm{in\,Fe}) \tag{2}$$

Furthermore, the effect of another alloying element j may be expressed as:

$$\ln \gamma_{\mathrm{Te}} = \epsilon_{\mathrm{Te}}^{j} \cdot X_j \tag{3}$$

Here, $\epsilon_{\mathrm{Te}}^{j}$ is Wagner's first-order interaction parameter of element j on tellurium. From Eqs. (1) and (3), the following relation can be obtained:

$$\ln X_{\mathrm{Te}} = \ln a_{\mathrm{Te}} - \epsilon_{\mathrm{Te}}^{j} \cdot X_j \tag{4}$$

Therefore, as long as the activity of tellurium, or the partial pressure of tellurium, is kept constant in a series of experiments, a linear relation between $\ln X_{\mathrm{Te}}$ and X_j with a slope of $-\epsilon_{\mathrm{Te}}^{j}$ is expected. Beyond the limit of dilute solution, a second-order parameter may be necessary:

$$\ln \gamma_{Te} = \epsilon_{Te}^{j} \cdot X_j + \rho_{Te}^{j} \cdot X_j^2 \tag{5}$$

$$\ln X_{Te} = \ln a_{Te} - \epsilon_{Te}^{j} \cdot X_j - \rho_{Te}^{j} \cdot X_j^2 \tag{6}$$

Results

The experimental results corresponding to the above relations for aluminum, sulfur, chromium, copper, and molybdenum are given in Fig. 2. The error bars in the graphs are calculated based only on the experimental errors of ICP-AES. Except for copper, almost all of the points can be fit with linear regression within error, suggesting that Eq. (4) is satisfied. A second-order term based on Eq. (6) is required to fit the data obtained for copper. Sulfur and copper are found to increase the solubility of tellurium, while aluminum, chromium, and molybdenum show opposite effects. From a microscopic point of view, the former case and the latter case correspond to attractive and repulsive interaction between tellurium and an alloying element, respectively. We shall discuss it in detail in the following section. From the coefficients obtained based on these fits, the following Wagner's interaction parameters can be obtained:

$$\epsilon_{Te}^{Al} = 5.1 \pm 0.5 \quad \epsilon_{Te}^{S} = -17.2 \pm 2.2 \quad \epsilon_{Te}^{Cr} = 1.5 \pm 0.7$$
$$\epsilon_{Te}^{Cu} = -9.9 \pm 0.5 \quad \rho_{Te}^{Cu} = 50 \pm 20 \quad \epsilon_{Te}^{Mo} = 18.5 \pm 1.6$$

A positive sign and a negative sign correspond to attractive and repulsive interaction, respectively. It is notable that sulfur has a relatively large effect, since resulfurized free-machining steel always has sulfur as an alloying element.

Discussion

Previously, we have proposed an empirical equation for the interaction parameters of various elements on oxygen and sulfur based on the standard Gibbs energy of formation of the corresponding oxide and sulfide [5]. The interaction parameters on tellurium obtained in the present study can also be explained in a similar fashion. Let us assume the following empirical equation, which is a modification of the modified quasichemical model proposed by Jacob and Alcock [6]:

$$RT \ln\left(1 - \frac{\epsilon_X^Y}{Z}\right) = A \cdot \left[\Delta G_{Fe-X}^{o} - \Delta G_{Y-X}^{o}\right] + B \cdot RT \ln \gamma_{Y\,in\,Fe}^{o} \tag{7}$$

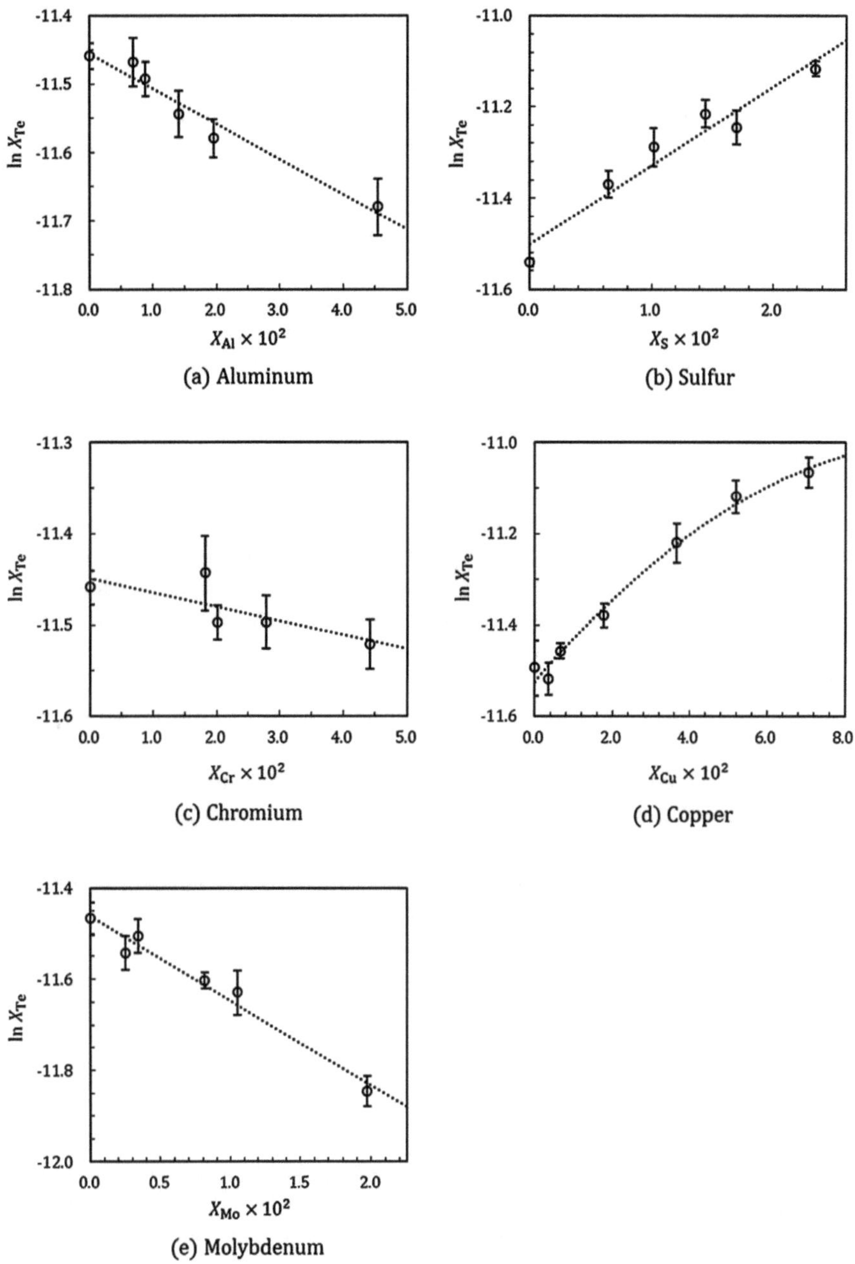

Fig. 2 Relations between the concentration of tellurium and those of **a** aluminum, **b** sulfur, **c** chromium, **d** copper, and **e** molybdenum in molten iron

Here, X is O, S, or Te; Y is Al, Cr, Mn, Ni, Cu, or Mo; R is the gas constant; T is the temperature (1873 K); and Z is the coordination number of molten iron, which is here assumed to be 10 based on an X-ray spectroscopic study [7]. The Raoultian activity coefficient of element Y in molten iron is represented by $\gamma^{\circ}_{\mathrm{Y\,in\,Fe}}$. Finally, $\Delta G^{\circ}_{\mathrm{Fe-X}}$ and $\Delta G^{\circ}_{\mathrm{Y-X}}$ are the standard Gibbs energies of reactions represented by Eqs. (8) and (9) at 1873 K, respectively:

$$\mathrm{Fe(l)} + \mathrm{X(g)} = \mathrm{FeX(s)} \tag{8}$$

$$a\mathrm{Y(l)} + \mathrm{X(g)} = b\mathrm{Y}_c\mathrm{X}_d\mathrm{(s)} \tag{9}$$

The values for $\gamma^{\circ}_{\mathrm{Y\,in\,Fe}}$ are taken from [8], and those for $\Delta G^{\circ}_{\mathrm{Fe-X}}$ and $\Delta G^{\circ}_{\mathrm{Y-X}}$ are taken from the FactSage thermochemical software [9–11]. The experimental values for $\epsilon_{\mathrm{X}}^{\mathrm{Y}}$ are taken from [12, 13], and this work. The dimensionless free parameters A and B (must not be less than zero) are optimized for each case of X, as shown in Table 1.

The left and right sides of the equation (corresponding to the experimental and predicted values, respectively) are compared for each case of X in Fig. 3. Considering the crudeness of the model, the agreement is satisfactory. Although the predictions are not quantitative, the results indicate that the above equation may be useful for a rough estimation of interaction parameters that have not yet been determined experimentally.

For the case of X = Te, $\epsilon_{\mathrm{Te}}^{\mathrm{Mo}} = 19$ cannot be explained by this model unless an unphysically large coordination number ($Z > 19$) is assumed, and therefore, it is excluded from the optimization process. One possible explanation for such a large interaction parameter is the size effect. According to the theoretical study by Ueno and Waseda based on the hard-sphere model [14], when the atomic sizes of X and Y are both larger than that of Fe, as may be the case for tellurium and molybdenum, $\epsilon_{\mathrm{X}}^{\mathrm{Y}}$ can be extremely large. Since the atomic size (the effective hard-sphere diameter) of tellurium in molten iron is currently unknown, it is difficult to confirm this hypothesis. Furthermore, $\epsilon_{\mathrm{X}}^{\mathrm{S}}$ is also excluded because the corresponding reaction expressed by Eq. (9) does not take place.

One may notice that the optimized parameters for oxygen are quite different from those for sulfur and tellurium. This might be explained by the differences in bonding mechanisms or solution types between these species. Taking into consideration the concepts behind the quasichemical model, parameter A can be considered as a measure of how X is affected when an Fe atom next to X is replaced by Y. Therefore, the larger value of A for oxygen may be because of ionic- or covalent-like interactions between oxygen and the surrounding atoms.

Table 1 Optimized parameters for each case

X	A	B
O	0.124	0.000
S	0.028	0.199
Te	0.033	0.255

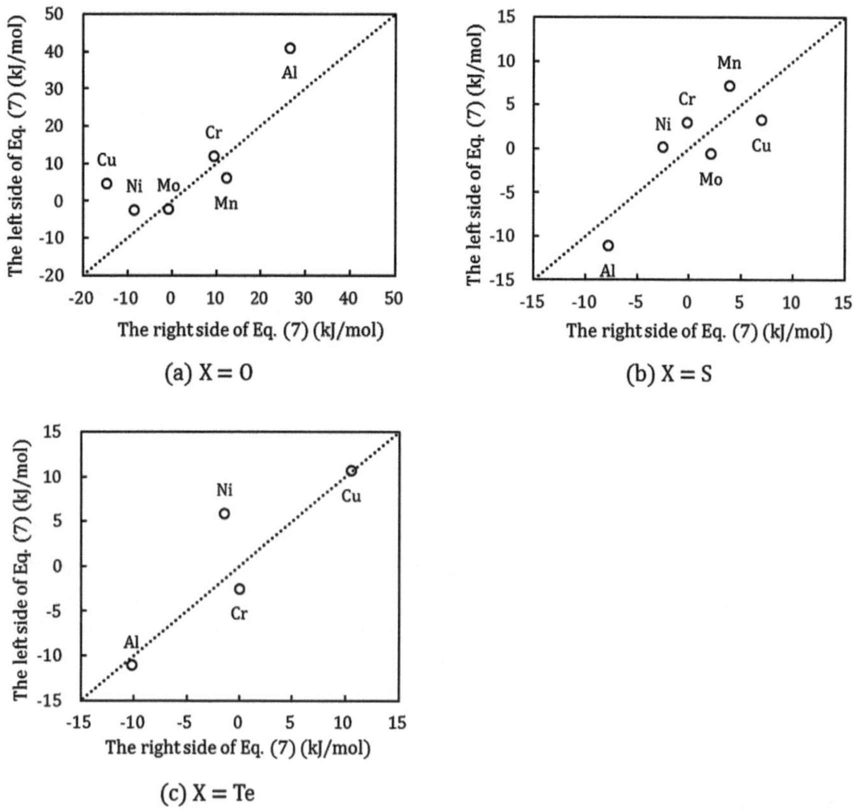

Fig. 3 Comparison between the left hand side of Eq. (7), calculated using experimental data, with the theoretical prediction from the expression on the right hand side for each case of **a** X = O, **b** X = S and **c** X = Te. *Dotted lines* represent the situation where the left hand side equals the hand right side

Such interactions are strongly affected by the types of atoms present at the nearest-neighbor sites. On the other hand, the smaller values of A for sulfur and tellurium may be attributed to the more metallic-like interactions that they take part in, which are less affected by the nearest-neighbors because of the screening effect of the conduction electrons between atoms.

Parameter B corresponds to how the interactions between Y and the surrounding Fe atoms are weakened when X is placed next to Y. The $B = 0$ value for oxygen may thus be explained by assuming that oxygen is an interstitial solute. In such a case, the number of Fe atoms around Y is virtually unchanged, even when X comes in contact with Y. Sulfur and tellurium are more likely to be substitutional, since their atomic sizes are larger than oxygen. In this scenario, the number of Fe atoms around Y decreases when X is placed in contact with the latter, resulting in a weakening of the interactions between Y and the Fe atoms. This results in non-zero values of B for sulfur and tellurium.

Conclusions

The Wagner's interaction parameters for aluminum, sulfur, chromium, copper, and molybdenum on tellurium in molten iron were determined experimentally via a vapor-liquid equilibration technique using the transpiration method to control the chemical potential of tellurium in the ambient atmosphere.

$$\epsilon_{\mathrm{Te}}^{\mathrm{Al}} = 5.1 \pm 0.5 \quad \epsilon_{\mathrm{Te}}^{\mathrm{S}} = -17.2 \pm 2.2 \quad \epsilon_{\mathrm{Te}}^{\mathrm{Cr}} = 1.5 \pm 0.7$$

$$\epsilon_{\mathrm{Te}}^{\mathrm{Cu}} = -9.9 \pm 0.5 \quad \rho_{\mathrm{Te}}^{\mathrm{Cu}} = 50 \pm 20 \quad \epsilon_{\mathrm{Te}}^{\mathrm{Mo}} = 18.5 \pm 1.6$$

The values obtained here, together with the values reported from previous studies for oxygen and sulfur, were qualitatively explained by an empirical equation based on the standard Gibbs energy for the formation of the corresponding tellurides, oxides, and sulfides. A hypothesis is presented for explaining the differences in the interaction parameters on oxygen, sulfur, and tellurium by taking into account their bonding mechanisms and solution types.

References

1. H. Yaguchi, N. Onodera, The effect of tellurium on the machinability of AISI 12L14+ Te steel. Trans. Iron Steel Inst. Japan **28**, 1051–1059 (1988)
2. T. Katoh et al., A study on resulfurized free-machining steel containing a small amount of Tellurium. DENKI-SEIKO **53**, 195–202 (1982)
3. H. Ipser, Vapor pressure methods: a source of experimental thermodynamic data. Berichte der Bunsengesellschaft für physikalische Chemie **102**, 1217–1224 (1998)
4. T. Ashino, K. Takada, K. Hirokawa, Determination of trace amounts of selenium and tellurium in high-purity iron by electrothermal atomic absorption spectrometry after reductive coprecipitation with palladium using ascorbic acid. Anal. Chim. Acta. **297**, 443–451 (1994)
5. Ueda, K. Morita, An empirical equation for interaction parameters of oxygen and sulfur in molten iron. 170st ISIJ Meeting, Fukuoka, Japan, 2015, 646
6. K.T. Jacob, C.B. Alcock, Quasi chemical equations for oxygen and sulphur in liquid binary alloys. Acta. Metall. **20**, 221–232 (1972)
7. Y. Waseda, S. Tamaki, The structures of 3D-transition metals in the liquid state. Philos. Mag. **32**, 273–281 (1975)
8. JSPS, 19th Committee of Steelmaking, ed., Recommended equilibrium values for steelmaking processes, 1984, 254
9. C.W. Bale et al., FactSage thermochemical software and databases. Calphad **26**, 189–228 (2002)
10. C.W. Bale et al., FactSage thermochemical software and databases—recent developments. Calphad **33**, 295–311 (2009)

11. C.W. Bale et al., Reprint of: FactSage thermochemical software and databases, 2010–2016. Calphad (2016). doi:10.1016/j.calphad.2016.07.004
12. M. Hino, K. Ito (eds.), *Thermodynamic data for steelmaking* (Tohoku University Press, Sendai, Japan, 2010), pp. 259–264
13. S. Ueda, K. Morita, Measurement of thermodynamic properties of tellurium in iron-based molten alloy using the transpiration method, Asia Steel International Conference 2015, Yokohama, Japan, 2015, 42–43
14. S. Ueno, Y. Waseda, Interaction parameters in multi-component metallic solutions by the hard sphere model. J. Japan Inst. Met. Mat. **50**, 788–796 (1986)

Thermodynamics of Simultaneous Desulfurization and Dephosphorization of Silicomanganese Alloy

Jongmin Jeong, Jaehong Shin, Joo Hyun Park, Chul-Woo Nam and Kyung-Ho Park

Abstract Because sulfur and phosphorus are very harmful in steel products, there are increasing requirements for refining these elements from ferroalloys industries. Hence, in the present study, simultaneous desulfurization (de-S) and reducing dephosphorization (Rde-P) from silicomanganese (SiMn) alloy were investigated using metal-slag equilibration method under highly reducing condition at 1773 K. Experiments were carried out by equilibration between SiMn alloy and MnO-based slag in graphite crucible under $p(O_2) = 10^{-18}$ atm. Slag composition was designed by changing basicity (CaO/SiO_2) and MnO content at fixed Al_2O_3 (=20%) and MgO (=5%) contents. De-S ratio increased with increasing basicity until C/S = 1.5 whereas Rde-P ratio was not significantly affected by slag basicity. This was because thermodynamic driving force of de-S reaction was greater than that of Rde-P reaction under present experimental conditions. The de-S and Rde-P ratios commonly increased with decreasing oxygen partial pressure.

Keywords Silicomanganese (SiMn) alloy · Reducing dephosphorization · Desulfurization · MnO-slags

Introduction

High manganese steels (HMnS) get attention because of their good mechanical properties such as high strength and good ductility [1]. However, one of the problems for producing HMnS is the low quality of Mn ferroalloy, i.e., high impurity levels. Sulfur and phosphorus in Mn ferroalloys have a significant effect on a deteriorating the quality of steel products. Nevertheless, a conventional oxidizing dephosphorization technique is not easily applicable to Mn ferroalloys

J. Jeong · J. Shin · J.H. Park (✉)
Department of Materials Engineering, Hanyang University, Ansan, Korea
e-mail: basicity@hanyang.ac.kr

C.-W. Nam · K.-H. Park
Korea Institute of Geoscience and Mineral Resources (KIGAM), Daejeon, Korea

S. Wang et al. (eds.), *Applications of Process Engineering Principles in Materials Processing, Energy and Environmental Technologies*,
The Minerals, Metals & Materials Series, DOI 10.1007/978-3-319-51091-0_48

because Si or Mn is preferentially oxidized compared to phosphorus under oxidizing conditions. Therefore, reducing dephosphorization (Rde-P) process is considered to produce high purity Mn ferroalloys.

The mechanism of Rde-P has been reported by several researchers [2]. Tabuchi and Sano investigated the thermodynamic behavior of phosphate and phosphide ions in the $BaO–BaF_2$ and the $CaO–CaF_2$ fluxes and concluded that the phosphide capacity of the latter was higher than that of the former system [3, 4]. Eric et al. [5] measured the phosphide capacity of the $CaO–SiO_2–MgO–Al_2O_3(–MnO)$ slags used for the FeMn smelting process at 1773 K and found that the phosphide capacity increased with increasing basicity.

Although several studies concerning the phosphide capacity of the lime-based flux systems have been investigated, there are few reports of the simultaneous removal of sulfur and phosphorus from SiMn alloy melts under strongly reducing atmosphere. Consequently, the aim of present study is to optimize the slag composition and other conditions for producing high purity SiMn alloy with lower contents of P and S.

Experimental Procedure

The SiMn alloys and the slags of various compositions were equilibrated using a super-kanthal electric furnace with a $MoSi_2$ heating element. The temperature was controlled within ±2 K using a B-type thermocouple and a PID controller, and was maintained at 1773 K throughout the experiments. Experimental apparatus is shown in Fig. 1. A SiMn alloy containing 0.1% P and 0.06% S was prepared using an induction furnace under an Ar-3%H_2 atmosphere, and the slags were prepared by mixing regent grade chemicals.

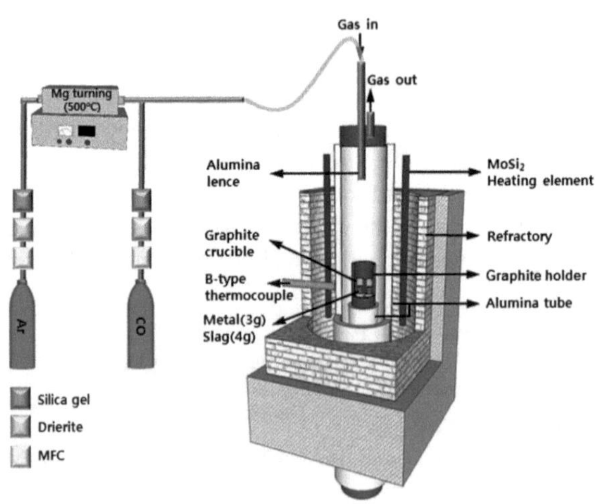

Fig. 1 Schematic diagram of the experimental apparatus

SiMn alloy (3 g) and slag (4 g) were placed in the graphite crucible. The oxygen partial pressure was controlled $p(O_2) = 10^{-19}$–10^{-18} atm at 1773 K by C–CO equilibrium reaction. Equilibrium was maintained for 12 h. After equilibration, the samples were quenched by Ar flushing and separated carefully from the crucible for chemical analysis. The alloy was analyzed by ICP-AES for Mn, Fe, Si, and P. The contents of oxides in the slag was analyzed with an XRF spectroscopy. The content of sulfur in the alloy and slag samples was determined using LECO combustion analyzer. Moreover, the XRD was used for the identification of slag phase.

Results and Discussion

After the equilibrating, the desulfurization (de-S) and reducing dephosphorization (Rde-P) ratio was evaluated as a function of slag basicity, viz. CaO/SiO_2 ratio. The influences of basicity and MnO content on the removal efficiency of sulfur and phosphorus are shown in Figs. 2 and 3, respectively. According to the results of current study, de-S ratio increases with increasing slag basicity until C/S = 1.5, whereas the influence of slag basicity on the Rde-P ratio is negligible. A difference between these two refining reactions are originated from the fact that the thermodynamic driving force of de-S reaction, Eq. (1), is more predominant than that of Rde-P reaction, Eq. (2), at 1773 K [6, 7].

$$[\mathrm{Ca}] + \tfrac{1}{2}\mathrm{S_2(g)} = (\mathrm{CaS}), \quad \Delta G^0 = -131000 + 24.82\ \mathrm{T(cal/mol)} \tag{1}$$

$$\tfrac{3}{2}[\mathrm{Ca}] + \mathrm{P_2(g)} = (\mathrm{Ca_{1.5}P_2}), \quad \Delta G^0 = -78520 + 26.77\ \mathrm{T(cal/mol)} \tag{2}$$

Because CaS was more stable than Ca_3P_2, Ca^{2+} ion preferentially participated in the de-S reaction, resulting in a lack of adequate Ca^{2+} for Rde-P reaction. Therefore, the Rde-P ratio had no relationship with slag basicity and thus the Rde-P ratio was not more than approx. 10%. A decrease in a de-S ratio at C/S > 1.5 was due to an

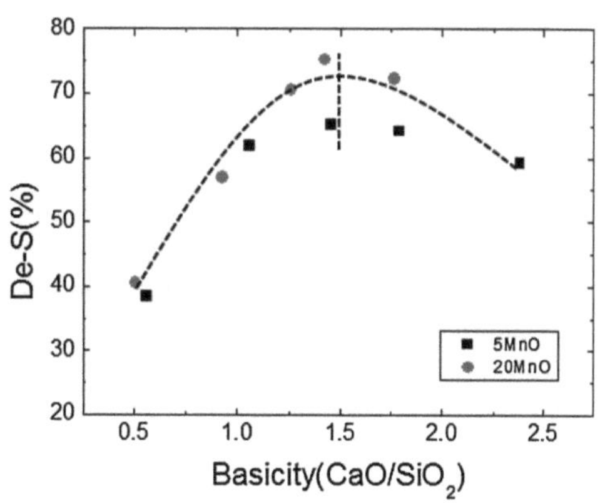

Fig. 2 Effect of slag basicity and MnO content on the desulfurization ratio at 1773 K

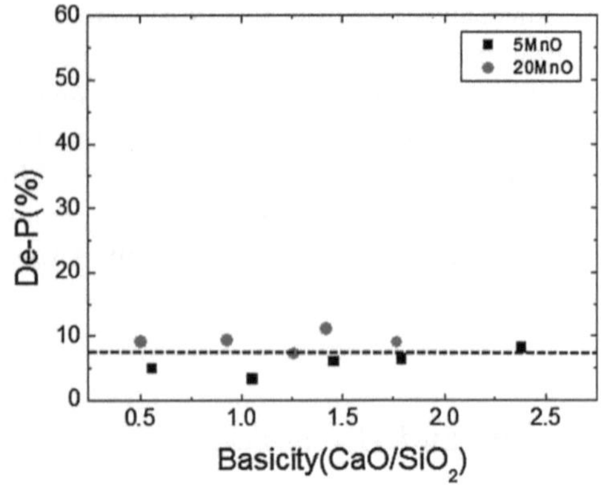

Fig. 3 Effect of basicity and MnO content on the dephosphorization ratio at 1773 K

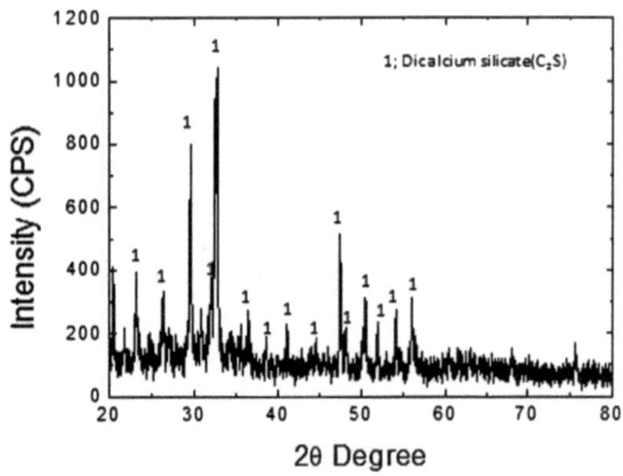

Fig. 4 X-Ray Diffraction analysis of the highly basic (C/S = 2.4) slag system

increase in solid fraction in the slag. The solid dicalcium silicate (Ca_2SiO_4) was precipitated, which was confirmed from an XRD analysis of the quenched slag samples as shown in Fig. 4.

Conclusions

The equilibrium reaction between SiMn alloy and $CaO–SiO_2–Al_2O_3–MgO–MnO$ slag was carried out at 1773 K. The de-S ratio was strongly affected by slag basicity, whereas Rde-P ratio was not affected by slag basicity under the present experimental conditions. The de-S ratio decreased at C/S > 1.5 because the effective liquid slag was reduced due to a precipitation of dicalcium silicate. For the Rde-P ratio, the more systematic experiments should be carried out regarding the operating parameters including oxygen partial pressure, temperature, etc.

References

1. Y. Lu, D.A. Molodov, G. Gottstein, ISIJ Int. **51**, 812–817 (2011)
2. H. Momokawa, N. Sano, Metall. Trans. B **13B**, 643–644 (1982)
3. S. Tabuchi, N. Sano, Metall. Trans. B **15B**, 351–356 (1984)
4. S. Tabuchi, N. Sano, Tetsu-to-Hagane **71**, 687–692 (1985)
5. B. Maramba, R.H. Eric, Min. Eng. **21**, 132–137 (2008)
6. M.W. Chase et al., J. Phys. Chem. Ref. Data **3**, 311 (1974)
7. I. Barin, O. Knacke, *Thermochemical Properties of Inorganic Substances* (Springer-Verlag, Berlin and New York, 1977) (supplement)

Reduction Behavior of CaO–Fe_2O_3–8 wt% SiO_2 System at 1123, 1173 and 1223 K with CO–N_2 Gas Mixtures

Chengyi Ding, Xuewei Lv, Kai Tang, Senwei Xuan, Yun Chen and Jie Qiu

Abstract The influence of SiO_2 on the reducibility of the CaO–Fe_2O_3–SiO_2 system was fully examined in this study. The isothermal reduction kinetics of CF and CF8S (CaO/Fe_2O_3 = 1:1, wt% (SiO_2) = 8) were investigated through thermogravimetric analysis at 1123, 1173, and 1223 K with 30% CO and 70% N_2 gas mixtures. The reduction of the samples with 8% SiO_2 was not only highly accelerated but also proceeded easily. Rate analysis revealed that CF and CF8S reduction occurs in two stages, the Fe_3O_4-to-FeO stage overlaps with the previous Fe_2O_3-to-Fe_3O_4 stage and tends to approach the following FeO-to-Fe stage with the addition of SiO_2. The apparent activation energy values of CF and CF8S reduction are 46.89 and 8.71 kJ mol^{-1}. Sharp analysis indicated that CF and CF8S reduction was expressed by the Avrami–Erofeev equation presenting a 2D shrinking layer reaction.

Keywords CF · SiO_2 · Reduction kinetics · Apparent activation energy · Model function

Introduction

In sintering production, the effect of SiO_2 on sinter quality and on the determination of blast furnace slag has been examined by various studies. The SiO_2–Fe_2O_3–CaO–Al_2O_3 (SFCA) phase is regarded as a desirable bonding phase for fluxed sinter, and SiO_2 content influences the phase composition and bonding phase mass of SFCA. Edstrom [1] in 1986 showed that low-SiO_2 sinter exhibits good metallurgical performance and reduces blast furnace slag in the iron-making process. SiO_2–Fe_2O_3–CaO (SFC) is believed to be the transition phase in the SFCA formation process and has been widely investigated [2, 3]. Hyunsik [4, 5] explored the effect

C. Ding · X. Lv (✉) · K. Tang · S. Xuan · Y. Chen · J. Qiu
College of Materials Science and Engineering, Chongqing University, Chongqing 400044, China
e-mail: lvxuewei@163.com

S. Wang et al. (eds.), *Applications of Process Engineering Principles in Materials Processing, Energy and Environmental Technologies*, The Minerals, Metals & Materials Series, DOI 10.1007/978-3-319-51091-0_49

of SiO_2 on the reduction kinetics of carbon composite pellets and found that SiO_2 decreases the reduction rate by lowering the surface area of samples. Taguchi et al. [6] analyzed phase transition in the CaO–2Fe_2O_3 system with SiO_2 addition. The reduction of samples in previous studies was chemically regarded as the solid–gas reduction of iron oxide [7–10]. The reduction route of iron oxide has been explored by many studies, but explanations with the model function derived by thermal kinetics are lacking. The current study attempts to improve the understanding of these aspects.

Experimental

Materials

Samples were prepared from $CaCO_3$ ($\geq$99.99%) and Fe_2O_3 (α-Fe_2O_3, $\geq$99.99%) at a molar ratio of 1:1 with the addition of 0 and 8 wt% SiO_2, as shown in Table 1. The powdery raw materials were uniformly mixed and then pressed into cylinder-shaped samples with 10 MPa pressure. The samples were roasted in a $MoSi_2$ furnace at 1173 K for 1 h to decompose $CaCO_3$ to CaO. Then, the temperature was increased to 1473 K for 10 h to allow the complete reaction of CF and CF8S generation. The samples were crushed and sieved into powder (<74 μm) for subsequent tests.

X-ray diffraction (XRD) analysis with Rigaku D/max2500/PC (Cu Kα) was conducted to confirm the phase composition of the roasted samples. Scanning was carried out at an angular range of 10°–90° and scan rate of 4°/min. MDI Jade 5.0 was utilized to analyze the intensity data obtained by XRD analysis. Figure 1 shows the XRD patterns of samples CF and CF8S.

The reaction of the CaO–Fe_2O_3–SiO_2 system in the roasting process can be described as follows:

$$\begin{aligned} Fe_2O_3 + CaO + SiO_2 = {} & CaO \cdot Fe_2O_3 + 2CaO \cdot SiO_2 \\ & + 3(CaO \cdot SiO_2) \cdot Fe_2O_3 + Fe_2O_3, \end{aligned} \quad (1)$$

where $3(CaO{\cdot}SiO_2){\cdot}Fe_2O_3$ is equal to the phase SFC indicated in the XRD pattern shown in Fig. 1. As the SiO_2 content added to the CaO–Fe_2O_3–SiO_2 system increases, the peak corresponding to CF weakens gradually, whereas that of C_2S and Fe_2O_3 strengthens. This result indicates that the CF content in the roasted phase decreases and C_2S and Fe_2O_3 increase. The increase in SFC content is not obvious.

Table 1 Chemical composition of samples CF and CF8S (wt%)

Samples	$CaCO_3$	Fe_2O_3	SiO_2
CF	38.46	61.54	0
CF8S	35.38	56.61	8

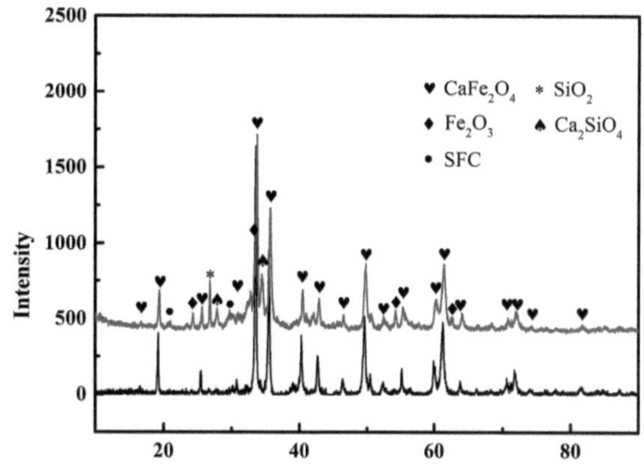

Fig. 1 XRD patterns of samples CF and CF8S

TGA

Isothermal analysis was carried out via thermogravimetric (TG) measurement in a Setaram analyzer (Model Setsys Evo TG-DTA 1750), as shown in Fig. 2. The samples (20 mg) in an alumina crucible were heated to 1123, 1173, and 1223 K in N_2 ($\geq$ 99.999%) atmosphere at 15 K/min from room temperature. Gas mixtures of 30% CO ($\geq$ 99.999%) and 70% N_2 were blown into the furnace at 20 ml/min for 150 min to enable complete reaction with the samples at the isothermal stage. To exclude the influence of the system error from the thermal analyzer and the buoyance force from the gas mixtures, a blank test was conducted under the same elimination conditions with no samples in alumina crucibles. Weight loss data were obtained during the isothermal reduction stage, from which the TG data of the blank test were deduced.

Thermal Analysis Kinetics

Reduction degree is defined as the ratio of removed oxygen mass at a fixed time t to the theoretically removed oxygen mass from iron oxide; it can be expressed as

$$\alpha = \frac{\Delta m_t}{\Delta m_0}, \tag{2}$$

where α is the reduction degree and Δm_t and Δm_0 refer to the removed oxygen mass at fixed time t and the theoretically removed oxygen mass from iron oxide, respectively. In the reduction of the $CaO–Fe_2O_3–SiO_2$ system, the removed oxygen is only from phases containing iron oxides.

The basic kinetic equation [11] that describes the relationship between reduction rate and time can be expressed as

Fig. 2 Schematic of the TG analyzer

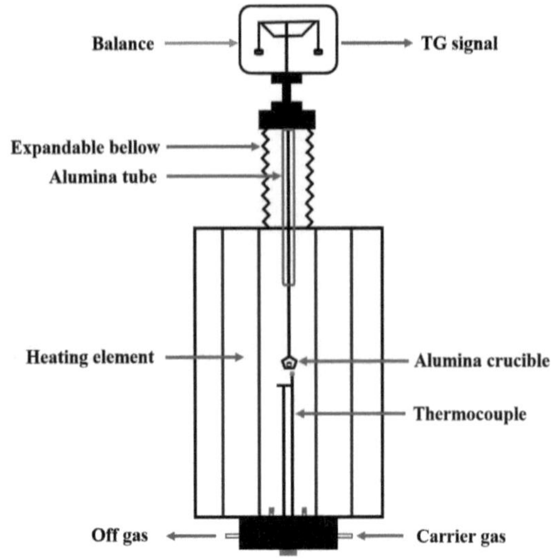

$$\frac{d\alpha}{dt} = k(T)f(\alpha), \tag{3}$$

where $d\alpha/dt$ is the reduction rate and $k(T)$ and $f(\alpha)$ are the rate constant and model function of the reduction reaction, respectively. $f(\alpha)$ is influenced by the reaction mechanism. $k(T)$ is determined by the Arrhenius equation as follows:

$$k(T) = A\exp\left(\frac{-E}{RT}\right), \tag{4}$$

where A is the pre-exponential factor, E is the apparent activation energy, and R is the gas constant [8.314 J/(mol K)]. Equation (3) can be further expressed as

$$\frac{d\alpha}{dt} = A\exp\left(\frac{-E}{RT}\right)f(\alpha). \tag{5}$$

Given that reduction degree α is a constant, ln $f(\alpha)$ remains unchanged. Therefore, activation energy can be calculated as

$$E = -R\frac{d\left(\ln\frac{d\alpha}{dt}\right)}{d\left(\frac{1}{T}\right)}. \tag{6}$$

The method to calculate activation energy can eliminate the limit from the model function and is called the model-free method [12]. $f(\alpha)$ is usually not easily obtained. $G(\alpha)$, which is the integral function of $f(\alpha)$, can be described as

$$G(\alpha) = \int_0^{\alpha} \frac{d\alpha}{f(\alpha)} = \int_0^{t} A \exp\left(\frac{-E}{RT}\right) dt = k(T)t. \tag{7}$$

Results and Discussion

Reduction Degree and Reduction Rate

The reduction degree of samples CF and CF8S with the change in time during the isothermal reaction stage is shown in Fig. 3. The reduction process completes early as temperature increases. The maximum reduction degree and its corresponding time for samples CF and CF8S at 1123, 1173, and 1223 K are shown in Table 2, where α_m and t_m are the maximum reduction degree and its reduction time, respectively. The results indicated that the reduction time for samples CF8S decreases than CF, and maximum reduction degree increases for CF8S than CF.

As shown in Fig. 2, the phases in the roasted samples mainly comprise SiO_2, CF, C_2S, CFS, and Fe_2O_3. Removed oxygen originates only from the iron oxides in CF, CFS, and Fe_2O_3. The reducibility of Fe_2O_3 is better than that of CF, and that of CFS is the worst among the three phases [13]. The presence of Fe_2O_3 significantly improves sample reducibility. The reduction rates of CF and CF8S at three temperatures are shown in Fig. 4. The reduction routes of iron oxides can be expressed as

$$Fe_2O_3(CF, SFC) \rightarrow Fe_3O_4 \rightarrow FeO \rightarrow Fe. \tag{8}$$

The theoretical reduction degree is fixed at 0.11 and 0.33 when Fe_2O_3 reduces to Fe_3O_4 and Fe_3O_4 reduces to FeO, respectively. The Fe_3O_4-to-FeO stage overlaps with the previous Fe_2O_3-to-Fe_3O_4 stage and tends to approach the following FeO-to-Fe stage with the addition of SiO_2.

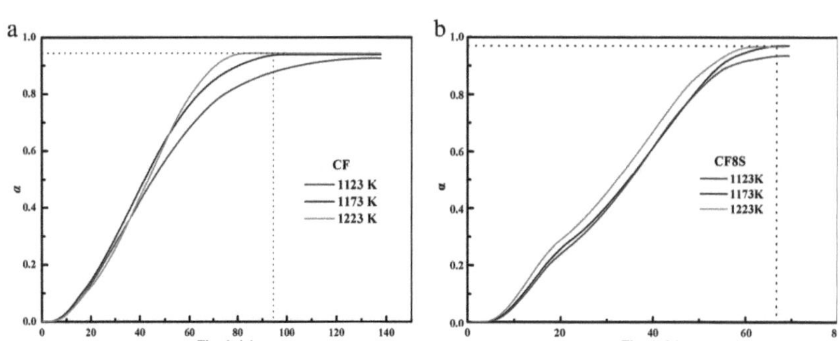

Fig. 3 Reduction degree of samples CF (**a**) and CF8S at 1123, 1173, and 1223 K (**b**)

Table 2 Maximum reduction degree of samples CF and CF8S at 1123, 1173, and 1223 K and the corresponding time

Samples	α_m			t_m/min		
	1123 K	1173 K	1223 K	1123 K	1173 K	1223 K
CF	0.92	0.93	0.94	137	100	84
CF8S	0.94	0.94	0.97	69	68	62

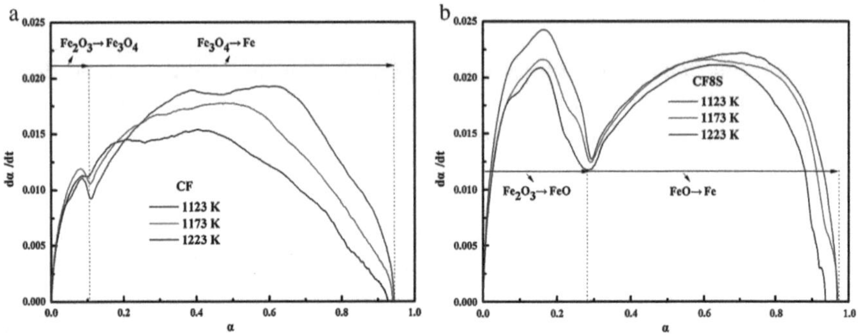

Fig. 4 Reduction rate of samples CF (**a**) and CF8S at 1123, 1173, and 1223 K (**b**)

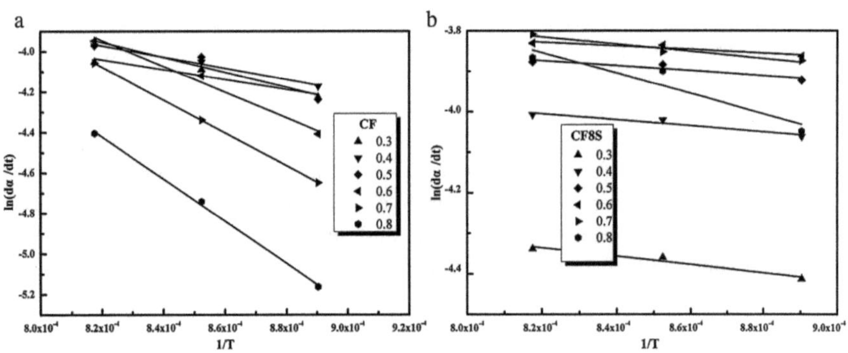

Fig. 5 ln(dα/dt) against 1/T of samples CF (**a**) and CF8S (**b**)

Apparent Activation Energy for Samples Reduction

Apparent activation energy was calculated by the slope of the plots of ln(dα/dt) against 1/T shown in Fig. 5, and the results are presented in Table 3. Given that sample reduction was combined with several reaction stages and gas diffusion, activation energy was regarded as the apparent value reflecting the comprehensive influence on the reduction process. Gas switching from pure N_2 to N_2 and CO mixtures led to the samples not being reduced by the target CO content during the

Table 3 Apparent activation energy of samples CF and CF8S

E/(kJ mol^{-1})							
α	0.3	0.4	0.5	0.6	0.7	0.8	Avg
CF	20.32	23.11	30.85	52.85	67.51	86.68	46.89
CF8S	8.58	6.24	5.18	3.78	7.46	21.01	8.71

initial stage. Consequently, the reduction stage before $\alpha = 0.3$ was not considered in this study for the calculation of apparent activation energy. Instead, the reduction during $\alpha = 0.3$–0.8 was considered for the activation energy calculation. The apparent activation energy values of CF and CF8S reduction are 46.89 and 8.71 kJ mol^{-1}. Sample reduction tends to be easy with the addition of 8% SiO_2 in the Fe_2O_3–CaO system.

Model Function Results

Model function $G(\alpha)$ was obtained to describe the relationship of reaction degree with time. Nine normal functions [14] expressing the solid-state reaction are listed in Table 4. Functions A_2 and A_3 were obtained from the nucleation process (crystallization) and can also be applied to the shrinking layer reaction. Considering that the model function cannot be directly obtained, Sharp [15, 16] defined non-dimensional parameter $y(\alpha)$ to target model function.

$$y(\alpha) = \frac{G(\alpha)}{G(0.5)} = \frac{kt}{kt_{0.5}} = \frac{t}{t_{0.5}}, \tag{9}$$

where $G(0.5)$ represents $G(\alpha)$ when $\alpha = 0.5$ and $t_{0.5}$ refers to the time when $\alpha = 0.5$. The curves of plots of α against $t/t_{0.5}$ derived by the nine solid reactions are called standard curves. Experimental data $[t_i/t_{0.5}(i = 1, 2, \ldots l), \alpha_i]$ were obtained from the TG curves of sample reduction. Target $G(\alpha)$ is clear, given that the experimental data match one of the corresponding standard curves. Sharp analysis was conducted when the transition of the reaction rate peak approximately occurred at $\alpha = 0.5$. According to the rate change of CF and CF8S reduction, $y(\alpha)$ was set to $t/t_{0.1}$ and $t/t_{0.3}$ for CF and CF8S, respectively.

Experimental data $[t_i/t_{0.1}(i = 0.1, 0.2, \ldots 0.8, 0.9), \alpha_i]$ and $[t_i/t_{0.3}(i = 0.1, 0.2, \ldots 0.8, 0.9), \alpha_i]$ of CF and CF8S reduction were obtained and are shown in Fig. 6. The results indicate that the experimental data $y(\alpha)$ values for CF and CF8S reduction mostly lie on the standard curve corresponding to function A_2. The reduction of CF and CF8S can only be described by the mechanism of 2D shrinking layer reaction.

Table 4 Model function $G(\alpha)$ for normal solid-state reactions

Mechanism	Functions	$G(\alpha)$
One-dimensional diffusion	$D_1(\alpha)$	$\alpha^2 = kt$
Two-dimensional diffusion (bidimensional particle shape)	$D_2(\alpha)$	$(1-\alpha)\ln(1-\alpha)+\alpha = kt$
Three-dimensional diffusion (tridimensional particle shape) Jander equation [17]	$D_3(\alpha)$	$\left[1-(1-\alpha)^{1/3}\right]^2 = kt$
Three-dimensional diffusion (tridimensional particle shape) Ginstling-Brounshtein equation [18]	$D_4(\alpha)$	$(1-2/3\alpha)-(1-\alpha)^{2/3} = kt$
Bimolecular decay law (instantaneous nucleation and unidimensional growth)	$F_1(\alpha)$	$-\ln(1-\alpha) = kt$
Phase boundary controlled reaction (contracting area, i.e., bidimensional shape)	$R_2(\alpha)$	$1-(1-\alpha)^{1/2} = kt$
Phase boundary controlled reaction (contracting volume, i.e., tridimensional shape)	$R_3(\alpha)$	$1-(1-\alpha)^{1/3} = kt$
Random instant nucleation and two-dimensional growth of nuclei (Avrami-Erofeev equation [19–21])	$A_2(\alpha)$	$[-\ln(1-\alpha)]^{1/2} = kt$
Random instant nucleation and three-dimensional growth of nuclei (Avrami-Erofeev equation)	$A_3(\alpha)$	$[-\ln(1-\alpha)]^{1/3} = kt$

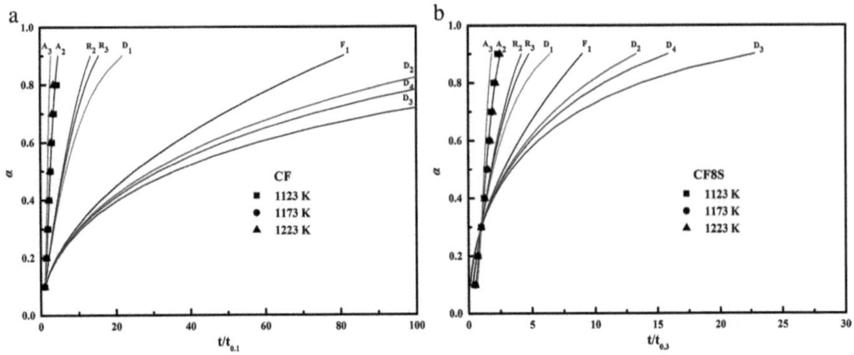

Fig. 6 Standard curves and experimental data based on Sharp analysis for CF (**a**) and CF8S (**b**)

Conclusions

The isothermal reduction kinetics of CF and CF8S was investigated via TG measurement with 30% CO and 70% N_2 gas mixtures. The following conclusions were obtained.

- CF and CF8S reduction occurs in two stages, the Fe_3O_4-to-FeO stage overlaps with the previous Fe_2O_3-to-Fe_3O_4 stage and tends to approach the following FeO-to-Fe stage with the addition of SiO_2.
- The apparent activation energy values of CF and CF8S reduction are 46.89 and 8.71 kJ mol^{-1}. Sample reduction tends to be easy with the addition of 8% SiO_2 in the Fe_2O_3–CaO system.
- Sharp analysis implied that reduction of CF and CF8S can only be described by the mechanism of 2D shrinking layer reaction.

Acknowledgements The authors are grateful for the financial support provided by the Natural Science Foundation of China (51544203).

References

1. S. Santen, J.O. Edstrom, Method of Manufacturing Silicon from Powdered Material Containing Silica. U.S. Patent, No. 4,439,410, 27 Mar 1984
2. X. Ding, X.M. Guo, The formation process of silico-ferrite of calcium (SFC) from binary calcium ferrite. Metall. Mater. Trans. B **45**(4), 1221–1231 (2014)
3. X. Ding, X.M. Guo, Study of SiO_2 involved in the formation process of silico-ferrite of calcium (SFC) by solid-state reactions. Int. J. Miner. Process. **149**, 69–77 (2016)
4. H. Park, V. Sahajwalla, Effect of alumina and silica on the reaction kinetics of carbon composite pellets at 1473 K. ISIJ Int. **54**(1), 49–55 (2014)
5. H. Park, V. Sahajwalla, Reduction behavior of carbon composite pellets including alumina and silica at 1273 K and 1373 K. ISIJ Int. **54**(6), 1256–1265 (2014)
6. N. Taguchi, T. Otomo, K. Tasaka, Effect of SiO_2 and Al_2O_3 additions on the reduction and expansion of synthetic materials based on $CaO_2·Fe_2O_3$. Tetsu-to-Hagane **73**(15), 1885–1892 (1987)
7. S.K. Dutta, A. Ghosh, Kinetics of gaseous reduction of iron ore fines. ISIJ Int. **33**(11), 1168–1173 (1993)
8. H.W. Kang, W.S. Chung, T. Murayama, Effect of iron ore size on kinetics of gaseous reduction. ISIJ Int. **38**(2), 109–115 (1998)
9. A. Unal, A.V. Bradshaw, Rate processes and structural changes in gaseous reduction of hematite particles to magnetite. Metall. Mater. Trans. B **14**(4), 743–752 (1983)
10. N.A. Warner, Reduction kinetics of hematite and the influence of gaseous diffusion. Trans. Metall. Soc. AIME **230**, 1631–1676 (1964)
11. R.C. Mccune, P. Wynblatt, Calcium segregation to a magnesium oxide (100) surface. J. Am. Ceram. Soc. **66**(2), 111–117 (1983)
12. S. Vyazovkin, C.A. Wight, Model-free and model-fitting approaches to kinetic analysis of isothermal and non-isothermal data. Thermochim. Acta **340**, 53–68 (1999)
13. J.O. Edstrom, The mechanism of reduction of iron oxides. J. Iron Steel Inst. **175**(3), 289 (1953)
14. J.D. Hancock, J.H. Sharp, Method of comparing solid–state kinetic data and its application to the decomposition of kaolinite, brucite, and $BaCO_3$. J. Am. Ceram. Soc. **55**(55), 74–77 (1972)
15. M.E. Brown, D. Dollimore, A.K. Galwey, *Reactions in the Solid State*, vol 22 (Elsevier, 1980)
16. J.H. Sharp, G.W. Brindley, B.N. Narahari, Numerical data for some commonly used solid state reaction equations. J. Am. Ceram. Soc. **49**(7), 379–382 (1966)

17. S.S. Tamhankar, L.K. Doraiswamy, Analysis of solid–solid reactions: a review. AIChE J. **25** (4), 561–582 (1979)
18. A.M. Ginstling, B.I. Brounshtein, On diffusion kinetics in chemical reactions taking place in spherical powder grains. Zhur. Priklad. Khim. **23** (1950)
19. M. Avrami, Kinetics of phase change. I general theory. J. Chem. Phys. **7**(12), 1103–1112 (1939)
20. M. Avrami, Kinetics of phase change. II transformation–time relations for random distribution of nuclei. J. Chem. Phys. **8**(2), 212–224 (1940)
21. M. Avrami, Granulation, phase change, and microstructure kinetics of phase change. III. J. Chem. Phys. **9**(2), 177–184 (1941)

A Review of Some Studies on Impurity Capacity Predictions in Molten Melts

Bora Derin

Abstract This study is a collection of both newly published and/or presented works on impurity capacity predictions in molten melts which were carried out by Derin et al. Some sulfide capacity prediction results of iron oxide containing ternary silicate melts and industrial lead and copper slags using the Reddy-Blander model and some impurity capacity predictions of multicomponent oxide and salt melts obtained using a Artificial Neural Network Approach are discussed.

Keywords Impurity capacity · Molten melts · Reddy-Blander · Neural network

Introduction

Inclusions containing sulfur, phosphorus, nitrogen, carbon etc. have to be controlled during many metal production processes due mainly to their undesirable effects on the metal properties. Better quality metal is possible with better slag design, which may influence the impurity content in the metal. Impurity capacity (C_i) models have been developed for decades to predict the ability of slag to remove the impurities in hot metal. These models are empirical, semi-empirical and/or theoretical and are used to estimate the capacities of sulfide, phosphorus, nitrogen, and carbon in molten melts [1–18]. For example, Sosinsky and Sommerville first proposed a sulfide capacity (Cs) prediction model using an optical basicity concept defined by Duffy and Ingram [1, 2]. Mori also used the optical basicity concept in phosphorus capacity prediction for $CaO–MgO–SiO_2–Fe_tO$ slags at 1873 K [3]. Moretti and Ottonello developed a sulfide capacity prediction model for simple and complex silicate melts using different polymeric solution theory [4]. KTH model was designed for the optimization of experimentally determined Cs values of simple systems in order to obtain those of multicomponent slags [5]. Selin et al. developed

B. Derin (✉)
Faculty of Chemistry and Metallurgy, Department of Metallurgical and Materials Engineering, Istanbul Technical University, Maslak, Istanbul, Turkey
e-mail: bderin@itu.edu.tr

S. Wang et al. (eds.), *Applications of Process Engineering Principles in Materials Processing, Energy and Environmental Technologies*,
The Minerals, Metals & Materials Series, DOI 10.1007/978-3-319-51091-0_50

a phosphate capacity prediction model using temperature and basicity index for $CaO–SiO_2–CaF_2$ slags [6]. Koyabashi et al. developed a simple linear formula for phosphate capacity predictions as functions of temperature for the slag composition for the $MnO–SiO_2–Fe_tO–MgO_{satd}$ systems [7]. Yang et al. developed a thermodynamic model for predicting the phosphate capacity of convertor slags based on the ion and molecule coexistence theory (IMCT) [8]. Maramba and Eric created the quadratic regression models to estimate phosphide capacity (C_P^{3-}) of ferromanganese smelting slags under highly reducing conditions [9]. Sasabe et al. developed a simple regression model for prediction of nitride capacities (C_N^{3-}) of molten $CaO–SiO_2–Al_2O_3$ slags containing TiO_2 or ZrO_2 [10]. Shin and Lee studied on the nitrogen solubility of $CaO–Al_2O_3–CaF_2$ melts and developed a formula based on the optical basicity concept for 1773 K [11]. Park and Min investigated the carbide capacities ($C_{C_2^{2-}}$) of $CaO–SiO_2–CaF_2$ ($–Na_2O$) slags at 1773 K and they presented a linear regression formula. Derin et al. used neural network approach for the prediction of impurity capacities in some molten melts [12, 13]. Reddy and Blander (RB) developed a model, a priori, based on a simple solution model and on knowledge of the chemical and solution properties of impurities and oxides [14–16]. The RB model was then modified by Pelton et al. [17]. Later, Kang and Pelton developed a thermodynamic model in the framework of the modified quasichemical model in the quadruplet approximation to permit the calculation of solubilities of various gaseous species (sulfide, sulfate, nitride, carbide, water, etc.) in molten slags [18]. This paper is a review of research on the predications of impurity capacities of some molten melts using the Reddy-Blander and a Neural Network Approach methods which have been carried out by Derin et al.

Theoretical Consideration

Reddy-Blander (RB) Model

The sulfide capacity, C_S, as an indicator of desulphurization power of molten slags, was first described by Fincham Richardson as:

$$C_S = (\mathrm{wt\%S})\left(\frac{P_{O_2}}{P_{S_2}}\right)^{1/2} \tag{1}$$

where, wt%S is weight percent of sulfur in the slag, PO_2 and PS_2 are the oxygen and sulfur partial pressures in system, respectively [19] RB model was described elsewhere in detail [14–16]. Shortly, it proposes the following definitions for the Cs calculations. For a binary $MO–SiO_2$ (M = Ca, Mg, Mn, Fe, etc.) system, the sulfide equilibrium reaction can be written as:

$$MO_{(l)} + 1/2S_{2(g)} = MS_{(l)} + 1/2O_{2(g)} \quad (2)$$

The equilibrium constant, K_M, for the above reaction is given by:

$$K_M = \left(\frac{a_{\text{MS}}}{a_{\text{MO}}} \sqrt{\frac{P_{\text{O}_2}}{P_{\text{S}_2}}} \right) \quad (3)$$

For a ternary system MO–NO–SiO_2, such as the FeO–MgO–SiO_2 system, the Flood-Grjotheim approximation is applied. The C_S value is then deduced from the sub-binary systems at constant silica mole fractions using the following equation:

$$\ln C_S = N_M \ln C_{S,M} + N_N \ln C_{S,N} \quad (4)$$

where N_M and N_N are electrical equivalent fractions, $N_M = \frac{X_{MO}}{X_{MO} + X_{NO}}$ and $N_N = \frac{X_{NO}}{X_{MO} + X_{NO}}$, $C_{S,M}$ and $C_{S,N}$ are sulfide capacities in the MO–SiO_2 and NO–SiO_2 binary systems at a fixed mole fraction of silica; and C_S is the sulfide capacity of the ternary MO–NO–SiO_2 system. When the multi-component melts contain more than one acid component such as MO–NO–SiO_2–$FeO_{1.5}$, then a subdivision is made i.e. MO–SiO_2–$FeO_{1.5}$ and NO–SiO_2–$FeO_{1.5}$. When $X_{SiO2} + X_{FeO1.5} \leq 0.33$

$$C_S = 100 W_S K_M a_{MO} \left(\frac{1 - 2X_{SiO_2} - 2X_{FeO_{1.5}}}{\overline{W}} \right) \quad (5)$$

When $X_{SiO2} + X_{FeO1.5} \geq 0.33$

$$C_S = 100 W_S K_M a_{MO} \left[\frac{X_{SiO_2} + X_{FeO_{1.5}}}{\overline{W}} \right] \left(\frac{\phi_S}{a_{\text{MS}}} \right) \quad (6)$$

Where, a_{MO} is the activity value of the sub-system with one basic component i.e. MO–SiO_2–$FeO_{1.5}$ at a fixed mole fractions of SiO_2 and $FeO_{1.5}$. In order to obtain total Cs value of the multi-component system, Eq. (4) is applied.

Neural Network Model

Neural network (NN) computation was described elsewhere [12, 13]. NN is a processing method and it imitates the features of the human brain that contains approximately 10 billion nerve cells (neurons). A human neuron which is a functioning unit of the nervous system consists of dendrite, soma, axon and synapse. Neurons communicate via input signals. A neuron accepts inputs associated with different weights from multiple neurons. The summation of the inputs (intensity) is multiplied by their associated weight. When the intensity of the signal is high enough to pass over a certain critical value (threshold), then an output signal is transmitted through axon and synapse to the next neuron. The state that the intensity of signal exceeds the threshold is called "ignition" and it can be expressed in a sigmoid function shown in Eq. (7) in the neural network concept.

$$y = f(x) = 1/[1 + \exp(-\eta \cdot x)] \tag{7}$$

Here, x is an input value and y is the output. η is a coefficient which determines the shape of the sigmoid curve. In the back propagation method, three layers-type neural network computation consists of an input layer, a middle (hidden) layer and an output layer. The units in the middle layer are connected with the input and output units. However, there are no connections within a layer. The final result "y" in the output layer is expressed by using the above sigmoid function as follows

$$a_k = f\left(\sum x_i \cdot W_{ki} - h_i\right) \tag{8}$$

$$y = f\left(\sum a_k \cdot V_k - h_k\right) \tag{9}$$

where x_i is an input value of unit i in the input layer, W_{ki} is a connection weight between unit i in the input layer and unit k in the middle layer, h_i is a critical value for unit i. y is the final output, V_k is a connection weight between unit k in the middle layer and the final output, h_k is a critical value for unit k. After values are applied to the units in the input layer, signals propagate through the middle layer to the output layer. Each link between neurons contains a unique weight value. A comparison is made between output values and the teaching values. The errors are calculated for each output unit and then propagated backwards through the network to correct the connection weights and the critical values in each unit. This "learning" process is repeated until the overall error value drops to acceptable levels. In the impurity capacity estimation studies (C_i), the computation was carried out by using SlagVis software designed by Research Center of Computational Mechanics Inc., Osaka University, and Sumitomo Metal Industries Ltd. to estimate some physical properties of multi-component slags using a neural network model. In our studies, the program was found applicable for the C_i predictions. [12, 13].

Result and Discussions

In first study, Reddy-Blander model was applied to predict the sulfide capacity (Cs) values of FeO_x–CaO–SiO_2, FeO_x–MnO_x–SiO_2 and FeO–MgO–SiO_2 ternary slags which were experimentally studied in the literature within the temperatures between 1673 and 1923 K [20]. The sulfide capacity values increased with increasing temperature but, decreased with increasing acidic SiO_2 content. The RB model was found in agreement for all slag systems at the temperature range of 1773–1923 K. A small discrepancy occurred just when the basic content is

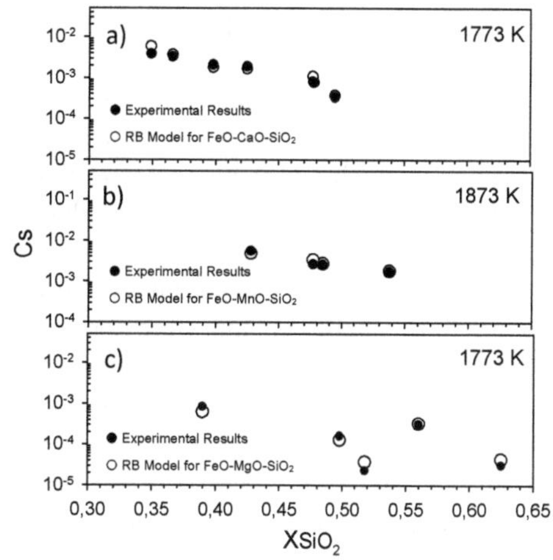

Fig. 1 Sulfide capacities comparison of model and experimental data for **a** FeO–CaO–SiO_2 system at 1773 K, **b** FeO–MnO–SiO_2 system at 1873 K, and **c** FeO–MgO–SiO_2 system at 1773 K [20]

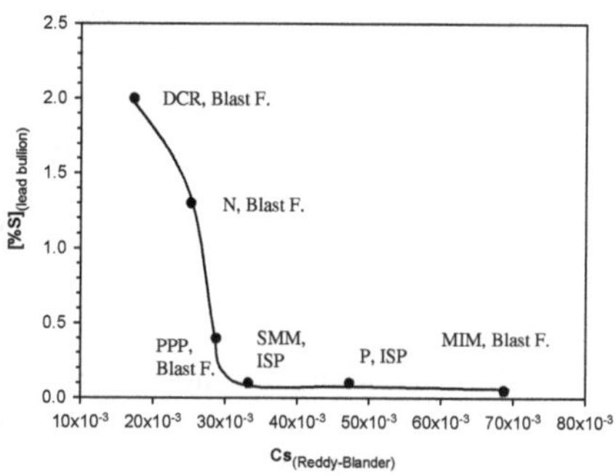

Fig. 2 The relationship between sulfur capacity of industrial slags and sulfur content of lead bullions [21]

relatively high at 1673 K for FeO_x–CaO–SiO_2 and FeO–MnO_x–SiO_2 slags. This may be due to high amounts of Fe^{+3} and Mn^{+3} in the slags resulting in a decrease in Cs values. Figure 1a–c shows some Cs comparisons of model and experimental data from the study [20]. Also, the predictions of sulfide capacities in industrial lead slags and interactions on sulfur content of lead bullions were the studied using the RB model [21]. The sulfide capacities (Cs) of 19 lead smelter slags indicated in the literature were calculated and effects of Cs values on sulfur content of lead bullions were discussed. It was found that, the calculated sulfide capacities for industrial lead slags changed with changing slag component and slag temperature. Lead smelting slag had a remarkable effect on removing sulfur from lead bullion. When, sulfide capacity of a slag increased, sulfur content in the lead bullion decreased. Sulfide capacity of a lead smelting slag was found to be one of the important parameter for removing sulfur from lead bullion. Figure 2 shows the relationship between Cs of industrial furnace slags and sulfur content in lead bullions. To make the importance of slag capacity clearly seen, both blast furnace and imperial smelting processes were selected for this figure, since they have similar process conditions involving only reduction step compare to direct reduction furnaces having oxidizing and reduction zone such as QSL or Kivcet reactors. The figure depicts that in industrial smelters, as sulfide capacity of a slag increases, sulfur content in the lead bullion decreases. Having different Cs values of slags, blast furnaces, for example, Mount Isa Mines (MIM, Cs: 6.87×10^{-2}), Noranda (N, Cs: 2.97×10^{-2}) and Doe Run (DRC, Cs: 1.73×10^{-2}) smelters produce lead bullions having ≤0.05, 1.3 and <2% sulfur [21].

In another study, the partial pressure ratios of sulfur and oxygen for several copper flash smelting plants were derived using the Reddy-Blander (RB) model and predicted dissolved sulfur contents available in the literature [22]. The sulfide capacities (Cs) of six different flash smelting plants ($FeO–CuO_{0.5}–CaO–MgO–FeO_{1.5}–AlO_{1.5}–SiO_2$) slags were calculated. The log (PO_2/PS_2) versus matte grades of various industrial processes was determined and shown in Fig. 3. The PO_2/PS_2 ratio increased with an increase of copper grade. It can be concluded that calculated partial pressure ratios of sulfur and oxygen, log (PO_2/PS_2), for the copper flash smelting plants, changed between −5.39 and −4.67 with increasing Cu content in matte. These values were especially comparable with Takeda's experimental data. Predicted pressure ratio values of Sridhar et al. were found to be much lower than our prediction as well as other experimental data described in the literature [22].

The neural network approach was applied to the experimental Cs values of 20 different molten oxide and salt systems up to five component system for the temperatures between 1273 and 1923 K [12]. It was also constructed iso-sulfide counters on ternary phase diagrams establishing a link among experimental Cs values of molten ternary and lower sub-systems. Figure 4a shows an NN constructed Cs contours for liquid region of $CaO–FeO–SiO_2$ melt at 1773 K. For this aim, experimentally obtained Cs values of molten state pure FeO, $CaO–SiO_2$, $FeO–SiO_2$, CaO–FeO, and $CaO–FeO–SiO_2$ melts were used. Another example shown in Fig. 4b is an NN constructed Cs contours for liquid region of $CaO–CaCl_2–CaF_2$ melt at 1273 K. For this aim, experimentally obtained Cs values of $CaO–CaCl_2$ and $CaO–CaCl_2–CaF_2$ melts were used [12].

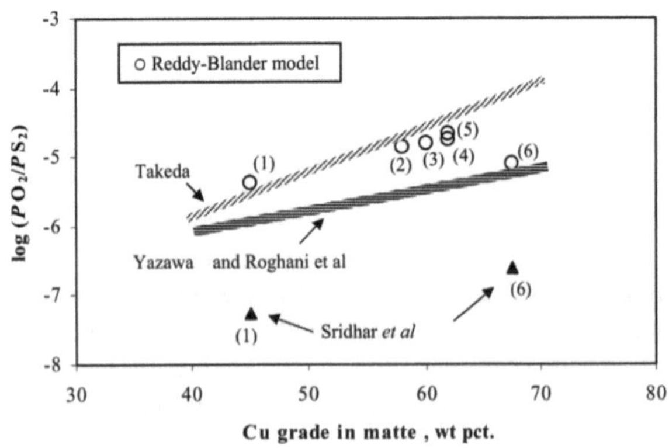

Fig. 3 Comparison of potential ratios for the matte grades from different smelting plants: *1* INCO *2* Rio-Tinto *3* Tamano *4* Norddeutsche *5* Phelps *6* Outokumpu [22]

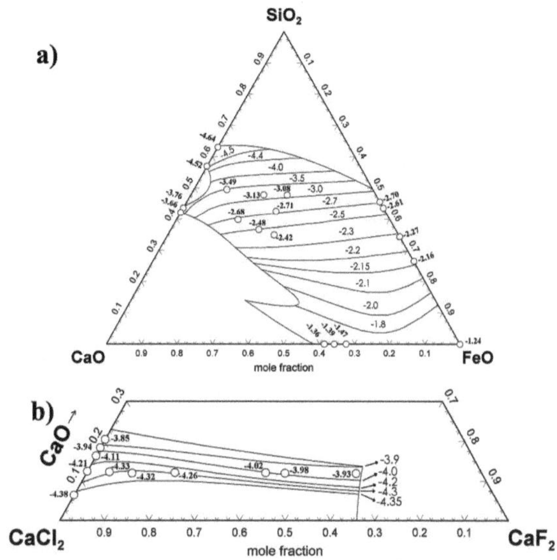

Fig. 4 Neural network predicted iso-sulfide capacity contours in liquid region of **a** $CaO–FeO–SiO_2$ melt system at 1773 K, and **b** $CaO–CaCl_2–CaF_2$ melt system at 1273 K [12]

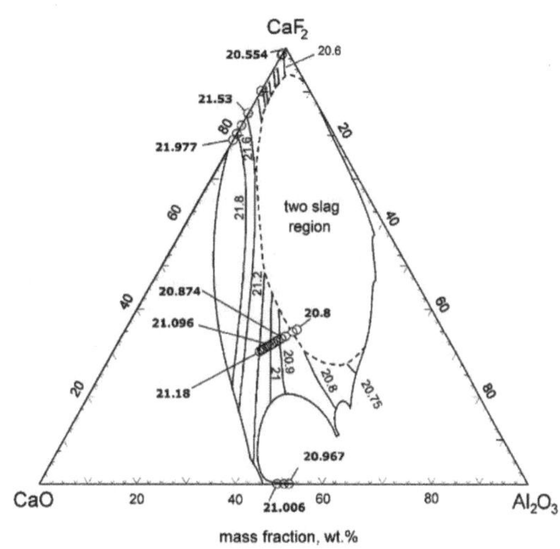

Fig. 5 Artificial neural network predicted iso-phosphate capacity contours in liquid region of $CaO–CaF_2–Al_2O_3$ melts at 1773 K [13]

In the final study, the capacities of phosphate, phosphide, nitride and carbide in binary and multi-component molten melts were estimated at different temperatures using the artificial neural network approach [13]. The experimental data from literature were introduced to the artificial neural network, then the calculated results were plotted against the experimental values for comparative purposes. The calculated results obtained through neural network computation agreed well with the experimental ones and were found to be more accurate than some model results. Figure 5 shows a plotted iso-phosphate capacity contours on the liquid region of $CaO–CaF_2–Al_2O_3$ ternary phase diagram at 1773 K generated by using NN results [13].

Conclusions

This paper is a review of research on impurity capacity predictions in molten melts which were carried out by Derin et al. It is concluded that, as a priori prediction method for impurity capacities, the RB model is a very useful tool for the optimization of impurity removal in the existing processes and for the development of new processes. The computation results were found to be in good agreement with the experimental values. It can be also concluded that neural network based computation is a very useful technique for predicting Ci values in molten melts, but attention needs to be paid to the quantity and accuracy of the experimental data.

Acknowledgments We acknowledge with great pleasure the financial support from Istanbul Technical University.

References

1. D.J. Sosinsky, I.D. Sommerville, The composition and temperature dependence of the sulfide capacity of metallurgical slags. Metall. Mater. Trans. B **17**, 331–337 (1986)
2. J.A. Duffy, M.D. Ingram, An interpretation of glass chemistry in terms of the optical basicity concept. J. Non-Cryst. Solids **21**(3), 373–410 (1976)
3. T. Mori, On the phosphorus distribution between slag and metal. Bull. J. Inst. Met. **23**(5), 354–361 (1984)
4. R. Moretti, G. Ottonello, Solubility and speciation of sulfur in silicate melts: the conjugated Toop-Samis-Flood-Grjotheim (CTSFG) model. Geochim. Cosmochim. Acta **69**(4), 801–823 (2005)
5. M.M. Nzotta, D. Sichen, S. Seetharaman, A study of the sulfide capacities of iron-oxide containing slags. Metall. Mater. Trans. B **30**(5), 909–920 (1999)
6. R. Selin, Y. Dong, Q. Wu, Uses of lime-based fluxes for simultaneous removal of phosphorus and sulphur in hot metal pretreatment. Scand. J. Metall. **19**(3), 98–109 (1990)
7. Y. Kobayashi, N. Yoshida, K. Nagai, Thermodynamics of phosphorus in the $MnO–SiO_2–Fe_tO$ system. ISIJ Int. **44**(1), 21–26 (2004)
8. X.M. Yang, C.B. Shi, M. Zhang, J.P. Duan, J. Zhang, A Thermodynamic model of phosphate capacity for $CaO–SiO_2–MgO–FeO–Fe_2O_3–MnO–Al_2O_3–P_2O_5$ slags equilibrated with molten steel during a top–bottom combined blown converter steelmaking process based on the ion and molecule coexistence theory. Metall. Mater. Trans. B **42**(5), 951–977 (2011)
9. B. Maramba, R.H. Eric, Phosphide capacities of ferromanganese smelting slags. Min. Eng. **21**, 132–137 (2008)
10. M. Sasabe, S. Yamashita, S. Shiomi, T. Tamura, H. Hosokawa, K. Sano, Nitride capacity of the molten $CaO–SiO_2–Al_2O_3$ system containing TiO_2 or ZrO_2 and equilibrated with Molten Si. Tetsu-to-Hagane **87**(12), 727–733 (2001)
11. W.Y. Shin, H.G. Lee, Nitride capacities of $CaO–Al_2O_3–CaF_2$ melts at 1773 K. ISIJ Int. **41**(3), 239–246 (2001)
12. B. Derin, M. Suzuki, T. Tanaka, Sulphide capacity prediction of Molten slags by using a neural network approach. ISIJ Int. **50**(8), 1059–1063 (2010)
13. B. Derin, E. Alan, M. Suzuki, T. Tanaka, Phosphate, phosphide, nitride and carbide capacity predictions of Molten melts by using an artificial neural network approach. ISIJ Int. **56**(2), 183–188 (2016)

14. R.G. Reddy, M. Blander, Modeling of sulfide capacities of silicate melts. Metall. Trans. B **18**, 591–596 (1987)
15. R.G. Reddy, M. Blander, Sulfide capacities of MnO–SiO_2 slags. Metall. Trans. B **20**, 137–140 (1989)
16. R.G. Reddy, Thermodynamic modeling of sulfide capacities of melts. Paper presented at 6th international conference on Molten slags, fluxes and salts, Stockholm, Sweden–Helsinki, Finland, 12–17 June, 2000
17. A.D. Pelton, G. Eriksson, A. Remero-Serrano, Calculation of sulfide capacities of multicomponent slags. Metall. Trans. B **24**(5), 817–825 (1993)
18. Y.B. Kang, A.D. Pelton, Thermodynamic model and database for sulfides dissolved in molten oxide slags. Metall. Meter. Trans. B **40**, 979–994 (2009)
19. C.J.B. Fincham, F.D. Richardson, The behaviour of sulphur in silicate and aluminate melts. Proc. R. Soc. Lond. A **223**, 40–62 (1954)
20. B. Derin, O. Yücel, R.G. Reddy, Sulfide capacity modeling of FeO_x–MO–SiO_2 (MO = CaO, MnO_x, MgO) melts. Miner. & Metall. Process. J. **28**(1), 33–36 (2011)
21. B. Derin, O. Yucel, R.G. Reddy, Predicting of sulfide capacities of industrial lead smelting slags. Paper presented at TMS, Sohn International Symposium on Advanced Processing of Metals and Materials: Volume 1-Thermo And Physicochemical Principles: Non-Ferrous High-Temperature Processing, Catamaran Resort. San Diego, California, USA, pp. 237–244, 27–31 Aug 2006
22. B. Derin, R.G. Reddy. Sulfur and Oxygen partial pressure ratios prediction in copper flash smelting plants using reddy-blander model. Paper presented at Yazawa International Symposium on Metallurgical and Materials Processing, TMS, San Diego, California, USA, pp. 625-632, 02–06 Mar 2003

Part X
Poster Session

Application of Sharp Analysis on Reduction Kinetics of Vanadium Titanium Magnetite Sintering Ore

Jinsheng Wang, Pingsheng Lai, Wei Lv, Xuewei Lv and Jie Dang

Abstract The isothermal reduction kinetics of vanadium titanium magnetite sintering ore was investigated by thermo-gravimetric analysis at a temperature range of 1123–1223 K under 30% CO + 70% N_2 gas mixtures. The apparent activation energy of reduction was calculated by iso-conversional method and its value is 36.7 kJ/mol. The model function describing the reduction mechanism was discussed by Sharp analysis and the results revealed that the reduction process is controlled first by random nucleation and nucleus growth mechanism (RNNGM) as reduction degree is lower than 0.5, followed by the mixed influence of RNNGM and phase boundary reaction when reduction degree is higher than 0.5.

Keywords Kinetics · Vanadium titanium magnetite · Gaseous reduction

Introduction

The vanadium titanium magnetite (VTM) is a kind of complex iron ore which contains a variety of valuable metals. On account of its enormous reserves and the decrease of high-grade iron ore, the status of VTM in iron and steelmaking in China and even in the world becomes more and more significant [1–6].

In the past few decades, many researches on the carbothermic reduction mechanism and kinetics of VTM have been carried out. Chen [7] et al. investigated the carbothermic reduction mechanism of VTM. They used the thermogravimetric-differential scanning calorimetry-quadrupole mass spectrometer method to analyze the kinetics of main reduction process of the VTM. The results indicated that the kinetic mechanism followed the principle of random nucleation and growth. However, in Liu's research [8], they divided the reduction process of vanadium and

J. Wang · P. Lai · W. Lv · X. Lv (✉) · J. Dang
College of Material Science and Engineering, Chongqing University, Chongqing, China
e-mail: lvxuewei@163.com

J. Wang
e-mail: jinshengwang126@163.com

S. Wang et al. (eds.), *Applications of Process Engineering Principles in Materials Processing, Energy and Environmental Technologies*,
The Minerals, Metals & Materials Series, DOI 10.1007/978-3-319-51091-0_51

titanium iron concentrate pellets containing carbon into two stages. Their results showed that the first stage was controlled by chemical reaction between the metal oxides and carbon, the later stage was controlled by the diffusion of reduced products.

Meanwhile, some researches concerned with gaseous reduction of VTM have been conducted as well. Qi [9] et al. used CO and H2 to reduce VTM respectively. The results showed that gaseous reduction process of VTM was controlled primarily by diffusion when temperature was below 1050 °C and mainly controlled by interfacial reaction when temperature is more than 1050 °C. But, the research about reduction mechanism of high-chromium vanadium–titanium magnetite under H2–CO–CO2 gas mixtures by Tang et al. [10] showed different results. It turned out that the reduction was controlled by the mixed internal diffusion and interfacial reaction at the initial stage where the interfacial reaction is dominant. As the reduction proceeds, the internal diffusion became the controlling step.

However, the information on gaseous reduction of VTM sintering ore is particularly sparse. In the current study, the kinetics of VTM sintering ore gaseous reduction by CO at a temperature range of 1123–1223 K was presented.

Experimental

Materials

The experimental samples used in this study were derived from homemade VTM sintering ore in laboratory. The main chemical composition is shown in Table 1. The samples were grinded and sieved to a size less than 200 mesh before TG experiments. High purity N_2 was used as a protective gas.

Procedure

The experiments were carried out in a way of isothermal reduction in Setaram Evo TG-DTA 1750 thermal analyzer. The sample was put into an alumina crucible and placed in the reactor which was pumped to a state of vacuum and then filled by N_2 to 1 atm so that there was no other residual air to influence experimental data. Then, the reactor was heated to 1123, 1173 and 1223 K respectively at a heating rate of 20 °C/min in N_2 atmosphere. When the temperature was stable, the N_2 was

Table 1 Main chemical position of VTM sintering ore, wt%

CaO	FeO	MgO	TFe	SiO_2	TiO_2	V_2O_5	Cr_2O_3	R
6.26	19.68	2.43	50.62	4.06	12.49	0.524	0.177	1.5

switched to reduction gas containing 30% CO and 70% N_2, and then the weight loss was monitored continuously. After reacting for a period of time until the weight loss stay constant, the reduction gas was switched to N_2 again, and the sample was cooled to the room temperature along with the furnace.

In all experimental runs, the mass of sample was 20 mg and a constant flow rate of all gas was kept as 20 ml/min, which was controlled by gas mass flowmeter.

Results and Discussion

TG Analysis

Figure 1 shows that the mass loss ratio and the reduction degree of VTM sintering ore changed with time at different temperature. The reduction degree was calculated by Eq. (1):

$$\alpha = \frac{\Delta O}{\sum O} = \frac{m_O - m_t}{\sum O} \quad (1)$$

where, ΔO is the actual mass of removed oxygen; ΣO is the theoretical mass of removable oxygen in the sample; *m0* is the initial mass of the sample and *mt* is the residual mass of the sample after reduced for time 't'.

It can be seen from Fig. 1 that the mass loss and reduction degree of samples increase with the processing of reduction until it becomes stable, which means the reduction comes to an end. Meanwhile, it is obvious to find that both the maximum mass loss and reduction degree of samples increase with the rising of reduction temperature, which indicates that it is favorable for reduction of VTM sintering ore at higher reduction temperature.

Figure 2 shows that the reduction rate changed along with reduction degree of samples at different temperature. It can be found that there are two peaks in all three curves, which reveals that the reduction process of VTM sintering ore can be

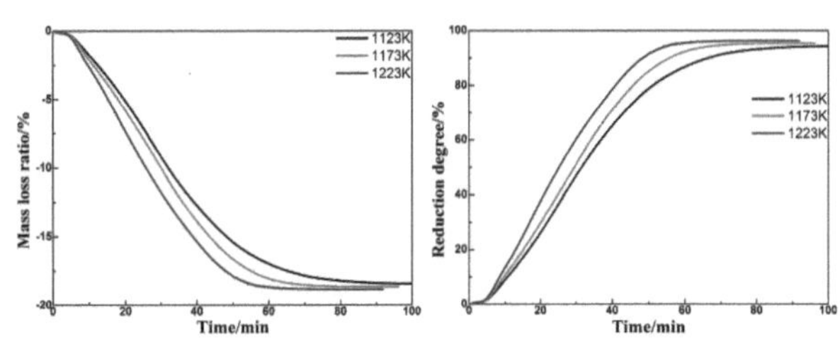

(a) (b)

Fig. 1 Mass loss ratio and reduction degree of VTM sintering ore

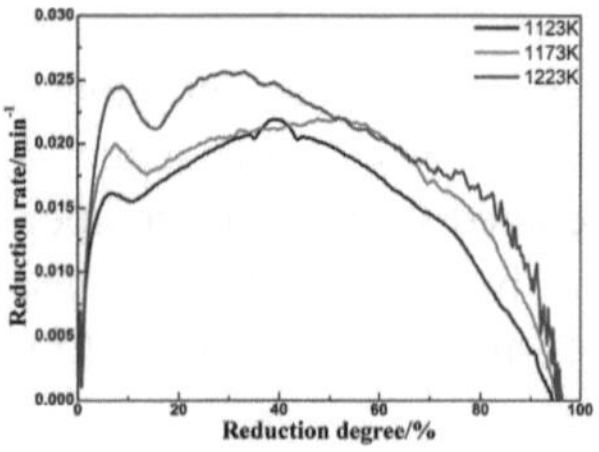

Fig. 2 The trend of reduction rate changing with reduction degree

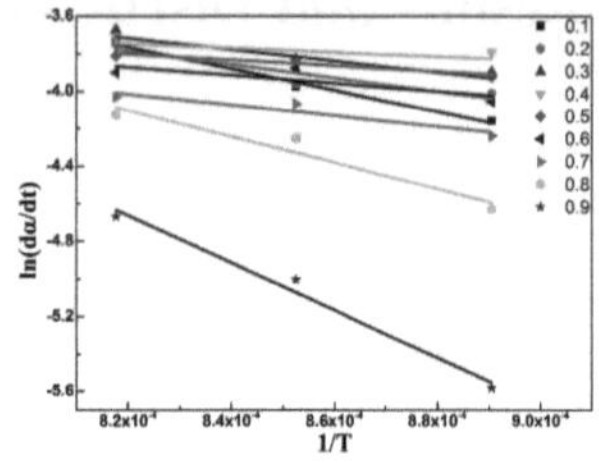

Fig. 3 The relationship between ln(dα/dt) and 1/T by iso-conversion method

divided into two stages. But, further XRD and SEM analysis of reduction products at different stages are necessary to verify the specific reduction path.

Figure 3 shows that the relationship between ln(dα/dt) and 1/T, where dα/dt is the reduction rate of the reaction and T is the reduction temperature. According to model-free approach, dynamical equation of non-uniform reaction can be expressed as Eq. (2):

$$d\alpha/dt = k(T)f(\alpha) \tag{2}$$

where, dα/dt is reduction rate; k(T) is reduction rate constant; f(α) is model of reaction mechanism function. Equation (3) can be obtained by substitute k(T) by Arrhenius formula [11]:

$$d\alpha/dt = A\ \exp\left(\frac{-E}{RT}\right)f(\alpha) \tag{3}$$

Taking the logarithm of Eq. (3) can obtain Eq. (4):

$$\ln(d\alpha/dt) = \ln A - \frac{E}{RT} + \ln f(\alpha) \tag{4}$$

Equation (4) indicates that the reaction activation energy of different reduction degree can be calculated by the slope(-E/R) of each fitting line in Fig. 3. The average apparent activation energy of the reduction process is 36.64 kJ/mol. The value is showed in Table 2.

Table 2 The reaction activation energy at different conversion

α	0.1	0.2	0.3	0.4	0.5	0.6	0.7	0.8	0.9	Average
E/(kJ/mol)	47.9	29.9	25.2	9.4	12.8	17.9	24.1	57.9	104.7	36.64

Kinetics Model of VTM Sintering Ore Reduction

Sharp [12] et al. proposed a method to conclude the reaction mechanism function:

$$G(\alpha) = kt = \frac{G(\alpha)_{\alpha=0.5}}{G(\alpha)_{\alpha=1.0}}\left(\frac{t}{t_{0.5}}\right) \quad (5)$$

where, G(α) is a mechanism function, t is time, t0.5 is the time when α is 0.5, G(α)α = 0.5 is the value of G(α) when α = 0.5, G(α)α = 1.0 is the value of G(α) when α = 1.0.

The experimental curve showing relationship between α and t/t0.5 is calculated from experimental data. And, the theoretical curves of α versus t/t0.5 can be acquired by commonly known G(α). The reaction mechanism function can be easily obtained by comparing these two kind of curves. Table 3 shows commonly used kinetic mechanism functions, and Fig. 4 shows the comparison of experimental curve of VTM sintering ore reduction with theoretical curves. It is clear that the reduction process is controlled first by random nucleation and nucleus growth mechanism (RNNGM) as reduction degree is lower than 0.5, followed by the mixed influence of RNNGM and phase boundary reaction when reduction degree is higher than 0.5.

Table 3 Commonly used kinetic mechanism functions [11, 13]

Symbol		Kinetic models	G(α)
Diffusion models	D1(α)	One-dimensional diffusion	α2 = kt
	D2(α)	Two-dimensional diffusion	(1 − α)ln (1 − α) + α = kt
	D3(α)	Three-dimensional diffusion	[1 − (1 − α)1/3]2 = kt
	D4(α)	Three-dimensional diffusion	(1 − 2/3α) − (1 − α) 2/3 = kt
Boundary reaction	F1(α)	Phase boundary reaction	1 − (1 − α)1/2 = kt
	R2(α)	Phase boundary reaction	1 − (1 − α)1/3 = kt
Nucleation models	R3(α)	Random nucleation and subsequent growth	−ln(1 − α) = kt
	A2(α)	Random nucleation and subsequent growth	[−ln(1 − α)]1/2 = kt
	A3(α)	Random nucleation and subsequent growth	[−ln(1 − α)]1/3 = kt

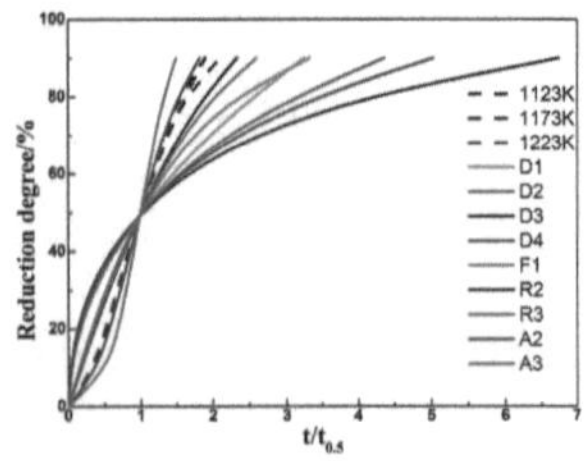

Fig. 4 The comparison of experimental curves and theoretical curves

Conclusions

The reduction kinetics of VTM sintering ore is experimentally determined, and the conclusions are acquired as follows:

1. The apparent activation energy of reduction was calculated by iso-conversional method and its value is 36.7 kJ/mol.
2. The reduction process is controlled first by random nucleation and nucleus growth mechanism (RNNGM) as reduction degree is lower than 0.5, followed by the mixed influence of RNNGM and phase boundary reaction when reduction degree is higher than 0.5.

Acknowledgments This work was supported by "National Key Basic Research Program of China (973 Program)" (Grant No. 2013CB632603).

References

1. J.L. Zhang et al., Reduction kinetics of vanadium titano-magnetite carbon composite pellets adding catalysts under high temperature. J. Iron. Steel Res. Int. **20**, 1–7 (2013)
2. D.U. Xing-Hong, Bin Xie, Tai Ping Lou, Research on solid reduction of vanadium–titanium magnetite. J. Northeast. Univ. **33**, 685–688 (2012)
3. S.S. Liu et al., Solid-state reduction kinetics and mechanism of pre-oxidized vanadium–titanium magnetite concentrate. Trans. Nonferrous Met. Soc. China **24**, 3372–3377 (2014)
4. Z. Hao et al., Comprehensive utilization of vanadium–titanium magnetite deposits in China has come to a new level. Acta Geologica Sinica 87, 286–287 (2013)
5. D.S. Chen et al., Solid state reduction of Panzhihua titanomagnetite concentrates with pulverized coal. Miner. Eng. **24**, 864–869 (2011)
6. J. Zhu, Beneficiation and comprehensive utilization of vanadium titanium magnetite. Metal Mine (2000)
7. S.Y. Chen et al., Carbothermic reduction mechanism of vanadium–titanium magnetite. J. Iron. Steel Res. Int. **23**, 409–414 (2016)
8. S. Li Liu et al., Kinetics of direction reduction on vanadium and titanium iron concentrate pellets containing carbon. J. Iron. Steel Res. **23**, 5–8 (2011)
9. D.G. Qi et al., The reduction mechanism study of vanadium–titanium magnetite. Chem. Eng. Metall. **2**, 16–30 (1980)
10. J. Tang et al., Preparation of oxidized pellets with high chromium vanadium–titanium magnetite. J. Northeast. Univ. **34**, 545–550 (2013)

11. J.D. Hancock, J.H. Sharp, Method of comparing solid-state kinetic data and its application to the decomposition of kaolinite, brucite, and $BaCO_3$. J. Am. Ceram. Soc. **55**, 74–77 (1972)
12. J.H. Sharp, G.W. Brindley, B.N. Narahari Achar, Numerical data for some commonly used solid state reaction equations. J. Am. Ceram. Soc. **49**, 379–382 (1966)
13. A. Khawam, D.R. Flanagan, Solid-state kinetic models: basics and mathematical fundamentals. J. Phys. Chem. B **110**, 17315–17328 (2006)

High Temperature Properties of Molten Nitrate Salt for Solar Thermal Energy Storage Application

Mehedi Bin Mohammad, Geoffrey Brooks and M. Akbar Rhamdhani

Abstract Molten alkali nitrates are used commercially as thermal storage fluids (HTF) for solar thermal electricity generation. Their range of operation is limited by the thermal stability and this limits the energy (steam-Rankine cycle) efficiency of these processes. In this study, the effect of atmosphere on the thermal stability of nitrates was investigated using simultaneous thermal analysis techniques (STA). The use of oxygen (O_2) as a blanket gas was found to delay the decomposition point of the ternary $NaNO_3$–KNO_3–$LiNO_3$ (13.23–57.14–29.63 weight ratio) mixture by 55 °C. By which operating range will increase 65 °C more compared to the binary solar salt. Higher heating rates increased the decomposition point but had less effect on premelting and melting properties. Nitrogen monoxide, and O_2 evolution after melting indicates concurrent reversible reactions. Similar phenomena were observed for single and binary nitrates. The eutectic melting point of ternary nitrate obtained from the STA agreed well with FactSage eutectic composition prediction.

Keywords Concentrated solar power · Heat transfer fluids · Thermal analysis · $LiNO_3$ · $NaNO_3$ · KNO_3

M.B. Mohammad (✉) · G. Brooks · M. Akbar Rhamdhani
Faculty of Science, Engineering and Technology, Swinburne University of Technology, Hawthorn, VIC 3122, Australia
e-mail: mmohammad@swin.edu.au

G. Brooks
e-mail: gbrooks@swin.edu.au

M. Akbar Rhamdhani
e-mail: arhamdhani@swin.edu.au

S. Wang et al. (eds.), *Applications of Process Engineering Principles in Materials Processing, Energy and Environmental Technologies*,
The Minerals, Metals & Materials Series, DOI 10.1007/978-3-319-51091-0_52

Background

Solar thermal power (STP) is a form of renewable energy that produces sustainable power using concentrated solar thermal energy [1, 2]. Concentrated solar power (CSP) plant's electricity generation is similar to conventional power plant [3] using conventional cycles [4], but instead of fossil fuel to supply heat to the boiler or heat exchanger, it uses concentrated solar radiation from solar field which is stored in thermal energy storage (TES) system [3, 5]. The various types of CSP systems, central receiver solar tower, parabolic trough, dish engine, and Fresnel system, are globally available technologies. Commercially the central receiver solar tower system, and the parabolic trough system use TES with heat transfer fluid (HTF) [6, 7]. A TES system consists of two tanks: a cold and hot storage tanks. A cold tank dispatch HTF and hot tank collect HTF from the solar field via pipe and pump accessories, as shown schematically in Fig. 1.

The use of molten alkali nitrate/nitrite and alkaline nitrate salts as a HTF and heat storage fluid is promising in CSP plants because of their negligible vapor pressure and optimum fluid velocity. The best-established HTF is a binary $NaNO_3$–KNO_3 (60–40 weight ratio) solar salt [8] which has melting point 220 °C and decomposition point at 565 °C [9]. The ternary $NaNO_3$–KNO_3–$LiNO_3$ salt (13.23–57.14–29.63 weight ratio) obtained from thermodynamic modeling has a melting point of 120.8, 100 °C lower liquidus temperature compared to the binary (melting: 222 °C [10]). The purpose of this study is to investigate the increase of the degradation limit of proposed ternary nitrate salt system by using oxygen as cover gas to obtain broader operating temperature range to potentially increase the steam-Rankine cycle efficiency solar thermal electricity plants.

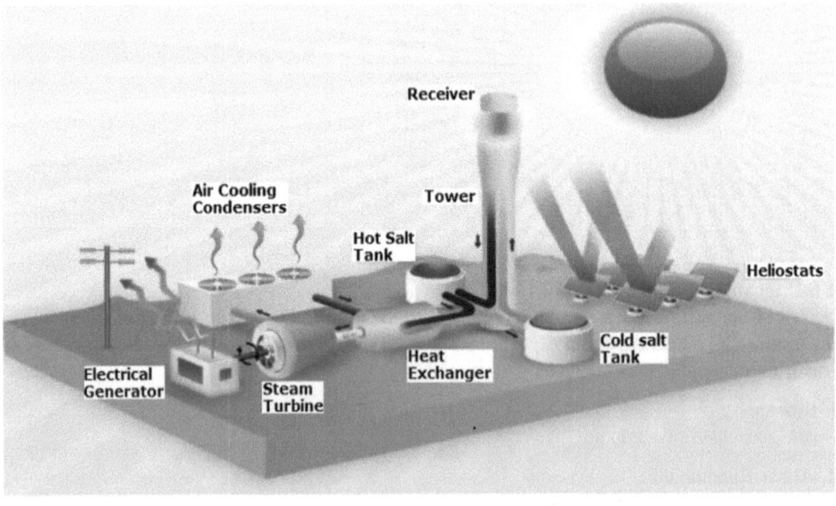

Fig. 1 Schematic diagram of Gemasolar power tower (central receiver tower) plant [11]

The main course of decomposition reactions are nitrate being dissociate to nitrite and liberating oxygen in reversible manner [12], as shown in Eq. 1. Then nitrite dissociates reversibly to oxides, and NO_x gasses [13], as shown in Eq. 2. Recent studies showed that, in alkali nitrate/nitrite (Hitec salt) system, pure oxygen as a blanket gas (partial pressure of oxygen, $P_{O_2} = 1$) alters the nitrite back to nitrate at higher temperatures by exothermic oxidation and delays the thermal degradation point [14]. The decomposition limits of many alkali nitrate systems reported were based on thermo gravimetric analysis (TGA) or TGA and differential thermal analysis (DTA)/differential scanning calorimetry (DSC) only and the fundamental decomposition reactions have not been reported in detail. To address these issues, the proposed ternary nitrate salt were studied using simultaneous thermal analysis techniques (STA) which includes simultaneous DTA/DSC-TGA and mass spectrometry (MS) under three gases (argon, air and O_2) and scanning rates (2.5, 5 and 10 °C/min) to observe decomposition behavior and identify thermal events accurately. To aid this evaluation the single $LiNO_3$, $NaNO_3$ and KNO_3 and binary solar salts were also subjected to STA.

$$MNO_3 \rightleftharpoons MNO_2 + \frac{1}{2}O_2 \quad (M = Li, Na, K) \tag{1}$$

$$\begin{aligned} &2MNO_2 \rightleftharpoons M_2O + NO_2 + NO \quad OR, \\ &2MNO_2 \rightleftharpoons M_2O + \frac{3}{2}O_2 + N_2 \end{aligned} \tag{2}$$

Experimental Methodology

A ternary nitrate mixture was prepared from reagent grade $LiNO_3$ (anhydrous 99.99% purity), $NaNO_3$ (anhydrous 99.995% purity) and KNO_3 (anhydrous 99.997% purity) from Sigma Aldrich® according to the composition 29.63$LiNO_3$:13.23$NaNO_3$:57.14KNO_3 weight% (37.35$LiNO_3$:13.53$NaNO_3$:49.12KNO_3 mol%) obtained from the thermodynamic analysis [15]. The binary solar salt was prepared as 60:40 weight ratio of $NaNO_3$, KNO_3. Salts were weighed using Mettler Toledo digital balance (±0.001 g weighing accuracy) and mixed without further purification. The 10 g binary and ternary salt mixtures were dehydrated in Pyrex crucible for 2 h and heated up to 250 °C in an oven furnace in an air atmosphere. The salt melts were kept in molten state at for 2 h to achieve sufficient homogenization. After heat treatment, the salt were cooled slowly by keeping the salts insulated and turning off the furnace. Then the quenched salt mixtures were crushed and ground using mortar and pestle and stored in a master container and placed in a Duran vacuum desiccator before subsequent STA testing.

A Setaram DSC/DTA-TGA Setsys Evolution 18 analyzer unit coupled with a Pfeiffer Quadstar-422 MS was used for thermal and evolved gas analysis.

The gaseous products were detected by MS simultaneously and were presented with DTA and TGA curves [16, 17]. In case of stability test the salts were heated from 50 to 800 °C using DTA in three atmospheres (argon, air and O_2). And in case of transition events detection the salts were heated at various heating-cooling rates from 50 to 30 °C above the melting point of respective sample in argon using DSC. A platinum-rhodium 6%/platinum-rhodium 30% protected B-type DTA/TGA thermocouple (150–1700 °C) was used for thermal stability measurements. A platinum/platinum-rhodium 10% S-type DSC/TGA transducer (−50 to 1600 °C) was used for transition event detection. The DSC/TGA sensor was calibrated against certified reference materials (CRM) following Standards procedures [18–24] to obtain corrected temperature and heat flow (HF) sensitivity coefficients. The CRMs, namely pure (purity of 4N or 99.99%) Indium, Lead, Aluminium, and Aurum were heat treated to observe the melting events. A built-in least square optimization routine [16] was used to calculate the temperature and HF correction coefficients. Errors associate with temperature uncertainty was ±4.9 °C, heat sensitivity variation was ±3.02 J/g and, noise level was ±6.7 mK. The method used for the detailed error calculations are described by Olivares [14].

Results and Discussion

The STA study showed that, the oxygen blanket gas delayed the degradation temperature of ternary molten nitrate by 55 °C at 10 °C/min heating rate compared to an inert argon atmosphere [25, 26]. Similar effects of oxygen atmosphere are seen in case of binary and single constituent nitrates [25, 26]. Figure 1 is showing decomposition behavior of binary solar salt at three atmospheres. Rapid degradation temperatures were observed at 575, 610 and 624 °C in argon, air and oxygen blanket gas respectively. Only NO gas was plotted for comparison. O_2 gas was unable to detect in air and oxygen atmospheres because of same AMU (atomic mass unit). N_2, O_2, NO, N_2O and NO_2 were the main evolved gases of nitrates decomposition. NO and O_2 was started to evolved at same temperature (490 °C) in-between melting and boiling temperatures for each sample suggesting that, the primary and secondary reversible reactions are concurrent and overlapping, as shown in Fig. 2

The higher heating rate of 10 °C/min resulted in a delayed degradation point compared to the lower heating rate of 2.5 °C/min as observed in argon for the ternary nitrate system. The liquidus and solidus temperature gap was higher in ternary and binary nitrates compared to single nitrates perhaps due to super cooling and/or inadequate calibration of heat flow curves (during cooling of DTA/DSC transducer). For similar reason, also the crystalline transition that was observed prior to melting did not appear during cooling for the binary and ternary nitrates. There was little effect of heating rates on pre-melting and melting (transition, fusion and crystallization) temperatures compared to decomposition temperatures for all samples.

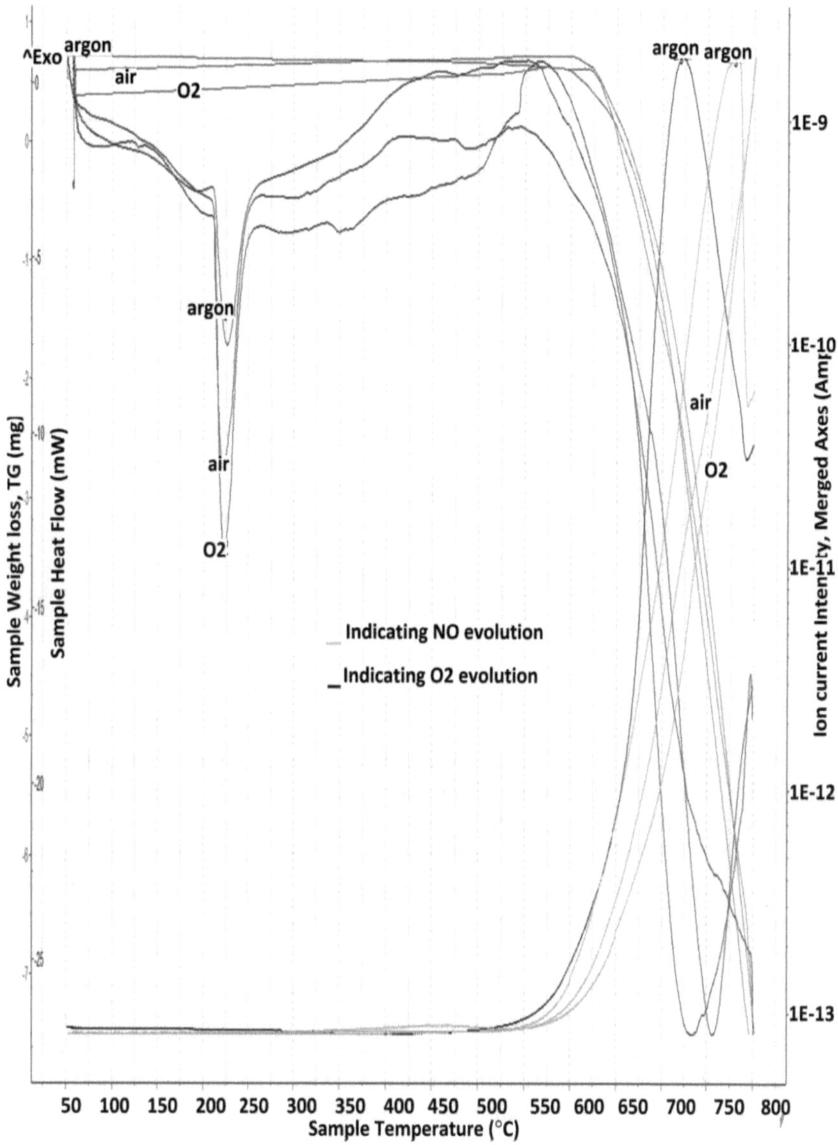

Fig. 2 Binary solar salt heated from 50 to 800 °C in argon, air and oxygen atmosphere (sweep velocity: 40 ml/min) at 10 °C/min heating rate. *Blue curves* are indicating sample heat flow; *green curves* indicating sample weight loss; and NO and O_2 gas evolution curves (ion current intensity) are shown in *pink* and *black* color. The curves in three atmospheres are artificially lined up in Y-axis for comparison

In order to avoid time consuming and costly experimentation, thermodynamic modeling using thermochemical package FactSage 7.0 [15] was performed to obtain eutectic composition (13.23–57.14–29.63 weight ratio) of ternary nitrate

Table 1 Binary phase diagrams comparison

Salt systems	Eutectic composition (mol%)	Eutectic temperature (°C)	Method [reference]
$NaNO_3$–KNO_3	0.5 KNO_3	224.85	DSC [27]
	0.5 KNO_3	220.85	DSC [30]
	0.5KNO_3	220.85	DSC and high temperature XRD [31]
	0.51 KNO_3	223	FactSage [15]
	0.5 KNO_3	220.85	This work calculated C_p as constant
	0.52 KNO_3	221.85	This work calculated varying C_p
$LiNO_3$–$NaNO_3$	0.537 $LiNO_3$	200.35	DSC [32]
	0.537 $LiNO_3$	195	FactSage [15]
	0.524 $LiNO_3$	192.8	This work calculated C_p as constant
	0.524 $LiNO_3$	190	This work calculated varying C_p
$LiNO_3$–KNO_3	0.45 $LiNO_3$	137.05	DSC [32]
	0.46 and 0.53 $LiNO_3$	136.85 and 146.85	DSC [33]
	0.424 $LiNO_3$	125	FactSage [15]
	0.44 $LiNO_3$	137	This work calculated C_p as constant
	0.424 $LiNO_3$	135	This work calculated varying C_p

system and verified experimentally by STA. The melting point obtained by DSC was 121 °C and it has excellent agreement with fusion point obtained by thermodynamic modeling (120.8 °C). The mathematical models of three binary systems $NaNO_3$–KNO_3, $NaNO_3$–$LiNO_3$ and KNO_3–$LiNO_3$ were formulated using a regular solution model and Gibbs energy minimization method [27, 28], as shown in Table 1. Kleppa's [29] empirical polynomial expression of binary alkali nitrate was used to obtain molar enthalpy of mixing, then by utilizing the Gibbs-Duhem equation the partial molar enthalpies of mixing of each component at both phases (liquid and solid) were obtained. The Gibbs free energy of both components were formulated, in Eqs. 3 and 4. The concentration of both components at liquid and solid phase for $NaNO_3$–KNO_3 binary system is shown in Eqs. 5 and 6 respectively. These four nonlinear coupled equations (Eqs. 3–6) were solved numerically using a contour plot technique. An algorithm was developed to find the convergence of real roots (x_A^l, x_B^l and T). The liquidus and solidus curves of mathematical modeling were fitted well against values obtained from the quasi chemical model used in FactSage and experimental value from literature, as shown in Table 1. Heat capacities as a function of temperatures have less effect on liquidus and solidus curves. This study showed that the regular solution model described alkali binary nitrates system well and could be useful method for predicting the phase diagrams for other alkali nitrates and alkali halide systems.

$$
\begin{aligned}
\Delta G^{\overline{A}}_{\overline{\mathrm{NaNO_3}}} = {} & \left(1-\frac{T}{T^{A}_{mp}}\right)\Delta H^{A}_{fus} + \Delta C_{PA}\left[T - T^{A}_{mp} - T\ln\frac{T}{T^{A}_{mp}}\right] \\
& + RT\left(\ln x^{l}_{A} - \ln x^{s}_{A}\right) + \left(a_l x^{l^2}_{B} + 2b_l x^{l}_{A} x^{l^2}_{B} + 2c_l x^{l}_{A} x^{l^3}_{B} - c_l x^{l^2}_{A} x^{l^2}_{B}\right) \\
& - \left(a_s x^{s^2}_{B} + 2b_s x^{s}_{A} x^{s^2}_{B} + 2c_s x^{s}_{A} x^{s^3}_{B} - c_s x^{s^2}_{A} x^{s^2}_{B}\right)
\end{aligned} \tag{3}
$$

$$
\begin{aligned}
\Delta G^{\overline{B}}_{\overline{\mathrm{KNO_3}}} = {} & \left(1-\frac{T}{T^{B}_{mp}}\right)\Delta H^{B}_{fus} + \Delta C_{PB}\left[T - T^{B}_{mp} - T\ln\frac{T}{T^{B}_{mp}}\right] \\
& + RT\left(\ln x^{l}_{B} - \ln x^{s}_{B}\right) + \left(a_l x^{l^2}_{A} + b_l x^{l^3}_{A} - b_l x^{l^2}_{A} x^{l}_{B} - c_l x^{l^2}_{A} x^{l^2}_{B}\right. \\
& \left. + 2c_l x^{l^3}_{A} x^{l}_{B}\right) - \left(a_s x^{s^2}_{A} + b_s x^{s^3}_{A} - b_s x^{s^2}_{A} x^{s}_{B} - c_s x^{s^2}_{A} x^{s^2}_{B} + 2c_s x^{s^3}_{A} x^{s}_{B}\right)
\end{aligned} \tag{4}
$$

$$x^{l}_{A} + x^{l}_{B} = 1 \tag{5}$$

$$x^{S}_{A} + x^{S}_{B} = 1 \tag{6}$$

Conclusions

In summary, at 10 °C/min scanning rate the salt started to degrade rapidly at 545 °C in argon, 571 °C in air and 600 °C in oxygen, respectively. In argon at a higher heating rate (10 °C/min), the salt degradation was delayed by 42 °C. Oxygen as a blanket gas delayed the decomposition point of ternary nitrate by 55 °C. Higher heating rates increased the decomposition point. O_2, N_2, NO, N_2O, and NO_2 were the main evolved gasses of decomposition detected by MS in an argon atmosphere. As observed in argon atmosphere, NO and O_2 evolution after melting and onward indicates concurrent overlapping reversible decomposition reactions. Similar phenomena were observed for single and binary nitrates. The ternary nitrate offered 65 °C more operating temperature compared to binary solar salt. The eutectic melting point of ternary nitrate obtained from STA agreed well with FactSage eutectic composition formulation. A regular solution model predicted experimental literature values well. The overall conclusion is that the use of oxygen as a blanket gas for alkali molten nitrates is likely to increase the Rankine cycle efficiency of a solar thermal power plant.

References

1. S.J. Wagner, E.S. Rubin, Economic implications of thermal energy storage for concentrated solar thermal power. Renew. Energy, 1–15 (2012)
2. K.H. Solangi, M.R. Islam, R. Saidur, N.A. Rahim, H. Fayaz, A review on global solar energy policy. Renew. Sustain. Energy Rev. **15**, 2149–2163 (2011)

3. A. Gil, M. Medrano, I. Martorell, A. Lázaro, P. Dolado, B. Zalba, L.F. Cabeza, State of the art on high temperature thermal energy storage for power generation. Part 1—Concepts, materials and modellization. Renew. Sustain. Energy Rev. **14**, 31–55 (2010)
4. P. Schwarzbözl, R. Buck, C. Sugarmen, A. Ring, M.J. Marcos Crespo, P. Altwegg, J. Enrile, Solar gas turbine systems: design, cost and perspectives. Sol. Energy **80**, 1231–1240 (2006)
5. W. Han, J. Hongguang, S. Jianfeng, L. Rumou, W. Zhifeng, Design of the first chinese 1 MW solar-power tower demonstration plant. Int. J. Green Energy **6**, 414–425 (2009)
6. R.W. Bradshaw, J.G. Cordaro, N.P. Siegel, Molten nitrate salt development for thermal energy storage in parabolic trough solar power systems (2009), pp. 615–624
7. H.J. Hou, Y.P. Yang, Y.H. Cui, S. Gao, Y.X. Pan, Assessment of concentrating solar power prospect in China, in *International Conference on Sustainable Power Generation and Supply, 2009, SUPERGEN '09*, pp. 1–5
8. L.L.C. HITEC® Heat Transfer Salt. Coastal Chemical Co., viewed 18 Sept 2014, http://www.coastalchem.com/
9. W. Delameter, N. Bergan, *Review of the Molten Salt Electric Experiment: A Solar Central Receiver Project* (Sandia National Labs, Livermore, CA (USA), 1986)
10. T. Bauer, D. Laing, R. Tamme, Recent progress in alkali nitrate/nitrite developments for solar thermal power applications, in *Molten Salts Chemistry and Technology*, MS9, Trondheim, Norway, June(2011), pp. 5–9
11. R.I. Dunn, P.J. Hearps, M.N. Wright, Molten-salt power towers: newly commercial concentrating solar storage. Proc. IEEE **100**, 504–515 (2012)
12. W. Benaissa, D. Carson, Oxidation properties of "solar Salt", in *45th Annual Loss Preventive Symposium 2011 (LPS)*, AIChE, Chicago, Illinois, USA, 2011
13. K.H. Stern, *High Temperature Properties and Thermal Decomposition of Inorganic Salts with Oxyanions* (CRC Press, Boca Raton, 2001)
14. R.I. Olivares, The thermal stability of molten nitrite/nitrates salt for solar thermal energy storage in different atmospheres. Sol. Energy **86**, 2576–2583 (2012)
15. FACTSAGE©, CRCT- ThermFact Inc. & GTT-Technologies, viewed 6 Aug 2014, http://www.factsage.com/
16. Calisto, Thermal analysis software—CALISTO, KEP technologies, viewed 4 Mar 2013, http://www.setaram.com/CALISTO.htm
17. Quadera, viewed 6 Aug 2014, http://www.pfeiffer-vacuum.com/know-how/mass-spectrometers-and-residual-gas-analysis/introduction-operating-principle/portfolio/data-analysis-systems/technology.action?chapter=tec4.1.4.2
18. H.K. Cammenga, W. Eysel, E. Gmelin, W. Hemminger, G.W. Höhne, S.M. Sarge, The temperature calibration of scanning calorimeters: part 2. Calibration Subst. Thermochim. Acta **219**, 333–342 (1993)
19. G.W.H. Höhne, H.K. Cammenga, W. Eysel, E. Gmelin, W. Hemminger, The temperature calibration of scanning calorimeters. Thermochim. Acta **160**, 1–12 (1990)
20. P.K. Gallagher, M.E. Brown, R.B. *Kemp, Handbook of Thermal Analysis and Calorimetry* (Elsevier, 1998)
21. F.D. Rossini, Thermogravimetry (TG) or thermogravimetric analysis (TGA), viewed on 10 Oct 2015, http://www.astm.org/Standards/E967.htm. Pure applied chemistry, 22 (1970) 577
22. S.M. Sarge, E. Gmelin, G.W. Höhne, H.K. Cammenga, W. Hemminger, W. Eysel, The caloric calibration of scanning calorimeters. Thermochim. Acta **247**, 129–168 (1994)
23. I.W. Group, R. Sabbah, A. Xu-wu, J.C. USA, M.P. Leitão, M. Roux, L. Torres, Reference materials for calorimetry and differential thermal analysis. Thermochim. Acta **331**, 93–204 (1999)
24. J. Emsley, *The Elements*, 3rd edn. (Oxford Press, Clarendon Press, 1989)
25. M.B. Mohammad, G.A. Brooks, M.A. Rhamdhani, High temperature study of single alkali nitrate salts: $LiNO_3$, $NaNO_3$, KNO_3; and binary alkali nitrate salt: $NaNO_3$-KNO_3. Thermochim. Acta (2016, Submitted)
26. B.G.A. Mohammad, M.A. Rhamdhani, Thermal analysis of molten ternary alkali nitrate: lithium-sodium-potassium nitrate. Renew. Energy (2016, Submitted)

27. C.M. Kramer, C.J. Wilson, The phase diagram of $NaNO_3$—KNO_3. Thermochim. Acta **42**, 253–264 (1980)
28. T. Elliott, *Gibbs Energy Modeling of Binary and Ternary Molten Nitrate Salt Systems* (2012)
29. O.J. Kleppa, R.B. Clarke, L.S. Hersh, Studies of fused salts. III. Heats of mixing in silver nitrate-alkali nitrate mixtures. J. Chem. Phys. **35**, 175–180 (1961)
30. O. Greis, K.M. Bahamdan, B.M. Uwais, The phase diagram of the system $NaNO_3 \cdot KNO_3$ studied by differential scanning calorimetry. Thermochim. Acta **86**, 343–350 (1985)
31. X. Zhang, J. Tian, K. Xu, Y. Gao, Thermodynamic evaluation of phase equilibria in $NaNO_3$-KNO_3 system. J. Phase Equilib. **24**, 441–446 (2003)
32. M.J. Maeso, J. Largo, The phase diagrams of $LiNO_3$-$NaNO_3$ and $LiNO_3$-KNO_3: the behaviour of liquid mixtures. Thermochim. Acta **223**, 145–156 (1993)
33. X. Zhang, K. Xu, Y. Gao, The phase diagram of $LiNO_3$–KNO_3. Thermochim. Acta **385**, 81–84 (2002)

Influence of Diluents Dosage on the Performance of High Solid Anti-corrosion Coating by Converter Dust

Jinglong Liang, Hui Li, Ramana G. Reddy and Yungang Li

Abstract High solid corrosion protective coatings were prepared by the dust of the converter with an average particle size of 59.33 μm. The effect of different amount of benzyl alcohol diluent on the performance of high solid anti-corrosion coatings was studied. With the increase of the amount of diluent, the viscosity of high solid anti-corrosion coatings reducing, the film apparent bubble content decreasing, and film resistant to salt water corrosion ability decreased, while acid corrosion capacity enhancement. When the dosage of diluents is 15%, the coating film is more resistant to acid corrosion.

Keywords Converter dust · Anti-corrosive coatings · Diluents · Anti-corrosion performance

There is about 4% of GDP losses annual caused by metal corrosion in China [1]. Therefore the research on the corrosion protection technology of metal materials is great significance to the development of the whole national economy [2, 3]. As a new type of metal anti-corrosion coating, high solid anti-corrosion coatings [4, 5] with less organic volatile components is conducive to environmental protection. It will become the development trend of anti-corrosion coatings technology because of the curing speed, impact resistance and superior construction performance [6, 7]. Converter dust contains a large amount of metal oxide [8] can be used as high solid anti-corrosive coating filler, and the smaller size reducing the pigment grinding process costs [9]. The technology of preparing high solid corrosion coating by converter dust is not only recycling the waste of converter steelmaking process, reducing the pollution of the environment, but also reduces the cost of the

J. Liang (✉) · H. Li · Y. Li
College of Metallurgy and Energy, North China University of Science and Technology, Tangshan 063009, Hebei, China
e-mail: 35138105@qq.com

R.G. Reddy
Department of Metallurgical and Materials Engineering, The University of Alabama, Tuscaloosa, AL 35487, USA

S. Wang et al. (eds.), *Applications of Process Engineering Principles in Materials Processing, Energy and Environmental Technologies*, The Minerals, Metals & Materials Series, DOI 10.1007/978-3-319-51091-0_53

anti-corrosion coating, and improves the process performance, although the use of low viscosity epoxy resin, the high solid anti-corrosion epoxy coating viscosity is still relatively large and cannot reach the effect of air spraying. It is necessary to add a certain amount of diluent, so that epoxy resin coating to reach the required viscosity [10]. Using Benzene methanol as a diluent mainly because of the non-toxic, high boiling point, not volatile. And because of the strong polarity, after curing benzyl alcohol can be very good compatibility with epoxy resin, will not appear migration phenomenon. In the experiment, the influence of different amount of diluent on the performance of the high solid anti-corrosion coatings was studied.

Experimental Method

Select the dust particle size of 59.33 μm for preparing the pigment and filler. The pigments and fillers are 100 g, pigment to binder ratio, namely the pigment and filler: epoxy resin = 1:1. The adding amount of diluent accounts for epoxy resin for 5, 10, 15, 20%, respectively. Pigments and fillers are made up with converter dust, a certain amount of talc, quartz powder and mica powder (the amount of 4% of the converter dust). Leveling agent, dispersant dosage generally account for the main agent of 0.5–1% (take 0.75%); Thixotropic agent, silane coupling agent generally account for main agent of 0.25–0.5% (take 0.3%). After the good coating equipped with on the surface of the silicon steel sheet, the coating film need 6.5 h of the surface dry and 19 h of the actual drying process.

NK-2 viscosity cup viscometer detector is used to conduct analysis and detection of viscosity, Zeiss microscope and Axovert 200MAT metallographic microscope are used to apparent detection of coating and study the influence of the amount of diluent on the performance of the coatings, and salt spray experiment box is used to study performance of the high solid anti-corrosive coating resistant to salt water corrosion. Neutral salt spray test: 5% NaCl solution; pH value of 7.0–7.2 (35 ± 2 °C); salt spray settlement amount is 1–2 mL/80 (cm^2 h); sample placement angle of 30°; continuous spray. Acid salt spray test: a 5% sodium chloride solution with adding some of glacial acetic acid, so that PH value of the solution is reduced to about 3, starting the salt spray test when the solution becomes acid.

Results and Discussion

Influence of Diluents Dosage on Coatings Viscosity

Figure 1 is the influence of diluents dosage on coatings viscosity. When the dosage of benzene was 5%, the viscosity was 244.66 Pa s; the dosage of benzyl alcohol increased to 20%, coating viscosity reduced to 243.50 Pa s. When the dosage of

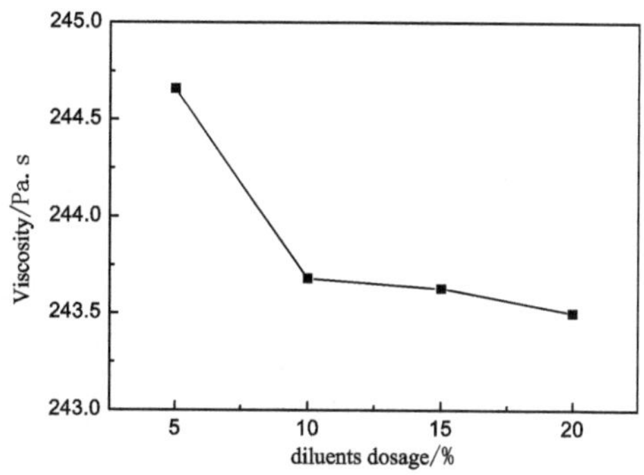

Fig. 1 Influence of diluents dosage on coatings viscosity

benzene methanol increased from 5 to 10%, the viscosity of the coating decreased from 244.66 to 243.68 Pa s, and decreased by 0.98 Pa s, and coating viscosity reduction is relatively obvious. Along with the amount of diluent benzyl alcohol increased from 5% to 20%, the viscosity of the coating decreases, and when the amount of the diluent is less than 10%, the viscosity decreased significantly with increasing of diluent dosage, but when the amount of the diluent is more than 10%, the viscosity changes little with increasing of diluent dosage. Because a certain amount of benzene methanol can promote the reaction between epoxy resin and amine, accelerate the reaction rate, so that the solid in the solid content decreased significantly, thereby reducing the viscosity of high solid anti-corrosion coatings. However, the addition of diluent can also reduce the mechanical properties and chemical properties of the cured, so control the dosage is needed.

The Influence of Diluents Dosage on Coating Surface

Coating surface prepared at different concentrations of diluent was observed by Zeiss microscope at a magnification of 500 times, as shown in Fig. 2. Figure 2a is the film prepared by a reference coating sample, and as a reference sample. Figure 2b–e is films for different experimental conditions, the black spots in the picture as the hole. Figure 2a shows that the surface of film prepared by reference coatings is not flat, more impurities and film have exposing the substrate white, indicating its poor hiding; Fig. 2b shows that when the diluent amount is 5%, the coating film pores small (<1 μm), and more stomata number per unit area; Fig. 2c, d shows when the amount of diluent is 10 and 15%, the number of holes on per unit area of the film decrease, but the porosity increased slightly (<5 μm); Fig. 2e the film stomata have aggregation, appears nonporous region. It is indicated that with the amount of diluent increased, the number of pores per unit area of the film decreased, and the area occupied by the pores increased; When the diluents dosage is 20%, the porosity on film decrease, and appears nonporous region. Comparative Fig. 2a shows that there are no impurities, but there is a small hole on the film

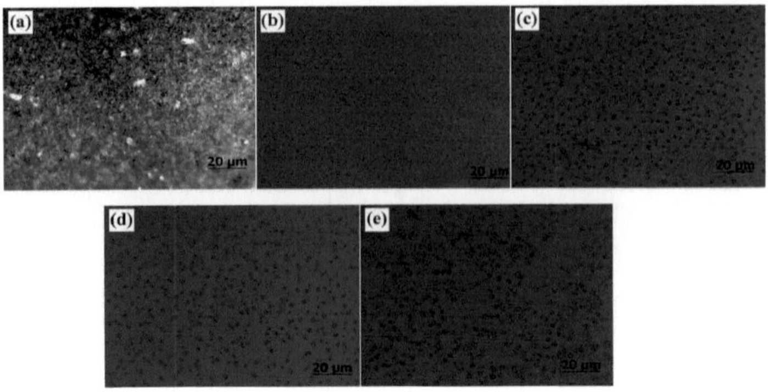

Fig. 2 Influence of diluents dosage on films appearance. **a** Reference sample, **b** 5%, **c** 10%, **d** 15%, **e** 20%

obtained by experiment. With the increase of the amount of diluent benzyl alcohol, the film surface is flat, the porosity of the film is reduced and becomes large. When the dosage of benzyl alcohol increases to 20%, the porosity of the film is easy to be gathered together, so that the corrosive substances can easily penetrate and damage the film performance. The formation of shrinkage hole on the film is mainly because of the existence of low surface tension material in the center of the shrinkage hole, and the presence of different surface tension with the surrounding material, so that coatings departing from the center to flow to form pores [11]. The viscosity of the coating itself, drying rate and thickness of film will affect the formation of shrinkage hole. And with the increase of the dosage of diluent, high solid coating viscosity increases, curing time of the extension, and the flow of the surrounding liquid is enhanced, so it is easier to form a larger shrinkage hole.

Influence of Diluents Dosage on Adhesion and Hardness of Films

Firstly, the metal substrates are treated by surface treatment [12] of sanding and solvent cleaning. The adhesion and hardness of the coating film prepared by different diluent dosage of benzene methanol were tested, and the results are shown in Table 1. The adhesion and hardness of prepared films change little with the dosage

Table 1 Influence of diluents dosage on adhesion and hardness of films

Experiment number	Residual lattice	Shedding rate/%	Hardness
1	98	2	5H
2	99	1	4H
3	97	3	5H
4	98	2	6H

of diluents increasing from 5 to 20%. Usually, to enhance the adhesion of film and substrate, need to add the accelerator, to strengthen the interaction force between the film and the substrate. Therefore, the dosage of diluent has almost no effect on adhesion and hardness of film.

Influence of Diluents Dosage on the Anti-corrosion Performance of the Film

Film Resistance of Salt Water

Figure 3 is the result of the NSS test of the film with different diluent dosage. Figure 3a for the film prepared by coating of reference sample, Fig. 3b–e for the films prepared by converter dust in experiment. In comparison, Figs. 2a and 3a can be known, for the reference sample, dew point corrosion point increase, the matrix is exposed and corrosion is serious; From Fig. 3b–e, the dosage of diluent is 5, 10 and 15%, the film is good, no obvious corrosion phenomenon, in which the dosage of diluent is 10% with the individual corrosion points may be caused by uneven coating. When the dosage of diluent is 20%, the film has obvious corrosion phenomenon, corrosion area by salt water of film is larger, but the corrosion degree of film is light, and the matrix is not damaged. The salt water resistance of the coating film is decreased with the increase of the dosage of diluent. Comparison of Fig. 3a–e, experiment preparation of film have good anti-corrosion performance and salt water corrosion resistance performance. The thickness and uniformity of the coating film has a great influence on the salt corrosion resistance of the film. With the increase of the dosage of diluent, the viscosity of high solid anti-corrosion coating is lower, which is not conducive to the formation of a layer of thick and uniform coating on the substrate surface, so the performance of corrosion resistance has a downward trend.

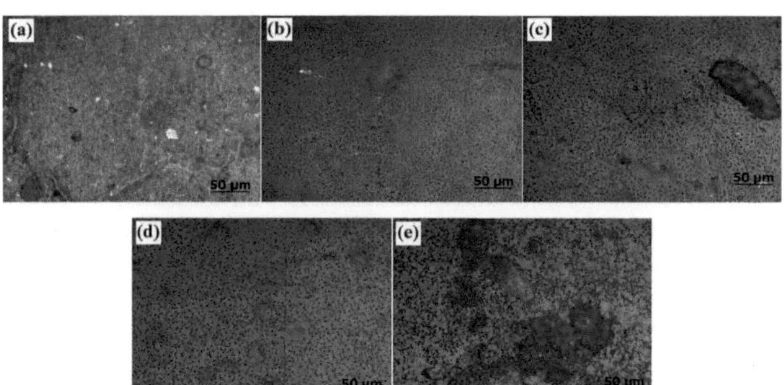

Fig. 3 NSS result of films made by different diluent dosage. **a** Reference sample, **b** 5%, **c** 10%, **d** 15%, **e** 20%

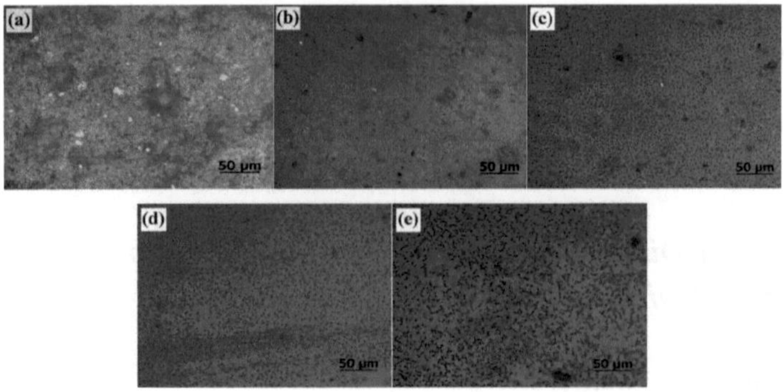

Fig. 4 ASS result of films made by different diluent dosage. **a** Reference sample, **b** 5%, **c** 10%, **d** 15%, **e** 20%

Film Corrosion Resistance to Acid

Figure 4 is the ASS test result of films prepared by different diluent dosage. Figure 4a as reference sample by the corrosion of acid spray testing, Fig. 4b–e is the result of acid spray corrosion of the film prepared by different diluent dosage. From Fig. 4a, it is known that there is a lot of corrosion spots on the reference coating film, the corrosion spots is large; and on the film appears obvious pits, the film is seriously corrosion; From Fig. 4b–e, with the increase of the dosage of diluent, the corrosion spots on the coating film is reduced, the film is more resistant to acid; When the dosage of diluent is 5%, the film has more corrosion spots. When the dosage is 10%, the unit area of the film is reduced, but the individual spots are corroded seriously; when the dosage is 20%, the film has less corrosion spots; when the dosage is 15%, the coating film is in good condition. Thus, the 15% is the most appropriate diluent dosage.

Conclusion

(1) The anti-corrosive film apparent bubble content gradually reduce with the increase of dosage of diluent, until disappear, thus the quality of the film is improved,and the corrosion resistance ability is enhanced.
(2) The viscosity of the anti-corrosion coating gradually decline with the increase of diluent concentration, thus the dosage more conducive to the air spraying coating.
(3) The dosage of diluent had little effect on the adhesion and hardness of films.
(4) With the increase of dosage of diluent, lowering the performance of resistant to salt water corrosion, and the ability of corrosion resistance to acid is enhanced, When the dosage of diluent is 15%, the acid corrosion resistance of coating film is the best.

Acknowledgements This work was financially supported by the National Natural Science Foundation of China (Grant No. 51401075, 51474093), and the National Natural Science Foundation of Hebei Province (Grant No. E2013209307).

References

1. W. Han, Research and development of anti-corrosion coating for metal, Intern. Combust. Engines **1**, 4–6 (2008)
2. J.G. Xu, G.F. Zhang, J. Wang et al., Preparation and properties of heat-reflective insulation thin layer anti-corrosion coating, New Chem. Mater. **8**,168–171 (2010)
3. X.S. Du, X.L. Yu, W.J. Yang et al., Research and development of anticorrosion coatings. Shanghai Coat. **47**(7), 26–29 (2009)
4. Z.Y. Zhan, Research and development of anti-corrosion coatings. Henan Chem. Ind. **28**(2), 19–21 (2011)
5. N.J. Zhu, S.Y. Pan, X.J. Lan et al., Preparation of high solid anti-corrosion epoxy coating for steel structure, Ningbo Chem. Ind. **3**, 22–25 (2009)
6. G. Lu, The trend of high solid anti-corrosion coatings development and product promotion and application. Chem. Ind. **31**(4), 35–40 (2013)
7. J. Xiao, High solid like paint application of energy conservation and environmental protection. Chem. Ind. **32**(2), 58–61 (2014)
8. J.C. Gao, M.X. Fang, Y. Wang et al., Extracting iron powder from revolving furnace sludge. Powder Metall. Technol. **20**(1), 24–27 (2002)
9. F. Hao, Y.G. Li, J.L. Liang, Anticorrosive coating production method with converter dust gas as filler. Energy Metall. Ind. **32**(3), 52–54 (2013)
10. X.G. Wang, N.B. Ma, J.B. Wu et al., Preparation of high solid epoxy anticorrosion coatings, China Coat. **29**(1), 27–30 (2014)
11. Y. Tang, S.P. Chen, Z.B. Hu et al., Water-based paint coating surface defects prevention and cure of shrinkage hole. Shanghai Coat. **5**, 31–35 (2005)
12. L. Xie, A. Freytag, M. Mutha, Novel adhesion promoter used for heavy duty coatings. Electroplating Finish. **30**(12), 53–56 (2011)

Permselectivity Study of Ion-Exchange Membranes in the Presence of Cu-HEDP Complexes from a Copper Plating Wastewater Treatment

J.M.S. Jesus, T. Scarazzato, J.A.S. Tenório and D.C.R. Espinosa

Abstract The development of separation process for the wastewater treatment in distinct industrial process have been applied for recover water and inputs. Electrodialysis is an example of separation process that have been studied for electroplating wastewater treatment. The use of ion-exchange membranes and an electric energy allows the selectivity separation of ions in a solution. The 1-hydroxyethane-1,1diphosphonic acid HEDP is an organic acid that is has been applied for the cyanide substitution in electroplating baths, and in this condition the electrodialysis has showed recover potential. The ion-exchange membrane used in an electrodialysis process contain organic sample are exposed for extreme conditions. These conditions can provide degradation structure and loss of permselectivity, which causes unusual membrane properties. The aim of this study is evaluate the influence of CuHEDP complexes in synthetic ion-exchange membranes using electrochemical test.

Keywords Ion-exchange membranes · Electrodialysis · HEDP · Permselectivity

Introduction

Ion-exchange membranes are the main tool for some separation process such as electrodialysis. However, the maintenance costs are high, because the development of membrane separation processes brings to the process a concern about the involved costs. The ion-exchange membranes are able to perform an ionic separation, with a degree of selectivity that depends on the membrane's properties. There are many kinds of ion-exchange membranes, such as cationic, anionic,

J.M.S. Jesus (✉) · T. Scarazzato · J.A.S. Tenório · D.C.R. Espinosa
Chemical Engineering Department, University of São Paulo, Av. Prof. Lineu Prestes, 580, Block 18, Butantã, São Paulo, Brazil
e-mail: julianams.silva@gmail.com

S. Wang et al. (eds.), *Applications of Process Engineering Principles in Materials Processing, Energy and Environmental Technologies*,
The Minerals, Metals & Materials Series, DOI 10.1007/978-3-319-51091-0_54

bipolar and monovalent membranes. All of them may be used, depending on the aim of the process and on the solution composition to be treated. Cationic ion-exchange membranes are permselective for cations, and anionic membrane is permeselective for anions. The permeseletivity is related to the flux of ionic species through the membrane [1–3].

Ion-exchange membranes are composed of fixed counter-ions, mobile counter-ions, mobile co-ions and a polymeric matrix. These components are part of the membrane structure, which characterizes the membrane morphology that can be classified as heterogeneous or homogenous. This distinction occurs because of the distribution of the ion exchange resin throughout the matrix. In separation process that uses ion-exchange membranes, there are three transport mechanisms: convection, diffusion and migration. All of them are responsible for the ionic transport from the bulk solution to the membranes [1, 4, 5].

The permselectivity of an ion-exchange membrane can be determined from the application of zero current and the measure of potential variation through the membrane. In this case, its necessary to use the same solution but in different concentrations for evaluating the flux of a specific ion based on potential variation [1, 4, 5].

Materials and Methods

Ion-Exchange Membranes

In this work, two pairs of ion exchange membranes were used: a cationic (HDX 100) and an anionic (HDX 200) membrane, both presenting a heterogeneous morphology. The results were compared with the specifications provided by the manufacturer for new membranes (Table 1).

Ageing

A solution simulating the wastewater from an alkaline electroplating bath composed of copper and an organic acid called diphosphonic acid (Table 2) that was used as ageing agent. This solution was diluted in a solution of 1% v/v from the strike bath. The evaluated membranes were immersed in the copper-HEDP solutions during 13, 26, 60 and 90 days.

Table 1 Factory specifications of HDX 100 and HDX 200 membranes [6]

Membrane	Permselectivity (%)/(KCl 0.1 − 0.2 M)
HDX 100	≥ 89.0
HDX 200	≥ 90.0

Table 2 Strike bath of CuHEDP

Composition	Concentration
Ions Cu^{2+}	4.5 $g.L^{-1}$
HEDP	105.0 $g.L^{-1}$
Potassium Chloride	7.0 $g.L^{-1}$
pH	~ 10.0

After these over time, the membranes were immersed in deionized water for twenty-four hours in order to remove some superficial dross. After that, the membranes were submitted to an electrochemical test to determinate their permselectivity.

Permselectivity

The analysis of the apparent permselectivity of the aged membranes was performed using the electromotive force method and in a reactor containing three compartments (Fig. 1).

The membrane's effective area was measure to be 4.90 cm^2. In addition, two graphite electrodes (working and counter electrode) and two Ag/AgCl reference electrodes were used, which were responsible for measuring the voltage drop through the membrane. A direct current was applied by using a potentiostat/galvanostat, PGSTAT 302 N (Metrohm). In order to evaluate the permselectivity of membranes, the solutions used in these experiments were NaCl 1.0 M and NaCl 0.5 M. An open circuit potential was applied during 180 s between the two graphite electrodes. The average value of the membrane potential was used to calculate the permselectivity, using Eqs. (1) and (2).

Fig. 1 Experimental setup for permselectivity test

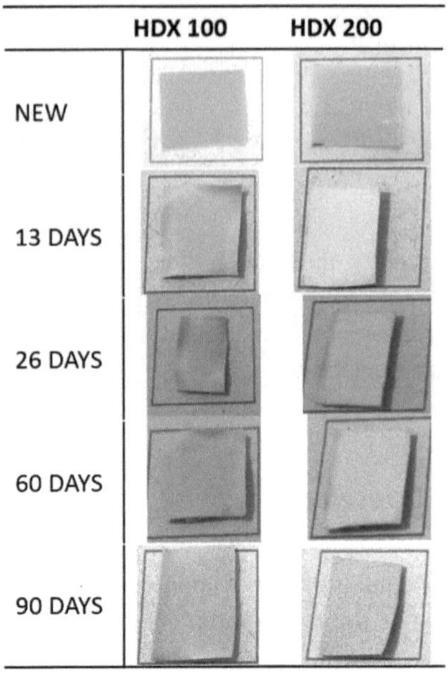

Fig. 2 Aged membranes HDX 100 and HDX 200 in solution of CuHEDP 1%

$$E_m = \left(1 - 2t_j^m\right)\frac{RT}{F}\ln\frac{\alpha_1}{\alpha_2} \quad (1)$$

$$P = \frac{t_j^m - t_j^s}{1 - t_j^s} \quad (2)$$

Equation (1) was used to calculate the counter-ion transport number in the membranes (t_j^m). In Eq. (1) E_m is the voltage drop, α_1 and α_2 are the molar concentrations of the electrolyte solutions, R is the gas constant, T the temperature in Kelvin and F is Faraday constant. In Eq. (2), (t_j^s) is the counter-ion transport number in the solution phase and P is the permselectivity. The experiments were conducted at 25.0 ± 0.4 °C (Fig. 2).

Results

Ageing

The HDX 100 and HDX 200 membranes showed color modifications in comparison with the new ones after 13, 26, 60 and 90 days of immersion in the solutions of CuHEDP 1% v/v.

It was noted color modifications between the membranes aged in CuHEDP 1% solution during all the ageing periods showed that the substances in the CuHEDP 1% had interacted with the membranes.

Permselectivity

HDX 100

By using the equations described, the permselectivity of the aged HDX 100 membrane was obtained (Table 3). All the results were considered such as modification when the variations between the parameters were higher than 5%.

It was observed that the permselectivity of the aged membrane in the CuHEDP 1% solution decreased in comparison with the new membrane in 13 and 26 days. The results show the effect of ageing under the evaluated conditions, which caused an increase in the permselectivity, therefor the flux of cations and anions increased in the membrane.

HDX 200

The permselectivity of HDX 200 membrane decreased after the ageing tests in both conditions (Table 4). This suggests that there was an interaction between the membranes and the ageing solutions. The decrease of the permselectivity indicates that modifications in the membrane transport properties may have occurred, beyond the ionic flux had increased in anionic membrane. Its results showed that the CuHEDP modified this property due to the ageing process.

Table 3 Permselectivity of the new HDX 100 new and aged membranes in solutions of CuHEDP 1% v/v

HDX 100	Permselectivity (%) (NaCl 1.0/0.5 M)
13 days	89.8
26 days	88.7
60 days	87.5
90 days	87.2

Table 4 Permselectivity of the new HDX 200 and aged membranes in solutions of CuHEDP 1% v/v

HDX 200	Permselectivity (%) (NaCl 0.1/0.5 M)
13 days	82.5
26 days	81.6
60 days	80.2
90 days	79.1

It is important to highlight that the permeselectivities of both membranes were obtained by means of an electrochemical test using NaCl solutions, since the ionic transport numbers for Cl^- and for Na^+ are already established in literature [7].

It was showed different behavior in permselectivity of HDX 100 and HDX 200 it is related by the fact that the HDX 200 (anionic membrane) had more influence and interaction with the anions of CuHEDP than that of the cationic membrane (HDX 100).

Conclusion

The interaction between the copper complexes and the evaluated membranes was studied by analyzing the modifications on the membrane electrical properties and color modifications. The obtained results of permselectivity and ageing suggests that the ageing test caused modifications in the structural and transport membrane properties. The HDX 200 membrane presented higher variation in its electrical properties than the HDX 100, due to the affinity with the complexes and the anionic species of HEDP.

References

1. H. Strathmann, *Ion-Exchange Membrane Separation Processes* (Elsevier, 2004), pp. 89–146
2. R. Baker, *Membrane Technology and Applications* (Wiley, California, 2004), pp. 393–422
3. W.S.W. Ho, K.K. Sirkar, Membrane Handbook (Springer Science Business Media, 2001), pp. 217–255
4. M. Mulder, *Basic Principles of Membrane Technology* (Kluwer Academic, Enschede, The Netherlands, 1996), pp. 380–385
5. T. Sata, Properties, characterization and microstructure of Ion Exchange Membranes, *Ion Exchange Membranes: Preparation, Characterization, Modification and Application* (Japan, Tokuyama, 2004), p. 314
6. T. Scarazzato, Wastewater treatment contains HEDP by electrodialysis. Master Degree dissertation. Polytechnic School of São Paulo, 2013, p. 114
7. D.R. Lide, in *CRC Handbook of Chemistry and Physics*, 84th edn, 2003–2004. Handbook of Chemistry and Physics, vol. 53 (2003), p. 2616

Treatment of Blast Furnace Gas Washing Water by Utilization of Coagulation Associated with Microwave

Junhong Zhang, Qinghai Pang, Zhijun He, Chen Tian and Tingfeng Wu

Abstract Microwave technology as new aided tool has been applied on waste water treatment and metallurgical fields. This paper explores the affected factors such as microwave power, microwave irradiation time, polyaluminium chloride (PAC) amount and addition of phosphoric acid on treatment of Blast Furnace gas washing water by utilization of coagulation associated with microwave. Through the orthogonal experiment, the optimum experimental parameters were obtained in the course of treatment of 300 ml Blast Furnace gas washing wastewater: 396 W microwave power, 2 min of microwave irradiation time, 0.12 ml Added PAC and 0.8 ml added phosphoric acid. It indicates the affected factors in descending order on removal of SS in BF gas's washing water: Added PAC level, added phosphoric acid level, time of microwave irradiation, and microwave power while those parameters that can affect suspended solids(SS)'s hardness reduction in descending order are added phosphoric acid level, added PAC level, time of microwave irradiation, and microwave power. Also, it demonstrates those parameters that can affect washing water's turbidity in descending order: Added PAC level, time of microwave irradiation, added phosphoric acid level, and microwave power. The results showed that PAC coagulation associated with microwave on treatment of Blast Furnace gas washing wastewater has several benefits: short treatment time, strong sedimentation velocity and other excellent characteristics, it can effectively remove suspended solids in wastewater, the SS hardness and turbidity can be reduced.

Keywords Microwave · Coagulation and sedimentation

J. Zhang (✉) · Q. Pang · Z. He · C. Tian
University of Science and Technology Liaoning, Anshan 114015, China
e-mail: Edikitty@126.com

T. Wu
Dalian University of Technology, Dalian 116024, China

S. Wang et al. (eds.), *Applications of Process Engineering Principles in Materials Processing, Energy and Environmental Technologies*,
The Minerals, Metals & Materials Series, DOI 10.1007/978-3-319-51091-0_55

Introduction

Washing water of Blast Furnace gas is main part of wastewater during the Blast Furnace iron-making process, and it has main features: large consumed water, high temperature, heavy pollution and high content of suspended solids, and a small amount of inorganic salts and other toxic substances such as phenols, cyanide, and heavy metals, etc. It should be cycled after treatment and it can achieve zero discharge of wastewater in iron-making process. Currently, due to utilization of advanced technologies, such as high-pressure blast furnace operations and pulverized coal injection, there is an increase in fine particles in washing water of Blast Furnace gas and its composition is greatly changed. Because of the surface of the fine particle suspension with a strong negative charge, it can relatively be easy to obtain a stable colloidal form, so it becomes more difficult to economically treat washing water of Blast Furnace gas.

Microwave is new kind of energy that can promote, catalyze and induce chemical reaction, and it has been used in the field of wastewater treatment. It is reported to have the good effect on water washing coal and coking wastewater treatment by microwave [1–6].

This paper describes the processing of washing water of Blast Furnace gas by PAC and phosphoric acid in the presence of microwave radiation, and it also discussed microwave effect, PAC and phosphoric acid's role and their mechanism on this kind of treatment. It tried to find new way to economically and efficiently treat and process washing water of Blast Furnace so that it is very vital to improve recycling rate of washing wastewater of blast furnace gas, reduce emissions and achieve efficient production on iron-making.

Experimental

Experimental Samples

The washing water of Blast Furnace gas from Anshan Iron and steel Company was used as the specimens in current study and their chemical composition is shown in Table 1.

Table 1 Main chemical composition of Blast Furnace gas washing wastewater

Index	COD/mg L^{-1}	Alk./ mg L^{-1}	SS/mg L^{-1}	NH3-N/mg L^{-1}	t.d/NTU	DKH/ mg L^{-1}	pH
Measured value	492	2323.2	2179	92.6	588	627.1	8.5

Experimental Device and Treatment Chemical Agent

Treatment chemical agent: PAC as a coagulant, phosphoric acid (0.845 g/mL) used to reduce the hardness, and chloride, EDTA disodium, chrome black T etc. for experimental effect with KH-6HMOA industrial microwave oven.

Experimental Procedure

Stirring of sample water with a glass rod was done to make even and weighing of 300 mL of the water as a specimen, and addition of the quantitative PAC and phosphoric acid into it and stirring; then putting into a microwave by setting up power, and radiation time. After processing it in microwave, taking the sample out and testing it.

Experimental Methodology

Microwave power specific selection: 132, 264 and 396 W.

Through the basic experiment, it was found the best effect can be obtained: adding 0.1 mL PAC in sample water, among the selected volumes of PAC was 0.08, 0.10 and 0.12 mL.

Under acidic conditions, it is very useful to add coagulated chemical agent for removal of pollutants and contaminants in the of presence of microwave field, and to reduce the water hardness. So, phosphoric acid level was chosen in three volmes: 0.6, 0.8 and 1.0 mL. The orthogonal experimental design method is shown in Table 2.

Table 2 Experimental methodology

Number	Microwave power/W	Irradiation time/min	PAC/mL	Phosphoric Acid/mL
1	132	3	0.08	0.6
2	132	2	0.10	0.8
3	132	2.5	0.12	1.0
4	264	3	0.10	1.0
5	264	2	0.12	0.6
6	264	2.5	0.08	0.8
7	396	3	0.12	0.8
8	396	2	0.08	1.0
9	396	2.5	0.10	0.6

Results and Discussion

According to Table 2 orthogonal design, the nine experiments were completed. The testing index included mass concentration of suspended solids, hardness and turbidity as experimental assessed indicators. In order to consider the impact of various factors on washing water treatment of Blast Furnace gas, the comprehensive assessed score method was employed. Comprehensive assessed score method is defined by the impact of each factor. And, the average value of the index range is defined as range proportion. The sum of range proportion is defined as the range sum, while the effect coefficient of comprehensive score is the ratio of the range proportion and the range sum. The influence coefficient of all indexes are added to equal to 1. According to the above method, the experimental results were shown in Tables 3 and 4, respectively.

From Tables 3 and 4 analysis, in the washing water treatment of blast furnace gas, the main factor is PAC effect on the comprehensive assessed index while microwave only played an aid role that can accelerate sedimentation rate, shorten the processing time. The optimal combination condition was selected for A1B2C2D1, i.e. 132 W Microwave Power, 3 min Irradiation Time, 0.10 mL added PAC, and 0.6 mL added Phosphoric Acid.

In the course of experimental treatment, the various factors strong effect on different index is not consistent; i.e. effective removal of SS in wash water of blast furnace gas in descending order: PAC Level, Phosphate Level, Microwave Irradiation Time, and Microwave power; effective hardness reduction factors in descending order: Phosphoric Level, PAC Level, Microwave Radiation Time, Microwave Power while effective turbidity reduction in descending order: PAC Level, Microwave Radiation Time, Phosphate Level, and Microwave Power.

In a summary, PAC is the most important factor in reducing the SS and turbidity, while increasing of the phosphoric acid level is a key factor on reduction of its hardness.

There is no significant effect on SS change when PAC dosage is 0.08 and 0.1 mL. When its dosage is 0.12 mL, the SS in wash water of blast furnace gas was significantly decreased, so in the course of treatment of wastewater, coagulant must reach a certain level so that it can show the coagulation effect. It indicates PAC as a coagulant, is the key factor on reduction of SS in washing water because it has the strong adsorption in water and it generates aggregation and precipitation of the physical and chemical reactions during its hydrolysis process.

As shown in Fig. 1, with the microwave power increasing, the SS would be decreased because microwave power can bring a considerable amount of heat and accelerate the removal rate of suspended solid chemical reactions in washing water of blast furnace gas. In addition, with the increase of absorbed microwave energy per PAC unit area, it can form many "hot spots" on its surface so that it can boost up the frequency of collision of suspended solids in wash water. From these points, microwave can impact on PAC removing rate of suspended solids in washing water of blast furnace gas by the chemical and physical processes. During this

Table 3 Result of orthogonal experiment

Number	Microwave power/W	Irradiation time/min	PAC/mL	Phosphoric Acid/mL	SS/mg L^{-1}	Hardness/mg L^{-1}	Turbidity/NTU
1	132	3	0.08	0.6	379	20.8	96.7
2	132	2	0.10	0.8	224	25	60.1
3	132	2.5	0.12	1.0	122	29.1	65.8
4	264	3	0.10	1.0	342	33.3	67.5
5	264	2	0.12	0.6	55	20.1	72.4
6	264	2.5	0.08	0.8	177	29	90
7	396	3	0.12	0.8	10	12.1	82.6
8	396	2	0.08	1.0	204	41.6	80.5
9	396	2.5	0.10	0.6	276	16.7	69

Table 4 Analysis of orthogonal experiment result

Number	SS/mg L^{-1}	Hardness/mg L^{-1}	Turbidity/NTU	Comprehensive score
1	379	20.8	96.7	220.99
2	224	25	60.1	135.25
3	122	29.1	65.8	83.35
4	342	33.3	67.5	201.56
5	55	20.1	72.4	45.75
6	177	29	90	115.77
7	10	12.1	82.6	20.63
8	204	41.6	80.5	132.98
9	276	16.7	69	161.26
Range	369	29.50	36.60	
Mean value	198.78	25.30	76.10	
Range ratio	185.63	116.6	48.10	350.33
Influence coefficient	52.99	33.33	13.68	100
	Microwave power/W	Irradiation time/min	PAC/mL	Phosphoric Acid/mL
Level 1	146.53	147.73	156.58	142.67
Level 2	121.03	104.66	156.60	90.55
Level 3	104.96	120.13	49.98	139.36
Range	41.57	43.07	106.62	52.12
Optimal levels	132	3	0.1	0.6
Factors to sort	4	3	1	2

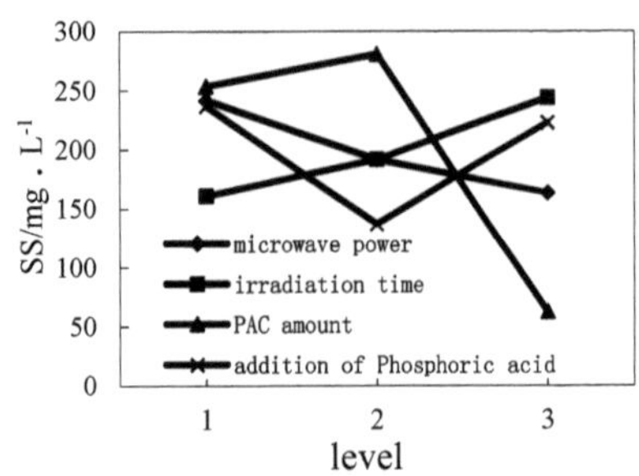

Fig. 1 Effect of various factors on SS

experiment, with the increasing irradiation time, contrarily, SS level is increased because the solubility of SS is rapidly reduced and precipitated with water evaporation due to heating effect of microwave radiation on washing water of blast furnace gas. Therefore, it is desirable to control the microwave power and its irradiation time in order to precipitate SS during wash water treatment.

With increasing of added phosphoric acid level, SS values are decreased first and then increased, but with the increase in the amount of added phosphoric acid, the water hardness would increase. The experiment showed the best level of adding phosphoric acid on removal of SS is 0.6 mL. Also, phosphoric acid is the most important factor in reducing water hardness because it can react with calcium carbonate ($CaCO_3$) in water and produce calcium phosphate with greater solubility to reduce the hardness of washing water.

Seen from Fig. 2, microwave power has little effect on the water hardness, but with the increase of irradiation time, water hardness would decrease. So, longer the radiation time, the lower water hardness. The hardness of washing water of blast furnace gas is mainly caused by the $CaCO_3$ amount, while the acidification of phosphoric acid is the main reason for reducing its hardness. With the increase of microwave radiation time, the various substances in washing water of blast furnace gas can absorb different levels of radiant heat. This causes high frequency movement of polar molecules in water which will weaken the hydrogen bonds between each other and the Van der Waals force; thereby that can break hydrogen bonds between molecules to promote the forward reaction of phosphoric acid for acidification. So, it can reduce the water hardness.

As seen from Fig. 3, PAC level can greatly impact on water turbidity and the 0.1 mL added PAC can get the best turbidity with different other factor levels.

Figures 4, 5, 6 and 7 show that the coagulation associated with microwave can effectively reduce suspended solids, hardness and turbidity in waste water. The removal rate of SS is between 82.6 and 99.5% while the removal rate of hardness is

Fig. 2 Effect of various factors on hardness

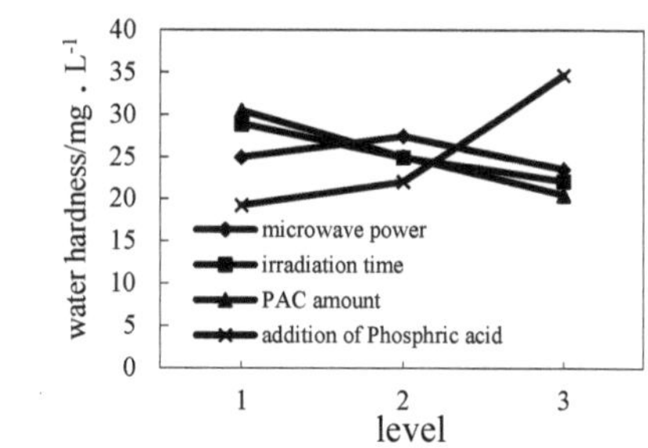

Fig. 3 Effect of various factors on turbidity

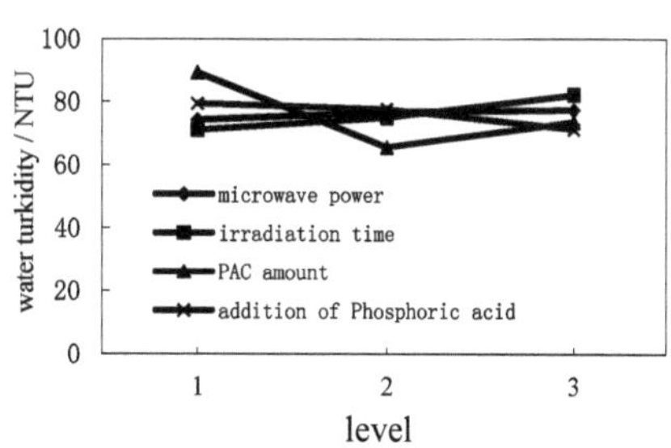

Fig. 4 Effect of microwave power on removal rate of pollutants

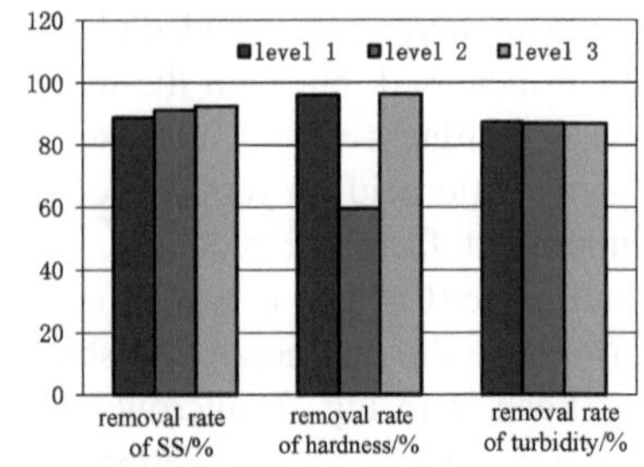

Fig. 5 Effect of irradiation time on removal rate of pollutants

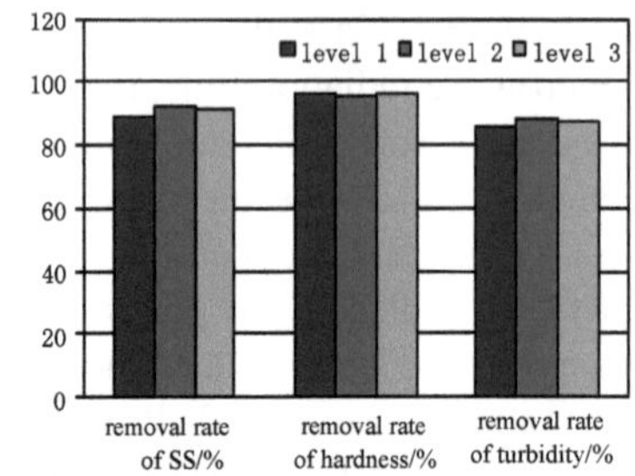

Fig. 6 Effect of PAC amount on removal rate of pollutants

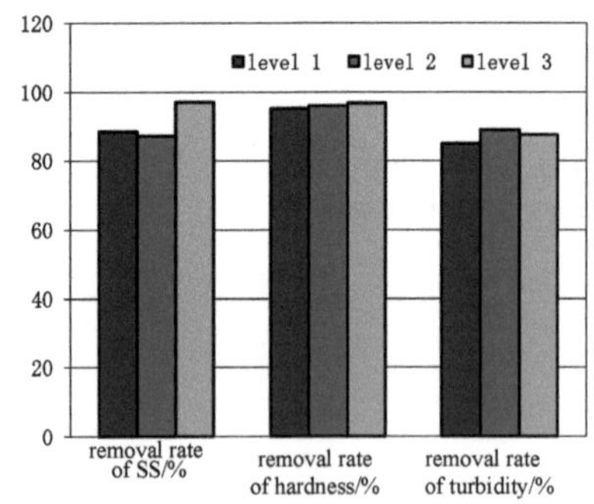

Fig. 7 Effect of Phosphoric acid on removal rate of pollutants

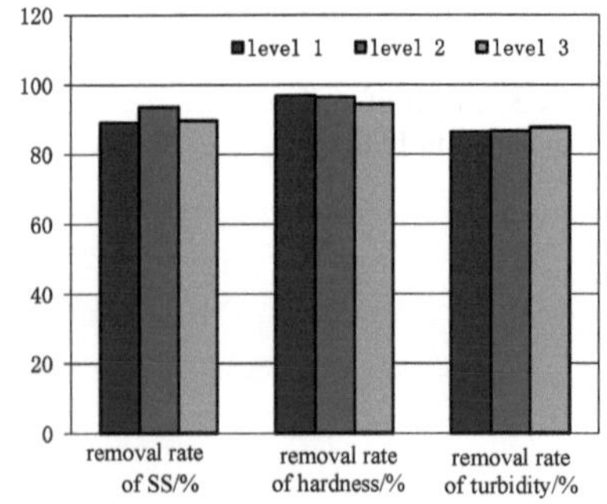

Table 5 Experiment result under the optimum combination condition

Index	SS/mg L^{-1}	Hardness/mg L^{-1}	Turbidity/NTU
Measured value	130	17.8	52.5
Removal rate/%	94.03	97.16	91.07

between 93.4 and 98.1% and its removal rate on turbidity is between 83.6 and 89.8%.

The optimum combination condition does not occur in the orthogonal experimental method, and the optimum condition is obtained by further experiments, its results are shown in Table 5.

Conclusion

The treatment of washing water of Blast Furnace gas through coagulation sedimentation aided by microwave has better effect and it has enhancement on coagulation and sedimentation and shorten the coagulation time.

During treatment of 300 mL wash water of Blast Furnace gas, the optimum parameters are: 132 W microwave power, 3 min, radiation time, 0.10 mL added PAC, and 0.6 mL phosphate. With this optimum condition of treatment, SS in wastewater dropped to 130 mg/L, the water hardness decreased to 17.8 mg/L, and water turbidity reduced to 52.5 NTU.

Acknowledgements This work is financially supported by funds sponsored by National Science Foundation of China (NO. 51474124, NO. 51504131, NO. 51504132 NO. 51674139 NO. 51604148) and Anshan Science and technology project (NO. 4104).

References

1. C.-L. Wang, J.-C. Gu, Application of microwave technology in water treatment. Adm. Tech. Environ. Monit. **19**(4), 36–39 (2007). (in Chinese)
2. X.-P. Qu, X.-H. Wu, X.-H. Lu et al., Treatment of coking wastewater by activated carbon combined microwave irradiation. J. Huazhong Univ. Sci. Technol. **22**(4), 79–81 (2005)
3. M.-X. Fan, Z. Zhang, Study on treatment of coke plant wastewater by microwave combined activated carbon and fenton reagent oxidation process. J. Hubei Univ. Technol. **26**(5), 36–39 (2011)
4. Z.-W. Deng, P. Yu, Y.-B. Luo, Study on removal of COD and colority from coking wastewater by microwave combined fenton reagent. Ind. Water Wastewater **40**(6), 40–43 (2009)
5. H.-C. Xu, X.-J. Xu, X.-H. Xu, Effect and mechanical analysis of washing coal water by microwave. J. Qingdao Technol. Univ. **29**(1), 60–62 (2008)
6. L. Lin, X.-H. Lu, Q.-Y. Li, Effect on removal of ammonia nitrogen and COD in coking wastewater by microwave combined activated carbon. J. Yangtze River Sci. Res. Inst. **28**(11), 19–21 (2011)

Author Index

S. Wang et al. (eds.), *Applications of Process Engineering Principles in Materials Processing, Energy and Environmental Technologies*,
The Minerals, Metals & Materials Series, DOI 10.1007/978-3-319-51091-0

Subject Index

S. Wang et al. (eds.), *Applications of Process Engineering Principles in Materials Processing, Energy and Environmental Technologies*,
The Minerals, Metals & Materials Series, DOI 10.1007/978-3-319-51091-0

MIX
Papier aus verantwortungsvollen Quellen
Paper from responsible sources
FSC® C105338

If you have any concerns about our products,
you can contact us on
ProductSafety@springernature.com

In case Publisher is established outside the EU,
the EU authorized representative is:
Springer Nature Customer Service Center GmbH
Europaplatz 3, 69115 Heidelberg, Germany

Printed by Libri Plureos GmbH
in Hamburg, Germany